Conformational Analysis of Polymers

Conformational Analysis of Polymers

Methods and Techniques for Structure-Property Relationships and Molecular Design

Yuji Sasanuma
Chiba University (retired)
Chiba, Japan

WILEY

Published by John Wiley & Sons, Inc., Hoboken, New Jersey. Published simultaneously in Canada.

Library of Congress Cataloging-in-Publication Data

Names: Sasanuma, Yuji, author.
Title: Conformational analysis of polymers : methods and techniques for structure-property relationships and molecular design / Yuji Sasanuma.
Description: Hoboken, NJ : Wiley, [2023] | Includes bibliographical references and index.
Identifiers: LCCN 2022049542 (print) | LCCN 2022049543 (ebook) | ISBN 9781119716358 (hardback) | ISBN 9781119716594 (adobe pdf) | ISBN 9781119716662 (epub)
Subjects: LCSH: Polymers–Conformation.
Classification: LCC QD381 .S287 2023 (print) | LCC QD381 (ebook) | DDC 547/.7–dc23/eng20230113
LC record available at https://lccn.loc.gov/2022049542
LC ebook record available at https://lccn.loc.gov/2022049543

Cover design: Wiley
Cover image: © Flavio Coelho/Getty Images

Set in 9.5/12.5pt STIXTwoText by Straive, Chennai, India

As my thanks to
those who led me to science,
those who have worked with me,
and
my father, mother, wife, and son

Contents

Preface

In the late 1920s and early 1930s, Carothers's team of DuPont reported the synthesis of some aliphatic polyesters [63–65]. Unfortunately, the polyesters melted at too low temperatures to be commercialized. Meanwhile, Lemoigne extracted an aliphatic polyester, poly((*R*)-3-hydroxybutyrate) (P3HB), from a bacterium, *Bacillus megaterium* [277, 278]. The equilibrium melting point of P3HB is as high as 203 °C. The asymmetric structure and steric hindrance of the methyl side chain restrict the degree of conformational freedom of the P3HB chain and, consequently, lead to the high melting point. Nowadays, P3HB attracts attention from polymer chemists and engineers because of its carbon neutrality, namely biosynthesis and biodegradability. The relationships between conformational characteristics and thermal properties of polymers are often discussed in this book.

In 1941, Whinfield and Dickson acquired a British patent on aromatic polyesters including poly(ethylene terephthalate) (PET), poly(trimethylene terephthalate) (PTT), and poly(butylene terephthalate) (PBT) [520]. The difference in structure among these polyesters is only the number of methylene units between the benzene rings. The high equilibrium melting point (262 °C) of PET is due partly to the intermolecular π–π interaction between the benzene rings. The aromatic polyesters are industrially manufactured all over the world. PET is molded to fibers, films, and bottles; PTT is so flexible as to be used for sports wears and carpets; and PBT is so superior in impact resistance as to be used in tooth and paint brushes, gears, and machine parts. The usages of the aromatic polyesters are mainly due to the mechanical properties. The crystalline Young's modulus in the chain-axis direction of PET is 182 GPa, that of PTT is as small as 7.1 GPa, and that of PBT (α form) is 20.8 GPa [266]. In the crystal, PET lies in a somewhat distorted all-trans structure, whereas PTT and PBT form bended tggt and distorted $g^+g^+tg^-g^-$ conformations, respectively. Here, t, g^+, and g^- represent trans, gauche$^+$, and gauche$^-$ states, respectively. In this book, the mechanical properties of polymers are frequently interpreted in terms of chain conformations.

As described above and will be shown later, higher order structures and thermal and mechanical properties of polymers depend on conformational characteristics, more fundamentally, the primary structure – what the component atoms and chemical bonds are and how they are arranged. Furthermore, scientists have faced with the fact that proteins composed of only 20 amino acids form unique secondary, tertiary, and quaternary structures and exhibit specific functions. Consequently, polymer scientists have reached the concept of molecular design: we may predict the primary structure(s) from which the desired higher order structures, physical properties, and functions are realized.

In Japan, the idea of molecular design was advocated early on. In 1972, a book consisting of three volumes, titled "Molecular Design of Polymers," was published by The Society of Polymer Science, Japan [239]. In the book, Kawai defined the molecular design as follows: First, one must reveal the correlations between the chemical structures and physical properties. Next, one actually synthesizes the polymer with the structure that is expected to show the desired properties. On the other hand, Kambara's molecular design is different: when a new polymer is suggested, namely its constituent atoms and chemical bonds are specified, to predict the structures and morphology to be formed therefrom is the molecular design. Kambara's concept is to predict the higher order structures from the primary structure, whereas Kawai's idea is more idealistic; the molecular design should be to propose the primary structure that actualizes such higher order structures, physical properties, and functions as desired. In order to realize either molecular design, as both Kawai and Kambara suggested, it is requisite to establish relationships between the primary structure, higher order structures, and, furthermore, if possible, properties and functions.

For that purpose, it is essential to elucidate the structures and properties of a single polymeric chain. From an experimental viewpoint, it is significant to determine bond conformations of the skeletal bonds via, for example, NMR and reveal the conformational characteristics and, furthermore, to investigate the configurational properties of the polymeric chain in the Θ state free from the excluded-volume effect via, for example, scattering methods. From a theoretical viewpoint, the free energies and geometrical parameters of individual conformations of a polymeric chain are evaluated by quantum chemistry, and the Boltzmann factors are summed over all conformations to yield the partition function from which various thermodynamic functions can be derived according to statistical mechanical theorems.

However, a polymer chain forms an enormous (astronomical) number of conformations. For example, under the rotational isomeric state (RIS) approximation based on the three states (t, g^+, and g^-), a single polyethylene chain of 100mer shows 3^{200} (approximately 2.7×10^{95}) conformations. The average $\langle M \rangle$ of a molecular parameter M_i, depending on the conformations, may be calculated from

$$\langle M \rangle = \sum_{i=1}^{N} M_i \exp\left(-E_i/RT\right)/Z$$

where E_i is the energy of conformation i, N is the total number of conformations, R is the gas constant, T is the absolute temperature, and Z is the partition function defined as

$$Z = \sum_{i=1}^{N} \exp\left(-E_i/RT\right)$$

To evaluate $\langle M \rangle$, one must solve the Schrödinger equation for each conformation, obtain its ground-state energy, and repeat this procedure N times; however, such colossal computations are impossible for the time being at least, apart from the remote future. Fortunately, statistical mechanics of chain molecules, designated as the RIS scheme, has been developed and formulated as matrix operations. According to the RIS scheme, one can exactly calculate the bond conformations, various configurational properties, and thermodynamic functions of unperturbed polymeric chains (lying in the Θ state). The early studies of the RIS scheme were summarized in Flory's book [141], the ensuing studies can be found in Mattice and Suter's book [307], and Rehahn, Mattice, and Suter's book collected almost all RIS models reported until the end of 1994 [383].

It is known that the configurational properties of unperturbed polymeric chains depend only on short-range intramolecular interactions [141]; therefore, the conformational energies may also be evaluated from a small model compound with the same bond sequence as that of the polymer. In order to derive the conformational energies from quantum chemical calculations, therefore, one need not treat the polymer itself but the small model compound instead. The MO computations on the model are sufficiently practical even if a high-level MO theory including electronic correlations is employed, together with large basis sets.

The RIS scheme can exactly characterize unperturbed polymeric chains in dilute solutions, amorphous phases, and melts. However, polymer scientists have also been desiring to acquire precise theoretical information on the structures and properties of solid-state polymers. Fortunately, the density functional theory (DFT) under periodic boundary conditions has enabled us to calculate the electronic structures of crystals [120]. The methodology can also be applied to polymer crystals and yield the following information: optimized crystal structure (lattice constants and atomic positions); intermolecular interaction energy corrected for the basis set superposition error (BSSE); thermodynamic functions; and vibrational spectroscopic frequency and intensity. The DFT calculations have shown that the chain conformation is also the principal factor in the structures and properties of polymer crystals.

Part I of this book describes the fundamental physical chemistry that is necessary to understand the characteristics of polymers and study their conformations. The contents are stereochemistry, polymer models, and the lattice model (the Flory–Huggins theory); molecular characteristics; solution properties; and rubber elasticity. This part is written so as to be understood without difficulty by graduate and undergraduate students who have learned general chemistry. Derivations of some important equations, which are mostly omitted from textbooks and original papers, are given as problems, and the answers are presented in Appendix B.

Part II explains the quantum chemistry used in conformational analysis, together with a wealth of practical applications. The contents are the Schrödinger equation, the Hartree–Fock method, electron correlations, DFT, dispersion-force correction, solvent effect, general statistical mechanics, NMR parameters, and periodic DFT for crystals.

Part III explains the RIS scheme including mathematical expressions and their derivations. The contents are the conventional RIS scheme, the refined RIS scheme, the RIS scheme including middle-range interactions, inversional–rotational isomeric state (IRIS) scheme, the RIS scheme with stochastic processes, and the RIS scheme with internally rotatable side chains.

Part IV introduces typical experimental methods for conformational analysis of polymers. The contents are broadly divided into two portions: NMR spectroscopy and scattering techniques. The former deals with NMR vicinal coupling constants to evaluate bond conformations and chemical shifts to determine stereo- and regiosequences. The latter half explains how to determine molecular characteristics, chain dimensions in solutions and melts via static light scattering (SLS), dynamic light scattering (DLS), and small-angle neutron scattering (SANS).

Part V exemplifies a number of studies on the following polymers: polyethers, polyamines, polyphosphines, polysulfides, polyselenides, alternating copolymers of amine, ether, and thioether units, polyesters, polycarbonates, nylon 4, substituted analogs of aromatic polyesters, polysilanes, and polyethylene.

Appendix A presents a FORTRAN source code for refined RIS calculations on polyethylene.

Appendix B presents answers of the problems given in the text.

In this book, a large number of mathematical symbols and physical quantities are used; therefore, the same alphabetic and Greek letters are often assigned to different parameters, and the same physical quantity is occasionally represented by different symbols. The dihedral angle is defined in two ways: the convention of polymer science, trans $\sim 0^\circ$ and gauche$^\pm$ $\sim \pm120^\circ$, and the IUPAC recommendation, trans $\sim 180^\circ$ and gauche$^\pm$ $\sim \mp60^\circ$. The former and latter are employed mainly in the RIS and MO calculations, respectively.

Chiba, Japan
August 2022

Yuji Sasanuma

Acknowledgments

The studies presented herein are mostly collaborations with the author's students and friends. The author is deeply indebted to the following people for their dedication and enthusiasm (honorific titles omitted): Taisuke Iwata, Akihiro Suzuki, Haruhisa Kato, Yasukazu Kato, Nobuyuki Miura, Hajime Ohta, Yuichi Ohta, Yugo Hayashi, Futoshi Nishimura, Misa Sawanobori, Makiko Yokoi, Hideto Ogura, Tetsushi Ono, Shinobu Takahashi, Ikuko Touma, Hiroaki Wakabayashi, Jun Fujii, Takayuki Iijima, Tomoyoshi Kaizuka, Yoshihiko Kuroda, Hiroki Matoba, Emi Miyazaki, Satoshi Hattori, Ken Hikino, Shinichi Imazu, Jun Arou, Kohji Nakata, Fumiko Teramae, Hiroki Yamashita, Ippei Hamano, Satoshi Ikeda, Kentaro Sugita, Ryota Kumagai, Tomohiro Miyai, Akinori Watanabe, Shunsuke Asai, Pei Na, Gou Suzuki, Kenta Tamura, Syugo Endo, Nobuaki Suzuki, Yoko Ogawa, Junichi Uchida, Daisuke Abe, Yoichi Hori, Masanao Matsumoto, Takumi Inada, Takayuki Ishibashi, Yuta Nonaka, Takuro Suzuki, Yuto Tanaka, Yuki Yamaguchi, Shiori Katsumata, Shogo Mori, Kou Moriya, Daichi Touge, Satoshi Uzawa, Yusuke Wagai, Tatsuya Ishii, Ryosuke Nagao, Masayuki Nagasawa, Takumi Wada, Yuichiro Fukuda, Kohei Miyamae, Tsubasa Sugita, Yuta Takahashi, Daiki Nagai, Masaki Otsuki, Taiga Kurita, Morihiro Takahashi, Syuto Tanaka, Hiromi Yamamoto, Daisuke Aoki, Somin Choi, Azumi Kawai, Kohyo Ono, Yusuke Kondo, Naofumi Yoshida, Yota Watabe, Shota Abe, Mitsutoshi Hoshide, Hiromu Kawasaki, Yuko Uesugi, Shunsuke Kouchi, Takayuki Morita, Ryota Ohmori, Tomoyuki Tomita, Naoki Hayashi, Sayaka Iwabuchi, Ikumi Saito, Shunsuke Shimomura, Naoki Hamamoto, Rena Ishida, Riko Kaidu, Yuma Yoshida, Shinnosuke Kogure, Ryosuke Ohnishi, Marino Yamamoto, Akira Kaito, Nobutaka Tanigaki, Yoshikazu Tanabe, Shinichi Kinugasa, Takashi Yarita, Muhammad A. Azam, Robert V. Law, and Joachim H. G. Steinke.

The author thanks the editorial staffs of Wiley, Ms. Skyler Van Valkenburgh, Dr. Andreas Sendtko, Mr. Jonathan T. Rose, Ms. Dimple Philip, Ms. Sabeen Aziz, and Ms. Devipriya Somasundaram for invitation to write this book, cordial support, and careful work.

About the Author

Yuji Sasanuma, PhD. Formerly Department of Applied Chemistry and Biotechnology, Graduate School and Faculty of Engineering, Chiba University, Japan

YS was born in Yokohama, Japan, in December 1956 and spent his boyhood in Kawasaki. He received bachelor's (1980, supervisors: Professor Ichitano Uematsu and Professor Junji Watanabe) and master's (1982, supervisors: Professor Akihiro Abe and Professor Isao Ando) degrees from Tokyo Institute of Technology (Department of Polymer Chemistry). Between 1982 and 1988, he worked in Toray Industries, Inc. (Toray Research Center, Inc.) and engaged in material characterization by X-ray diffraction and scattering. In 1988, he returned to Tokyo Institute of Technology to study orientational and conformational characteristics of liquid crystals with Professor Akihiro Abe. After he received doctor's degree (1992, chief examiner: Professor Shintaro Sasaki) from Tokyo Institute of Technology, he moved to National Institute of Materials and Chemical Research (Department of Polymer Physics) in Tsukuba, where he developed a measurement system (equipment and software) for small-angle X-ray scattering and studied conformational characteristics and configurational properties of polyethers (1993–1997). In 1997, he moved to Chiba University and continuously grappled with conformational analysis of polymers. In March 2022, he retired from Chiba University at the mandatory age of 65.

Acronyms

3HB	(*R*)-3-hydroxybutyrate
ABAMA	*N*-acetyl-γ-aminobutyric acid *N*′-methylamide
AM	amorphous
BDMePE	1,2-bis(dimethylphosphino)ethane
BDT	2-*tert*-butyl-1,3-dithiane
BGDA	butylene glycol diacetate
BMePE	1,2-bis(methylphosphino)ethane
BMePhPE	1,2-bis(methylphenylphosphino)ethane
BMSeE	1,2-bis(methylseleno)ethane
BMSeM	1,2-bis(methylseleno)methane
BMSeP	1,3-bis(methylseleno)propane
BMTE	1,2-bis(methylthio)ethane
BMTP	1,2-bis(methylthio)propane
BSSE	basis set superposition error
CC	coupled cluster
cc-pVQZ	Dunning's correlation consistent basis sets (quadruple)
cc-pVXZ	Dunning's correlation consistent basis sets
CCSD(T)	coupled cluster single–double and perturbative triple
CGST	continuous set of gauge transformations
CH	cyclohexane
CHC	cyclohexene carbonate
CHO	cyclohexene oxide
CI	configuration interaction
CID	configuration interaction with all double substitutions
CISD	configuration interaction with all single and double substitutions
cis-DMDO	*cis*-2,6,-dimethyl-1,4-dioxane
CONTIN	constrained regularization method for inverting data
COSMO	conductor-like screening model
COSY	correlation spectroscopy
CP	counterpoise
CP/MAS	cross-polarization / magic-angle spinning
D2	Grimme's D2 dispersion-force correction

D3	Grimme's D3 dispersion-force correction
DCA	dichloroacetic acid
DCB	*o*-dichlorobenzene
DCM	dichloromethane
DEPT	distortionless enhancement by polarization transfer
DFT	density functional theory
DFT-D	density functional theory with dispersion-force correction
di-MEDA	*N*, *N*′-dimethylethylenediamine
DLS	dynamic light scattering
DMB	1,4-dimethoxybutane
DMCC	di(methoxycarbonyloxy)cyclohexane
DME	1,2-dimethoxyethane
DMEDT	2-(1,1-dimethylethyl)-1,4-dithiane
DMF	*N*, *N*′-dimethyl formamide
DMP	1,2-dimethoxypropane
DMPA	2,2-dimethoxy-2-phenylacetophenone
DMS	dimethyl succinate
DMSO	dimethyl sulfoxide
DMT	dimethyl terephthalate
DPCM	dielectric polarizable continuum model
DSC	differential scanning calorimeter
DSO	diamagnetic spin–orbit
EDC	ethylene dichloride
EGDA	ethylene glycol diacetate
EGDB	ethylene glycol dibenzoate
EGDMS	ethylene glycol di(methyl succinate)
ELATE	online application for analysis and visualization of elastic tensors
EV	excluded volume
EVE	excluded-volume effect
FC	Fermi contact
FMB	1-fluoro-4-methoxybutane
GIAO	gauge-independent atomic orbital method
HBS	hydrogen bond strength
HF	Hartree–Fock
HFIP	1,1,1,3,3,3-hexafluoro-2-propanol
H–H	head-to-head
H–K	Hohenberg–Kohn
h-PMS	hexagonal poly(methylene sulfide) crystal
HSAB	hard and soft acids and bases
HSQC	heteronuclear single-quantum correlation
H–T	head-to-tail
IEF-PCM	polarizable continuum model using the integral equation formalism variant
IGLO	individual gauges for localized orbitals
IRIS	inversional-rotational isomeric state
IUPAC	International Union of Pure and Applied Chemistry

LA	lactic acid
LAOCOON	least-squares adjustment of calculated on observed NMR spectra
LC	low crystallinity
LCAO	linear combination of atomic orbitals
LCST	lower critical solution temperature
LD	low-power decoupling
LDA	local-density approximation
LM	lattice model
LS	light scattering
LSDA	local spin-density approximation
M2ONH	*N*, *N′*-(ethane-1,2-diyl) dibenzamide
M2OO	ethane-1,2-diyl dibenzoate
M2OS	*S*, *S′*-(ethane-1,2-diyl) dibenzothioate
M2SNH	*N*, *N′*-(ethane-1,2-diyl) dibenzothioamide
M2SO	*O*, *O′*-(ethane-1,2-diyl) dibenzothioate
M2SS	ethane-1,2-diyl dibenzodithioate
M3ONH	*N*, *N′*-(propane-1,3-diyl)dibenzamide
M3OO	propane-1,3-diyl dibenzoate
M3OS	*S*, *S′*-(propane-1,3-diyl) dibenzothioate
M3SNH	*N*, *N′*-(propane-1,3-diyl) dibenzothioamide
M3SO	*O*, *O′*-(propane-1,3-diyl) dibenzothioate
M3SS	propane-1,3-diyl dibenzodithioate
MAA	methyl 2-acetoxyacetate
MAH	methyl 6-acetoxyhexanoate
MD	molecular dynamics
MEMA	*N*-(2-methoxyethyl)methylamine
MEMS	2-methoxyethyl methyl sulfide
MO	molecular orbital
MOAA	2-methoxy-2-oxoethyl 2-acetoxyacetate
MP	Møller–Plesset
MP2	the Møller–Plesset expansion truncated at second-order
MR	monomer ratio (of PS to PMEI)
MTEMA	*N*-(2-methylthioethyl)methylamine
MTT	2-methyl-1,3,5-trithiane
NBO	natural bond orbital
NIS	neutron inelastic scattering
NOB	not observed
P(*N*-tosylEI-ES)	poly(*N*-tosylethylene imine-*alt*-ethylene sulfide)
P(EI-EO)	poly(ethylene imine-*alt*-ethylene oxide)
P(EI-ES)	poly(ethylene imine-*alt*-ethylene sulfide)
P(EO-ES)	poly(ethylene oxide-*alt*-ethylene sulfide)
P2HB	poly(2-hydroxybutyrate)
P2ONH	poly(ethylene terephthalamide)
P2OO	poly(ethylene terephthalate)
P2OS	poly(ethylene dithioterephthalate)
P2SNH	poly(ethylene terephthalthioamide)

P2SO	poly(ethylene thionoterephthalate)
P2SS	poly(ethylene tetrathioterephthalate)
P3HB	poly((*R*)-3-hydroxybutyrate)
P3ONH	poly(trimethylene terephthalamide)
P3OO	poly(trimethylene terephthalate)
P3OS	poly(trimethylene dithioterephthalate)
P3SNH	poly(trimethylene terephthalthioamide)
P3SO	poly(trimethylene thionoterephthalate)
P3SS	poly(trimethylene tetrathioterephthalate)
P5OS	poly(pentamethylene dithioterephthalate)
PAS	principal-axis system
PBS	poly(butylene succinate)
PBT	poly(butylene terephthalate)
PCHC	poly(cyclohexene carbonate)
PCL	poly(ϵ-caprolactone)
PCM	polarized continuum model
PDB	Protein Data Bank
PDBS	poly(di-*n*-butylsilane)
PDHS	poly(di-*n*-hexylsilane)
PDMS	poly(dimethylsilane)
PDO	1,3-propanediol
PE	polyethylene
PEC	poly(ethylene carbonate)
PEI	poly(ethylene imine)
PEN	poly(ethylene-2,6-naphthalate)
PEO	poly(ethylene oxide)
PEOX	poly(2-ethyl-2-oxazoline)
PES	poly(ethylene sulfide) or poly(ethylene succinate)
PESe	poly(ethylene selenide)
PET	poly(ethylene terephthalate)
PGA	poly(glycolic acid)
PH	phenol
Pip	piperidinium
PLA	poly(lactic acid)
PMEI	poly(*N*-methylethylene imine)
PMEI-PS	PMEI and PS latex
PMePP	poly(1-methylphosphirane)
PMO	poly(methylene oxide)
PMPL	poly(DL-β-methyl β-propiolactone)
PMPrS	poly(methyl-*n*-propylsilane)
PMS	poly(methylene sulfide)
PMSe	poly(methylene selenide)
PMTMI	poly(*N*-methyltrimethylene imine)
PO	propylene oxide
PPC	poly(propylene carbonate)
PPhPP	poly(1-phnylphosphirane)

PPO	poly(propylene oxide)
PPP	polyphosphine
PPS	poly(propylene sulfide)
PS	polystyrene
PSO	paramagnetic spin–orbit
PST	pulse saturation transfer
PTetMO	poly(tetramethylene oxide)
PTMI	poly(trimethylene imine)
PTrMO	poly(trimethylene oxide)
PTrMS	poly(trimethylene sulfide)
PTrMSe	poly(trimethylene selenide)
PTT	poly(trimethylene terephthalate)
PyONH	poly(alkyl terephthalamide)
PyOS	poly(alkyl dithioterephthalate)
PySNH	poly(alkyl terephthalthioamide)
PySO	poly(alkyl thionoterephthalate)
PySS	poly(alkyl tetrathioterephthalate)
PyTS4	poly(alkyl tetrathioterephthalate)
QST2	synchronous transit-guided quasi-Newton method with two molecule specifications
RIS	rotational isomeric state
RMSE	root-mean-square error
S4TPA	tetrathioterephthalate acid
S4TPA-Pip	tetrathioterephthalate acid complexed with piperidinium
SANS	small-angle neutron scattering
SAXS	small-angle X-ray scattering
SC	single chain or semicrystalline
SCF	self-consistent field
SD	spin-dipole or standard deviation
S–F	Stockmayer–Fixman
SLS	static light scattering
SS(V)PE	surface and simulation of volume polarization for electrostatics
SVPE	surface and volume polarization for electrostatics
TetCE	tetrachloroethane
TetMGDB	tetramethylene glycol dibenzoate
tetra-MEDA	N,N,N',N'-tetramethylethylenediamine
TFA	trifluoroacetic acid
TFE	trifluoroethanol
THF	tetrahydrofuran
TMS	tetramethylsilane
TOCOSY	total correlation spectroscopy
TriCPH	trichlorophenol
TriMGDB	trimethylene glycol dibenzoate
T–T	tail-to-tail
UCST	upper critical solution temperature
VESTA	visualization for electronic and structural analysis

Part I

Fundamentals of Polymer Physical Chemistry

Conformational Analysis of Polymers: Methods and Techniques for Structure-Property Relationships and Molecular Design, First Edition. Yuji Sasanuma.

1

Stereochemistry of Polymers

1.1 Configuration

It is well known that the direction, position, and length of a side chain, arrangements of the side chains, and connection ways of monomeric units significantly influence the conformational characteristics and spatial shape of the polymer. Such structural characteristics of polymers are generally termed *configurations*. In Oxford Advanced Learner's Dictionary, the word "configuration" is explained as follows: an arrangement of the parts of something or a group of things; the form or shape that this arrangement produces. If the monomeric unit has a chiral center, it is assigned to either *R* or *S* enantiomer. Whether the monomeric unit is *R* or *S* will be due to the chirality of the monomer itself and the mechanism of polymerization. Besides, the chain dimension is often termed *spatial configuration*, which will be determined by the conformational sequence along the polymeric chain. It should be noted that the word "configuration" has been used in different senses in polymer chemistry.

Figure 1.1 represents vinyl polymers such as polypropylene (R = CH_3), poly(vinyl chloride) (R = Cl), and polystyrene (R = C_6H_5). When the polymeric chain in the all-trans form is put on the paper so that the methine carbon is located below as in Figure 1.1, and if the side chain R appears on the front or back side of the paper, the arrangement is defined as *d* or *l* form, respectively. As illustrated on the right of Figure 1.1, if the polymeric chain of all-*d* configuration (above) is rotated around the central arrow by 180°, it will be the all-*l* structure (below). Such *d* and *l* definitions are temporary, thus designated as Flory's pseudoasymmetry [140]. The configurational relation between two adjacent units, diad, is termed as follows: *meso*, *dd* and *ll*; *racemo*, *dl* and *ld*. The diad will stay invariant even after the rotation; accordingly, it may be preferable to use *meso* and *racemo* rather than *d* and *l*. If a polymeric chain includes only *meso* diads, the configurational regularity, tacticity, is referred to as isotactic. If only *racemo* diads are included,

Conformational Analysis of Polymers: Methods and Techniques for Structure-Property Relationships and Molecular Design, First Edition. Yuji Sasanuma.

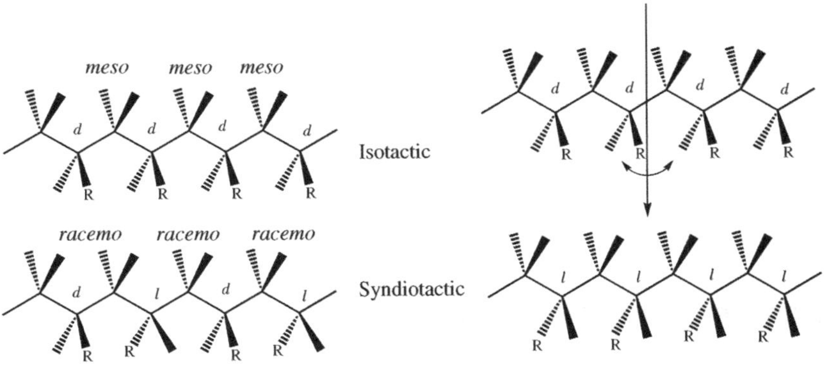

Figure 1.1 Vinyl polymer $[-CH_2CHR-]_x$ with the definition of *d* and *l* forms based on Flory's pseudoasymmetry and *meso* and *racemo* diads.

the tacticity is termed syndiotactic. When no configurational regularity is formed, the polymeric chain is atactic.

As illustrated in Figure 1.1, isotactic all-trans vinyl polymers have two skeletal C—C bonds in the monomeric unit and stick the side chains out only on one side of the paper, while the syndiotactic chains show the side chains on the front and back sides alternately. In contrast, tacticities of polymers with odd numbers of skeletal bonds in the monomeric unit, also being defined according to the *meso* and *racemo* diads, appear to be opposite to those of vinyl polymers: the side groups of the isotactic chain appear alternately, and those of the syndiotactic chain are found on the same side (see Figure 1.2).

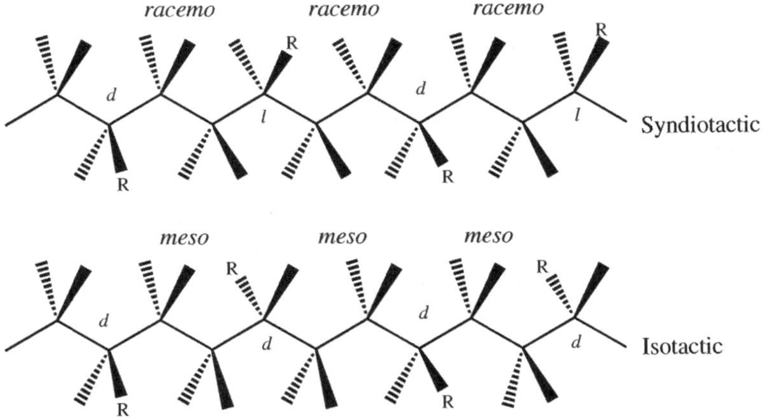

Figure 1.2 Iso- and syndiotactic polymeric chains with three skeletal bonds in the monomeric unit.

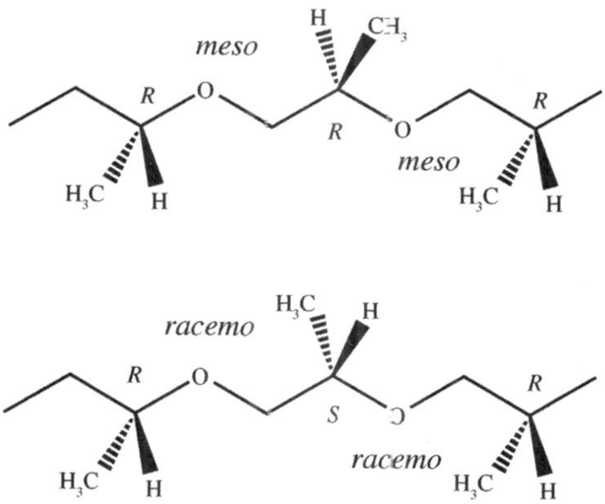

Figure 1.3 Portions of iso- and syndiotactic poly(propylene oxide)s with the definition of the *meso* and *racemo* diads.

So far, the configurations of polymers have been discussed according to Flory's pseudoasymmetry. On the other hand, the diads and tacticities of the polymeric chains containing the absolutely asymmetric carbon are defined with respect to the *R* or *S* centers. In Figure 1.3, the *meso* and *racemo* diads of poly(propylene oxide) (PPO) are illustrated: *meso*, *RR* and *SS*; *racemo*, *RS* and *SR*. All *meso* and all *racemo* chains are also designated as isotactic and syndiotactic, respectively. Inasmuch as the repeating unit of PPO has three skeletal bonds, the *meso* and *racemo* diads are defined similarly as those in Figure 1.2.

1.2 Connection Type of Monomeric Units

Poly(propylene carbonate) (PPC) has three types of connections between the monomeric units. If the $CH(CH_3)$ and CH_2 groups are termed "head" and "tail," respectively, the three connections are represented as head-to-head (H—H, —CH(CH_3)—OC(=O)O—CH(CH_3)—), head-to-tail (H—T, —CH(CH_3)—C(=O)O—CH_2—), and tail-to-tail (T—T, —CH_2—C(=O)O—CH_2—) (see Figure 1.4). In addition, since the monomeric unit of PPC has an asymmetric carbon and assigned to either (*R*)- or (*S*)-optical isomer, the *meso* and *racemo* diads may be formed (Figure 1.5) [415].

1.3 Nitrogen Inversion

Commercially available poly(ethylene imine) (PEI) is mostly prepared by ring-opening polymerization of aziridine, and such PEIs tend to branch at the

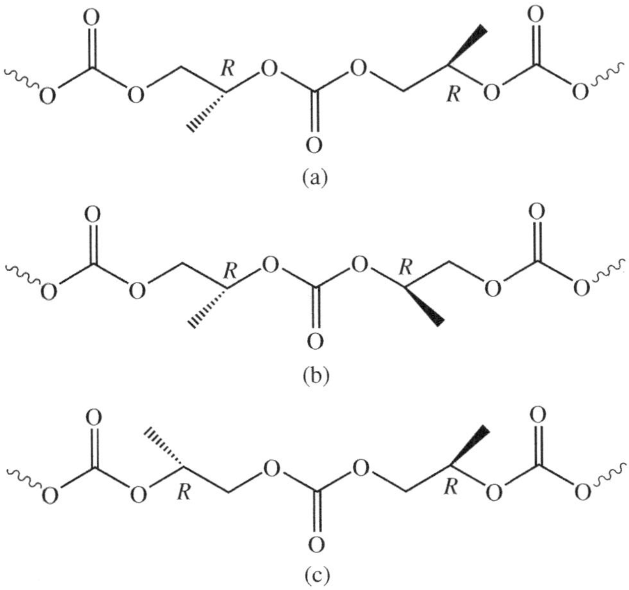

Figure 1.4 Three connection types between the repeating units of poly(propylene carbonate) (PPC): (a) head-to-tail (H—T); (b) head-to-head (H—H); and (c) tail-to-tail (T—T). All the repeating units here are (*R*)-form. Source: Adapted with permission from reference [415]. Copyright 2017 American Chemical Society.

nitrogen site. Therefore, to obtain linear PEI ($[—CH_2—CH_2—NH—]_x$), hydrolysis of, for example poly(2-ethyl-2-oxazoline) has been employed. Inasmuch as it forms aggregates with DNA and acts as a gene-delivery polymer, PEI has attracted attention in the field of gene therapy.

Since nitrogen is trivalent, the nitrogen atom of PEI may possibly be a center of the pseudoasymmetry. Therefore, *meso* and *racemo* diads and, furthermore, tacticities may be defined for the linear PEI chain. In the 1980s, however, it was found that amines have a unique nature termed nitrogen inversion [55], which counterchanges the positions of the substitute and lone pair of the nitrogen atom. Even if the substitute is as bulky as the *tert*-butyl group, the nitrogen inversion rapidly occurs at room temperature.

Figure 1.6 shows the nitrogen inversion of *N*,*N*′-dimethylethylenediamine, a model compound of PEI [403]. The nitrogen inversion always switches the configuration between *meso* and *racemo*, and the transition state has a higher Gibbs free energy (activation energy) by approximately 4 kcal mol^{-1} than the *meso* and *racemo* states. In the transition state, the lone pair seems as if it were a p orbital, and the two carbon, nitrogen, and hydrogen atoms are seen to be coplanar.

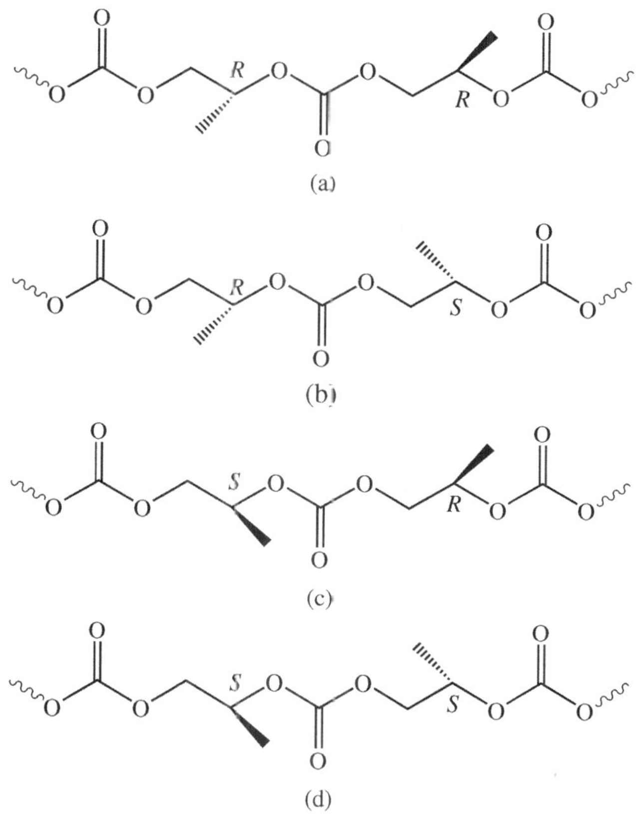

Figure 1.5 Diads of PPC: (a) *RR* (*meso*); (b) *RS* (*racemo*); (c) *SR* (*racemo*); and (d) *SS* (*meso*). Source: Adapted with permission from reference [415]. Copyright 2017 American Chemical Society.

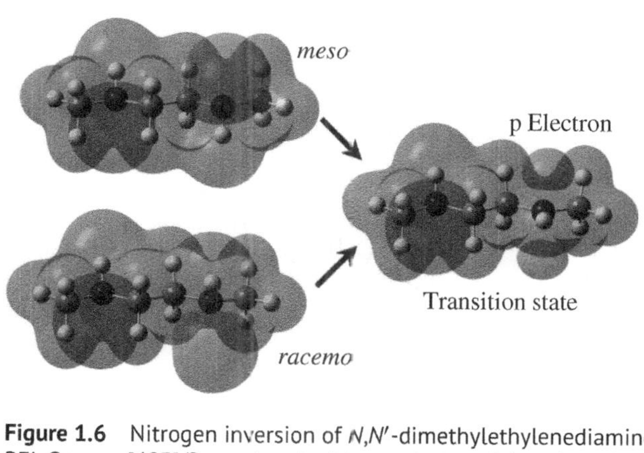

Figure 1.6 Nitrogen inversion of *N*,*N*′-dimethylethylenediamine, a model compound of PEI. Source: [403]/Reproduced with permission of American Chemical Society.

1.4 Conformation

Polymeric chains in solutions and melts always change the spatial configuration owing to internal rotations around single bonds. Figure 1.7 shows Newman projections of three stable staggered states, namely, trans, gauche$^+$, and gauche$^-$ conformations of *n*-alkanes and polyethylene (PE). In the trans conformation, the two carbon atoms are located opposite to each other. If the distant carbon atom is rotated clockwise (counterclockwise) by about 120°, the new conformation is gauche$^+$ (gauche$^-$). The hypothesis that such staggered conformations may represent all rotational states, designated as the rotational isomeric state (RIS) approximation, has often been adopted in polymer chemistry.

The dihedral angle of the trans state is defined as either 0° or 180°. The former definition (ϕ) has been traditionally used in polymer science, and the latter (φ) is recommended by the International Union of Pure and Applied Chemistry (IUPAC) [308]. According to the former, the gauche$^+$ (notation: g$^+$) and gauche$^-$ (g$^-$) states are, respectively, positioned at $\phi \approx 120°$ and $-120°$, and according to the latter, the gauche$^+$ (notation: G) and gauche$^-$ ($\overline{\text{G}}$) states are, respectively, located at $\varphi \approx 60°$ and $-60°$. It should be noted that the g$^+$ and g$^-$ conformations correspond to the $\overline{\text{G}}$ and G states, respectively.

Figure 1.8 shows three staggered conformations around the $CH(CH_3)—CH_2$ bond of poly((*R*)-propylene oxide) (PPO). In symmetric chains such as PE, the gauche$^+$ and gauche$^-$ conformations are equivalent (mirror images of each other in the Newman projection), whereas those of PPO are nonequivalent owing to the asymmetric methine carbon atom.

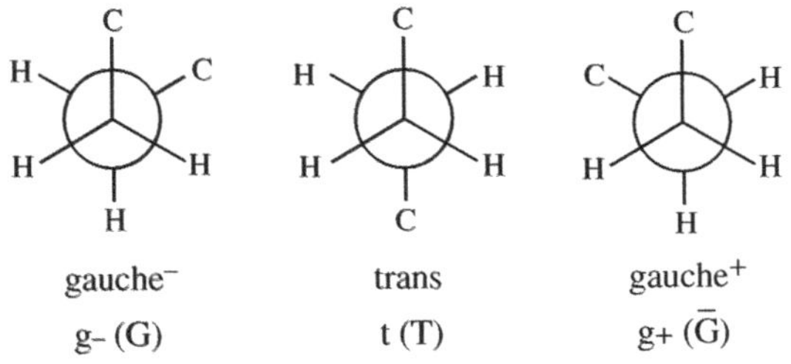

Figure 1.7 Rotational isomeric states of *n*-alkanes and polyethylene.

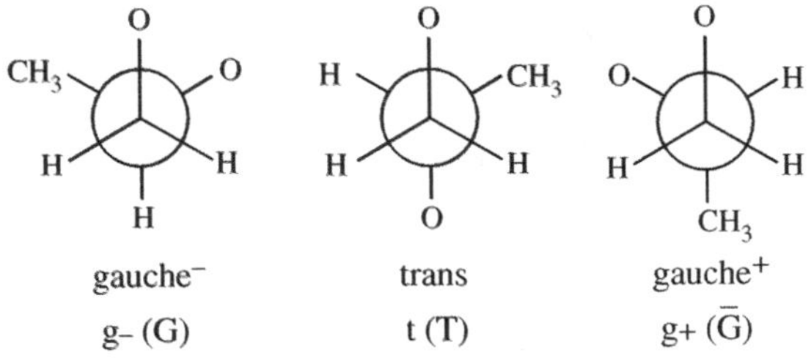

Figure 1.8 Rotational isomeric states of poly((*R*)-propylene oxide) (PPO).

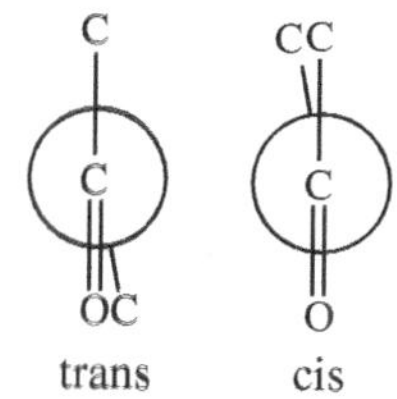

Figure 1.9 Rotational isomeric states around the C(=O)—O bond of esters.

Figure 1.9 depicts the preferred conformations around the C(=O)—O bond of esters. In contrast to the CH_2—CH_2 bond, the C(=O)—O bond adopts either of two eclipsed forms: trans or cis. In aliphatic esters, the cis state is higher in Gibbs free energy by 7–8 kcal mol^{-1} than trans [418]; thus, it can be assumed that the bond lies exclusively in the trans conformation.

1.5 Secondary Structure

The polymeric chain forms some secondary structure owing to its conformational characteristics and intramolecular interactions. A typical secondary structure is the helix. Figure 1.10 shows the helical structures of poly(methylene oxide) (PMO), poly(methylene sulfide) (PMS), and poly(methylene selenide)

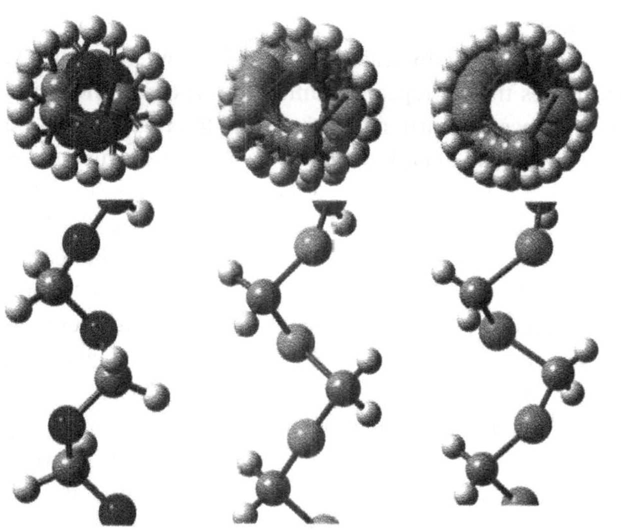

Figure 1.10 Helical structures of (a) PMO (9/5. 17.4 Å), (b) PMS (17/9, 36.5 Å), and (c) PMSe (21/11, 46.3 Å), where the values in the parentheses represent the helical pitches. Source: [421]/Reproduced with permission of American Chemical Society.

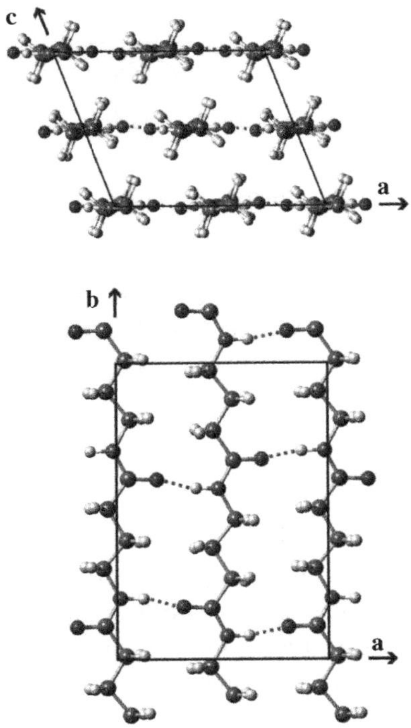

Figure 1.11 All-trans zigzag (α) form of nylon 4 in the crystal. The dotted line represents an N—H· · ·O=C hydrogen bond. Source: [165], Figure 6 (p. 9549)/American Chemical Society/Licensed under CC BY 4.0.

(PMSe). The helical pitches of PMO, PMS, and PMSe are 17.4 Å (9/5) [57, 458, 465, 466, 470, 503], 36.5 Å (17/9) [61], and 46.3 Å (21/11) [60], respectively, where the m/n notation means that m repeating units form n turns; therefore, 1.80, 1.89, and 1.91 units exist per one turn, respectively. All the three helices lie in all-gauche structures. The gauche energies relative to the trans states of the CH_2—X (X = O, S, and Se) bonds were evaluated to be −1.4 to −1.5 kcal mol^{-1}, −1.0 to −1.4 kcal mol^{-1}, and about −1.0 kcal mol^{-1}, respectively [7, 153, 421, 423]. The heteroatoms, O, S, and Se, are chalcogens. As the atomic number increases, the helix becomes loose, and the gauche stability is reduced. The PMO chain crystallizes to form another helix (2/1) that is more stable than 9/5 at low temperatures [58, 166, 251]. PMO is so highly crystalline as to be used as an engineering plastic [290], and its equilibrium melting point is as high as 479.2 K [293].

Another representative secondary structure of polymers is the all-trans zigzag structure. Figure 1.11 shows the all-trans form of nylon 4 [156, 165]. Between the neighboring chains arranged antiparallelly (directions: one NH→CO and the other CO→NH), N—H· · ·O=C hydrogen bonds are formed to stabilize the sheet structure.

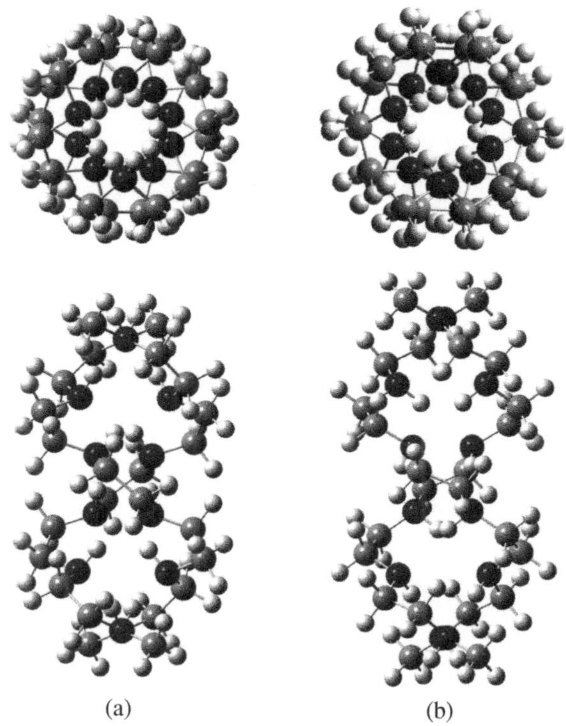

Figure 1.12 Double-stranded helices composed of two methyl-capped ethylene imine pentamers: (a) anhydrous crystal structure [71] and (b) isotactic structures optimized by density functional calculations at the B3LYP/6-31G(d) level. Source: [403], Figure 15 (p. 9181)/Reproduced with permission of American Chemical Society.

1.6 Double Helix

Two helical chains are coupled, and interchain hydrogen bonds are formed to reinforce the associated structure. Such structures were found in DNA and termed double helix. Linear PEI also forms a double helix in an anhydrous environment (Figure 1.12) [71]. The helical pitch is five units per turn, and the NH—CH_2—CH_2—NH bond sequence of the repeating unit lies in a tgt conformation of the *meso* configuration, which is the most stable state of the isolated PEI chain [403]. As described above, the NH site of PEI performs the nitrogen inversion, and directions of the N—H bond and lone pair may be adjusted so as to lower the conformational energy, form the interchain hydrogen bond, and, consequently, minimize the Gibbs free energy of the double helix, where only *meso* diads exist; therefore, the tacticity is isotactic. It is obvious that the nitrogen inversion enables PEI to realize the unique double helix.

2

Models for Polymeric Chains

2.1 Spatial Configuration of Polymeric Chain

The spatial configuration of a polymeric chain has usually been represented by the mean-square end-to-end distance $\langle r^2 \rangle$, which can be expressed with bond vector $\boldsymbol{l}_j$ (j, bond symbol, see Figure 2.1):

$$\langle r^2 \rangle = \left\langle \left(\sum_{i=1}^{n} \boldsymbol{l} \right) \cdot \left(\sum_{j=1}^{n} \boldsymbol{l}_j \right) \right\rangle = nl^2 + 2\sum_{i<j}^{n} \langle \boldsymbol{l}_i \cdot \boldsymbol{l}_j \rangle \tag{2.1}$$

where $\langle \ \rangle$ stands for a weight average, and n is the number of skeletal bonds. This relation is always valid regardless of what model is employed to represent the polymeric chain. The first term on the right-hand side expresses the sum of scalar products between the identical bond vectors, and the second term is that of scalar products between different bond vectors (Figure 2.2). The former and latter are independent of and dependent on the polymer model, respectively. In the following sections, some representative models are introduced, and the $\langle r^2 \rangle$ will be formulated individually.

2.2 Freely Jointed Chain

The freely jointed chain, the simplest model with the concept of chemical bond, has a fixed bond length l, whereas no restrictions are imposed on the bond angle and dihedral angle. The average of the scalar product between $\boldsymbol{l}_j$ and $\boldsymbol{l}_{j+k}$ is expressed by

$$\langle \boldsymbol{l}_i \cdot \boldsymbol{l}_{i+k} \rangle = l^2 \langle \cos(\pi - \theta_i) \cdots \cos(\pi - \theta_{i+k-1}) \rangle \tag{2.2}$$

where θ_j is the bond angle of bond j, i.e. angle between bonds j and $j+1$. Each bond angle varies freely; therefore, the average is simplified to

$$\langle \cos(\pi - \theta_i) \cdots \cos(\pi - \theta_{i+k-1}) \rangle = (-\langle \cos\theta \rangle)^k \tag{2.3}$$

Conformational Analysis of Polymers: Methods and Techniques for Structure-Property Relationships and Molecular Design, First Edition. Yuji Sasanuma.

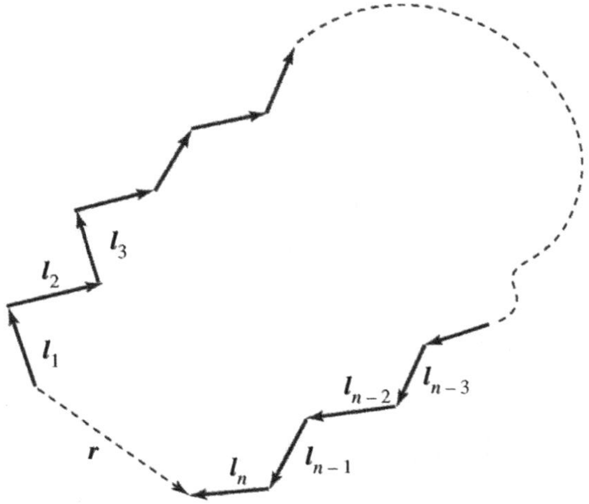

Figure 2.1 Bond vectors $\boldsymbol{l}_j$s of a polymeric chain. $\sum_{j=1}^{n}\boldsymbol{l}_j = \boldsymbol{r}$, where $\boldsymbol{r}$ is the end-to-end vector.

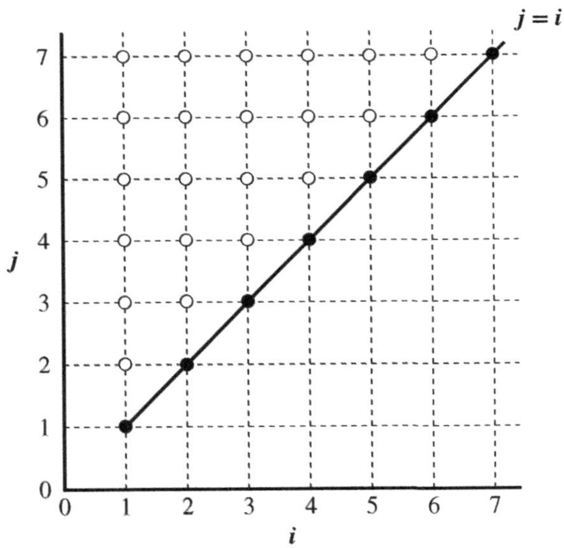

Figure 2.2 i and j pairs. The filled circles represent $i = j$ pairs for the calculation of the first term of Eq. (2.1), and the open circles express $i < j$ pairs to calculate the second term of Eq. (2.1).

Since θ is uniformly distributed between 0 and 2π, the average can be expressed by an integral:

$$\langle \cos\theta \rangle = \int_{-\pi}^{\pi} \cos\theta \, d\theta = [\sin\theta]_{-\pi}^{\pi} = 0 \tag{2.4}$$

Therefore, the second term of Eq. (2.1) is null, and the $\langle r^2 \rangle$ value of the freely jointed chain is

$$\langle r^2 \rangle = nl^2 \tag{2.5}$$

The chain dimension has often been scaled by the sum of squared bond lengths

$$\langle r^2 \rangle / nl^2 \tag{2.6}$$

This quantity was designated as *characteristic ratio*, which means a parameter characterizing the polymeric chain in terms of spatial configuration. The characteristic ratio of the freely jointed chain is unity.

2.3 Freely Rotating Chain

The bond length and bond angle of the freely rotating chain are fixed, whereas the dihedral angles are freely variable independently of each other. The second term of Eq. (2.1) can be expressed as

$$\begin{aligned} 2\sum_{i<j}^{n} \langle \boldsymbol{l}_i \cdot \boldsymbol{l}_j \rangle &= 2\sum_{i=1}^{n-1}\sum_{k=1}^{n-i} \langle \boldsymbol{l}_i \cdot \boldsymbol{l}_{i+k} \rangle = 2l^2 \sum_{k=1}^{n-1} (n-k)(-\cos\theta)^k \\ &= 2nl^2 \left\{ -\frac{\cos\theta}{1+\cos\theta} + \frac{\cos\theta\,[1-(-\cos\theta)^n)]}{n(1+\cos\theta)^2} \right\} \end{aligned} \tag{2.7}$$

where the following relation is used:

$$\langle \boldsymbol{l}_j \cdot \boldsymbol{l}_{j+k} \rangle = l^2 \cos^k(\pi - \theta) = l^2(-\cos\theta)^k \tag{2.8}$$

The scalar product between bonds j and $j + 1$ is $\langle \boldsymbol{l}_j \cdot \boldsymbol{l}_{j+1} \rangle = l^2 \cos(\pi - \theta)$. Bond $j + 2$ is rotatable freely and distributed uniformly on the surface of the circular cone as illustrated in Figure 2.3, and hence, the average direction of bond $j + 2$ coincides with that of bond $j + 1$, and the average magnitude of bond $j + 2$ is equal to the height of the cone. Therefore, we have $\langle \boldsymbol{l}_j \cdot \boldsymbol{l}_{j+2} \rangle = l^2 \cos^2(\pi - \theta)$. By repeating the coordinate transformation, we can obtain Eq. (2.8).

The characteristic ratio of the freely rotating chain is

$$\langle r^2 \rangle / nl^2 = \frac{1-\cos\theta}{1+\cos\theta} + \frac{2\cos\theta\,[1-(-\cos\theta)^n)]}{n(1+\cos\theta)^2} \tag{2.9}$$

For the infinitely long chain ($n \to \infty$), the second term of Eq. (2.9) will vanish; therefore,

$$\langle r^2 \rangle / nl^2 = \frac{1-\cos\theta}{1+\cos\theta} \tag{2.10}$$

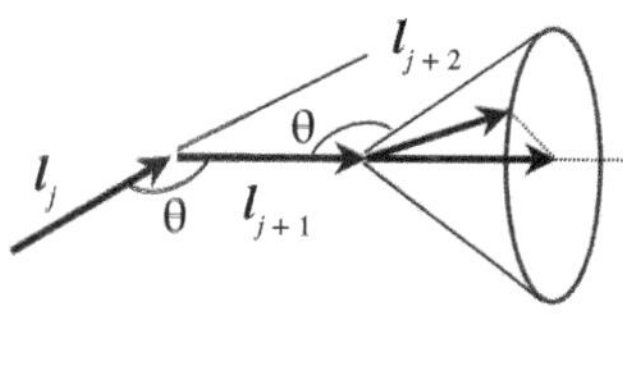

Figure 2.3 Bond vectors of the free rotating chain (see text).

When $\theta = 109.5°$ (tetrahedral sp^3-type bond), the characteristic ratio is 2.00, double that of the freely jointed chain.

Problem 1 Derive Eqs. (2.7) and (2.9).

2.4 Simple Chain with Rotational Barrier

Figure 2.4 schematically illustrates the potential curve of the rotation around the central CH_2—CH_2 bond of *n*-butane. The characteristic ratio of the simple chain with the rotational barrier is given by

$$\langle r^2 \rangle / nl^2 = \frac{1 - \cos\theta}{1 + \cos\theta} \frac{1 + \langle \cos\phi \rangle}{1 - \langle \cos\phi \rangle} \tag{2.11}$$

where the average cosine of dihedral angle ϕ is calculated from

$$\langle \cos\phi \rangle = \frac{\int_0^{2\pi} \cos\phi \, \exp\left[-E(\phi)/RT\right] d\phi}{\int_0^{2\pi} \exp\left[-E(\phi)/RT\right] d\phi} \tag{2.12}$$

Here, $E(\phi)$ is the potential energy as a function of ϕ, R is the gas constant, and T is the absolute temperature. If, under the rotational isomeric state (RIS) approximation, the potential curve is represented by only three staggered (locally minimum) states, i.e. trans, gauche$^+$, and gauche$^-$ conformations (Figure 1.7), the integral of Eq. (2.12) may be replaced with

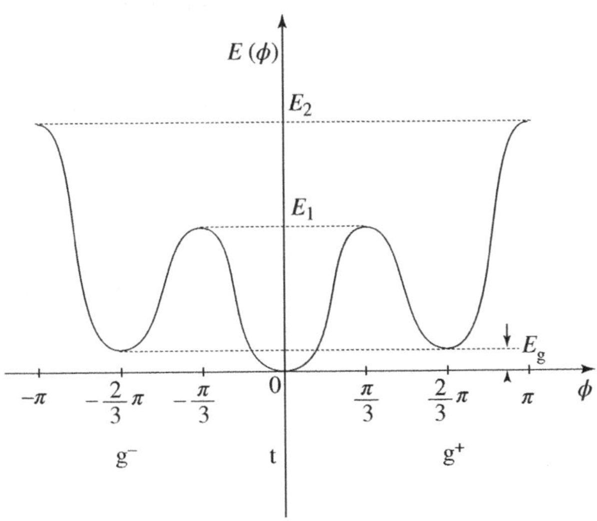

Figure 2.4 Potential curve for the rotation around the central CH_2—CH_2 bond of *n*-butane.

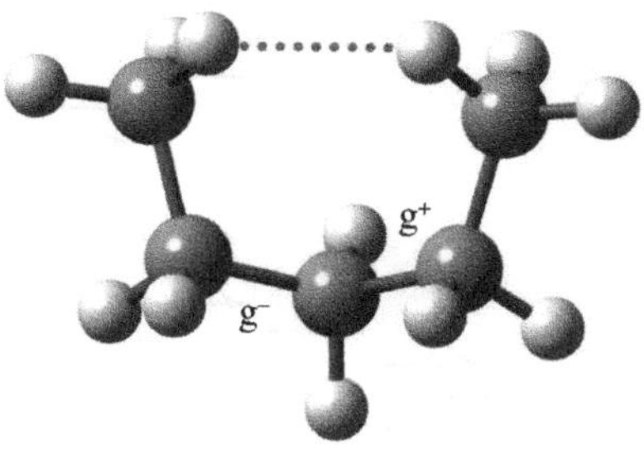

Figure 2.5 Pentane effect, a repulsive interaction between the terminal methyl groups separated by four bonds, when the two central bonds adopt $g^{\pm}g^{\mp}$ conformations.

$$\langle \cos\phi \rangle = \frac{\cos 0 + \cos(2/3)\pi \exp(-E_{g+}/RT) + \cos(-2/3)\pi \exp(-E_{g-}/RT)}{1 + \exp(-E_{g+}/RT) + \exp(-E_{g-}/RT)} \tag{2.13}$$

where the trans, gauche$^+$, and gauche$^-$ states are assumed to be located at $\phi = 0$, $(2/3)\pi$, and $-(2/3)\pi$ and have energies of 0, E_{g+}, and E_{g-}, respectively. With $E_{g+} = E_{g-} = 0.50$ kcal mol^{-1} [6, 141, 383], the $\langle \cos\phi \rangle$ value is obtained as 0.304, and the characteristic ratio can be calculated therefrom to be 3.8. This $\langle r^2 \rangle / nl^2$ value is still smaller than that (from 6 to 8) observed from polyethylene.

It has been suggested that the discrepancy between the calculated and observed $\langle r^2 \rangle / nl^2$ ratios here is due to the so-called pentane effect (Figure 2.5), which raises the interaction energy by as much as 2–3 kcal mol^{-1}. When two adjacent C—C bonds adopt $g^{\pm}g^{\mp}$ conformations, the repulsion occurs between the methyl (or methylene) groups separated by four C—C bonds. This effect was introduced into the RIS scheme, by which the experimental $\langle r^2 \rangle / nl^2$ value was well reproduced. The details of the RIS scheme will be explained later.

2.5 Gaussian Chain

Let us consider a one-dimensional random walk along the x-axis (see Figure 2.6). The numbers of steps in the plus (x_+) and minus (x_-) directions are defined as n_+ and n_-, respectively. The total number (n_x) of steps is

$$n_x = n_{x+} + n_{x-} \tag{2.14}$$

The position (x) after the n_x steps is

$$x = (n_{x+} - n_{x-})\, l \tag{2.15}$$

where l is the step width. From Eqs. (2.14) and (2.15), we derive

$$n_{x+} = \frac{n_x}{2}\left(1 + \frac{x}{n_x l}\right) \tag{2.16}$$

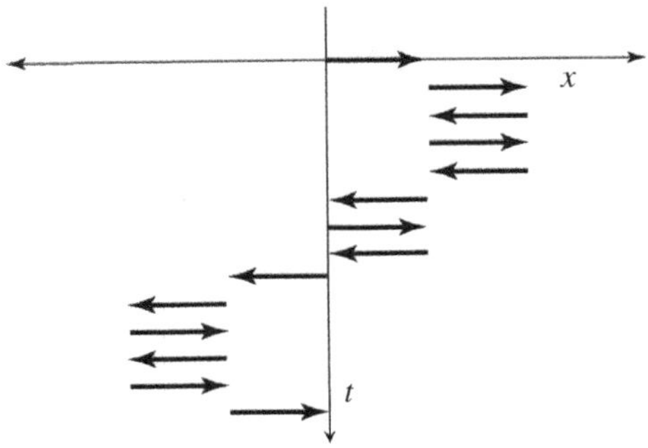

Figure 2.6 An example of one-dimensional random walks. One step width is l. The horizontal and vertical axes represent the one-dimensional position (x) and time (t), respectively. In this picture, after 14 steps, the walker has returned to the origin ($x = 0$)!.

and

$$n_{x-} = \frac{n_x}{2}\left(1 - \frac{x}{n_x l}\right) \tag{2.17}$$

For the random walk, probabilities of walks in the plus and minus directions are the same as 1/2 (Bernoulli trial); therefore, the probability that the walker stays at x after n_x steps is given by

$$p(x, n_x) = \frac{n_x!}{n_{x+}!n_{x-}!}\left(\frac{1}{2}\right)^{n_x} \tag{2.18}$$

With Stirling's approximation

$$\ln n_x! \approx n_x \ln n_x - n_x \tag{2.19}$$

Equation (2.18) can be free from the factorials to be

$$p(x, n_x) = \left(\frac{1}{2\pi n_x l^2}\right)^{1/2} \exp\left(-\frac{x^2}{2n_x l^2}\right) \tag{2.20}$$

This expression is a Gaussian function.

The above results can be extended to those of a three-dimensional random walk by

$$n_x = n_y = n_z = \frac{n}{3} \tag{2.21}$$

and

$$r^2 = x^2 + y^2 + z^2 \tag{2.22}$$

The x-, y-, and z-direction walks can be considered to be independent of each other; therefore,

$$p(r, n) = p(x, n_x)\cdot p(y, n_y)\cdot p(z, n_z) = \left(\frac{3}{2\pi n l^2}\right)^{3/2} \exp\left(-\frac{3r^2}{2nl^2}\right) \tag{2.23}$$

Equation (2.23) is also a Gaussian function. Figure 2.7 shows an example of the three-dimensional random walks. The thick arrows represent the walker's trace

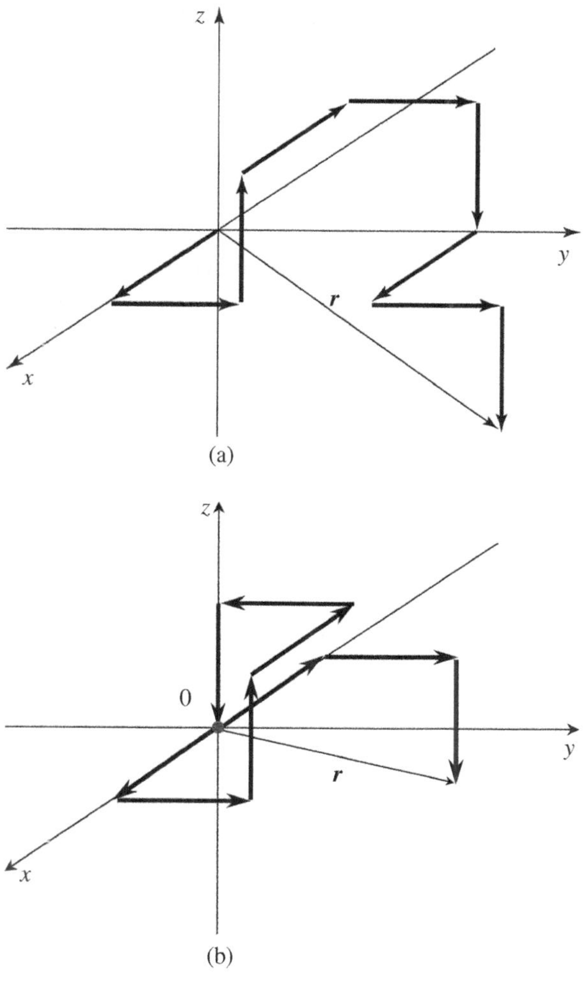

Figure 2.7 Examples of three-dimensional random walks: (a) self-avoiding walk and (b) simple random walk without the excluded-volume effect (the origin is occupied twice). ***r*** is the end-to-end vector.

and seem as if they were a chain molecule with a bond length of l and a bond angle of 90°. The arrows are results of Bernoulli trials in which the plus and minus steps occur at the same probability; accordingly, there is the possibility that the walker may return to and pass through the same point, which seems as if the arrow chain were lacking in volume. Therefore, the polymeric chain expressed by Eq. (2.23) is termed, generally, Gaussian chain, sometimes, phantom chain, and, often, ideal chain. Here, the word *ideal* means that the polymeric chain is free from the excluded-volume effect, which will be elucidated later. On the other hand, a polymeric chain that never passes the same point again is called self-avoiding chain.

The position (r_{max}) of the maximum of the radial distribution of the Gaussian function can be obtained from

$$\frac{d\left[4\pi r^2 p(r,n)\right]}{dr} = 0 \tag{2.24}$$

to be

$$r_{max} = \left(\frac{2}{3}nl^2\right)^{1/2} \tag{2.25}$$

The $\langle r^2 \rangle$ value of the Gaussian function is

$$\langle r^2 \rangle = \int_0^\infty r^2 p(r,n) 4\pi r^2 dr = nl^2 \tag{2.26}$$

The characteristic ratio is unity in agreement of that of the free jointed chain.

Problem 2 Derive Eq. (2.25) from Eq. (2.24).

Problem 3 Derive Eq. (2.26).

3

Lattice Model

3.1 Lattice Model of Small Molecules

As a simple thermodynamic model for solutions, the lattice model (LM) is well known. In this section, the LM for small molecules is explained. The LM is based on the following hypotheses:

(1) Either a solute or a solvent molecule occupies a single cell.
(2) The cell size is invariant irrespective of the rate between the solvent and solute.

Figure 3.1 schematically illustrates an LM. The open and filled circles represent solvent and solute molecules, respectively. If the numbers of solvent and solute molecules are defined as n_1 and n_2, respectively, and every cell is filled with either a solute or a solvent molecule, then the total number (n) of the cells is equal to

$$n = n_1 + n_2 \tag{3.1}$$

and the number (W) of *configurations*, namely the way to fill all the cells, can be calculated from

$$W = \frac{n!}{n_1!\, n_2!} \tag{3.2}$$

According to Boltzmann's entropy formula, the statistical entropy is obtained from

$$S = k \ln W = k \ln \frac{n!}{n_1!\, n_2!} \tag{3.3}$$

where k is Boltzmann's constant equal to R/N_A, with R and N_A being the gas constant and Avogadro's number, respectively. With the aid of Stirling's approximation,

$$\ln n! = n \ln n - n \tag{3.4}$$

the entropy becomes free from the factorials, being simplified to

$$S = -k\left(n_1 \ln X_1 + n_2 \ln X_2\right) \tag{3.5}$$

Conformational Analysis of Polymers: Methods and Techniques for Structure-Property Relationships and Molecular Design, First Edition. Yuji Sasanuma.

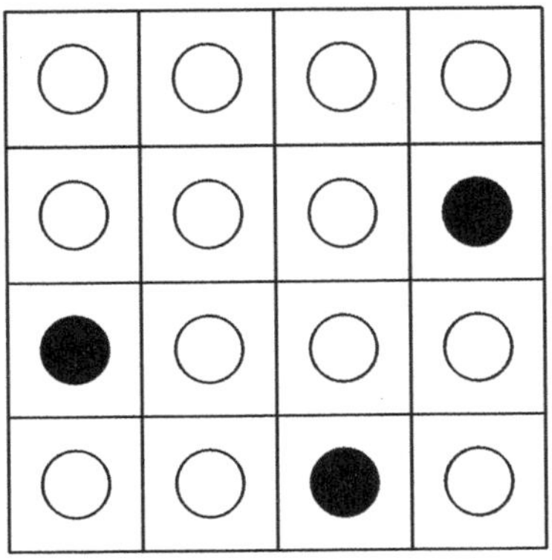

Figure 3.1 Lattice model (LM) for solutions. The open and filled circles stand for solvent and solute molecules. The numbers of solvent and solute are n_1 and n_2, respectively. Here, $n_1 = 13$ and $n_2 = 3$.

where

$$X_1 = \frac{n_1}{n} \tag{3.6}$$

and

$$X_2 = \frac{n_2}{n} \tag{3.7}$$

Here, X_1 and X_2 are termed molar fractions. Equation (3.5) can be rewritten as

$$S = -R\left(X_1 \ln X_1 + X_2 \ln X_2\right) \tag{3.8}$$

The S value corresponds to the configurational-entropy difference between the solution and the state where the solvent and solute molecules are separated, thus being henceforth referred to as mixing entropy and expressed as ΔS_{mix}.

Problem 4 Derive Eq. (3.5) from Eq. (3.3) with Stirling's approximation.

3.2 Flory–Huggins Theory

3.2.1 Entropy of Polymeric Chain

In this section, the LM for polymer solutions, designated as the Flory–Huggins theory [134, 135, 138, 205, 206], is explained. For polymer solutions, the above assumption (1) is invalid because a polymeric chain is much larger than a solvent molecule and hence cannot be pushed into a single cell. Therefore, the concept of *segment* was introduced: the polymeric chain is divided into a number of segments equivalent to a solvent molecule in size. An example of the LM is illustrated in Figure 3.2. The numbers of solvent molecules and polymeric chains are, respectively, n_1 and n_2, and the polymeric chain comprises x segments. After all n_2 polymeric chains are packed into $x \times n_2$ cells, the remaining n_1 cells will be

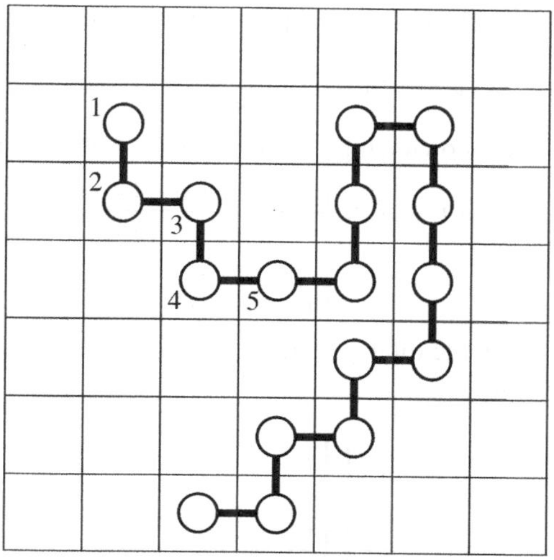

Figure 3.2 Lattice model (LM) for a polymer solution. The first to the *i*th polymeric chains have been already put in the lattice (not shown). The open circles represent polymer segments of the $(i + 1)$th chain, being connected to the adjacent segments. After all n_2 chains are put, the remaining vacant cells are filled with n_1 solvent molecules.

filled with the solvents; therefore, in all of the n cells, either a solvent or a polymer segment exists:

$$n = n_1 + xn_2 \tag{3.9}$$

Now, it is assumed that i polymeric chains have been placed already, and the $(i + 1)$th chain will be put in vacant cells. The number of vacant cells is $n - xi$; therefore, the first segment of the $(i + 1)$th chain can be put in any one of the $(n - xi)$ remaining cells. In the lattice, a cell is surrounded with z cells (z: coordination number). The second segment may be put in one of the z cells enclosing the first one if it is empty because the first and second segments are connected. The number of cells ready for the second segment may be $z(1 - f_i)$, given that the probability of vacancy, $1 - f_i$, is expressed as

$$1 - f_i = \frac{n - xi}{n} \tag{3.10}$$

where f_i $(= xi/n)$ corresponds to the overall segment occupancy. The third to the terminal xth segments are assumed to have the same possibility of $(z - 1)(1 - f_i)$ because one of the z cells has been already occupied, and hence, $(z - 1)$ is used in place of z. Even if one of the z cells is filled with the preceding segment, f_i is essentially invariant from the second to the xth segment because n would be as huge as comparable to, for example Avogadro's number, whereas x may be of the order of degree of polymerization $(n \gg x)$. The number of configurations of the $(i + 1)$th chain can be calculated to be

$$\begin{aligned} \nu_{i+1} &= (n - xi) \times [z(1 - f_i)] \times \left[(z - 1)(1 - f_i)\right]^{x-2} \\ &= (n - xi)z(z - 1)^{x-2}(1 - f_i)^{x-1} \\ &\approx (n - xi)(z - 1)^{x-1}(1 - f_i)^{x-1} \end{aligned} \tag{3.11}$$

The substitution of Eq. (3.10) into Eq. (3.11) leads to

$$v_{i+1} = (n - xi)^x \left(\frac{z-1}{n}\right)^{x-1} \tag{3.12}$$

The following ingenious approximation may be allowable because of $n \gg x$:

$$\begin{aligned}(n-xi)^x &\approx \frac{(n-xi)(n-xi-1)(n-xi-2)\cdots(n-xi-x+1)(n-xi-x)\cdots 1}{(n-xi-x)(n-xi-x-1)\cdots 1} \\ &= \frac{(n-xi)!}{[n-x(i+1)!]}\end{aligned} \tag{3.13}$$

Then, the total number of configurations to pack all n_2 polymeric chains into the lattice is

$$\begin{aligned}W &= \frac{1}{n_2!}\prod_{i=1}^{n_2} v_i \\ &= \frac{1}{n_2!}\prod_{i=1}^{n_2}\frac{[n-x(i-1)]!}{(n-xi)!]}\left(\frac{z-1}{n}\right)^{x-1} \\ &= \frac{n!}{(n-xn_2)!\ n_2!}\left(\frac{z-1}{n}\right)^{(x-1)n_2}\end{aligned} \tag{3.14}$$

Stirling's approximation removes the factorials:

$$\begin{aligned}S &= k\ln W \\ &= -k\left[n_1\ln\frac{n_1}{n_1+xn_2} + n_2\ln\frac{n_2}{n_1+xn_2} - n_2(x-1)\ln\frac{z-1}{e}\right]\end{aligned} \tag{3.15}$$

When $n_1 = 0$, namely without the solvent, Eq. (3.15) becomes

$$S_{\text{dis}} = kn_2\left[\ln x + (x-1)\ln\frac{z-1}{e}\right] \tag{3.16}$$

The S_{dis} term, designated as disordering entropy, represents the entropy difference of the polymeric chains in the disordered state and in the perfect regular arrangement. The difference between Eqs. (3.16) and (3.15) corresponds to the mixing entropy (ΔS_{mix})

$$S - S_{\text{dis}} = \Delta S_{\text{mix}} = -k(n_1\ln v_1 + n_2\ln v_2) \tag{3.17}$$

where

$$v_1 = \frac{n_1}{n_1+xn_2} \tag{3.18}$$

and

$$v_2 = \frac{xn_2}{n_1+xn_2} \tag{3.19}$$

are volume fractions of the solvent and polymer, respectively. The ΔS_{mix} term is similar in form to that for small molecules: If $x = 1$, which means that one solvent = one segment, is substituted into Eq. (3.17), the resultant expression corresponds to Eq. (3.5).

Problem 5 Derive Eqs. (3.15)–(3.17) from Eq. (3.14).

Equation (3.15) can be converted to

$$\begin{aligned} W &= \left[\left(\frac{n}{n_1}\right)^{n_1}\left(\frac{n}{xn_2}\right)^{n_2}\right] \times \left[(z-1)^{(x-1)n_2}\left(\frac{x}{e^{x-1}}\right)^{n_2}\right] \\ &= [\text{mixing}] \times [\text{disordering}] \\ &= [\text{intermolecular}] \times [\text{intramolecular}] \end{aligned} \tag{3.20}$$

The total number of the configurations, W, can be divided into two: intermolecular (mixing) and intramolecular (disordering) factors. The former is due to the mixing of polymers and solvents, whereas the latter includes no solvent parameters, thus being dependent only on the characteristics of the polymeric chain and further broken down into two subfactors: $(z-1)^{(x-1)n_2}$ and $(x/e^{x-1})^{n_2}$ [142]. The former subfactor represents the total number of potential configurations of the n_2 polymeric chains, and the latter subfactor considerably reduces the former number due to the restricted chain packing owing to $x \ll e^{x-1}$. The *mixing* may be interpreted as intermingling of a polymeric chain with solvents (in solutions) or with segments of other polymer chains (in melts and amorphous states). The *disordering* suggests changes in the polymeric chain from well-arranged to random states, namely from the perfect crystal to solutions, melts, or amorphous states. The entropy difference per segment between the two states, i.e. the entropy of fusion per segment, can be estimated as [137]

$$\frac{S_{\text{dis}}}{xn_2} \approx k \ln \frac{z-1}{e} \tag{3.21}$$

Inasmuch as the entropy function can be clearly divided into the inter- and intramolecular terms, configurational properties of polymers may be less affected by the environment regardless of whether it is placed in solution, melt, or amorphous phase [145, 146].

3.2.2 Enthalpy of Mixing

When a polymer segment comes in contact with a solvent molecule, the phenomenon can be expressed by the following reaction formula:

$$\frac{1}{2}\,[\text{solvent} \cdots \text{solvent}] + \frac{1}{2}\,[\text{solute} \cdots \text{solute}] \longrightarrow [\text{solvent} \cdots \text{solute}] \tag{3.22}$$

Then, the difference (Δw_{12}) in interaction energy may be written as

$$\Delta w_{12} = w_{12} - \frac{1}{2}(w_{11} + w_{22}) \tag{3.23}$$

where w_{11}, w_{22}, and w_{12}, respectively, stand for interaction energies between solvents, between segments, and between solvent and segment. If Δw_{12} is negative, null, or positive, the mixing would be exothermic, athermal, or endothermic, respectively. According to the LM, the enthalpy of mixing can also be formulated as

$$\begin{aligned}\Delta H_{\text{mix}} &= \Delta w_{12} \times x(z-2)\, n_2 \times v_1 \\ &= \Delta w_{12}\, x(z-2)\, n_2 \frac{n_1}{n_1 + xn_2} \\ &= \Delta w_{12}(z-2)\, n_1 \frac{xn_2}{n_1 + xn_2} \\ &= \Delta w_{12}(z-2)\, n_1 v_2 \end{aligned} \tag{3.24}$$

The derivation of Eq. (3.24) can be explained as follows: One contact between the segment and solvent generates the heat of Δw_{12}. One segment faces $(z-2)$ cell where solvent molecules may possibly exit at the probability of v_1, except for the terminal segments.[1] One polymeric chain has x segments, and the lattice includes n_2 chains.

Flory defined the χ parameter as [138, 142][2]

$$\chi_1 \equiv \frac{\Delta w_{12}\, z}{kT} \tag{3.25}$$

With the χ parameter, Eq. (3.24) is rewritten as

$$\Delta H_{\text{mix}} = kT\chi_1 n_1 v_2 \tag{3.26}$$

From Eqs. (3.17) and (3.26), the Gibbs free energy of mixing is derived to be [138]

$$\begin{aligned}\Delta G_{\text{mix}} &= \Delta H_{\text{mix}} - T\Delta S_{\text{mix}} \\ &= kT\left(n_1 \ln v_1 + n_2 \ln v_2 + \chi_1 n_1 v_2\right)\end{aligned} \tag{3.27}$$

3.2.3 Chemical Potential

The difference ($\Delta\mu_1$) in chemical potential between pure solvent (μ_1^0) and solution (μ_1) can be obtained from the derivative of the Gibbs free energy with respect to

1 Because only the terminal segments face $(z-1)$ cells, $x(z-2)+2$ is more accurate than $x(z-2)$; however, the term, 2, is removed to simplify Eq. (3.24).

2 Strictly, instead of z, $(z-2)$ should be used in Eq. (3.25). However, Flory defined χ_1 as above. This is because z is unknown, and hence, the adaptation of $z-2 \rightarrow z$ does not cause any problems.

the number of solvents [138]:

$$\left(\frac{\partial \Delta G_{mix}}{\partial n_1}\right)_{T,P,n_2} = kT\left[\ln(1 - v_2) + \left(1 - \frac{1}{x}\right)v_2 + \chi_1 v_2^2\right] \tag{3.28}$$

It should be noted that the chemical potential is defined per mole; therefore, k of Eq. (3.28) is replaced with R:

$$\Delta\mu_1 = \mu_1 - \mu_1^0 = RT\left[\ln(1 - v_2) + \left(1 - \frac{1}{x}\right)v_2 + \chi_1 v_2^2\right] \tag{3.29}$$

Problem 6 Derive Eq. (3.28) from Eq. (3.27). Note that both v_1 and v_2 are functions of n_1.

On the basis of Eq. (3.29), colligative properties of polymer solutions can be formulated. For example, the vapor pressure is given by

$$\ln\frac{p_1}{p_1^0} = \frac{\Delta\mu_1}{RT} = \ln(1 - v_2) + \left(1 - \frac{1}{x}\right)v_2 + \chi_1 v_2^2 \tag{3.30}$$

where p_1 and p_1^0 are vapor pressures of the solution and pure solvent, respectively. When $x = 1$ and $v_2 \ll 1$, namely the dilute solution of small solutes, $p_1/p_1^0 = 1 - v_2 = v_1$, which corresponds to Raoult's law.

Equation (3.30) can be changed with a series

$$\ln(1 - v_2) = -\left[v_2 + \frac{1}{2}v_2^2 + \cdots\right] \tag{3.31}$$

to

$$\frac{\Delta\mu_1}{RT} = -\frac{v_2}{x} - \left(\frac{1}{2} - \chi_1\right)v_2^2 \cdots \tag{3.32}$$

The osmotic pressure π is given by

$$\pi V_1 = -\Delta\mu_1 \tag{3.33}$$

where V_1 is the molar volume of solvent. From Eqs. (3.32) and (3.33), it follows

$$\pi = \frac{RT}{V_1}\left[\frac{v_2}{x} + \left(\frac{1}{2} - \chi_1\right)v_2^2 + \frac{v_2^3}{3} + \cdots\right] \tag{3.34}$$

Inasmuch as $v_2 = c\bar{v}$ and $v_2/xV_1 = c\bar{v}/xV_1 = c/M$, with $\bar{v}$ being the specific volume, Eq. (3.34) can be rewritten as

$$\frac{\pi}{c} = RT\left[\frac{1}{M} + \frac{\bar{v}^2}{V_1}\left(\frac{1}{2} - \chi_1\right)c + \frac{\bar{v}^3}{3v_1}c^2 + \cdots\right] \tag{3.35}$$

where c is the concentration in the unit of gram per unit volume. The osmotic pressure can also be expressed as the virial expansion:

$$\frac{\pi}{c} = RT\left[\frac{1}{M} + A_2 c + A_3 c^2 \cdots\right] \tag{3.36}$$

where A_2 and A_3 are the second and third virial coefficients, respectively. Comparison between Eqs. (3.35) and (3.36) leads to

$$A_2 \propto \frac{\bar{v}^2}{V_1}\left(\frac{1}{2} - \chi_1\right) \tag{3.37}$$

3.2.4 Excluded-Volume Effect I

According to the LM, the mixing free energy of polymer segments and solvent molecules in a small volume δV may be expressed as [138]

$$\Delta G_{\mathrm{M}} = kT\left[\delta n_1 \ln(1 - v_2) + \chi_1 \delta n_1 v_2\right] \tag{3.38}$$

where δn_1 is the number of solvent molecules included in the δV volume, and $v_2 = xn_2/(\delta n_1 + xn_2)$. Because there is no polymer *molecule* in δV, Eq. (3.38) is obtained by substituting $n_2 = 0$ into Eq. (3.27). The *excess chemical potential*, $(\mu_1 - \mu_1)_{\mathrm{E}}$, is derived from the derivative of Eq. (3.38) with respect to δn_1:

$$(\mu_1 - \mu_1)_{\mathrm{E}} = RT\left[\ln(1 - v_2) + v_2 + \chi_1 v_2^2\right] \tag{3.39}$$

which can also be obtained by substituting $x = \infty$ into Eq. (3.29) and rewritten as a series:

$$(\mu_1 - \mu_1)_{\mathrm{E}} = -RT\left[\left(\frac{1}{2} - \chi_1\right)v_2^2 + \frac{1}{3}v_2^3 + \cdots\right] \tag{3.40}$$

In place of χ_1 and 1/2, Flory introduced different parameters, κ_1 and ψ_1 [138]:

$$\Delta\overline{H_1} = RT\kappa_1 v_2^2 \tag{3.41}$$

and

$$\Delta\overline{S_1} = R\psi_1 v_2^2 \tag{3.42}$$

where $\Delta\overline{H_1}$ and $\Delta\overline{S_1}$ are the relative partial molar heat content and relative partial molar configurational entropy, respectively. The reason why κ_1 and ψ_1 were defined is that the entropy of dilution was found to differ widely for each combination of polymer and solvent. Ideally, κ_1 (enthalpy term) and ψ_1 (entropy term) are equal to χ_1 and 1/2, respectively, and hence satisfy

$$\kappa_1 - \psi_1 = \chi_1 - \frac{1}{2} \tag{3.43}$$

In order to express the second virial coefficient, A_2, as a function of temperature, the following parameter, being termed "Θ point" (in the unit K^{-1}), was introduced [152]:

$$\Theta = \frac{\kappa_1 T}{\psi_1} \tag{3.44}$$

Accordingly,

$$\psi_1 - \kappa_1 = \psi_1 \left(1 - \frac{\Theta}{T}\right) \tag{3.45}$$

Adoption of $(1/3)v_2^3 \approx 0$ in Eq. (3.40) leads to

$$(\mu_1 - \mu_1)_{\mathrm{E}} \approx -RT\psi_1 \left(1 - \frac{\Theta}{T}\right) v_2^2 \tag{3.46}$$

From Eqs. (3.37), (3.43), and (3.45), the second virial coefficient is expressed by

$$A_2 \propto \frac{\bar{v}^2}{V_1} \psi_1 \left(1 - \frac{\Theta}{T}\right) \tag{3.47}$$

When $T = \Theta$, A_2 becomes null, and hence, Eq. (3.36) obeys van't Hoff's law. As will be shown later, the ideal state is designated as unperturbed state or Θ state and corresponds to the null excluded volume (EV) (β).

The polymer segments are serially connected with each other; therefore, even in a dilute solution, segments may interact with other nonbonded segments of the same molecule. Such a long-range intramolecular interaction will usually enlarge or occasionally reduce the overall dimension of the polymeric chain, being termed *the excluded-volume effect* (EVE). If the EVE renders the chain dimension α times as large as that of the ideal chain in the unperturbed state, the EVE generates the excess mixing free energy (ΔG_{M}), and its derivative with respect to α was derived to be [138, 152]

$$\frac{\partial \Delta G_{\mathrm{M}}}{\partial \alpha} = -6C_{\mathrm{M}} kT\psi_1 \left(1 - \frac{\Theta}{T}\right) M^{1/2} \alpha^{-4} \tag{3.48}$$

where

$$C_{\mathrm{M}} = \frac{27}{2^{5/2}\pi^{3/2}} \frac{\bar{v}^2}{N_{\mathrm{A}} V_1} \left(\frac{\overline{r_0^2}}{M}\right)^{-3/2} \tag{3.49}$$

Here, N_{A} is Avogadro's number, $\bar{v}$ is the specific volume, V_1 is the molar volume of solvent, and $\overline{r_0^2}$ is the mean-square end-to-end distance of the unperturbed polymer. In addition, the expansion or shrink due to the EVE induces rubber-like elastic recovery, and its free energy is expressed by [136]

$$\Delta G_{\mathrm{el}} = kT \left[\frac{3(\alpha^3 - 1)}{2} - \ln \alpha^3\right] \tag{3.50}$$

which yields

$$\frac{\partial \Delta G_{\mathrm{el}}}{\partial \alpha} = 3kT(\alpha - \alpha^{-1}) \tag{3.51}$$

The equilibrium condition

$$\frac{\partial \Delta G}{\partial \alpha} = \frac{\partial \Delta G_{\mathrm{M}}}{\partial \alpha} + \frac{\partial \Delta G_{\mathrm{el}}}{\partial \alpha} = 0 \tag{3.52}$$

leads to

$$\alpha^5 - \alpha^3 = 2C_M\psi_1\left(1 - \frac{\Theta}{T}\right)M^{1/2} \tag{3.53}$$

$$= 2C_M\left(\psi_1 - \kappa_1\right)M^{1/2} \tag{3.54}$$

Equation (3.54) represents the characteristic features of the EVE. The α value increases with molecular weight and $\psi_1(1 - \Theta/T)$, i.e. $\psi_1 - \kappa_1$; therefore, the better solvent, the large chain dimension, and the poorer solvent, the smaller chain dimension. If $T = \Theta$, then $\alpha = 1$, and the polymeric chain, being free from the EVE, will behave as if it could overlap or interpenetrate one another freely ("phantom chain").

To discuss the swelling phenomenon of polymer molecules with solvents, Flory also derived the following expression [136]:

$$\alpha^5 - \alpha^3 \propto x_s^{-1} \tag{3.55}$$

where x_s is the number of lattice cells occupied by a solvent molecule, i.e. the size of solvent. On the basis of this relation, Flory asserted

> If the solvent is itself polymeric such that x_s is very large, the term on the left approaches zero and α approaches unity... Thus, polymeric molecules will occupy spatial configurations coinciding with those calculated in the random flight approximation, undistorted by interference effects... Expansion of the molecule achieves no net improvement in the configurational freedom of the system as a whole. This deduction is important for it justifies use of the random flight approximation in the treatment of concentrated polymer systems.

Therefore, he suggested that the polymeric chain existing in amorphous states, rubbers, and melts would be free from the EVE and hence lies in the unperturbed (Θ) state. This deduction has been demonstrated both theoretically and experimentally [146].

The $\alpha^5 - \alpha^3$ law on the EVE was derived in a different way (see, for example [112]). When a polymeric chain exhibits the EVE, its spatial distribution will deviate from the ideal state that may be expressed as the Gaussian chain:

$$p_0(r, n) = \left(\frac{3}{2\pi nl^2}\right)^{\frac{3}{2}} \exp\left(-\frac{3r^2}{2nl^2}\right) 4\pi r^2 \tag{3.56}$$

where $p_0(r, n)$ is the probability at which the polymeric chain of n bonds (segments) has an end-to-end distance of r, and l is the bond length (segment size). The EVE prevents distant segments from overlapping each other and reduces the probability by $p(r)$:

$$\begin{aligned} p(r) &= \left(1 - \frac{v_c}{r^3}\right)^{\frac{n(n-1)}{2}} \\ &= \exp\left[\frac{1}{2}n(n-1)\ln\left(1 - \frac{v_c}{r^3}\right)\right] \end{aligned} \tag{3.57}$$

where v_c is the EV between a pair of segments. Equation (3.57) is based on the following suppositions. The reduction in the probability due to an EV may be estimated to be v_c/V_p, where V_p is the volume occupied by the polymeric chain and may be approximated by r^3. There is the possibility that ${}_nC_2 = (1/2)n(n-1)$ pairs of segments form the EV in a polymeric chain. With approximations of $\ln(1 - v_c/r^3) \sim -v_c/r^3$ and $n(n-1) \simeq n^2$, Eq. (3.57) can be simplified to

$$p(r) \propto \exp\left(-\frac{n^2 v_c}{2r^3}\right) \tag{3.58}$$

The probability of the real chain exhibiting the EVE may be expressed by

$$\begin{aligned} p(r,n) &= p(r) \cdot p_0(r,n) \\ &\propto \exp\left(-\frac{n^2 v_c}{2r^3} - \frac{3r^2}{2nl^2}\right) 4\pi r^2 \end{aligned} \tag{3.59}$$

The condition of the maximum probability is given by

$$\frac{d \ln p(r,n)}{dr} = \frac{d}{dr}\left(-\frac{n^2 v_c}{2r^3} - \frac{3r^2}{2nl^2} + 2\ln r + \ln 4\pi\right) = 0 \tag{3.60}$$

which leads to

$$\frac{3n^2 v_c}{2r^3} - \frac{3r^2}{nl^2} + 2 = 0 \tag{3.61}$$

Here, let the expansion coefficient α be defined as

$$\alpha = \frac{r}{r_0} \tag{3.62}$$

Then, Eq. (3.61) can be rewritten as

$$\alpha^5 - \alpha^3 = \frac{9\sqrt{6}}{16}\frac{v_c}{l^3} n^{\frac{1}{2}} \tag{3.63}$$

Equation (3.63) also represents the $\alpha^5 - \alpha^3$ law and indicates that the $\alpha^5 - \alpha^3$ term is proportional to v_c and $n^{1/2}$ ($\propto M^{1/2}$). If the expansion is so large as to satisfy $\alpha^5 \gg \alpha^3$, then

$$r \propto \left(\frac{n^{1/2} v_c}{l^3}\right)^{1/5} r_0 \propto n^{\frac{1}{10}+\frac{1}{2}} \propto n^{3/5} \propto M^{3/5} \tag{3.64}$$

The end-to-end distance is proportional to $M^{3/5}$.

3.2.5 Excluded-Volume Effect II

The equation of state of the ideal gas is well known:

$$PV = nRT \tag{3.65}$$

where P is pressure, V is volume, and n is molar quantity. At high pressures, however, the gas molecules frequently collide (interact) with each other, and the molecular volume may not be neglected. For the real gas, van der Waals equation of state was proposed:

$$\left(P + \frac{n^2 a}{V^2}\right)(V - nb) = nRT \tag{3.66}$$

where the (an^2/V^2) and nb terms represent the intermolecular interaction and molecular volume, respectively, and a and b are the coefficients. If the gas molecule is assumed to be spherical, the EV where other molecules cannot access would be $(4/3)\pi a_0^3 = (4/3)\pi(2r_c)^3 = 8 \times (4/3)\pi r_c^3 = 8 \times V_c$, i.e. eight times as large as the core volume (V_c) (Figure 3.3).

The interaction energy between segments i and j of a polymeric chain may be expressed similarly to that of van der Waals force as

$$\begin{aligned} V_{ij} &= \infty \qquad (r < a_0) \\ &= -k_0\, U(r) \quad (r \geq a_0) \end{aligned} \tag{3.67}$$

In the spatial region of $r < a_0$, i.e. $V < (4/3)\pi a_0^3$, the infinite repulsion acts, whereas an attractive force tails away in the region of $r \geq a_0$ (see Figure 3.4). According to the cluster expansion [133, 446, 549, 552], the binary cluster integral for a pair of segments, a measure of the EV, may be obtained from

$$\begin{aligned} \beta &= \int_0^\infty \left[1 - \exp\left(-\frac{V_{ij}}{kT}\right)\right] 4\pi r^2 dr \\ &= \frac{4}{3}\pi a_0^3 - \int_{a_0}^\infty \left[1 - \exp\left(-\frac{V_{ij}}{kT}\right)\right] 4\pi r^2 dr \\ &\simeq \frac{4}{3}\pi a_0^3 - \frac{k_0}{kT}\int_{a_0}^\infty U(r) 4\pi r^2 dr \end{aligned}$$

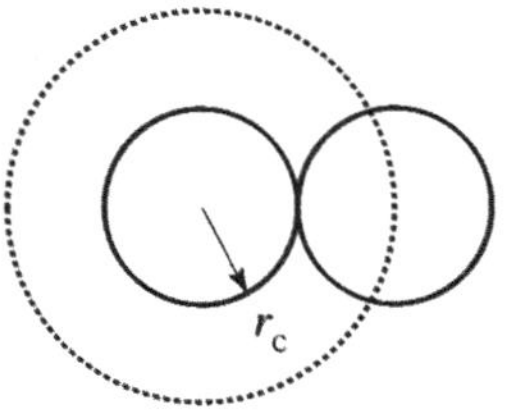

Figure 3.3 Excluded volume of a gas molecule, represented by the dotted sphere whose volume is eight times as large as the core volume expressed by the solid sphere. r_c is the core radius.

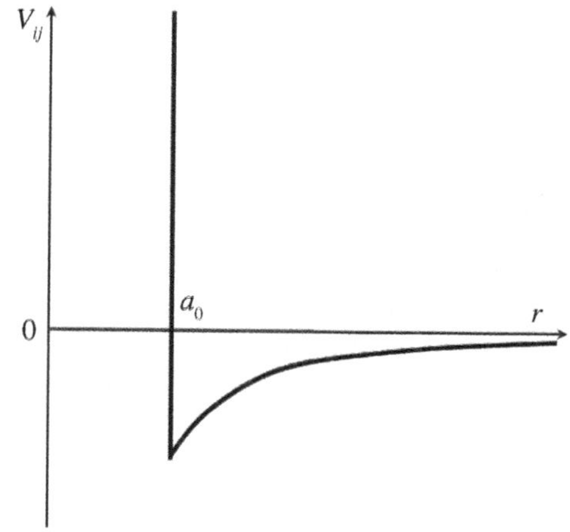

Figure 3.4 Nonbonded interaction (V_{ij}) assumed to act between polymer segments, expressed by Eq. (3.67)

$$
\begin{aligned}
&= V_c \left\{ 1 - \frac{k_0}{kT} \left(\frac{3}{a_0^3} \int_{a_0}^{\infty} U(r) r^2 dr \right) \right\} \\
&\equiv 2\psi_1 \left(1 - \frac{\Theta}{T} \right) V_c \\
&\simeq 2\psi_1 \left(1 - \frac{\Theta}{T} \right) V_1 \\
&\simeq 2\psi_1 \left(1 - \frac{\Theta}{T} \right) \frac{V_s^2}{V_1}
\end{aligned} \tag{3.68}
$$

where an approximation of $\exp(-V_{ij}/kT) = 1 - V_{ij}/kT + \cdots$ is used, and V_c, V_s, and V_1 are molecular volumes of the core, solvent, and segment, respectively.

The abovementioned $\alpha^5 - \alpha^3$ law can be rewritten with β as

$$
\begin{aligned}
\alpha^5 - \alpha^3 &= 2C_M \psi_1 \left(1 - \frac{\Theta}{T} \right) M^{\frac{1}{2}} \\
&\equiv Cz
\end{aligned} \tag{3.69}
$$

Here, the coefficient C was evaluated by Flory to be $3^{3/2}/2 = 2.60$ [149], whereas Zimm, Fixman, and Stockmayer recommended to adopt $C = 4/3$ [133, 445, 552], and, later, Stockmayer suggested that $C = 1.28$ would be rather better [446], and z is defined as

$$
\begin{aligned}
z &= \left(\frac{3}{2\pi \langle r^2 \rangle_0} \right)^{\frac{3}{2}} \beta n^2 \\
&= \left(\frac{3}{2\pi l^2} \right)^{\frac{3}{2}} \beta n^{\frac{1}{2}}
\end{aligned} \tag{3.70}
$$

with $\langle r^2 \rangle_0$ being the end-to-end distance of the unperturbed chain, e.g. the Gaussian chain: $\langle r^2 \rangle_0 = nl^2$.

The $\langle r^2 \rangle$ or $\langle S^2 \rangle$ value of a polymer under a given condition (solvent and temperature) can be predicted provided that both $\langle r^2 \rangle_0$ or $\langle S^2 \rangle_0$ and α values are known. This principle is termed *the two-parameter theory*. A number of mathematical representations for α as a function of z have been proposed so far. For the details, see, e.g. Yamakawa's [533] and Fujita's [163]. Of them, formulae derived by Domb, Barrett, and Lax [116, 272] should be mentioned here.

Domb et al. have employed the Domb–Joyce model [117] in which the excluded-volume interaction is varied between the phantom and self-avoiding chains, enumerated the end-to-end distances of three-dimensional lattices such as simple cubic, body-centered cubic, face-centered cubic, and diamond lattices, and derived the expansion coefficient as a function of z:

$$\alpha_r^2 = \left[1 + 10z + \left(\frac{70}{9}\pi + \frac{10}{3}\right)z^2 + 8\pi^{\frac{3}{2}}z^3\right]^{\frac{2}{15}} \tag{3.71}$$

where α_r is the expansion coefficient for the mean-square end-to-end distance,

$$\alpha_r^2 = \frac{\langle r^2 \rangle}{\langle r^2 \rangle_0} \tag{3.72}$$

The expansion coefficient α_S for the mean-square radius of gyration can be derived from

$$\frac{\alpha_S^2}{\alpha_r^2} = 0.933 + 0.067 \exp\left[-\left(0.85z + 1.39z^2\right)\right] \tag{3.73}$$

where

$$\alpha_S^2 = \frac{\langle S^2 \rangle}{\langle S^2 \rangle_0} \tag{3.74}$$

The expansion coefficients are defined separately for r^2 and S^2. Debye derived the following relation between $\langle r^2 \rangle_0$ and $\langle S^2 \rangle_0$ [103]:

$$\langle r^2 \rangle_0 = 6\langle S^2 \rangle_0 \tag{3.75}$$

The α_r^2 expression was further modified to [272]

$$\alpha_r^2 = \left[1 + 20z + 155.54z^2 + 591.86z^3 + 325z^4 + 1670z^6\right]^{\frac{1}{15}} \tag{3.76}$$

Equations (3.71), (3.73), and (3.76) were shown to be consistent with Monte Carlo simulations for self-avoiding chains with 16–4096 bonds filled into cubic lattices [20].

Problem 7 Calculate the $\langle r^2 \rangle_0/nl^2$ values of samples 2 and 3 in Table 3.1 according to Eqs. (3.73) and (3.76).

3.2.6 Phase Equilibrium

In general, as temperature decreases, the chemical potential of the polymer solution increases, and the solution may be divided into two phases and turn opaque. The conditions for the phase equilibrium are expressed by

$$\mu_1 = \mu_1' \tag{3.77}$$

and

$$\mu_2 = \mu_2' \tag{3.78}$$

where μ and μ' represent the chemical potentials of the separated dilute and concentrated phases, respectively. The critical temperature T_c, at which the phase diagram shows the maximum (viz., $v_2 = v_2'$), is designated as the upper critical solution temperature (UCST) and fulfills the following conditions [135, 138]:

$$\left(\frac{\partial \Delta\mu_1}{\partial v_2}\right)_{T,P} = 0 \tag{3.79}$$

and

$$\left(\frac{\partial^2 \Delta\mu_1}{\partial v_2^2}\right)_{T,P} = 0 \tag{3.80}$$

where $\Delta\mu_1$ is given by Eq. (3.29). Equations (3.79) and (3.80) yield

$$\chi_{1c} = \frac{1}{2}\left(1 + \frac{1}{x^{1/2}}\right)^2 \simeq \frac{1}{2} + \frac{1}{x^{1/2}} \tag{3.81}$$

and

$$v_{2c} = \frac{1}{1 + x^{1/2}} \tag{3.82}$$

Equation (3.81) represents that T_c approaches the Θ point ($\chi_{1c} \rightarrow 1/2$) as x, i.e. the molecular weight increases. Equation (3.82) indicates that the concentration of the critical point decreases with increasing molecular weight.

The definition of the Θ point leads to

$$-\psi_1\left(1 - \frac{\Theta}{T_c}\right) = \chi_{1c} - \frac{1}{2}$$
$$-\psi_1\left(1 - \frac{\Theta}{T_c}\right) = \frac{(1 + x^{1/2})^2}{2x} - \frac{1}{2}$$
$$-\psi_1\left(1 - \frac{\Theta}{T_c}\right) = \frac{1}{2x} + \frac{1}{x^{1/2}} \tag{3.83}$$

Equation (3.83) can be written as

$$\frac{1}{T_c} = \frac{1}{\Theta}\left[1 + \frac{1}{\psi_1}\left(\frac{1}{x^{1/2}} + \frac{1}{2x}\right)\right] \tag{3.84}$$

This equation shows again that T_c approaches Θ as x increases. To be more accurate, the Θ point may be estimated from the reciprocal of the ordinate intercept of a $1/T_c$ vs. $(1/M^{1/2} + 1/2M)$ plot because of $x \propto M$.

In contrast to UCST, the phase separation occasionally occurs as temperature increases. The critical point, at which the phase diagram exhibits the minimum, is termed lower critical solution temperature (LCST); however, the theoretical treatments for LCST are out of scope of this chapter.

Problem 8 Derive Eqs. (3.81) and (3.82) from Eqs. (3.79) and (3.80).

3.3 Intrinsic Viscosity

The viscosity η of a polymer solution is often expressed by the Huggins equation [207]:

$$\frac{\eta_{sp}}{c} = \frac{\eta - \eta_s}{\eta_s c} = [\eta] + k'[\eta]^2 c + \cdots \tag{3.85}$$

where η_s is the solvent viscosity, and η_{sp} and $[\eta]$ are termed specific viscosity and intrinsic viscosity, respectively. The $[\eta]$ value is determined by extrapolation to $c = 0$ of a η_{sp}/c vs. c plot for dilute solutions.

The intrinsic viscosity is related to the molecular weight M by the Houwink–Mark–Sakurada equation (see, for example [163]):

$$[\eta] = KM^{\nu} \tag{3.86}$$

The exponent ν, depending on the polymer, solvent, and temperature, lies ordinarily between 0.5 and 0.8 and increases with goodness of the solvent, whereas that of the Θ solution is equal to 1/2:

$$[\eta]_{\Theta} = KM^{1/2} \tag{3.87}$$

Under the EVE, the $[\eta]$ is expressed as [150]

$$[\eta] = [\eta]_{\Theta}\ \alpha_{\eta}^3 \tag{3.88}$$

where α_{η} is the viscosity expansion coefficient. For example, Kurata and Yamakawa expressed α_{η} as a function of z [265]:

$$\alpha_{\eta}^3 = 1 + 1.55z \tag{3.89}$$

Barrett [36, 37] derived

$$\alpha_{\eta}^3 = \left(1 + 3.8z + 1.9z^2\right)^{0.3} \tag{3.90}$$

Yamakawa revised Eq. (3.89) to [533]

$$\alpha_{\eta}^3 = 1 + 1.05z \qquad (0 < \alpha_{\eta}^3 < 1.6) \tag{3.91}$$

$$= 1.05 + 0.87z \qquad (0 < \alpha_{\eta}^3 < 2.5) \tag{3.92}$$

Flory and Fox proposed the relationship between K and the unperturbed chain dimension [150]:

$$K = \Phi_\Theta \left(\frac{\langle r^2 \rangle_0}{M} \right)^{3/2} \tag{3.93}$$

$$= \Phi_\Theta \left(\frac{6\langle S^2 \rangle_0}{M} \right)^{3/2} \tag{3.94}$$

where Φ_Θ is termed the viscosity constant. Notwithstanding the "viscosity constant," in fact, the proposed Φ_Θ values are widely scattered over $(1.8\text{–}2.9)\times10^{23}$ mol^{-1} [163, 533].

3.3.1 Stockmayer–Fixman Plot

In the community of polymer chemistry, the following statement is well established: when configurational properties characteristic of a given polymer are investigated, the polymeric chain should lie in the unperturbed state free from the EVE. The unperturbed state (Θ point), where the second virial coefficient becomes null, may be found via, e.g. static light scattering (SLS) experiments. As Eq. (3.84) suggests, the UCST of a polymer with infinite molecular weight may correspond to the Θ point ($\chi_c = 1/2$). Therefore, the cloud point of a polymer of extremely high molecular weight would give a clue to the Θ point. Ordinarily, it is difficult to find the Θ point, and it is not certain whether it lies between the melting and boiling points of the solvent. Therefore, methods to predict the unperturbed chain dimension of a given polymer from viscosity measurements for its good solution have been proposed.

The excluded-volume parameter z is rewritten as

$$z = \left(\frac{3}{2\pi} \right)^{3/2} BA^{-3}M^{1/2} \tag{3.95}$$

where

$$A = \left(\frac{\langle r^2 \rangle_0}{M} \right)^{1/2} \tag{3.96}$$

and

$$B = \frac{\beta}{M_0^2} \tag{3.97}$$

with M_0 being the formula mass of the segment (usually, the monomeric unit). From Eqs. (3.87)–(3.89), (3.93), (3.95)–(3.97), it follows

$$[\eta] = KM^{1/2} + 0.51\Phi_\Theta BM \tag{3.98}$$

that is,

$$\frac{[\eta]}{M^{1/2}} = K + 0.51\Phi_\Theta BM^{1/2} \tag{3.99}$$

Equation (3.99) suggests that the extrapolation of an $[\eta]/M^{1/2}$ vs. $M^{1/2}$ plot yields K at $M^{1/2} = 0$. From the K value, the unperturbed chain dimension $\langle r^2\rangle_0/M$ can be evaluated with an appropriate Φ_Θ value. This procedure is designated as the Stockmayer–Fixman plot [447] and allows us to derive the $\langle r^2\rangle_0/M$ value from intrinsic viscosities using good solvents but requires us to prepare several samples of different molecular weights.

If Eqs. (3.91) and (3.92) are used in place of Eq. (3.89), it follows that [533]

$$\frac{[\eta]}{M^{1/2}} = 1.05K + 0.287\Phi_\Theta BM^{1/2} \qquad (0 < \alpha_\eta^3 < 1.6) \tag{3.100}$$

$$\frac{[\eta]}{M^{1/2}} = K + 0.346\Phi_\Theta BM^{1/2} \qquad (0 < \alpha_\eta^3 < 2.5) \tag{3.101}$$

Equations (3.100) and (3.101) also indicate that the $[\eta]/M^{1/2}$ vs. $M^{1/2}$ plot yields the K value at $M^{1/2} = 0$.

Exercise

Poly((*R*)-3-hydroxybutyrate) (P3HB, $[\text{—C(=O)OCH(CH}_3\text{)CH}_2\text{—}]_x$) is a biodegradable polyester produced by microorganisms (see, for example [115]). Fujita's group carried out viscosity measurements for the trifluoroethanol (TFE) solution of P3HB at 25 °C and determined the $\langle S^2\rangle_0/M_w$, K, and β values to be, respectively, 1.05×10^{-17} cm^2, 1.00×10^{-1} cm^{3}g^{-1}, and 159×10^{-24} cm^3 from the plots of Eqs. (3.100) and (3.101) [199]. The samples were fully fractionated so as to be considered monodisperse: $M_w = M_n = M$.

(1) Evaluate the characteristic ratio $\langle r^2\rangle_0/nl^2$ of P3HB from the K value with $\Phi_\Theta = 2.00\times10^{23}$ (mol^{-1}) [322], bond lengths of C(=O)—O = 1.356 Å, O—CH(CH_3) = 1.455 Å, CH(CH_3)–CH_2 = 1.530 Å, and CH_2—C(=O) = 1.520 Å, and the molar mass of the repeating unit = 86.1 [407].

(2) SLS experiments on P3HB in TFE at 25 °C yielded the data shown in Table 3.1. Evaluate the characteristic ratios $\langle r^2\rangle_0/nl^2$ of sample 1 from the $\langle S^2\rangle_0/M_w$ and β values with Eq. (3.69) and Eqs. (3.73) and (3.76) and compare the data thus derived with that obtained in (1).

Answers

(1) $nl^2 = xL$, where x is the degree of polymerization, and L is the sum of square bond lengths in the repeating unit: $L = 1.356^2 + 1.455^2 + 1.530^2 + 1.520^2 = 8.61\text{Å}^2 = 8.61\times10^{-16}\text{cm}^2$. The characteristic ratio can be calculated from the K value:

Table 3.1 Static light scattering data on P3HB in TFE at 25 °C.

Sample number	$M_w \times 10^{-4}$ (g mol^{-1})	$A_2 \times 10^4$ (cm^3 mol g^{-2})	$\langle S^2 \rangle^{1/2}$ (Å)
1	910 ± 20	6.20 ± 0.2	2560 ± 50
2	761 ± 30	6.28 ± 0.2	2320 ± 50
3	434 ± 20	6.46 ± 0.4	1710 ± 50

Source: Adapted from [322].

$$\begin{aligned}\frac{\langle r^2\rangle_0}{nl^2} &= \frac{6\langle S^2\rangle_0}{nl^2}\\ &= \frac{(K/\Phi_\Theta)^{2/3}xM_0}{xL}\\ &= \frac{(K/\Phi_\Theta)^{2/3}M_0}{L}\\ &= \frac{(1.00\times10^{-1}/2.00\times10^{23})^{2/3}\times 86.1}{8.61\times10^{-16}}\\ &= 6.30\end{aligned}$$

(2) To evaluate the excluded-volume parameter z according to Eq. (3.95), the A and B values must first be calculated by

$$A^2 = \frac{6\langle S^2\rangle_0}{M} = 6\times1.05\times10^{-17} = 6.30\times10^{-17}$$

and

$$B = \frac{\beta}{M_0^2} = 159\times10^{-24}/86.1^2 = 2.14\times10^{-26}$$

Therefore,

$$\begin{aligned}z &= \left(\frac{3}{2\pi}\right)^{3/2} BA^{-3}M^{1/2}\\ &= \left(\frac{3}{2\pi}\right)^{3/2}\times 2.14\times10^{-26}\times(6.30\times10^{-17})^{-3/2}\times(910\times10^4)^{1/2}\\ &= 42.6\end{aligned}$$

For the z value, Eq. (3.69) yields $\alpha = 2.65$, and hence, $\alpha^2 = 7.00$.[3] The unperturbed mean-square radius of gyration can be calculated from the α^2 values by

3 Equation (3.69) can be solved numerically by, e.g. Newton's or Bairstow's method. You will find the algorithms and source codes in textbooks on computer numerical analyses.

$$\langle S^2\rangle_0 = \frac{\langle S^2\rangle}{\alpha^2} = \frac{2560^2}{7.00} = 9.36\times 10^5$$

The degree of polymerization of sample 1 is calculated from

$$x = \frac{M_\mathrm{w}}{M_0} = \frac{910\times 10^4}{86.1} = 1.06\times 10^5$$

The characteristic ratio can be obtained as

$$\frac{\langle r^2\rangle_0}{nl^2} = \frac{6\langle S^2\rangle_0}{xL} = \frac{6\times 9.36\times 10^5}{1.06\times 10^5\times 8.61} = 6.15$$

If in place of Eq. (3.69), Eqs. (3.73) and (3.76) are used, then

$$\begin{aligned}\alpha_r^2 &= \left[1 + 20\times 42.6 + 155.54\times 42.6^2 + 591.86\times 42.6^3 + 325\times 42.6^4\right.\\ &\quad \left.+ 1670\times 42.6^6\right]^{\frac{1}{15}}\\ &= 7.36\end{aligned}$$

and

$$\alpha_S^2 = \left\{0.933 + 0.067\exp\left[-\left(0.85\times 42.6 + 1.39\times 42.6^2\right)\right]\right\}\times 7.36 = 6.86$$

Therefore,

$$\langle S^2\rangle_0 = \frac{\langle S^2\rangle}{\alpha_S^2} = \frac{2560^2}{6.86} = 9.56\times 10^5$$

which yields

$$\frac{\langle r^2\rangle_0}{nl^2} = \frac{6\times 9.56\times 10^5}{1.06\times 10^5\times 8.61} = 6.28$$

The $\langle r^2\rangle_0/nl^2$ values obtained in Exercises (1) and (2) satisfactorily agree with each other. Equations (3.73) and (3.76) in particular gave exact agreement.

4

Rubber Elasticity

Metals exhibit the stress–strain curves as shown in Figure 4.1, where the strain increases linearly with the stress, at least, between the origin and point *a* (elastic limit). The linear relation is known as Hooke's law. Figure 4.2 shows the stress–strain curve of rubbers, where the strain of as much as 400% may be possible; the extraordinarily large and recoverable deformation is a characteristic feature of rubber elasticity. Elastic moduli of ordinary materials range from 10^9 to 10^{11} Pa, whereas those of elastomer are as small as 10^5–10^6 Pa.

The molecular requirements for rubber elasticity are (1) that the chains have a high degree of flexibility and mobility and (2) that the chains must be joined into a network structure. Some unique properties of rubbers are experimentally observed: a rapid (adiabatic) deformation renders the rubber warm; when temperature is raised, it follows a decrease in length at constant force or an increase in force at constant length. The former behavior is dissimilar to, but the latter is similar to those of a compressed gas, if the force (f) and length (L) are assumed to correspond to pressure (p) and volume (V), respectively. An increase in temperature enlarges the volume at constant pressure or the pressure at constant volume because of $pV = RT$.

4.1 Thermodynamics of Rubber Elasticity

The first law of thermodynamics is written as

$$dU = dQ - dW \tag{4.1}$$

where U is the internal energy, Q is the heat supplied by the surroundings, and W is the work that the rubber does. In a reversible process

$$dQ = T\,dS \tag{4.2}$$

Conformational Analysis of Polymers: Methods and Techniques for Structure-Property Relationships and Molecular Design, First Edition. Yuji Sasanuma.

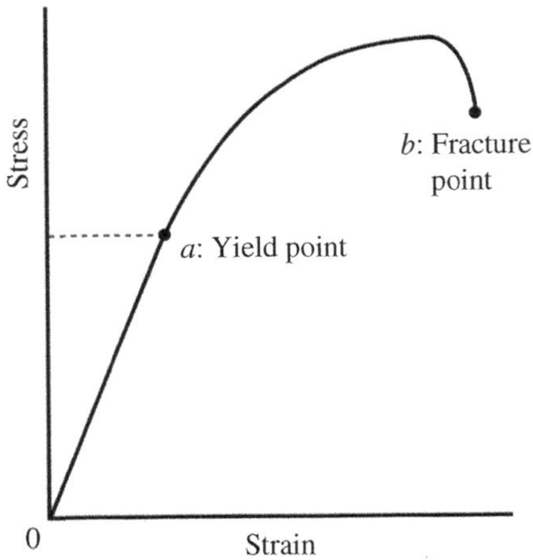

Figure 4.1 Typical stress–strain curve of metals. Under the yield point *a*, Hooke's law, the linearity between stress and stress, holds.

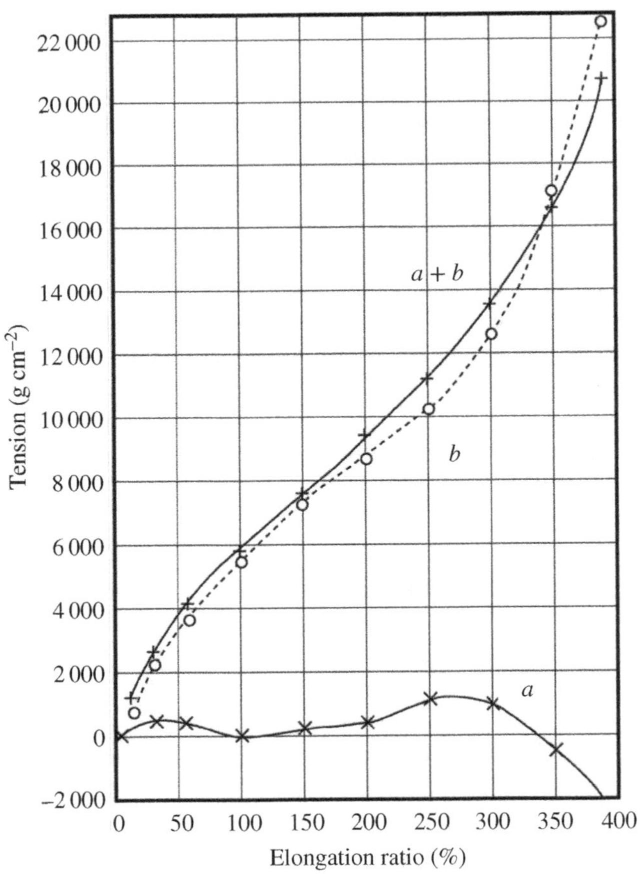

Figure 4.2 Stress (tension)–strain (elongation ratio) curve of a rubber. *a*, energy contribution, *b*, entropy contribution, and *a* + *b*, observation.

For a rubber, dW is divided into terms due to changes in volume (dV) and length (dL)

$$dW = p\,dV - f\,dL \tag{4.3}$$

Substitution of Eqs. (4.2) and (4.3) into Eq. (4.1) leads to

$$dU = T\,dS - p\,dV + f\,dL \tag{4.4}$$

Under isovolumetric and isothermal conditions, from Eq. (4.4), we obtain

$$f = \left(\frac{\partial U}{\partial L}\right)_{V,T} - T\left(\frac{\partial S}{\partial L}\right)_{V,T} \tag{4.5}$$

The first and second terms on the right-hand side are due to the energy and entropy, respectively. The Helmholtz free energy of a rubber may be expressed by

$$dF = -S\,dT - p\,dV + f\,dL \tag{4.6}$$

According to Schwarz's theorem,

$$\frac{\partial}{\partial L}\left(\frac{\partial F}{\partial T}\right) = \frac{\partial}{\partial T}\left(\frac{\partial F}{\partial L}\right) \tag{4.7}$$

that is

$$-\left(\frac{\partial S}{\partial L}\right)_{V,T} = \left(\frac{\partial f}{\partial T}\right)_{V,L} \tag{4.8}$$

From Eqs. (4.5) and (4.8), we obtain

$$f = \left(\frac{\partial U}{\partial L}\right)_{V,T} + T\left(\frac{\partial f}{\partial T}\right)_{V,L} \tag{4.9}$$

In the experiment, the tension may be directly measured, but it would be impossible to obtain the first and second terms on the right side as they are. For that reason, the following approximation was derived (see, for example Appendix A of Chapter 11 of [138]):

$$\left(\frac{\partial f}{\partial T}\right)_{V,L} \approx \left(\frac{\partial f}{\partial T}\right)_{p,\alpha} \tag{4.10}$$

where α is the elongation,

$$\alpha = \frac{L}{L_0} \tag{4.11}$$

It may be possible to obtain the right-side term of Eq. (4.10) experimentally. In Figure 4.3, the observed values of tension are plotted against temperature under different αs. The slope of the f vs. T plot corresponds to the $(\partial f/\partial T)_{p,\alpha}$ value; therefore, the energy contribution, $(\partial U/\partial L)_{V,T}$, can be evaluated from the difference between f and $T(\partial f/\partial T)_{p,\alpha}$. The three terms, f, $(\partial U/\partial l)_{V,T}$, and $T(\partial f/\partial T)_{p,\alpha}$, are plotted against elongation percentage in Figure 4.2, from which

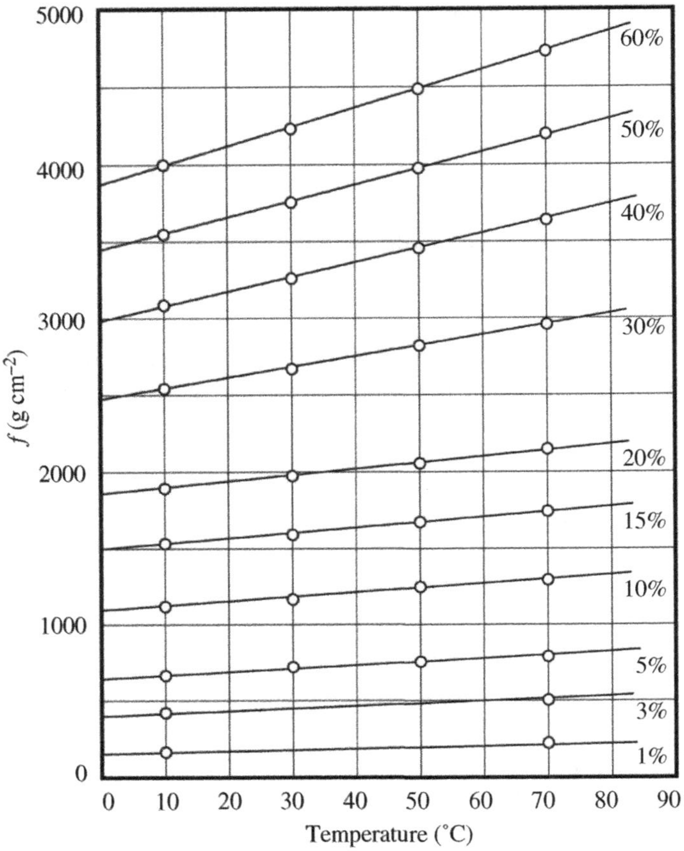

Figure 4.3 Temperature dependence of tension of a rubber. The percentage is the elongation ratio, which is defined as $(L - L_0)/L_0 \times 100$.

it can be seen that the tension comes almost entirely from the entropy term. Over 300%, however, the energy term drops, which may be due to partial crystallization. When a rubber satisfies the following condition

$$\left(\frac{\partial U}{\partial L}\right)_{V,T} \approx 0 \tag{4.12}$$

it is called the ideal rubber, and hence the rubber elasticity is often termed entropy elasticity.

The tension can also be written in a total differential form:

$$df = \left(\frac{\partial f}{\partial L}\right)_{V,T} dL + \left(\frac{\partial f}{\partial T}\right)_{V,L} dT \tag{4.13}$$

At constant load ($df = 0$),

$$\left(\frac{\partial f}{\partial L}\right)_{V,T} dL + \left(\frac{\partial f}{\partial T}\right)_{V,L} dT = 0 \tag{4.14}$$

Accordingly,

$$\left(\frac{\partial L}{\partial T}\right)_f = -\left(\frac{\partial f}{\partial T}\right)_{V,L} \Big/ \left(\frac{\partial f}{\partial L}\right)_{V,T} \tag{4.15}$$

From Eq. (4.8), we can derive

$$\left(\frac{\partial L}{\partial T}\right)_f = \left(\frac{\partial S}{\partial L}\right)_{V,T} \Big/ \left(\frac{\partial f}{\partial L}\right)_{V,T} \tag{4.16}$$

Inasmuch as $(\partial f/\partial L)_{V,T}$ is positive and $(\partial S/\partial L)_{V,T}$ is negative, the left side of Eq. (4.16) is negative. Equation (4.16) represents that, at constant force, an increase in temperature yields a decrease in length, which is a characteristic feature of rubbers.

4.2 Adiabatic Stretching: Gough–Joule Effect

When a rubber string undergoes a rapid elongation, its temperature will be raised. This property is known as the Gough–Joule effect. The characteristic of elastomers is elucidated here. The entropy of a rubber can be expressed in a total differential form of length and temperature as

$$dS = \left(\frac{\partial S}{\partial L}\right)_{p,T} dL + \left(\frac{\partial S}{\partial T}\right)_{p,L} dT \tag{4.17}$$

The isobaric specific heat (C_p) is

$$C_p = \left(\frac{\partial H}{\partial T}\right)_p = T\left(\frac{\partial S}{\partial T}\right)_F \tag{4.18}$$

Therefore,

$$\left(\frac{\partial S}{\partial T}\right)_p = \frac{C_p}{T} \tag{4.19}$$

Inasmuch as the Gibbs free energy is given by

$$dG = -SdT + Vdp + fdL \tag{4.20}$$

a Maxwell relation can be derived from

$$\frac{\partial}{\partial l}\left(\frac{\partial G}{\partial T}\right) = \frac{\partial}{\partial T}\left(\frac{\partial G}{\partial l}\right) \tag{4.21}$$

that is

$$-\left(\frac{\partial S}{\partial L}\right)_{p,T} = \left(\frac{\partial f}{\partial T}\right)_{p,L} \tag{4.22}$$

Substitution of Eqs. (4.19) and (4.22) into Eq. (4.17) yields

$$dS = -\left(\frac{\partial f}{\partial T}\right)_{p,L} dL + \frac{C_p}{T} dT \tag{4.23}$$

In an adiabatic process, we have

$$dQ = TdS = 0 \tag{4.24}$$

that is

$$dS = 0 \tag{4.25}$$

Accordingly, Eq. (4.23) can be rewritten as

$$dT = \frac{T}{C_p}\left(\frac{\partial f}{\partial T}\right)_{p,L} dL \tag{4.26}$$

When the rubber string is stretched from L_0 to L, the accompanying temperature change (ΔT) is given by

$$\Delta T = \int_{L_0}^{L} \frac{T}{C_p}\left(\frac{\partial f}{\partial T}\right)_{p,L} dL \tag{4.27}$$

Because $(\partial f/\partial T)_{p,L} > 0$, $T > 0$, and $C_p > 0$, ΔT is also positive, we feel the rubber to be warm.

4.3 Phenomenological Theory: Affine Model

A rubber will be elongated by a force and shrink if released from the force. To exhibit such properties, the rubber polymer must be flexible, mobile, and cross-linked [302]. The deformation has often been represented by the affine transformation, which preserves collinearity, being composed of operations such as translation, scale, shear, and rotation and expressed by

$$\boldsymbol{x} \to A\boldsymbol{x} + B \tag{4.28}$$

where A includes scale, shear, and rotation, and B corresponds to translation. Even under the deformation, the network structure is expected to be kept. This is the reason why the affine transformation is adopted as the model for rubbers. If the polymeric chain is assumed to be a Gaussian chain, the distribution is expressed by

$$w(r, n) \propto \exp\left(-\frac{3r^2}{2\langle r^2\rangle_0}\right) \tag{4.29}$$

where the relation $\langle r^2\rangle_0 = nl^2$ of the Gaussian chain is used. The Helmholtz free energy per ν chains may be given as

$$F(T) = -\nu kT \ln w(r, n) = \left(\frac{3\nu kT}{2\langle r^2\rangle_0}\right) r^2 + C(T) \tag{4.30}$$

where $C(T)$ is a constant at a specified temperature. When the rubber is stretched from $\boldsymbol{r}_0 = (x, y, z)$ to $\boldsymbol{r} = (\alpha_x x, \alpha_y y, \alpha_z z)$, the free energy change (ΔF) is

$$\Delta F = \left(\frac{3\nu kT}{2\langle r^2\rangle_0}\right)\left[\left(\alpha_x^2 x^2 + \alpha_y^2 y^2 + \alpha_z^2 z^2\right) - \left(x^2 + y^2 + z^2\right)\right] \tag{4.31}$$

If $\langle r^2\rangle_0$ is expressed as $\langle r^2\rangle_0 = \left(\langle x^2\rangle, \langle y^2\rangle, \langle z^2\rangle\right)$, we have

$$\Delta F = \left(\frac{3\nu kT}{2\langle r^2\rangle_0}\right)\left[\left(\alpha_x^2 - 1\right)\langle x^2\rangle + \left(\alpha_y^2 - 1\right)\langle y^2\rangle + \left(\alpha_z^2 - 1\right)\langle z^2\rangle\right] \tag{4.32}$$

As an affine transformation, we define

$$\alpha_x = \frac{L_x}{L_{x_0}},\ \alpha_y = \frac{L_y}{L_{y_0}},\ \alpha_z = \frac{L_z}{L_{z_0}} \tag{4.33}$$

where the subscript 0 indicates the natural length. The three-dimensional isotropism of the Gaussian chain leads to

$$\langle x^2\rangle = \langle y^2\rangle = \langle z^2\rangle = \frac{\langle r^2\rangle}{3} \tag{4.34}$$

Therefore,

$$\Delta F = \left(\frac{\nu kT}{2}\right)\frac{\langle r^2\rangle}{\langle r^2\rangle_0}\left(\alpha_x^2 + \alpha_y^2 + \alpha_z^2 - 3\right) \tag{4.35}$$

Because $\langle r^2\rangle$ is close to $\langle r^2\rangle_0$, Eq. (4.35) may be changed to

$$\Delta F = \left(\frac{\nu kT}{2}\right)\left(\alpha_x^2 + \alpha_y^2 + \alpha_z^2 - 3\right) \tag{4.36}$$

Also, if the rubber is isovolumetrically deformed in the x-direction, then $\alpha_x = \alpha > 1$ and $\alpha_y = \alpha_z = \alpha^{1/2}$, and consequently, Eq. (4.36) may be changed to

$$\Delta F = \left(\frac{\nu kT}{2}\right)\left(\alpha^2 + 2\alpha^{-1} - 3\right) \tag{4.37}$$

The tension is related to the Helmholtz free energy by

$$f = \left(\frac{\partial \Delta F}{\partial L}\right)_{T,V} = \frac{1}{L_0}\left(\frac{\partial \Delta F}{\partial \alpha}\right)_{T,V} = \left(\frac{\nu kT}{L_0}\right)\left(\alpha - \alpha^{-2}\right) \tag{4.38}$$

Therefore, the stress (σ) is

$$\sigma = \frac{f}{A} = \left(\frac{\nu kT}{L_0 A}\right)\left(\alpha - \alpha^{-2}\right) = \left(\frac{\nu}{V}\right)kT\left(\alpha - \alpha^{-2}\right) \tag{4.39}$$

where A is the cross-sectional area of the rubber, and ν/V corresponds to the density of the network. Figure 4.4 shows an $\alpha - \alpha^{-2}$ vs. α plot, which is similar in shape to the f vs. elongation ratio curve in Figure 4.2. The equation of state of the ideal gas is well known as

$$p = \frac{N}{V}kT \tag{4.40}$$

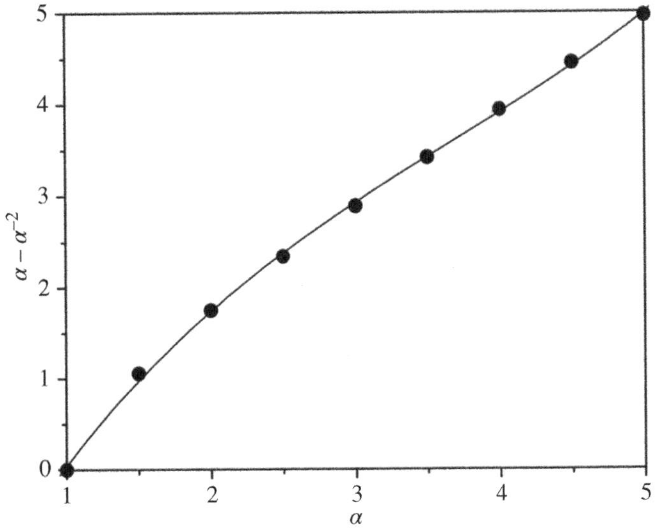

Figure 4.4 $\alpha - \alpha^{-2}$ vs. α plot.

where N is the number of gas molecules. Inasmuch as σ and ν correspond to p and N, respectively, Eq. (4.39) is close to Eq. (4.40) except for the $(\alpha - \alpha^{-2})$ factor due to the affine transformation resulting from the network structure.

4.4 Temperature Dependence of Chain Dimension in Rubber

The energy term of Eq. (4.5) is

$$f_e = \left(\frac{\partial U}{\partial L}\right)_{V,T} \tag{4.41}$$

On the fixed-length and isovolumetric conditions, the f_e/f ratio can be proved to be expressed as

$$\frac{f_e}{f} = -T\left[\frac{\partial \ln(f/T)}{\partial T}\right]_{V,L} \tag{4.42}$$

From Eq. (4.35), we can obtain

$$f = \left(\frac{\nu kT}{L_0}\right)\frac{\langle r^2\rangle}{\langle r^2\rangle_0}\left(\alpha - \alpha^{-2}\right) \tag{4.43}$$

From Eqs. (4.42) and (4.43), the following relation can be derived [86, 148, 296, 298]:

$$\frac{f_e}{f} = T\frac{d\ln\langle r^2\rangle_0}{dT} \tag{4.44}$$

Because, on the fixed-length and isovolumetric conditions, α and $\langle r^2\rangle = \langle V\rangle^{2/3}$ are invariant, only $\langle r^2\rangle_0$ can be considered to be a function of temperature.

Equation (4.44) means that the contribution of internal energy to the rubber elasticity can be estimated from the temperature dependence of the unperturbed chain dimension (characteristic ratio), which can be calculated by the rotational isomeric state (RIS) scheme. Equation (4.44) suggests that, even in the network structure, the rubber elasticity stems exclusively from intramolecular factors. The f_e/f values have been reported for a variety of polymers. For example, for polyethylene, it is −0.4, whereas for polybutadiene, polydimethylsiloxane, and natural rubbers they are 0.10–0.17, 0.13–0.30, and 0.12–0.18, respectively [86, 148, 296, 298]. In general, semicrystalline polymers show negative f_e/fs, whereas rubber polymers will give positive ones.

The f_e/f difference between polyethylene and rubber polymers can be explained in terms of conformational characteristics. The CH_2—CH_2 bond of polyethylene prefers the trans conformation; thus, an increase in temperature enhances the fraction of unstable gauche conformations. Consequently, the polyethylene chain will be more distorted and reduce the chain dimension. Therefore, the temperature coefficient of the characteristic ratio is observed to be negative, and the f_e/f value also becomes negative. The energy term of $f = f_e + f_S$ (f_S: entropy term) works against the entropy elasticity because f_S is positive. On the other hand, rubber polymers generally prefer distorted conformations and will be moved to more unstable extended states with increasing temperature; therefore, their temperature coefficients of the characteristic ratio are positive, and the energy term assists the entropy elasticity.

Problem 9 Prove Eqs. (4.42) and (4.44).

Part II

Quantum Chemistry

The structure and properties of molecules are attributed to their electronic structures. The same is true of macromolecules. Inasmuch as, for example, a 100mer polyethylene (PE) chain may exhibit an enormous number (3^{200}) of conformations even under the three-state rotational isomeric state (RIS) approximation (trans, gauche$^+$, and gauche$^-$ conformations), its properties $\langle M \rangle$s are the results of the weight-average

$$\langle M \rangle = \sum_i M_i e^{-E_i/RT} \Big/ \sum_i e^{-E_i/RT}$$

where M_i and E_i are the property and energy of conformation i, respectively. Intramolecular interactions of a polymeric chain in the unperturbed state mostly stay within several chemical bonds; therefore, a small compound with the same bond sequence as that of the polymer can be adopted as a model to provide us with information on intramolecular interactions, geometrical parameters, and conformational energies of the polymer. The model compound must include, at least, a repeating unit of the polymer, and its ends are preferably terminated by inert atomic groups. If a functional group such as hydroxy and carbonyl groups is put in the terminal, some fictitious interaction such as a hydrogen bonding would appear.

In the RIS scheme, geometrical parameters and conformational energies of the polymer are needed. To acquire such data, molecular mechanics and molecular orbital (MO) calculations are used, and otherwise, they may be extracted from, for example, X-ray diffraction data. However, the experimental data are usually either weight averages over all conformations or those of a specific structure. From the experiments, the structural data cannot be determined for each conformation.

Conformational Analysis of Polymers: Methods and Techniques for Structure-Property Relationships and Molecular Design, First Edition. Yuji Sasanuma.

In a paper of 1974, Flory stated [144]

> Current configurational statistical theory of polymer chains is founded upon classical statistical mechanics. Justification for the intuitively satisfactory premise has been called into question on the grounds that the frequencies of some of modes of motion within the chain are well beyond the classical limit. This undeniable fact appears to have been constructed to imply that a comprehensive theory of macromolecular configurations should be developed in the terms of quantum statistical mechanics. The ultimate rectitude of the latter is incontestable, but drastic approximations are necessary in order to implement its application to a macromolecule in the liquid state. Measures adopted to circumvent the difficulties confronting rigorous application of quantum statistical mechanics may entail errors far more serious than any approximations involved in resort to classical statistical mechanics.

Undoubtedly, he understood the significance of quantum chemistry. Unfortunately, however, the quantum chemistry of those days was not precise enough to reproduce small energy differences <1 kcal mol^{-1} between conformations. Afterward, quantum chemistry has been advanced significantly to dispel Flory's concerns. It has become possible to carry out quantum chemical calculations for a polymeric chain staying in a particular conformation; however, it is still far from reality to achieve the weight average over 3^{200} conformations. Instead, for long polymeric chains, molecular force field calculations have often been employed, and the potential function $V(r)$ is expressed as the sum of bond, angle, dihedral, van der Waals, and electrostatic energy terms [66]:

$$V(r) = \sum_{\text{bonds}} k_b(l - l_{eq})^2 + \sum_{\text{angles}} k_a(\theta - \theta_{eq})^2 + \sum_{\text{dihedrals}} \frac{V_n}{2}[1 + \cos(n\varphi - \gamma)] + \sum_{i<j}^{\text{atoms}} \left(\frac{A_{ij}}{R_{ij}^{12}} - \frac{B_{ij}}{R_{ij}^{6}} \right) + \sum_{i<j}^{\text{atoms}} \frac{q_i q_j}{\epsilon R_{ij}}$$

where l is bond length, θ is bond angle, φ is dihedral angle, R_{ij} is the distance between atoms i and j, q_i is the partial charge of atom i, ϵ is dielectric constant, and the subscript "eq" represents the equilibrium state. The parameters, k_b, k_a, V_n, γ, A_{ij}, and B_{ij}, are determined so as to reproduce either experimental observations or MO calculations. An additional term representing hydrogen bonds has occasionally been introduced in the force field, whereas MO calculations will often predict something unexpected, for example weak attractions such as C—H$\cdots$O of polyethers [398, 399, 405, 412, 413], N—H$\cdots$N of polyamines [403, 417], $\pi \cdots \pi$ of aromatic polyesters [400, 414, 419], and dipole–dipole of poly(ethylene sulfide) [412] and poly(glycolic acid) [422]. As will be described later, such weak interactions strongly influence the conformational characteristics and configurational properties of polymers.

The geometrical parameters and conformational energies of the polymer are extracted from optimized conformers of the model compound(s). To investigate the conformational equilibrium, the Gibbs free energy, being derived from the electronic energy and thermochemical term, is preferably used as the conformational energy. The thermochemical term must be calculated at the same level of theory with the same basis set as those used in the geometrical optimization [154, 159].

Flory also pointed out the importance of solvent effects [144]:

> The motions of the macromolecules are subject to influences of the neighboring molecules, consisting of solvent or of other polymer molecules, which invariably make up its environment. Conformational modes of the macromolecule are especially vulnerable to restrictions imposed by its neighbors. Hence, treatment of the system as a whole is obligatory. The Hamiltonian representing the system must include kinetic energies of the surrounding molecules as well as the intermolecular potential energies involving them. Under the term of quantum mechanics, the kinetic energies of the neighboring species are not separable.

The solvent effect has been introduced in the MO calculations, and the electronic energy perturbed by solvents can also be evaluated [56, 483]. From the Gibbs energy thus obtained, therefore, the conformational energies and configurational properties of polymeric chains dissolved in the solvents will be reproduced. Inasmuch as the MO calculations are so compatible with the RIS scheme as to be directly introduced thereinto, the RIS scheme combined with quantum chemistry can realize the quantum statistical mechanics that Flory advocated as the ideal.

Herein, the computational procedures for quantum chemical calculations on molecules and crystals are premised on the usage of the Gaussian series [159] and the CRYSTAL series [121, 122], respectively.

5

Ab Initio Molecular Orbital Theory

5.1 Schrödinger Equation

The starting point of quantum chemistry is the Schrödinger equation:

$$\left[-\frac{\hbar^2}{2m}\left(\frac{\partial^2}{\partial x^2}+\frac{\partial^2}{\partial y^2}+\frac{\partial^2}{\partial z^2}\right)+V\right]\Psi(\mathbf{r},t)=i\hbar\frac{\partial\Psi(\mathbf{r},t)}{\partial t} \tag{5.1}$$

Equation (5.1) expresses the motion of a particle with a mass m at a position $\mathbf{r}=(x,y,z)$ and a time t under a potential V. $\hbar$ is Planck's constant divided by 2π, i is the imaginary unit, and $\Psi(\mathbf{r},t)$ is the wave function. Since $\Psi(\mathbf{r},t)$ can be divided into the position and time parts, the Schrödinger equation independent of time can be written as

$$\left(-\frac{\hbar^2}{2m}\nabla^2+V\right)\Psi(\mathbf{r})=E\Psi(\mathbf{r}) \tag{5.2}$$

where $\nabla^2=\partial/\partial x^2+\partial/\partial y^2+\partial/\partial z^2$, and E is the eigenvalue of energy, and Ψ is the eigenfunction. Equation (5.2) is simply expressed as

$$\mathcal{H}\Psi=E\Psi \tag{5.3}$$

where $\mathcal{H}$ $(=-\hbar^2/2m\nabla^2+V)$ is designated as the *Hamiltonian* operator.

The Hamiltonian for the molecular system is formulated as [273, 368, 512]

$$\mathcal{H}=-\sum_{i}^{\text{electrons}}\frac{\hbar^2}{2m_e}\nabla_i^2-\sum_{\mathrm{A}}^{\text{nuclei}}\frac{\hbar^2}{2m_{\mathrm{A}}}\nabla_{\mathrm{A}}^2$$
$$-\sum_{i}^{\text{electrons}}\sum_{\mathrm{A}}^{\text{nuclei}}\frac{e^2Z_{\mathrm{A}}}{r_{i\mathrm{A}}}+\sum_{i>j}^{\text{electrons}}\frac{e^2}{r_{ij}}+\sum_{\mathrm{A}>\mathrm{B}}^{\text{nuclei}}\frac{e^2Z_{\mathrm{A}}Z_{\mathrm{B}}}{r_{\mathrm{AB}}} \tag{5.4}$$

where i and j indicate electrons, A and B stand for nucleus, m_e and m_{A} are masses of the electron and nucleus, respectively, r is the distance, and Z is the atomic number. The first and second terms of Eq. (5.4) correspond to kinetic energies of electron and nucleus, respectively, and the third, fourth, and fifth terms represent

Conformational Analysis of Polymers: Methods and Techniques for Structure-Property Relationships and Molecular Design, First Edition. Yuji Sasanuma.

potential energies between electron and nucleus, between electrons, and between nuclei, respectively.

A proton is 1836 times as heavy as an electron; therefore, the nuclear motion is much slower than the electronic motion, the second term of Eq. (5.4) can be removed from the Hamiltonian of electrons, and the fifth term may be calculated from the average r_{AB}s and added after the energies of the first, third, and fourth terms are obtained. This hypothesis is known as the Born–Oppenheimer approximation. According to the hypothesis, the Hamiltonian of the electron can be simplified and expressed in the atomic units as

$$\mathcal{H} = -\sum_{i}^{\text{electrons}} \frac{1}{2}\nabla_i^2 - \sum_{i}^{\text{electrons}} \sum_{\text{A}}^{\text{nuclei}} \frac{Z_{\text{A}}}{r_{i\text{A}}} + \sum_{i>j}^{\text{electrons}} \frac{1}{r_{ij}} \tag{5.5}$$

In the atomic units, the following quantities are set equal to unity: mass, m_e ($= 9.10938 \times 10^{-31}$ kg); charge, $|e|$ ($= 1.60218 \times 10^{-19}$ C); length, the Bohr radius a_0 ($= 5.29177 \times 10^{-11}$ m); and energy, Hartree ($= 4.35974 \times 10^{-18}$ J).

5.2 Wave Function

The wave function of a single electron (denoted as 1) is a product of a spatial function ($\phi(1)$) and a spin function ($\alpha(1)$ or $\beta(1)$) because the electron adopts the α and β spin states. Inasmuch as the electron is a Fermion, the wave function for many electrons, $\Psi(1,2,\ldots,n)$, must be asymmetric. Therefore, if the coordinates of two electrons are interchanged, the sign of the wave function will be changed:

$$\Psi(1,2,3,\ldots,i,\ldots,j,\ldots,n) = -\Psi(1,2,3,\ldots,j,\ldots,i,\ldots,n) \tag{5.6}$$

where n is the number of electrons. In addition, since the square of the wave function corresponds to the probability density of electrons, the Ψ function is normalized:

$$\int |\Psi^2| \, dV = n \tag{5.7}$$

An expression to satisfy the nature of the electron is the Slater determinant:

$$\Psi(1,2,3,\ldots,n) = \frac{1}{\sqrt{n!}} \begin{vmatrix} \phi_1(1)\alpha(1) & \phi_1(1)\beta(1) & \cdots & \phi_{n/2}(1)\alpha(1) & \phi_{n/2}(1)\beta(1) \\ \phi_1(2)\alpha(2) & \phi_1(2)\beta(2) & \cdots & \phi_{n/2}(2)\alpha(2) & \phi_{n/2}(2)\beta(2) \\ \vdots & \vdots & & \vdots & \vdots \\ \phi_1(n)\alpha(n) & \phi_1(n)\beta(n) & \cdots & \phi_{n/2}(n)\alpha(n) & \phi_{n/2}(n)\beta(n) \end{vmatrix} \tag{5.8}$$

where $1/\sqrt{n!}$ is the normalization factor, and the ϕ_1 function depends on the space coordinates of electron 1.

It has been assumed that the ϕ_1 function may be given as a linear combination of atomic orbitals (LCAO) approximation:

$$\phi_i = \sum_{\mu} c_{\mu i} \chi_{\mu} \tag{5.9}$$

where χ_μ is termed basis set, which expresses the spacial distribution of the electron. The variational principle shows that the solution (Ψ) of Eq. (5.3) minimizes the eigenvalue E. Therefore, to solve the Schrödinger equation of the polyelectronic system under the LCAO approximation is to find the combination of $c_{\mu i}$s that minimizes the E value.

5.3 Basis Set

The eigenfunctions of the hydrogen atom were exactly derived from the Schrödinger equation and termed the Slater functions that decay as a function $\exp(-\zeta r)$ of the distance (r) from the nucleus, with ζ being a coefficient. However, because the product of two Gaussian functions can be expressed as a single Gaussian, it has been attempted that a number of Gaussian functions are linearly combined to approximate the Slater-type function and used as a basis set.

For example, a split-valence basis set, 6-31G, has an inner *contracted* shell composed of six *primitive* Gaussian functions and an outer valence with a contracted function of three primitive Gaussian functions and a single primitive Gaussian [110, 190, 381]. Such a basis set may be expressed as

$$\chi_\mu = \sum_{p} d_{p\mu}\, g_p(\zeta, r) \tag{5.10}$$

where $g_p(\zeta, r)$ is the primitive Gaussian function:

$$g_p(\zeta, r) = cx^n y^m z^l e^{-\zeta r^2} \tag{5.11}$$

The parameters, (x^n, y^m, z^l), were determined for each orbital type (s, p, d, ...) and each atom, the exponent ζ expresses the spatial distribution, and the coefficient c is the normalization constant evaluated from

$$\int g_p^2(\zeta, r)\ dV = 1 \tag{5.12}$$

The coefficients, $d_{p\mu}$s, were determined by the spin-restricted open-shell Hartree–Fock (HF) calculations to minimize the self-consistent field (SCF) energy of the atomic ground state.

In the 6-31G(d,p) basis set, polarization functions are added to the 6-31G set: a primitive d-type Gaussian for heavier atoms than hydrogen and a p-type Gaussian for hydrogen. The mixing of the polarization function will better represent the distorted distribution of the electron cloud. Furthermore, especially for chemical species spreading electrons away from the nuclear centers, such as anions and lone pairs, diffuse functions are effective. When the diffuse functions are added to atoms except hydrogen, the basis set is denoted as 6-31+G, and, when furthermore added to hydrogen, the basis set is 6-31++G.

Dunning's correlation-consistent basis sets denoted as cc-pVXZ (X = D, T, Q, 5, and 6) have often used [123]. The cc-pVXZ basis sets are composed of the sp-type and polarization functions, whose primitive Gaussians were determined to reproduce atomic correlation effects of the HF and single and double excitations. The X of cc-pVXZ is D = doublet, T = triplet, Q = quadruple, and so on, representing the size of the basis sets. For example, the cc-pVQZ basis set of oxygen comprises one contracted sp-type set of 12 s and 6 p primitive Gaussians and three (3d, 2f, and 1g) polarization functions. The aug-cc-pVXZ basis sets include diffuse functions as well as the cc-pVXZ components. A large number of basis sets have been and will be developed for various purposes and applied to different molecules [154].

5.4 Hartree–Fock Method

The HF energy for the polyelectronic system is

$$E = 2\sum_{i}^{n/2} H_{ii} + \sum_{i}^{n/2}\sum_{j}^{n/2} \left(2J_{ij} - K_{ij}\right) \tag{5.13}$$

where H_{ii} is the one-electron core Hamiltonian, and J_{ij} and K_{ij} are the coulomb and exchange integrals, respectively [273, 368, 512]. These terms are given by

$$H_{ii} = \int \phi_i(1)\left(-\frac{1}{2}\nabla^2 - \sum_{A}^{M}\frac{Z_A}{r_{1A}}\right)\phi_i(1)d\tau_1 \tag{5.14}$$

$$J_{ij} = \iint \phi_i^*(1)\phi_j^*(2)\frac{1}{r_{12}}\phi_i(1)\phi_j(2)d\tau_1 d\tau_2 \tag{5.15}$$

and

$$K_{ij} = \iint \phi_i^*(1)\phi_j^*(2)\frac{1}{r_{12}}\phi_j(1)\phi_i(2)d\tau_1 d\tau_2 \tag{5.16}$$

If the one-electron orbital energy is defined as

$$\epsilon_i = H_{ii} + \sum_{j}^{n/2} \left(2J_{ij} - K_{ij}\right) \tag{5.17}$$

the HF energy is also expressed by

$$E = 2\sum_{i}^{n/2} \epsilon_i - \sum_{i}^{n/2}\sum_{j}^{n/2} \left(2J_{ij} - K_{ij}\right) = \sum_{i}^{n/2} \left(\epsilon_i + H_{ii}\right) \tag{5.18}$$

5.5 Roothaan–Hall Equation

The HF energy can be derived from the Roothaan–Hall equation:

$$\sum_{\nu} \left(F_{\mu\nu} - \epsilon_i S_{\mu\nu}\right) c_{\nu i} = 0 \tag{5.19}$$

where $F_{\mu\nu}$ is the element of the Fock matrix and given by

$$F_{\mu\nu} = H_{\mu\nu} + \sum_{\lambda}\sum_{\sigma} P_{\lambda\sigma} \left[(\mu\nu|\lambda\sigma) - \frac{1}{2}(\mu\lambda|\nu\sigma)\right] \tag{5.20}$$

The two-electron integral is defined as

$$(\mu\nu|\lambda\sigma) = \iint \phi_\mu(1)\phi_\nu(1)\frac{1}{r_{12}}\phi_\lambda(2)\phi_\sigma(2)d\tau_1 d\tau_2 \tag{5.21}$$

The element of the charge density matrix is defined as

$$P_{\lambda\sigma} = 2\sum_{i}^{\text{occupied}} c_{\lambda i}^{*} c_{\sigma i} \tag{5.22}$$

where the summation is performed only for the occupied orbitals.

The Roothaan–Hall matrix equation is rewritten as

$$\mathbf{FC} = \mathbf{SCE} \tag{5.23}$$

where **F** is the Fock matrix, **C** is the orbital coefficient matrix, **E** is the diagonal orbital energy matrix, and **S** is the overlap integral matrix whose element is given by

$$S_{\mu\nu} = \int \phi_\mu(1)\phi_\nu(1)d\tau_1 \tag{5.24}$$

The Roothaan–Hall matrix equation will be solved self-consistently in the following manner:

1. Calculate the integrals of the Fock matrix.
2. Calculate the overlap integrals, diagonalize the **S** matrix, and derive the $\mathbf{S}^{-1/2}$.
3. Set the initial orbital coefficients and form the **P** matrix.
4. Form the **F** matrix.
5. Calculate $\mathbf{F}^{\tau} = \mathbf{S}^{-1/2}\mathbf{F}\mathbf{S}^{-1/2}$.
6. Solve the secular equation $|\mathbf{F}^{\tau} - \mathbf{E}\mathbf{I}| = 0$ by diagonalizing $\mathbf{F}^{\tau}$ and obtain the eigenvalues **E** and eigenvectors $\mathbf{C}^{\tau}$.
7. Calculate the molecular orbital coefficients **C** from $\mathbf{C} = \mathbf{S}^{-1/2}\mathbf{F}^{\tau}$.
8. Calculate a new density matrix **P** from **C**.
9. Check the convergence by comparison between the previous and current cycles in **P** and **E**. If the convergence conditions are satisfied, complete the calculation. Otherwise, return to step 4 and repeat the procedures.

5.6 Electron Correlation

The HF method adopts a single Slater determinant. Strictly, however, the wave function for polyelectronic systems cannot be exactly expressed by only the single determinant. The difference between the exact and HF energies is defined as the correlation energy:

$$E_{\mathrm{corr}} = E_{\mathrm{exact}} - E_{\mathrm{HF}} \tag{5.25}$$

A technique to evaluate the correction energy is the configuration interaction (CI) method [369], which deals with excited states as well as the ground state. The wave function is formulated as a linear combination of the determinants for the HF (Ψ_{HF}) and excited states:

$$\Psi = C_{\mathrm{HF}}\Psi_{\mathrm{HF}} + \sum_{ia} C_{i\to a}\Psi_{i\to a} + \sum_{ijab} C_{ij\to ab}\Psi_{ij\to ab} + \cdots \tag{5.26}$$

where $\Psi_{i\to a}$ corresponds to the single excitation from the ith occupied to the ath virtual orbitals, and $\Psi_{ij\to ab}$ represents the double excitation from the ith and jth occupied to the ath and bth virtual orbitals. The system energy is minimized to determine the coefficients, Cs, in Eq. (5.26) by a variational approach. Inasmuch as the full CI including all excitations is essentially impossible, instead, the selected excitations are adopted in the CI calculations. The HF wave function does not mix directly with the single excitations (Brillouin's theorem):

$$\langle \Psi_{\mathrm{HF}} | \hat{H} | \Psi_{i\to a} \rangle = 0 \tag{5.27}$$

Accordingly, of all the excitations, the double excitations $\Psi_{ij\to ab}$ would be the most significant; however, the single excitations interact with the double ones, which interact with Ψ_{HF}. Therefore, the single excitations also affect the system energy indirectly. Under the Gaussian program, the computations with the $\Psi_{\mathrm{HF}} + \Psi_{ij\to ab}$ and $\Psi_{\mathrm{HF}} + \Psi_{i\to a} + \Psi_{ij\to ab}$ functions will be performed using configuration interaction with all double substitutions (CID) and configuration interaction with all single and double substitutions (CISD) keywords, respectively.

A different technique to evaluate the electron correlation energy is the coupled cluster (CC) method [38]. The wave function of the CC theory is given by an exponential cluster operator ($e^{\hat{T}}$):

$$|\Psi\rangle = e^{\hat{T}} |\Psi_{\mathrm{HF}}\rangle \tag{5.28}$$

The cluster operator can produce all the excitations, being expressed by

$$\hat{T} = \hat{T}_1 + \hat{T}_2 + \cdots \tag{5.29}$$

where $\hat{T}_1$ and $\hat{T}_2$ are the operators of all single and all double excitations, respectively. The exponential operator may be expanded as a Taylor series. For example, up to the double excitations, the exponential operator is written as

$$e^T = 1 + \hat{T} + \frac{1}{2}\hat{T}^2 + \cdots = 1 + \hat{T}_1 + \hat{T}_2 + \frac{\hat{T}_1^2}{2} + \hat{T}_1\hat{T}_2 + \frac{\hat{T}_2^2}{2} + \cdots \tag{5.30}$$

Since the number of occupied orbitals is finite, the possible excitations are limited; therefore, the CC computations are not so expensive (time consuming) as the CI ones. The CC method including the single and double excitations with a perturbation contribution of the triple excitations [518], whose keyword of the Gaussian program is coupled cluster single–double and perturbative triple (CCSD(T)), yields essentially exact electronic energies, thus being accepted as the *gold standard* of quantum chemistry.

Compared with the CI and CC methods, the Møller–Plesset (MP) perturbation theory has been widely accepted as the most practical approach to evaluate the electron correlation energy [94, 324]. The MP method assumes the exact Hamiltonian (H) to be

$$H = H^{(0)} + \lambda V \tag{5.31}$$

where $H^{(0)}$ is the zeroth-order Hamiltonian expressed by the sum of Fock operators, and λV is the perturbation. The eigenfunction Φ and eigenvalue E are expanded in terms of λ:

$$\Psi = \Psi^{(0)} + \lambda\Psi^{(1)} + \lambda^2\Psi^{(2)} + \lambda^3\Psi^{(3)} + \cdots \tag{5.32}$$

and

$$E = E^{(0)} + \lambda E^{(1)} + \lambda^2 E^{(2)} + \lambda^3 E^{(3)} + \cdots \tag{5.33}$$

where $E^{(1)}$, $E^{(2)}$, and $E^{(3)}$ are the first-, second-, and third-order corrections to the energy, respectively. If Eqs. (5.32) and (5.33) are substituted in Eq. (5.31) and the terms are arranged according to the power of λ, it follows

$$\left(H^{(0)} - E^{(0)}\right)\Psi^{(0)} = 0 \tag{5.34}$$

$$\left(H^{(0)} - E^{(0)}\right)\Psi^{(1)} = \left(E^{(1)} - V\right)\Psi^{(0)} \tag{5.35}$$

$$\left(H^{(0)} - E^{(0)}\right)\Psi^{(2)} = \left(E^{(1)} - V\right)\Psi^{(1)} + E^{(2)}\Psi^{(0)} \tag{5.36}$$

and so on. From these relations, the individual energies are derived to be

$$E^{(0)} = \langle\Psi^{(0)}|H^{(0)}|\Psi^{(0)}\rangle = \sum_i \epsilon_i \tag{5.37}$$

$$E^{(1)} = \langle\Psi^{(0)}|V|\Psi^{(0)}\rangle \tag{5.38}$$

$$E^{(2)} = \sum_{s} \frac{|\langle \Psi^{(0)} | V | \Psi^{s} \rangle|^2}{E^{(0)} - E^{s}} \tag{5.39}$$

$$E^{(0)} + E^{(1)} = \langle \Psi^{(0)} | H^{(0)} + V | \Psi^{(0)} \rangle = E^{\mathrm{HF}} \tag{5.40}$$

and

$$E^{\mathrm{MP2}} = E^{(0)} + E^{(1)} + E^{(2)} = E^{\mathrm{HF}} + E^{(2)} \tag{5.41}$$

where MP2 means the second-order MP perturbation. As mentioned above, Brillouin's theorem gives

$$\langle \Psi^{(0)} | V | \Psi^{(1)} \rangle = 0 \tag{5.42}$$

Accordingly, Eq. (5.39) does not include the single as well as the triple and higher excitations; thus, only the double excitations remain: $\Psi^{s} = \Psi_{ij \to ab}$. Consequently, Eq. (5.39) can be rewritten as

$$E^{(2)} = \sum_{i<j}^{n/2} \sum_{a<b} \frac{|\langle \Psi^{(0)} | V | \Psi_{ij \to ab} \rangle|^2}{\epsilon_i + \epsilon_j - \epsilon_a - \epsilon_b} \tag{5.43}$$

Because the virtual orbitals (a and b) are higher in energy (ϵ) than the occupied ones (i and j), the denominator of Eq. (5.43) is negative, and hence $E^{(2)}$ is always negative. This means that the electronic correlation reduces the energy of the poly-electronic system ($E^{\mathrm{MP2}} < E^{\mathrm{HF}}$).

6

Density Functional Theory

As described in Chapter 5, the electron correction must be included in the molecular orbital (MO) calculations to treat chemical problems accurately. However, the configuration interaction (CI) and coupled cluster (CC) computations are so expensive as to be applied only to small molecules, and the MP2 method is more practical but also needs much computational time. Instead of such ab initio MO methods, density functional theory (DFT) calculations have often been employed to investigate the structures, properties, and reactions of various molecules [253, 356]. The DFT is based on the Hohenberg–Kohn (H–K) theorems [201]: in the first theorem, the external potential $V_{\text{ext}}(\mathbf{r})$ is a unique function of electron density $\rho(\mathbf{r})$; therefore, the $V_{\text{ext}}(\mathbf{r})$ term determines the Hamiltonian $\hat{H}$, and hence, the many-particle ground state is also a unique function of $\rho(\mathbf{r})$; in the second theorem, the ground-state energy can be evaluated from the electron density that minimizes the functionals. Here, the functional means a function of function. Since $\rho(\mathbf{r})$ is a function of the variable $\mathbf{r}$, the ground-state energy $E[\rho(\mathbf{r})]$ is a functional of the function $\rho(\mathbf{r})$. The DFT proposed originally by Thomas and Fermi [130, 476] has been underpinned by the H–K theorems and established nowadays as the Kohn–Sham method [254].

The first H–K theorem shows that the ground-state energy is determined only from $\rho(\mathbf{r})$. However, the functional of the kinetic energy based on $\rho(\mathbf{r})$ is too crude to be used for chemical problems. Therefore, Kohn and Sham introduced the orbital ψ_i (designated as the Kohn–Sham orbital) related to the electron density by

$$\rho(\mathbf{r}) = \sum_{i}^{n} \sum_{\sigma}^{\alpha,\beta} |\psi_i(\mathbf{r}, \sigma)|^2 \tag{6.1}$$

where σ represents the spin state, that is either α or β. In the closed-shell case, i.e. $\rho_\alpha(\mathbf{r}) = \rho_\beta(\mathbf{r})$, it follows that $\rho_\alpha(\mathbf{r}) = \rho_\beta(\mathbf{r}) = (1/2)\rho(\mathbf{r})$. The electron density must be normalized as

$$\int \rho(\mathbf{r}) d\mathbf{r} = n \tag{6.2}$$

Conformational Analysis of Polymers: Methods and Techniques for Structure-Property Relationships and Molecular Design, First Edition. Yuji Sasanuma.

where n is the number of electrons. The Kohn–Sham orbital equations are

$$\left[-\frac{1}{2}\nabla^2 + v_{\text{eff}}(\mathbf{r})\right]\psi_i = \epsilon_i\psi_i \tag{6.3}$$

and

$$v_{\text{eff}}(\mathbf{r}) = v(\mathbf{r}) + \int \frac{\rho(\mathbf{r}')}{|\mathbf{r}-\mathbf{r}'|}d\mathbf{r}' + v_{\text{xc}}(\mathbf{r}) \tag{6.4}$$

where $v(\mathbf{r})$ is the Coulomb potential due to nuclei A's, acting on electron i

$$v(\mathbf{r}) = -\sum_{\text{A}}\frac{Z_{\text{A}}}{r_{i\text{A}}} \tag{6.5}$$

and $v_{\text{xc}}(\mathbf{r})$ is the exchange-correlation potential

$$v_{\text{xc}}(\mathbf{r}) = \frac{\delta E_{\text{xc}}[\rho]}{\delta\rho(\mathbf{r})} \tag{6.6}$$

with $E_{\text{xc}}[\rho]$ being the exchange-correlation energy. The total energy is given by

$$E = \sum_i^n \epsilon_i - \frac{1}{2}\int \frac{\rho(\mathbf{r})\rho(\mathbf{r}')}{|\mathbf{r}-\mathbf{r}'|}d\mathbf{r}d\mathbf{r}' - \int v_{\text{xc}}(\mathbf{r})\rho(\mathbf{r})d\mathbf{r} \tag{6.7}$$

where

$$\sum_i^n \epsilon_i = \sum_i^n \left\langle \psi_i \left| -\frac{1}{2}\nabla^2 + v_{\text{eff}}(\mathbf{r}) \right| \psi_i \right\rangle \tag{6.8}$$

To solve the Kohn–Sham equations, Eqs. (6.1), (6.3), and (6.4) must be satisfied self-consistently, namely similar to the HF equations: (1) The initial orbitals (ψ_is) are guessed from the molecular parameters. (2) The Fock operator is formulated according to

$$\begin{aligned} F_{\mu\nu} = &-\frac{1}{2}\int \chi_\mu(\mathbf{r}_1)\nabla^2\chi_\nu(\mathbf{r}_1)d\mathbf{r}_1 - \int \chi_\mu(\mathbf{r}_1)\sum_{\text{A}}^{M}\frac{Z_{\text{A}}}{r_{1\text{A}}}\chi_\nu(\mathbf{r}_1)d\mathbf{r}_1 \\ &+ \iint \chi_\mu(\mathbf{r}_1)\frac{\rho(\mathbf{r}_2)}{r_{12}}\chi_\nu(\mathbf{r}_1)d\mathbf{r}_1 d\mathbf{r}_2 + \int \chi_\mu(\mathbf{r}_1)v_{\text{xc}}(\mathbf{r}_1)\chi_\nu(\mathbf{r}_1)d\mathbf{r}_1 \end{aligned} \tag{6.9}$$

where χ_μ and χ_ν are the basis sets defined in Eq. (5.10). (3) Equation (6.3) is solved with the derived Fock operator. (4) The exchange-correlation energies are calculated with the orbitals to yield the total energy of Eq. (6.7). The current ψ_is and E are compared with those obtained previously. If the agreement between the previous and current cycles is satisfactory, the convergence is attained. Otherwise, the computations will be repeated from step 2.

6.1 Exchange and Correlation Functionals

A significant problem still remains: the mathematical expressions of the exchange-correlation functional must be given. A number of the exchange $E_X(\rho)$ and correlation $E_C(\rho)$ functionals, being mostly composed of complicated equations, have been proposed. Of them, some typical functionals are introduced here. The fundamental exchange functional is based on the local-density approximation (LDA, in terms of $\rho(\mathbf{r})$, referred also to as the local spin-density approximation, LSDA, in terms of $\rho_\sigma(\mathbf{r})$), which assumes that the functional may be expressed as a function of electron density alone. For example, the Dirac–Slater exchange functional is expressed by a set of equations [109, 436, 488]

$$E_X^{\mathrm{LSDA}}\left[\rho_\alpha, \rho_\beta\right] = \int d^3\mathbf{r}\rho\epsilon_x(\rho, \zeta) \tag{6.10}$$

$$\epsilon_x(\rho, \zeta) = \epsilon_x^0(\rho) + \left[\epsilon_x^1(\rho) - \epsilon_x^0(\rho)\right] f(\zeta) \tag{6.11}$$

$$\epsilon_x^0(\rho) = \epsilon_x(\rho, 0) = C_x\rho^{1/3} \tag{6.12}$$

$$\epsilon_x^1(\rho) = \epsilon_x(\rho, 1) = 2^{1/3}C_x\rho^{1/3} \tag{6.13}$$

$$C_X = \frac{3}{4}\left(\frac{3}{\pi}\right)^{1/3} \tag{6.14}$$

$$f(\zeta) = \frac{(1+\zeta)^{4/3} + (1-\zeta)^{4/3} - 2}{2\left(2^{1/3} - 1\right)} \tag{6.15}$$

and

$$\zeta = \frac{\rho_\alpha - \rho_\beta}{\rho_\alpha + \rho_\beta} \tag{6.16}$$

The Becke exchange energy functional (abbreviated as B88 or B), including a gradient-corrected term of x_σ as well as the LSDA functional, is expressed as [39]

$$E_X^{\mathrm{B88}}\left[\rho_\alpha, \rho_\beta\right] = E_X^{\mathrm{LSDA}}\left[\rho_\alpha, \rho_\beta\right] - \sum_\sigma^{\alpha,\beta} \int \rho_\sigma^{4/3} \frac{bx_\sigma^2}{1 + 6bx_\sigma \sinh^{-1} x_\sigma} d^3\mathbf{r} \tag{6.17}$$

where

$$x_\sigma = \frac{|\nabla\rho_\sigma|}{\rho_\sigma^{4/3}} \tag{6.18}$$

and b is an empirical parameter optimized to be 0.0042.

A number of the correlation functionals have also been formulated. For example, Lee, Yang, and Parr proposed the following functional (abbreviated as LYP) [276]:

$$E_{\mathrm{c}}^{\mathrm{LYP}}\left[\rho_\alpha, \rho_\beta\right] = -a \int d^3\mathbf{r} \frac{\gamma(r)}{1 + d\rho^{-1/3}} \left\{ \rho + 2b\rho^{-5/3} \left[2^{2/3} C_{\mathrm{F}} \rho_\alpha^{8/3} + 2^{2/3} C_{\mathrm{F}} \rho_\beta^{8/3} \right.\right.$$
$$\left. - \rho t_{\mathrm{w}} + \frac{1}{9}\left(\rho_\alpha t_{\mathrm{w}}^\alpha + \rho_\beta t_{\mathrm{w}}^\beta\right) + \frac{1}{18}\left(\rho_\alpha \nabla^2 \rho_\alpha + \rho_\beta \nabla^2 \rho_\beta\right) \right]$$
$$\left. \times \exp\left(-c\rho^{-1/3}\right) \right\} \tag{6.19}$$

where

$$\gamma(\mathbf{r}) = 2\left[1 - \frac{\rho_\alpha^2(\mathbf{r}) + \rho_\beta^2(\mathbf{r})}{\rho^2(\mathbf{r})}\right] \tag{6.20}$$

$$t_{\mathrm{w}} = \frac{1}{8}\left[\frac{|\nabla \rho(\mathbf{r})|^2}{\rho(\mathbf{r})} - \nabla^2 \rho\right] \tag{6.21}$$

$$C_{\mathrm{F}} = \frac{3}{10}(3\pi^2)^{2/3} \tag{6.22}$$

$a = 0.04918$, $b = 0.132$, $c = 0.2533$, and $d = 0.349$.

The B3LYP functional [40, 444], a hybrid exchange-correlation functional, in which the LSDA, Hartree–Fock, and B88 exchange terms are combined with the LYP and Vosko–Wilk–Nusair (VWN) (or PW91) correlation terms so as to reproduce various experimental data accurately, was proposed in the following form:

$$E_{\mathrm{xc}}^{\mathrm{B3LYP}} = (1 - a_0)E_{\mathrm{x}}^{\mathrm{LSDA}} + a_0 E_{\mathrm{x}}^{\mathrm{HF}} + a_{\mathrm{x}}^{\mathrm{B88}} + a_{\mathrm{c}} E_{\mathrm{c}}^{\mathrm{LYP}} + (1 - a_{\mathrm{c}})E_{\mathrm{c}}^{\mathrm{VWN}} (\text{or } E_{\mathrm{c}}^{\mathrm{PW91}}) \tag{6.23}$$

where a_0, a_{x}, and a_{c}, termed Becke's three parameters, were semiexperimentally optimized by a linear least-squares fit to 56 atomization energies, 42 ionization potentials, 8 proton affinities, and 10 first-row total atomic energies to be $a_0 = 0.20$, $a_{\mathrm{x}} = 0.72$, and $a_{\mathrm{c}} = 0.81$. The mathematical expressions of the VWN [514] and PW91 (Perdew 1991) [362] correlation functions are too complicated and lengthy to be shown here; for the details, see the original articles. The B3LYP functionals have provided reliable geometrical parameters, spectroscopic data, and thermochemical data so as to be most extensively used in a variety of quantum chemical calculations [154, 444].

In the DFT, the exchange and correlation functionals are functions of electron density, which is also expressed as a function of spatial coordinates; therefore, a large number of integrals are numerically calculated. The three-dimensional space surrounding the molecule is expressed by a grid pattern, which has n_{rad} radial shells and n_{ang} angular points per shell, and the numerical computations are carried out at the individual grid points and summed to yield the integrals. The fineness of the grid, (n_{rad}, n_{ang}), depends on the required computational accuracy: for example, coarse (35, 110), standard (50, 194), fine (75, 302), and ultrafine (99, 590).

6.2 Dispersion-force Correction

The DFT method, based on the Kohn–Sham equation, represents a mean-field electronic structure and hence lacks long-range electronic correlation effects such as the London dispersion interactions [177]. To compensate for the inherent defect, semiempirical dispersion energies of atom pairs have often been added to the DFT energy.

For instant, Grimme's D2 approach adopts the following expression for the dispersion correction [175, 177]:

$$E_{\text{DFT-D}} = E_{\text{DFT}} + E_{\text{disp}}^{\text{D2}} \tag{6.24}$$

where $E_{\text{DFT-D}}$ is the dispersion-corrected energy, and E_{DFT} is the energy as obtained from the DFT calculation. The $E_{\text{disp}}^{\text{D2}}$ energy, the correction term, is expressed similarly to the empirical van der Waals potential:

$$E_{\text{disp}}^{\text{D2}} = -s_6 \sum_{i=1}^{N_{\text{at}}-1} \sum_{j=i+1}^{N_{\text{at}}} \frac{C_6^{ij}}{R_{ij}^6} f_{\text{dmp}}(R_{ij}) \tag{6.25}$$

where s_6 is the global scaling factor, N_{at} is the number of atoms included in the system, C_6^{ij} $(=\sqrt{C_6^i C_6^j})$ is the dispersion coefficient for atom pair ij, and R_{ij} is the distance between atoms i and j. Here, the damping factor is given as

$$f_{\text{dmp}}(R_{ij}) = \left\{1 + \exp\left[-d\left(\frac{R_{ij}}{R_r} - 1\right)\right]\right\}^{-1} \tag{6.26}$$

where d represents the steepness of the damping function, and R_r is the sum of van der Waals radii (R_{vdW}s) of atoms i and j. The parameters are listed in Table 6.1.

Table 6.1 Parameters for Grimme's D2 dispersion correction.

		R_{vdW} (Å)	
	C_6^{X} [a]	Original[a]	Modified[b]
H	0.14	1.001	1.3013
C	1.75	1.452	1.70
N	1.23	1.397	1.55
O	0.70	1.342	1.52
S	5.57	1.683	1.80

a) Reference [175].
b) References [87, 375, 416].

Of them, van der Waals radii offered originally by Grimme [175] have been partly modified so as to reproduce experiments better (Table 6.1) [87, 375, 416]. The modified R_{vdW}s are equal to those offered by Bondi [45], except for that of hydrogen. The B3LYP calculations with the D2 correction have yielded satisfactory results consistent with various experiments [165, 166, 266, 422].

Furthermore, semiempirical methods such as Grimme's D3 method [176, 177] and the Austin–Petersson–Frisch functional with an empirical add-on (APF-D) [29] to correct dispersion forces have been proposed.

7

Solvent Effect

In quantum chemistry, the solvation is, in principle, based on the reaction field and the polarized continuum model (PCM) [93, 483]: On the assumption that the solute has a detailed molecular structure, the charge distribution of the solute polarizes the surrounding solvents that are represented as a dielectric continuum (permittivity ϵ), which in turn perturb the solute charge distribution.

In the PCM, the solute is placed in a cavity, which preferably reflects the molecular shape as accurately as possible. In the early stage, a spherical cavity was adopted [346], but afterward, more realistic models have been proposed: a cavity whose surface is shaped by a constant electron density (isodensity surface) [155] and a cavity composed of connected spheres whose centers are located at the atomic nuclei and radii somewhat larger (e.g. ×1.2) [91] than the van der Waals radii of Bondi [45] and Pauling [357]. The PCMs have been formulated in terms of the surface charge density [483]:

$$\sigma(\mathbf{s}) = \frac{\epsilon - 1}{4\pi\epsilon}\frac{\partial}{\partial \mathbf{n}}\left(V_{\mathrm{M}} + V_{\sigma}\right) \tag{7.1}$$

where $\mathbf{s}$ is the position vector pointing the cavity surface, $\mathbf{n}$ is the unit vector normal to the surface, V_{M} is the electronic potential due to the solute charge distribution, and V_{σ} is the surface electronic potential. To facilitate the computation, the cavity surface is divided into small elements (termed "tesserae") (see Figure 7.1), and the point charge of a tessera i is given by $q_i = \sigma_i a_i$, with a_i being its area. Then, the surface integral may be replaced by a summation:

$$V_{\sigma}(\mathbf{r}) = \int_{\Gamma} \frac{\sigma(\mathbf{s})}{|\mathbf{r} - \mathbf{s}_i|} d^2 s \simeq \sum_i \frac{\sigma(\mathbf{s}) a_i}{|\mathbf{r} - \mathbf{s}_i|} = \sum_i \frac{q_i}{|\mathbf{r} - \mathbf{s}_i|} \tag{7.2}$$

where $\mathbf{r}$ is the position vector of an arbitrary point in the space, and $\mathbf{s}_i$ is that of the tessera i on the cavity surface.

An early PCM, termed dielectric polarized continuum model (DPCM) [318], supposes that the solute charge distribution stays within the cavity, which can be

Conformational Analysis of Polymers: Methods and Techniques for Structure-Property Relationships and Molecular Design, First Edition. Yuji Sasanuma.

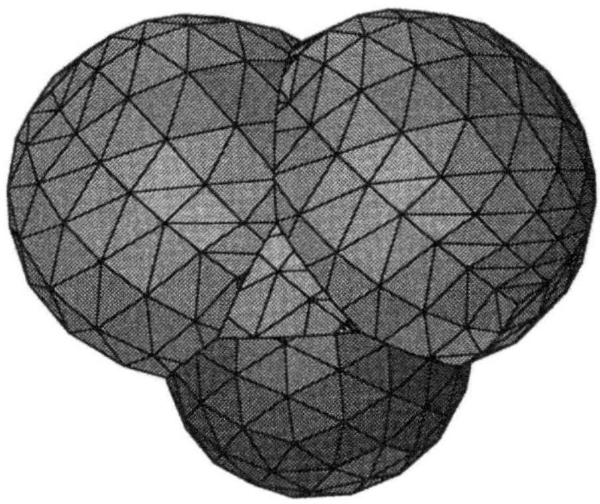

Figure 7.1 A cavity surface for acetone, produced by the program GEPOL [435] and divided into tesserae of 0.2 Å^2 area. Source: Cossi and Barone [90], Figure 2 (p. 10616)/ Reproduced with permission of American Chemical Society.

expressed with the Gauss theorem as

$$\int_{\Gamma} \sigma(\mathbf{s}) d\mathbf{s} = -\frac{\epsilon - 1}{4\pi\epsilon} \int_{\Gamma} \mathbf{E}(\mathbf{s}) \cdot \mathbf{n}(\mathbf{s}) d\mathbf{s} = -\frac{\epsilon - 1}{\epsilon} Q_{\mathrm{M}} \tag{7.3}$$

where $\mathbf{E}(\mathbf{s}) \cdot \mathbf{n}(\mathbf{s})$ is the magnitude of the electric field normal to the surface, and Q_{M} is the total charge of the solute (null for neutral solutes). From the viewpoint of the quantum theory, however, Eq. (7.3) is insufficient because the electronic charge distribution of the solute lies outside the cavity, which is termed the *outlying charge*.

In the conductor-like screening model (COSMO) [247], the permittivity of the solvent continuum is set to infinity ($\epsilon \to \infty$), which corresponds to that of a conductor, and the total potential $V(\mathbf{r})$ cancels out on the cavity surface. Consequently, the apparent surface charge is determined from the local electrostatic potential instead of the normal component of the gradient. To compensate for the hypothesis, the scale function $f(\epsilon)$ was introduced:

$$\sigma(\mathbf{s}) = f(\epsilon)\, \sigma^*(\mathbf{s}) \tag{7.4}$$

where

$$f(\epsilon) = \frac{\epsilon - 1}{\epsilon + k} \tag{7.5}$$

σ^* is the ideal unscreened charge density, and k is an empirical parameter: $k = 0.5$ is suggested to be appropriate for neutral solutes and $k = 0$ for charged solutes.

In the integral equation formalism of polarized continuum model (IEFPCM) [56, 312, 313], the electrostatic potential is divided into two terms, V_{M} and V_{R}, where V_{M} and V_{R} are generated by the solute in vacuo and by the reaction potential, respectively. The V_{R} term is defined as

$$V_{\mathrm{R}}(x) = \int_{\Gamma} \frac{\sigma(y)}{|x - y|} dy \tag{7.6}$$

and fulfill the boundary conditions:

$$-\nabla^2 V_{\mathrm{R}}(x) = 0 \quad \text{for both in and out} \tag{7.7}$$

$$V_{\mathrm{R}}^{\mathrm{in}}(x) - V_{\mathrm{R}}^{\mathrm{out}}(x) = 0 \tag{7.8}$$

and

$$\lim_{x \to \infty} V_{\mathrm{R}}(x) = 0 \tag{7.9}$$

where "in" and "out" mean the inside and outside of the cavity, respectively. The $V_{\mathrm{R}}(x)$ potential is continuous across the surface and decays toward $x \to \infty$, thus acting as an outlying charge.

In the surface and volume polarization for electrostatics (SVPE) model [78], the $V_{\mathrm{R}}(x)$ term is further divided into two potentials: $V_{\sigma}(x)$ due to the surface polarization charge density and $V_{\beta}(x)$ due to the apparent volume polarization charge density, that is

$$V_{\mathrm{R}}(x) = V_{\sigma}(x) + V_{\beta}(x) \tag{7.10}$$

where

$$V_{\beta}(x) = -\frac{\epsilon - 1}{\epsilon} \int_{\mathrm{ext}} \frac{\rho_{\mathrm{M}}(y)}{|x - y|} dy \tag{7.11}$$

Here, $\rho_{\mathrm{M}}(y)$ represents the solute charge distribution, and the integral is carried out over all points located outside the cavity; therefore, the exterior charge density may be considered to be an outlying charge. However, the numerical integral over the exterior volume to calculate $V_{\beta}(x)$ is so time consuming that, instead, a modified model, surface and simulation of volume polarization for electrostatics (SS(V)PE) [79], was proposed. The SS(V)PE model is based on the poof that all direct and indirect effects of the volume polarization can be exactly simulated from an additional surface polarization.

Afterward, it was shown that the SS(V)PE model is identical to the IEFPCM [80] and also suggested that no significant difference can be found between COSMO and SS(V)PE/IEFPCM if the correct scaling factor k of COSMO in Eq. (7.5) is chosen [246].

The solvent effect can be introduced to quantum chemical calculations by

$$\left[\hat{H}^0 + \hat{V}_{\mathrm{R}}\right] \Psi = E\Psi \tag{7.12}$$

where the superscript "0" stands for the vacuum, and $\hat{V}_{\mathrm{R}}$ is the perturbation due to the solvation. For a solute in vacuo, therefore, the Schrödinger equation is expressed as

$$\hat{H}^0 \Psi^0 = E^0 \Psi^0 \tag{7.13}$$

For example, according to the IEFPCM [56, 312, 313, 483], the Fock operator of the Hartree–Fock (HF) method is expressed as

$$\tilde{\mathbf{F}} = \left[\mathbf{h} + \frac{1}{2}(\mathbf{i}+\mathbf{j})\right] + [\mathbf{G}(\mathbf{P}) + \mathbf{X}(\mathbf{P})] \tag{7.14}$$

where **P** is the one-electron density matrix, **h** and **G(P)** are, respectively, the matrices of one- and two-electron integrals of the solute in vacuo, and the other three terms, **i**, **j**, and **X(P)**, represent the solvent effects. The Fock matrix including the solvation terms will also be treated self-consistently as described in Section 5.5. Finally, the internuclear repulsive potential, omitted from the Hamiltonian according to the Born–Oppenheimer approximation, is also added to the obtained electronic energy. In the density functional theory (DFT) computations, the solvation terms are added to the Kohn–Sham matrix. In the electron-correlation calculations, for example the MP2 energy may be evaluated from the perturbation with the solvated HF orbitals.

As an example, conformational internal energies (ΔE_ks) of poly(glycolic acid) (PGA, Figure 7.2), evaluated with the IEFPCM from its dimeric model compound,

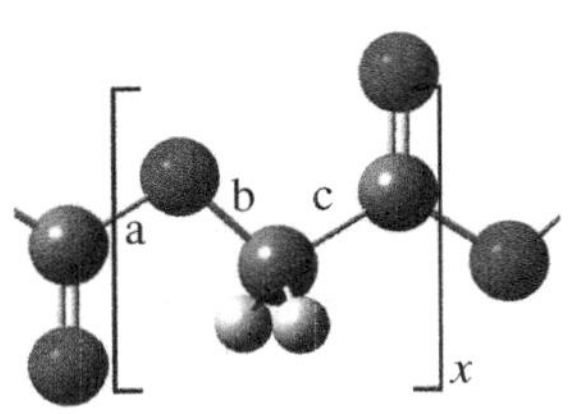

Figure 7.2 Poly(glycolic acid) (PGA). The skeletal bonds are designated as shown.

Table 7.1 Conformational internal energies (ΔE_ks) of PGA at 25 °C.

Bond[a)]			ΔE_k[b)] (kcal mol^{-1})		
a	**b**	**c**	**Gas**	**Methanol**	**DMSO**
t	t	t	2.27	1.75	1.74
t	t	s	1.58	1.40	1.39
t	g$^{\pm}$	t	0.00	0.00	0.00
t	g$^{\pm}$	s	0.67	0.45	0.44

a) See Figure 7.2. Abbreviations: t, trans; g$^{\pm}$, gauche$^{\pm}$; s, synperiplanar.
b) Determined from a dimeric model compound, 2-methoxy-2-oxoethyl 2-acetoxyacetate, at the MP2/6-311+G(2d,p)//B3LYP/6-311+G(2d,p) level with the IEFPCM. Relative to the tg$^{\pm}$t conformation.

Source: Sasanuma et al. [422], Table 1 (p. 3732)/Reproduced with permission of American Chemical Society.

2-methoxy-2-oxoethyl 2-acetoxyacetate, placed in vacuo (gas), methanol, and dimethyl sulfoxide (DMSO) at the MP2/6-311+G(2d,p)//B3LYP/6-311+G(2d,p) level, are listed in Table 7.1. In the later chapters, a lot of examples of solvent effects on conformational energies and configurational properties and the thermodynamic parameters will be presented and discussed.

8

Statistical Thermodynamics for Quantum Chemistry

In chemistry, the "energy" is strictly classified into internal energy, enthalpy, Helmholtz free energy, and Gibbs free energy; nevertheless, the electronic energy that the quantum chemistry yields corresponds to none of them. For the sake of comparison with experiment, the electronic energy should preferably be converted to one of the thermochemical quantities, and statistical mechanics fulfills the role. The theoretical basis and methods required therefore are described here [309, 336].

The molecular motions are divided into translation, rotation, and vibration. The solution of the Schrödinger equation for each motion is given as a function of the quantum number. The molecular partition function is defined as the sum of the Boltzmann factors corresponding to the energy levels:

$$z = \sum_i g_i e^{-\beta \epsilon_i} \tag{8.1}$$

where i stands for the energy level, g_i is its degeneracy degree, and

$$\beta = \frac{1}{kT} = \frac{1}{(R/N_A)T} \tag{8.2}$$

with k, R, N_A, and T being the Boltzmann constant, the gas constant, the Avogadro constant, and the absolute temperature, respectively. In this section, the lowercase z stands for the molecular partition function, and the uppercase Z is the partition function of the system composed of N molecules.

The internal energy E is given by

$$E = -\left(\frac{\partial \ln Z}{\partial \beta}\right)_V = kT^2\left(\frac{\partial \ln Z}{\partial T}\right)_V \tag{8.3}$$

and the entropy is derived from

$$S = \frac{E}{T} + k \ln Z \tag{8.4}$$

Conformational Analysis of Polymers: Methods and Techniques for Structure-Property Relationships and Molecular Design, First Edition. Yuji Sasanuma.

The Helmholtz free energy is given by

$$F = -kT \ln Z \tag{8.5}$$

and the pressure can be derived from

$$P = -\left(\frac{\partial F}{\partial V}\right)_T = kT\left(\frac{\partial \ln Z}{\partial V}\right)_T \tag{8.6}$$

The z function for the whole molecular motions is the product of those for the individual motions:

$$z = z_{\text{trans}} \cdot z_{\text{rot}} \cdot z_{\text{vib}} \cdot z_{\text{elect}} \tag{8.7}$$

where the subscripts trans, rot, vib, and elect represent translational, rotational, and vibrational motions and electronic excitation, respectively.

8.1 Translational Motion

In the translational motion, since the N molecules cannot be distinguished from each other, the partition function of the system is related to z_{trans} by

$$Z_{\text{trans}} = \frac{1}{N!} z_{\text{trans}}^N \tag{8.8}$$

The z_{trans} function is expressed as

$$z_{\text{trans}} = \left(\frac{2\pi mkT}{h^2}\right)^{3/2} V = \left(\frac{2\pi mkT}{h^2}\right)^{3/2} \frac{RT}{P} \tag{8.9}$$

where m is the mass of the molecule, h is Planck's constant, and V is the volume, which is derived from the equation of state of the ideal gas, $PV = RT$.

The internal energy per mole for the translation is given by

$$E_{\text{trans}} = RT^2 \frac{1}{z_{\text{trans}}} \left(\frac{\partial z_{\text{trans}}}{\partial T}\right)_V = RT^2 \frac{3}{2T} = \frac{3}{2}RT \tag{8.10}$$

Because Stirling's approximation leads to

$$\ln Z_{\text{trans}} = \ln\left(\frac{z_{\text{trans}}^N}{N!}\right) = N \ln z_{\text{trans}} - \ln N! = N \ln z_{\text{trans}} - N \ln N + N \tag{8.11}$$

the entropy per mole for the translation can be expressed as

$$\begin{aligned} S_{\text{trans}} &= \frac{E_{\text{trans}}}{T} + k \ln Z_{\text{trans}} \\ &= \frac{3}{2}R + k\left(N_{\text{A}} \ln z_{\text{trans}} - \ln N_{\text{A}} + N_{\text{A}}\right) \end{aligned}$$

$$= R\left(\frac{5}{2} + \ln\frac{z_{\text{trans}}}{N_A}\right)$$

$$= R\left\{\frac{5}{2} + \ln\left[\left(\frac{2\pi mkT}{h^2}\right)^{3/2}\frac{RT}{N_A P}\right]\right\} \tag{8.12}$$

8.2 Rotational Motion

The rotational partition function Z_{rot} is related to the molecular partition function by

$$Z_{\text{rot}} = z_{\text{rot}}^N \tag{8.13}$$

The z_{rot} function of linear molecules is expressed as

$$z_{\text{rot}} = \sum_J (2J+1)\exp\left[-\beta hc\tilde{B}\, J(J+1)\right] \tag{8.14}$$

where J (= 0, 1, 2, …) is the rotational quantum number, c is the speed of light, and $\tilde{B}$ is the rotational constant defined by

$$\tilde{B} = \frac{h}{8\pi^2 c I_B} \tag{8.15}$$

where I_B is the principal moment of inertia around the b-axis. The summation of Eq. (8.14) may be replaced with an integral

$$\begin{aligned} z_{\text{rot}} &= \int_0^\infty (2J+1)\exp\left[-\beta hc\tilde{B}J(J+1)\right] dJ \\ &= -\frac{1}{\beta hc\tilde{B}}\int_0^\infty \frac{d}{dJ}\exp\left[-\beta hc\tilde{B}J(J+1)\right] dJ \\ &= \left[-\frac{1}{\beta hc\tilde{B}}\exp\left[-\beta hc\tilde{B}J(J+1)\right]\right]_0^\infty \\ &= \frac{1}{\beta hc\tilde{B}} \\ &= \frac{8\pi^2 IkT}{h^2} \end{aligned} \tag{8.16}$$

A new parameter, symmetry number (σ_{rot}), is introduced into z_{rot}:

$$z_{\text{rot}} = \frac{8\pi^2 IkT}{\sigma_{\text{rot}} h^2} \tag{8.17}$$

σ_{rot} is 1 for asymmetric molecules such as CO or 2 for symmetric molecules such as N_2 and O_2. This is because the symmetric molecules have a twofold rotation (C_2) axis perpendicular to the molecular axis.

The rotational internal energy per mole of linear molecules is given by

$$E_{\text{rot}} = N_A kT^2\left(\frac{\partial \ln z_{\text{rot}}}{\partial T}\right)_V = N_A kT^2 \frac{1}{T} = RT \tag{8.18}$$

and the rotational entropy per mole is expressed as

$$S_{\text{rot}} = \frac{E_{\text{rot}}}{T} + k \ln Z_{\text{rot}} = N_{\text{A}}k + N_{\text{A}}k \ln z_{\text{rot}} = R\left(1 + \ln z_{\text{rot}}\right) \tag{8.19}$$

The z_{rot} function of nonlinear molecules is

$$z_{\text{rot}} = \frac{1}{\sigma_{\text{rot}}}\left(\frac{kT}{hc}\right)^{3/2}\left(\frac{\pi}{\tilde{A}\tilde{B}\tilde{C}}\right)^{1/2} \tag{8.20}$$

where $\tilde{A}$, $\tilde{B}$, and $\tilde{C}$ are the rotation numbers for the principal axes of inertia (a, b, and c axes). The symmetry number depends on the molecular symmetry, for example 3 for NH_3 and 12 for C_6H_6. The rotational internal energy per mole of nonlinear molecules is given by

$$E_{\text{rot}} = N_{\text{A}}kT^2\left(\frac{\partial \ln z_{\text{rot}}}{\partial T}\right)_V = N_{\text{A}}kT^2\frac{3}{2T} = \frac{3}{2}RT \tag{8.21}$$

and the rotational entropy per mole is

$$S_{\text{rot}} = \frac{E_{\text{rot}}}{T} + k \ln Z_{\text{rot}} = \frac{3}{2}R + N_{\text{A}}k \ln z_{\text{rot}} = R\left(\frac{3}{2} + \ln z_{\text{rot}}\right) \tag{8.22}$$

8.3 Vibrational Motion

The number of fundamental vibrational modes of linear and nonlinear molecules are, respectively, $3n - 5$ and $3n - 6$, where n is the number of atoms in the molecule, which can be explained as follows: The n gaseous atoms have $3n$ degrees of freedom (3 per atom), the translation takes 3 ones, and the rotation possesses 2 (linear) or 3 (nonlinear) ones; therefore, $3n - 5$ or $3n - 6$ degrees of freedom are left for the vibration.

The potential energy based on the harmonic oscillator is expressed as

$$\epsilon_v = \left(v + \frac{1}{2}\right)h\nu \tag{8.23}$$

where $v\ (= 0, 1, 2, \ldots)$ is the vibrational quantum number, and ν is the frequency (Figure 8.1). When the bottom of the potential well is set at the zero point of energy, the molecular partition function for the vibration is expressed as

$$\begin{aligned} z_{\text{vib}} &= \sum_{v=0} \exp\left[-\beta\left(v + \frac{1}{2}\right)h\nu\right] \\ &= e^{-\beta(1/2)h\nu} + e^{-\beta(3/2)h\nu} + e^{-\beta(5/2)h\nu} + \cdots \\ &= \frac{e^{-\beta(1/2)h\nu}}{1 - e^{-\beta h\nu}} \\ &= \frac{e^{-h\nu/(2kT)}}{1 - e^{-h\nu/(kT)}} \end{aligned} \tag{8.24}$$

Figure 8.1 Harmonic oscillator potential; r_{eq} is the equilibrium position; the zero-point energy of the jth vibrational mode is $(1/2)h\nu_j$; and the total zero-point energy is (E_0) $(1/2)\sum_j h\nu_j$.

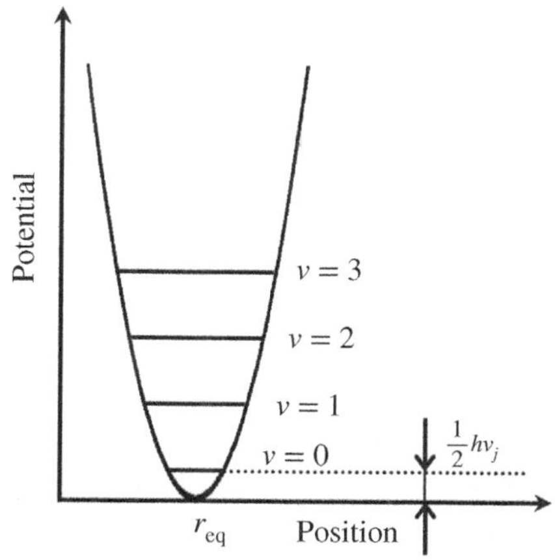

The vibrational partition function Z_{vib} is related to z_{vib} by

$$Z_{vib} = z_{vib}^N \tag{8.25}$$

Therefore, the vibrational internal energy per mole is derived as

$$\begin{aligned} E_{vib} &= -\left(\frac{\partial \ln Z_{vib}}{\partial \beta}\right)_V \\ &= N_A h\nu \left[\frac{1}{2} + \frac{1}{e^{\beta(h\nu)} - 1}\right] \\ &= R\left(\frac{h\nu}{k}\right)\left[\frac{1}{2} + \frac{1}{e^{h\nu/(kT)} - 1}\right] \end{aligned} \tag{8.26}$$

and the vibrational entropy per mole is

$$\begin{aligned} S_{vib} &= \frac{E_{vib}}{T} + k \ln Z_{vib} \\ &= R\left(\frac{h\nu}{kT}\right)\left[\frac{1}{2} + \frac{1}{e^{(h\nu)/(kT)} - 1}\right] + N_A k \ln \frac{e^{-h\nu/(2kT)}}{1 - e^{-h\nu/(kT)}} \\ &= R\left(\frac{h\nu}{kT}\right)\left[\frac{1}{2} + \frac{1}{e^{(h\nu)/(kT)} - 1}\right] + R\left\{-\frac{1}{2}\left(\frac{h\nu}{kT}\right) - \ln\left[1 - e^{-h\nu/(kT)}\right]\right\} \\ &= R\left\{\left(\frac{h\nu}{kT}\right)\frac{1}{e^{(h\nu)/kT} - 1} - \ln\left[1 - e^{-h\nu/(kT)}\right]\right\} \end{aligned} \tag{8.27}$$

Since there are $3n - 5$ or $3n - 6$ vibrational modes, the total vibrational partition function can be written as the product of the individuals, and the total internal energy (E_{vib}^{total}), zero-point energy (E_0), and entropy (S_{vib}^{total}) are the sums of individuals:

$$z_{vib}^{total} = \prod_j \frac{e^{-h\nu_j/(2kT)}}{1 - e^{-h\nu_j/(kT)}} \tag{8.28}$$

$$E_{\text{vib}}^{\text{total}} = R\sum_j \left(\frac{h\nu_j}{k}\right)\left[\frac{1}{2} + \frac{1}{e^{h\nu_j/(kT)} - 1}\right] \tag{8.29}$$

$$E_0 = \frac{1}{2}\sum_j h\nu_j \tag{8.30}$$

and

$$S_{\text{vib}}^{\text{total}} = R\sum_j \left\{\left(\frac{h\nu_j}{kT}\right)\frac{1}{e^{(h\nu_j)/kT} - 1} - \ln\left[1 - e^{-h\nu_j/(kT)}\right]\right\} \tag{8.31}$$

where j indicates the vibrational mode.

In quantum chemistry, the vibrational frequencies are calculated as follows [343]. The molecular structure is fully optimized at a given level of theory with proper basis sets. The Hessian matrix composed of the second partial derivatives of the potential (V) with respect to displacements ($\xi_{Al}, \xi_{Bm}, \ldots$) of atoms (A and B) in the l- and m-axis directions of the Cartesian coordinates in equilibrium (subscript 0 of Eq. (8.32)) is calculated at the same level of theory with the same basis sets. The Hessian matrix, whose size is $3n \times 3n$ (n: the number of atoms), is given by

$$H_{\text{A};l,\text{B};m} = \left(\frac{\partial^2 V}{\partial \xi_l \partial \xi_m}\right)_0 \tag{8.32}$$

and converted to that in the mass weighted Cartesian coordinates:

$$W_{\text{A};l,\text{B};m} = \frac{H_{\text{A};l,\text{B};m}}{\sqrt{M_\text{A} M_\text{B}}} \tag{8.33}$$

where M_A and M_B are the masses of atoms A and B, respectively. The $W_{\text{A};l,\text{B};m}$ matrix is diagonalized, and the square roots of the $3n$ eigenvalues correspond to the fundamental frequencies of the molecule. Of them, the lowest frequencies close to zero are due to the translational and rotational motions, and the remaining $3n - 5$ or $3n - 6$ data are physically meaningful. The Hessian matrix is further converted to that in the internal coordinates and diagonalized. The eigenvalues (λ_js) thus obtained yield the vibrational frequencies (ν_js):

$$\lambda_j = 4\pi^2\nu_j^2 \tag{8.34}$$

8.4 Electronic Excitation

The molecular partition function for the electron excitation is expressed as

$$z_{\text{elect}} = \sum_i g_i e^{-\epsilon_i/(kT)} = g_0 e^{-\epsilon_0/(kT)} + g_1 e^{-\epsilon_1/(kT)} + g_2 e^{-\epsilon_2/(kT)} + \cdots \tag{8.35}$$

where g_i represents the degeneracy degree of the ith energy level. When the energy (ϵ_0) of the ground state is set equal to zero, the energy differences between the excited and the ground states is much larger than the thermal energy, kT, and hence

$$z_{\text{elect}} = g_0 \tag{8.36}$$

In addition, if $g_0 = 1$, then $z_{\text{elect}} = 1$,

$$E_{\text{elect}} = 0 \tag{8.37}$$

and

$$S_{\text{elect}} = 0 \tag{8.38}$$

8.5 Thermochemistry

The internal energies derived above are summarized to be the total one:

$$E_{\text{total}} = E_{\text{trans}} + E_{\text{rot}} + E_{\text{vib}} + E_{\text{elect}} \tag{8.39}$$

and the total entropy is

$$S_{\text{total}} = S_{\text{trans}} + S_{\text{rot}} + S_{\text{vib}} + S_{\text{elect}} \tag{8.40}$$

If the electronic energy (E_{ee}) evaluated from quantum chemical calculations is added to E_{total}, the sum corresponds to the internal energy defined in thermodynamics [344]:

$$E = E_{\text{ee}} + E_{\text{total}} \tag{8.41}$$

and the thermodynamical enthalpy is

$$H = E + PV = E + RT \tag{8.42}$$

because the molecule is assumed to be gaseous. The Gibbs free energy can be evaluated from

$$G = H - TS_{\text{total}} \tag{8.43}$$

As described above, with the help of the statistical mechanics, the quantum chemical quantities can be connected to the thermodynamic functions at a given temperature and pressure and compared with the corresponding experimental data.

Conformational energies of polyethylene (PE) were derived from molecular orbital (MO) calculations on n-butane and n-pentane placed in the gas phase at 25 °C and 1 atm and in n-hexadecane at 25 °C and the Θ point (140 °C). The molecular geometries were optimized at the B3LYP/6-311+G(2d,p) level, and the thermochemical terms, E_{total} and S_{total}, were evaluated at the same level.

The electronic energies (E_{ee}) were calculated at the coupled cluster single–double and perturbative triple (CCSD(T))/cc-pVQZ and the MP2/cc-pVQZ levels. The Gibbs free energies of the individual conformers, obtained from Eq. (8.43), are shown in Table 8.1. It is known that the CCSD(T)/cc-pVQZ computation is extremely expensive (time consuming) but accurate. In order to evaluate the configurational properties of PE at the Θ point, the ΔG values were calculated at 140 °C, which is higher than the boiling points of *n*-butane (−0.5 °C) and *n*-pentane (36 °C). The gauche form of *n*-butane has a higher free energy by 0.73 kcal mol^{-1} than that of the trans state. The gauche energy is almost independent of the solvent and the temperature. This is partly because the solvent, *n*-hexadecane, is nonpolar (dielectric constant 2.04). The $g^{\pm}g^{\mp}$ conformations of *n*-pentane show a large ΔG ($\simeq$ 4 kcal mol^{-1}) owing to the pentane effect (see Figure 2.5). The CCSD(T) and MP2 methods yield almost the same ΔGs, which suggests the reliability of the MP2 calculations.

As another example, Gibbs free energies of ethylene glycol dibenzoate (EGDB, Figure 8.3), a model compound of poly(ethylene terephthalate) (PET, see Figure 12.3), are presented. The ΔG values of conformations in the oxyethylene chain intervening between two benzoate rings were calculated at the

Table 8.1 Gibbs free energies (ΔGs) of *n*-butane and *n*-pentane in the gas phase and *n*-hexadecane at 25 and 140 °C.

		ΔG[a] (kcal mol^{-1})					
		CCSD(T)[b]			MP2[c]		
		Gas	*n*-hexadecane		Gas	*n*-Hexadecane	
Conformation[d]		25 °C	25 °C	140 °C[e]	25 °C	25 °C	140 °C[e]
			n-Butane				
t		0.00	0.00	0.00	0.00	0.00	0.00
$g^{\pm}$		0.73	0.74	0.75	0.73	0.73	0.75
			n-Pentane				
t	t	0.00	0.00	0.00	0.00	0.00	0.00
t	$g^{\pm}$	0.70	0.71	0.72	0.69	0.69	0.70
$g^{\pm}$	$g^{\pm}$	1.12	1.13	1.13	1.06	1.07	1.07
$g^{\pm}$	$g^{\mp}$	3.96	3.96	4.47	4.06	4.06	4.57

a) Relative to the t (*n*-butane) or tt (*n*-pentane) conformation.
b) At the CCSD(T)/cc-pVQZ//B3LYP/6-311+G(2d,p) level with the IEFPCM.
c) At the MP2/cc-pVQZ//B3LYP/6-311+G(2d,p) level with the IEFPCM.
d) See Figure 8.2.
e) The Θ temperature of polyethylene in hydrocarbon solvents.

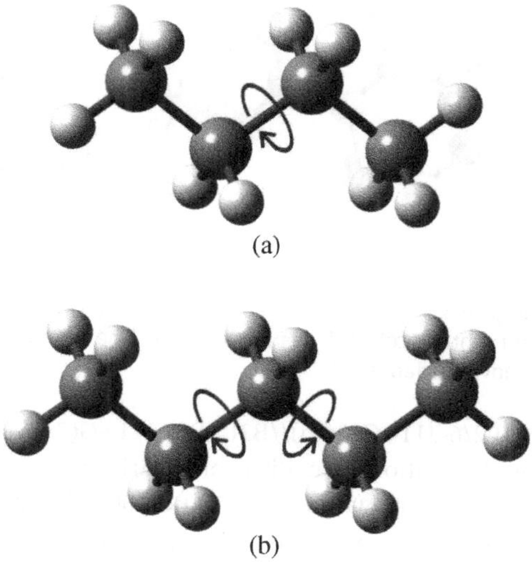

Figure 8.2 (a) *n*-Butane and (b) *n*-pentane. The round arrows represent the internal rotations shown in Table 8.1.

Table 8.2 Gibbs free energies (Δ*G*s) of EGDB in the gas phase and benzene at 25 and 250 °C.

			ΔG[a] (kcal mol^{-1})		
			Gas	Benzene	
Conformation[b]			25 °C	25 °C	250 °C[c]
t	t	t	0.00	0.00	0.00
t	t	g±	0.50	0.55	0.80
t	g±	t	−1.12	−1.24	−0.98
t	g±	g±	−0.96	−0.94	−0.82
t	g±	g∓	−0.65	−0.69	−0.58
g±	t	g±	1.09	1.16	1.28
g±	t	g∓	0.41	0.61	1.08
g±	g±	g±	−0.84	−0.73	−0.19
g±	g±	g∓	−0.61	−0.59	−0.59
g±	g∓	g±		(absent)[d]	

a) Relative to the ttt conformation. At the MP2/6-311+G(2d,p)//B3LYP/6-311+G(2d,p) level with the IEFPCM.
b) See Figure 8.3.
c) SANS experiments determined the chain dimension of amorphous PET at 250 °C [171].
d) The local minimum of the potential was not found by the geometrical optimization.

Source: [400], Table 4 (p. 2859)/Reproduced with permission of American Chemical Society.

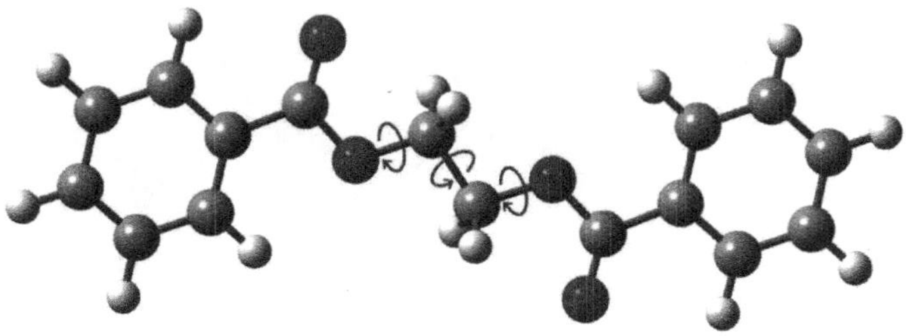

Figure 8.3 Ethylene glycol dibenzoate (EGDB), a model compound of poly(ethylene terephthalate) (PET, see Figure 12.3). The round arrows represent the internal rotations shown in Table 8.2.

MP2/6-311+G(2d,p)//B3LYP/6-311+G(2d,p) level with the solvation of integral equation formalism of polarized continuum model (IEFPCM) [400]. The temperature was set at 25 and 250 °C. Small-angle neutron scattering (SANS) experiments were carried out for amorphous PET to determine the radius of gyration of the unperturbed PET chain at 250 °C [171]. As seen from Table 8.2, almost all conformations but $g^{\pm}g^{\mp}g^{\pm}$ were geometrically optimized. The solvent effect on the ΔG values is seen to be small, whereas the temperature dependence is significant. The characteristic ratios of the PET chain at 25 and 250 °C were evaluated from the rotational isomeric state (RIS) calculations to be 2.63 and 2.84, respectively. The latter value falls within the range of the SANS experiment: 2.7 (Zimm plot) and 3.1 (Kratly plot).

9

NMR Parameters

Nuclear magnetic resonance (NMR) is the most powerful experimental tool for structural determination and conformational analysis of polymers and organic molecules. The most significant data provided by NMR are chemical shifts and indirect spin–spin coupling constants. The quantum chemical methods to calculate these NMR parameters are explained here.

In a magnetic field, the nuclei of the spin quantum number I will split the energy state into $2I + 1$ levels; therefore, ^{1}H and ^{13}C nuclei of $I = 1/2$, being included in almost all organic compounds, exhibit two spin states: lower α (up, +1/2) and higher β (down, −1/2). The energy difference between the α and β spins is proportional to the strength of the applied magnetic field $\mathbf{B_0}$ (the Zeeman effect), corresponding to the energy ($h\nu$) of a radio frequency (ν). At the thermal equilibrium, nuclear populations of the α and β states obey the Boltzmann distribution. The magnetic moment ($\boldsymbol{\mu}_A$) of nucleus A is related to its spin angular momentum ($\mathbf{I}_A$) and magnetogyric ratio (γ_A) as

$$\boldsymbol{\mu}_A = \hbar\gamma_A\mathbf{I}_A \tag{9.1}$$

and precesses at the Larmor frequency ($\omega_A = \gamma_A B_0$). A nonequilibrium state of the nuclear populations can be generated by applying the alternating radio frequency that matches the Larmor frequency, and the energy absorption occurs owing to the transition from the α to β level. This phenomenon is NMR. The electrons of the molecule shield the external magnetic field, and consequently, a nucleus designated here as A experiences a magnetic field (effective field $\mathbf{B}_{eff,A}$) different from $\mathbf{B}_0$.

As will be expressed in the second term of Eq. (9.7), the indirect spin–spin coupling is an interaction between the magnetic moments via intervening electrons and hence independent of the strength of B_0. The spin–spin coupling constants are represented as ^{n}Js, with n being the number of bonds lying between two coupled nuclei: ^{1}J, direct coupling; ^{2}J, geminal coupling; and ^{3}J, vicinal coupling. In

Conformational Analysis of Polymers: Methods and Techniques for Structure-Property Relationships and Molecular Design, First Edition. Yuji Sasanuma.

general, the magnitude of ${}^nJ$ decreases with increasing n. The ${}^nJ$ ($n \geq 4$) is called long-range coupling and usually negligibly small. For more details of NMR, consult proper textbooks [88, 158, 241, 389].

9.1 Chemical Shift

The chemical shift is defined as

$$\delta_A = 10^6 \times \frac{\nu_A - \nu_{ref}}{\nu_{ref}} \simeq 10^6 \times \frac{\nu_A - \nu_{ref}}{\nu_0} \tag{9.2}$$

where ν_{ref} is the frequency from a reference, usually tetramethylsilane (TMS) for ${}^1H$ and ${}^{13}C$, and

$$\nu_A = \frac{\gamma_A B_{eff,A}}{2\pi} = \frac{\gamma_A B_0}{2\pi}\left(1 - \sigma_A\right) \tag{9.3}$$

with σ_A being the shielding constant. ${}^1H$ and ${}^{13}C$ NMR signals of most molecules are observed within the ranges of 0–10 ppm and 0–300 ppm, respectively; however, negative δ_A values are rarely observed.

The shielding tensor is expressed as a 3×3 matrix defined with reference to the molecular axis system (x, y, z):

$$\sigma_A = \begin{bmatrix} \sigma_{xx} & \sigma_{xy} & \sigma_{xz} \\ \sigma_{yx} & \sigma_{yy} & \sigma_{yz} \\ \sigma_{zx} & \sigma_{zy} & \sigma_{zz} \end{bmatrix} = \begin{bmatrix} \sigma_Z & 0 & 0 \\ 0 & \sigma_Y & 0 \\ 0 & 0 & \sigma_Z \end{bmatrix} \tag{9.4}$$

The shielding tensor can be diagonalized and converted to that in the principal axis system (X, Y, Z). The shielding constant is given by

$$\sigma_A = \frac{1}{3}\left(\sigma_X + \sigma_Y + \sigma_Z\right) \tag{9.5}$$

The chemical shift, δ_A, is related to σ_A as

$$\delta_A = (\sigma_{ref} - \sigma_A) \times 10^6 \tag{9.6}$$

The Hamiltonian of the NMR phenomenon can be expressed as [95, 193]

$$H = -\sum_A \mathbf{B}_0^T\left(1 - \sigma_A\right)\boldsymbol{\mu}_A + \frac{1}{2}\sum_{A,B\ (A\neq B)} \boldsymbol{\mu}_A^T\left(\mathbf{D}_{AB} + \mathbf{K}_{AB}\right)\boldsymbol{\mu}_B \tag{9.7}$$

where $\mathbf{D}_{AB}$ and $\mathbf{K}_{AB}$ are the classical dipolar coupling tensor and reduced indirect nuclear spin–spin coupling tensor, respectively. The latter is related to the indirect spin–spin coupling tensor by

$$\mathbf{J}_{AB} = h\frac{\gamma_A}{2\pi}\frac{\gamma_B}{2\pi}\mathbf{K}_{AB} \tag{9.8}$$

In the Taylor series of the electronic energy in terms of $\mathbf{B}_0$ and $\boldsymbol{\mu}$, the second-order derivatives are given as

$$\left[\frac{d^2E\left(\mathbf{B}_0, \boldsymbol{\mu}\right)}{d\mathbf{B}_0\, d\boldsymbol{\mu}_\mathrm{A}}\right]_{\mathbf{B}_0=0,\boldsymbol{\mu}=0} = \boldsymbol{\sigma}_\mathrm{A} - 1 \tag{9.9}$$

and

$$\left[\frac{d^2E\left(\mathbf{B}_0, \boldsymbol{\mu}\right)}{d\boldsymbol{\mu}_\mathrm{A}\, d\boldsymbol{\mu}_\mathrm{B}}\right]_{\mathbf{B}_0=0,\boldsymbol{\mu}=0} = \mathbf{D}_\mathrm{AB} + \mathbf{K}_\mathrm{AB} \tag{9.10}$$

When the molecule is placed in isotropic media and performing rapid motions, the tensors are averaged to be scalar quantities, and in addition, the $\mathbf{D}_\mathrm{AB}$ term vanishes; therefore,

$$\sigma_\mathrm{A} = \frac{1}{3}\,\mathrm{tr}\left(\boldsymbol{\sigma}_\mathrm{A}\right) \tag{9.11}$$

and

$$K_\mathrm{AB} = \frac{1}{3}\,\mathrm{tr}\left(\mathbf{K}_\mathrm{AB}\right) \tag{9.12}$$

The magnetic field $\mathbf{B}$ is related to the vector potential by

$$\mathbf{B}(\mathbf{r}) = \nabla \times \mathbf{A}(\mathbf{r}) \tag{9.13}$$

and

$$\mathbf{A}_\mathrm{O}(\mathbf{r}) = \frac{1}{2}\mathbf{B} \times (\mathbf{r} - \mathbf{O}) \tag{9.14}$$

where $\mathbf{r}$ is the position vector, and $\mathbf{O}$ represents the origin of the vector potential (gauge origin). Therefore, $\mathbf{A}$ depends on the gauge position, but $\mathbf{B}$ does not. If a scalar function g is introduced, and a different origin $\mathbf{O}'$ is set, a new vector potential can be defined as

$$\mathbf{A}_{\mathrm{O}'}(\mathbf{r}) = \mathbf{A}_\mathrm{O}(\mathbf{r}) + \nabla g(\mathbf{r}) \tag{9.15}$$

Since, for any g function,

$$\nabla \times \nabla g = 0 \tag{9.16}$$

always holds, the vector potential (i.e. the gauge origin) required for the Hamiltonian cannot be determined uniquely. To solve this gauge problem and derive the shielding tensor, for example, the gauge-including atomic orbital method adopts such atomic orbitals as [74, 524, 525]

$$\chi_\mu(\mathbf{B}) = \exp\left[-\frac{i}{2c}\left(\mathbf{B} \times \mathbf{R}_\mu\right) \cdot \mathbf{r}\right] \chi_\mu(\mathbf{0}) \tag{9.17}$$

where c is the speed of light, $\mathbf{R}_\mu$ is the position vector of the χ_μ basis set, and $\chi_\mu(\mathbf{0})$ is the usual field-independent basis set. With the basis sets of Eq. (9.17), the shielding tensors have been calculated within the framework of the Hartree–Fock and density functional theory (DFT) schemes. In addition, the individual gauges for localized orbitals (IGLO) method [267] and the continuous set of gauge transformations (CGST) model [242, 243] have also been proposed for chemical-shift calculations.

9.1.1 Example: Determination of Reaction Process from NMR Chemical Shifts

If two oxygen atoms of the repeating unit of poly(ethylene terephthalate) (PET) are replaced by sulfur, the new polymer is poly(ethylene tetrathioterephthalate) ($P2TS_4$, see Figure 9.1). $P2TS_4$ was synthesized by ionic polycondensation [12] between 1,2-dibromoethane and tetrathioterephthalate acid complexed with piperidinium (S_4TPA-Pip)[68, 506], as illustrated in Figure 9.1. Initially, the molar ratio of tetrathioterephthalate acid (S_4TPA) to piperidinium (Pip) was presumed to be 1 : 4 from integrated intensities of its ^1H NMR; however, the chemical structure of negative tetravalent tetrathioterephthalate could not be depicted according to the conventional valence-bond concept. To determine the stoichiometry

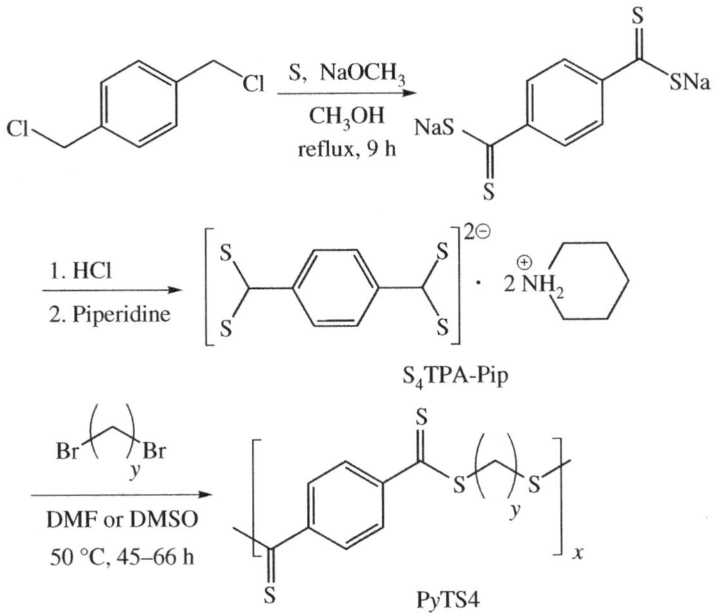

Figure 9.1 Synthesis of poly(ethylene tetrathioterephthalate) ($P2TS_4$) via tetrathioterephthalate acid complexed with piperidinium (S_4TPA-Pip). Here, $y = 2$. Source: Reproduced from reference [10] with permission from the Royal Society of Chemistry.

and clarify the polymerization process, ^{1}H and ^{13}C NMR chemical shifts of the related molecules were calculated by DFT at the B3LYP/6-311+G(3df,3pd)// B3LYP/6-311+G(2d,p) level [10].

Figure 9.2 shows an ^{1}H NMR spectrum observed from S_4TPA-Pip dissolved in dimethyl sulfoxide (DMSO) at room temperature. As shown in the spectrum, the observed peaks were assigned, and the relative intensities were evaluated. In the region of 7.0–8.2 ppm, a number of small peaks can be seen around the intense signal at 7.9 ppm of the aromatic protons. Only from the intense main peak, the S_4TPA to Pip ratio can be estimated to be 1 : 4; however, if the small

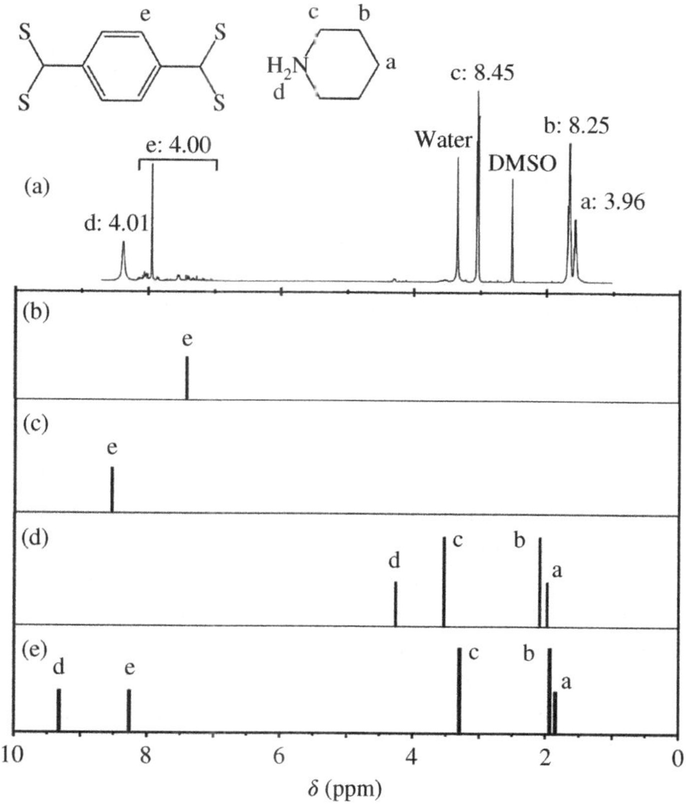

Figure 9.2 (a) Observed and (b–e) calculated ^{1}H NMR spectra of (a, e) the complex (abbreviated as S_4TPA-Pip) of tetrathioterephthalate acid (S_4TPA) with piperidinium (Pip), (b) S_4TPA^{4-}, (c) S_4TPA^{2-}, and (d) Pip. The spectrum e was calculated for the optimized complex structure shown in Figure 9.4. The signals are assigned as shown, and the numerical values in (a) represent the integrated intensities. Source: Reproduced from reference [10] with permission from the Royal Society of Chemistry.

Table 9.1 Elemental analysis of S_4TPA-Pip.

	Elemental composition (%)					
	C	H	N	S	Cl	Total
Calcd (S_4TPA:Pip = 1 : 2)	53.96	7.04	6.99	32.01	0.00	100.00
As measured	51.84	6.68	6.50	30.55	0.17	95.74
Normalized[a]	54.15	6.98	6.79	31.91	0.18	100.01

a) The "as measured" compositions were multiplied by a factor of 100/95.74 to be the "normalized" data.

Source: Reproduced from reference [10] with permission from the Royal Society of Chemistry.

signals are also included, the integrated intensity increases twofold to be 4.0 as written in the spectrum, and hence, the S_4TPA to Pip ratio may possibly be 1 : 2.

Furthermore, S_4TPA-Pip was subjected to elemental analysis, and the results are shown in Table 9.1. The composition as obtained is given on the line of "as measured," and the total percentage of the five elements was 95.74%. Accordingly, all the elemental compositions were multiplied by a factor of 100/95.74 to be modified to "normalized" data. The calibrated values are in good agreement with the "calculated" data based on S_4TPA: Pip = 1 : 2;. As shown above, it seems acceptable that one S_4TPA anion forms a complex with exactly two Pip cations. The small NMR signals may be due to partial degradation of S_4TPA by oxygen and water dissolved in the solvent, although solidified S_4TPA-Pip appears to be stable even in the air.

^{1}H and ^{13}C NMR chemical shifts of negative divalent and tetravalent S_4TPAs dissolved in DMSO were calculated by the gauge-independent atomic orbital method (GIAO) method at the B3LYP/6-311+G(3df,3pd) level with the solvent effect of the polarizable continuum model using the integral equation formalism variant (IEFPCM). The ^{1}H and ^{13}C NMR chemical shifts (in δ ppm) with reference to that (0 ppm) of TMS are plotted in Figures 9.2 and 9.3, respectively. The chemical shifts of TMS were calculated at the same level. In the experimental ^{1}H NMR spectrum, proton e appears at 7.9 ppm (Figure 9.2a) close to the mean value of the calculated δs of S_4TPA^{4-} (7.42 ppm, Figure 9.2b) and S_4TPA^{2-} (8.54 ppm, Figure 9.2c). In Figure 9.3a, the S—C—S carbon (designated as carbon G) of S_4TPA^{4-} shows a peak at 175 ppm, whereas δ_C of S_4TPA^{2-} is 281 ppm, relatively close to the experimental value (251 ppm). The calculated ^{1}H and ^{13}C NMR spectra of Pip are plotted in Figures 9.2d and 9.3d, respectively. The superposition ^{13}C NMR spectra of S_4TPA^{2-} (Figure 9.3c) and 2Pip$^+$ (Figure 9.3d) are close to the observation, whereas this is not the case with the ^{1}H NMR; proton d (NH_2) is observed at a much lower magnetic field (8.3 ppm).

Accordingly, we attempted to calculate the chemical shifts of the S_4TPA-Pip complex. The S_4TPA^{2-} and two Pip$^+$ molecules were initially set as optimized

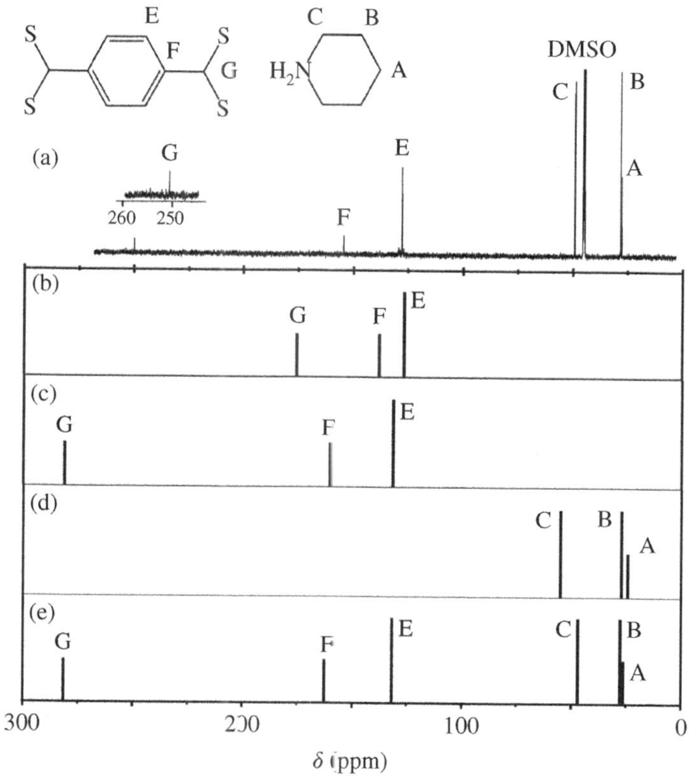

Figure 9.3 (a) Observed and (b–e) calculated ^{13}C NMR spectra with broadband proton decoupled of (a, e) S_4TPA-Pip, (b) S_4TPA^{4-}, (c) S_4TPA^{2-}, and (d) Pip. The signals are assigned as shown. The spectrum e was calculated for the optimized complex structure shown in Figure 9.4. Source: Reproduced from reference [10] with permission from the Royal Society of Chemistry.

individually, and the N—N—H and S—C—S groups were arranged so as to face each other. The geometrical optimization and chemical-shift calculation for the S_4TPA-Pip complex in DMSO were carried out. Figure 9.4 shows the optimized structure, and Figures 9.2e and 9.3e schematically illustrate the calculated NMR spectra. The ^{13}C NMR spectrum is similar to the superposition, whereas the calculated ^1H NMR (e) shows proton d at 9.3 ppm, thus being different from the simple superposition and in fairly good agreement with the experiment (a). The sulfur atoms of S_4TPA probably exert a strong electron-withdrawing effect on the NH_2 protons of Pip. From the above discussion, it could be concluded that S_4TPA forms a complex with two Pip molecules.

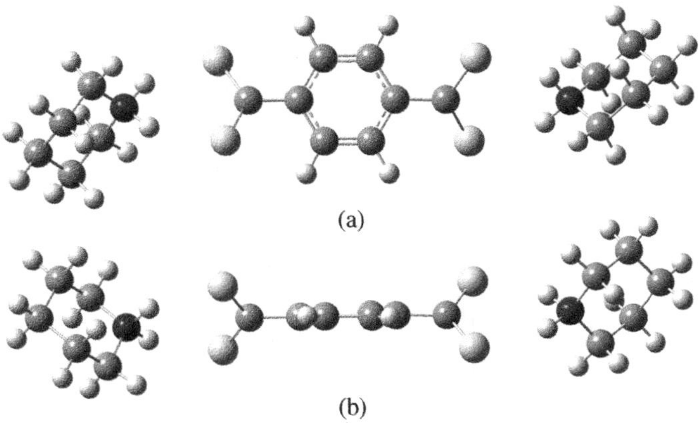

Figure 9.4 Optimized structure of the S_4TPA-Pip complex: (a) top and (b) side views with respect to the benzene ring. The H−N−H and S−C−S triangles are coplanar; the N−H and C−S lengths and N−H···S−C distance are, respectively, 1.04, 1.71, and 2.32 Å; the N−N−H and S−C−S angles are, respectively, 100.1° and 125.1°; and the two S−C−S triangles make an angle of 78°. Source: Reproduced from reference [10] with permission from the Royal Society of Chemistry.

9.2 Indirect Spin–Spin Coupling Constant

The indirect spin–spin coupling constant comprises four terms representing the following interactions: diamagnetic spin–orbit (DSO); paramagnetic spin–orbit (PSO); Fermi contact (FC); and spin–dipole (SD) [192, 379]. Of them, the FC interaction [131] is the most dominant, and its operator is expressed by [14, 192, 195, 453]

$$\mathbf{h}_{\mathrm{A}}^{\mathrm{FC}} = \frac{8\pi\alpha^2}{3}\sum_{i}\delta\left(\mathbf{r}_{i\mathrm{A}}\right)\mathbf{s}_i \tag{9.18}$$

where α is Sommerfeld's fine structure constant, δ is the Dirac delta function, $\mathbf{r}_{i\mathrm{A}}$ is the distance vector between electron i and nucleus A, and $\mathbf{s}_i$ is the electron spin in the unit of $\hbar$. Since Eq. (9.18) includes the delta function, the electron population of the s orbital at the nucleus ($\mathbf{r}_{i\mathrm{A}} = \mathbf{0}$) influences the FC term. For this reason, basis sets modified by adding tighter s functions to the original basis have occasionally been employed to calculate the spin–spin coupling constants [105]. It is suggested that the B3LYP functional will provide accurate values of coupling constants [195].

The vicinal coupling constant, ^3J, is often used for conformational analysis because the ^3J value can be expressed as a function of the dihedral angle (ϕ) of the central bond:

$$^3J(\phi) = A\cos^2 + B\cos\phi + C \tag{9.19}$$

This function is known as the Karplus equation [232, 233] and has the maximum $(A - B + C)$ at $\phi = 180°$ and minimum at $\phi = \arccos(-B/2A)$ (around 90°); therefore, it follows that the trans (antiperiplanar) coupling, ^3J (J_T), is larger than the gauche (synclinal) coupling, (J_G). The significant difference between J_T and J_G enables us to determine bond conformations (trans:gauche ratio) around the central bond between the coupled spins.

9.2.1 Example 1: Calculation of Vicinal Coupling Constants of Cyclic Compound

Figure 9.5 shows the ^{1}H NMR spectra of 2-(1,1-dimethylethyl)-1,4-dithiane (DMEDT) and 2-(1,1-dimethylethyl)-1,4,-dithiane-5,5,6,6-d_4 (DMEDT-d_4), whose Newman projections are depicted in Figure 9.6 [412]. In Figure 9.5, the observed spectra (above) are compared with those (below) simulated by the gNMR program [53]. The bulky *tert*-butyl group prevents the dithiane ring from flip-flopping and hence renders all the protons except *t*-Bu ones magnetically nonequivalent. Therefore, DMEDT provides vicinal coupling constants, J_Ts and J_Gs, that are useful for conformational analysis of the S—C—C—S bond sequences of, for example, poly(ethylene sulfide) [412] and poly(propylene sulfide) [404]. The deuterated

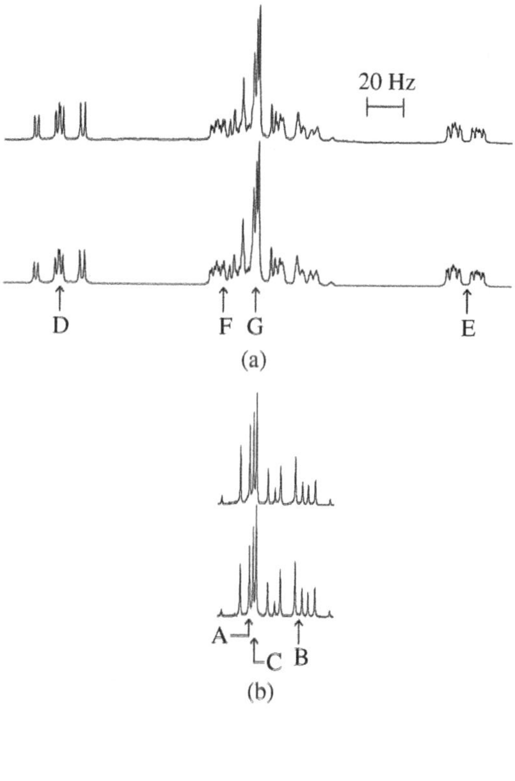

Figure 9.5 Observed (above) and calculated (below) ^{1}H NMR spectra of (a) DMEDT and (b) DMEDT-d_4. The capital letters (A–G) corresponds to the hydrogen atoms (see Figure 9.6), and the arrows indicate the chemical shifts. Source: [412], Figure 6 (p. 3751)/Reproduced with permission from American Chemical Society.

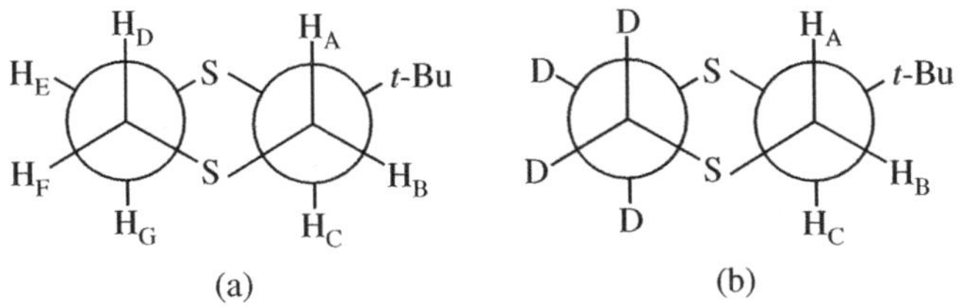

Figure 9.6 Newman projections of (a) DMEDT and (b) DMEDT-d_4. Source: [412], Figure 3 (p. 3750)/Reproduced with permission from American Chemical Society.

DMEDT (DMEDT-d_4) has only three protons bonded to the ring, and its ^{1}H NMR spectra are easy to analyze; therefore, DMEDT-d_4 was also synthesized and subjected to NMR measurements to yield chemical shifts and coupling constants of the three protons (A–C). In the next step, the ^{1}H NMR spectra of DMEDT were simulated with the acid of the data on the three protons to yield the NMR parameters of the other protons (D–G). Of them, only vicinal ^{1}H–^{1}H coupling constants are listed in Table 9.2 and compared with those calculated with the Gaussian program [159] at the B3LYP/6-311++G(3df,3pd)//B3LYP/6-311++G(3df,3pd) level. Although somewhat discrepancies are found between the calculated and observed values, the experimental tendencies are well reproduced.

Table 9.2 Vicinal coupling constants of DMEDT, obtained from NMR experiments and DFT calculations.[a)]

Medium/term[b)]	$^3J_{AC}$	$^3J_{BC}$	$^3J_{DE}$	$^3J_{DG}$	$^3J_{EF}$	$^3J_{FG}$
		NMR experiment				
Cyclohexane-d_{12}	10.85	1.85	2.41	12.04	4.12	2.44
Benzene-d_6	10.87	1.89	2.41	12.00	4.09	2.44
Chloroform-d	10.89	1.84	2.43	12.06	4.09	2.47
		DFT calculation[c)]				
DSO	−2.08	0.39	0.03	−2.44	−0.09	0.04
PSO	1.80	−0.23	0.09	2.13	0.18	0.09
FC	9.56	1.58	2.27	10.72	3.69	2.24
SD	0.03	0.09	0.10	0.01	0.09	0.09
Total	9.31	1.83	2.49	10.42	3.87	2.46

a) In Hz.
b) Abbreviations: DSO, diamagnetic spin–orbit; PSO, paramagnetic spin–orbit; FC, Fermi contact; SD, spin–dipole.
c) At the B3LYP/6-311++G(3df,3pd) level.

When the chemical-shift difference ($\Delta\sigma$) between protons is much larger than the spin–spin coupling between them (J), i.e. $\Delta\sigma/J \gg 1$, it is said that they are weakly coupled. Then, the splitting intensity obeys the binomial coefficients, namely Pascal's triangle: (1 : 1), (1 : 2 : 1), (1 : 3 : 3 : 1), etc. The J value may be readily read from the spectrum. In contrast, when $\Delta\sigma/J \leq 1$, the protons are strongly coupled; the spectrum is so complicated as to be simulated with a computer. In Figure 9.5, the chemical shifts of protons A, B, C, F, and G are seen to be close to each other, and hence, the four hydrogen atoms, D, E, F, and G, were replaced with deuterium so as to facilitate the analysis. However, two-dimensional NMR techniques such as the correlation spectroscopy (COSY), total correlation spectroscopy (TOCSY), and/or heteronuclear single-quantum correlation (HSQC) [51, 88, 188, 241, 507] methods may be helpful for the assignment, and the J-spectra with absorption-mode lineshapes method [359] separates the overlapping one-dimensional spectrum into individual ^{1}H spectra, from which the NMR parameters may be obtained.

9.2.2 Example 2: Derivation of Karplus Equation and Its Application

Figure 9.7 graphically illustrates the dihedral-angle (φ^{OC}) dependence of the vicinal coupling constant $^3J_{COCH}$, that is the Karplus equation [542, 543], which was calculated at 10° intervals of φ^{OC} for isopropyl methyl carbonate

$$(CH_3)_2(^1H{-})C{-}O{-}^{13}C({=}O)OCH_3$$

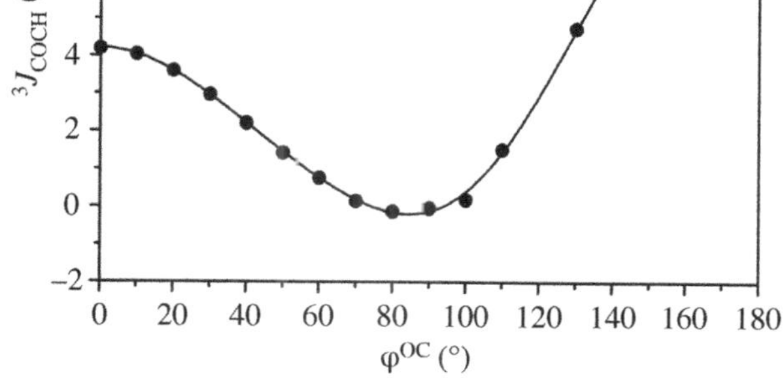

Figure 9.7 Vicinal coupling constant, $^3J_{COCH}$, as a function of dihedral angle φ^{OC} around the O–C bond of the carbonate group. The filled circles stand for the calculated data, and the solid line represents the Karplus equation of Eq. (9.20). Source: [542], Figure 6 (p. 9366)/Reproduced with permission of American Chemical Society.

Its geometry was optimized at the B3LYP/6-311+G(2d,p) with φ^{OC} fixed at the given position, and the $^3J_{COCH}$ value there was calculated at the B3LYP/6-311++G(3df,3pd) level. The Karplus equation was formulated by a nonlinear least-squares method so as to express the $^3J_{COCH}$ vs. φ^{OC} plot. The ordinary Karplus equation is given as a cosine function up to the cos($2\varphi^{OC}$) or $\cos^2(\varphi^{OC})$ term [233]; however, the Karplus equation here was optimized as a cosine function of six terms up to cos($5\varphi^{OC}$) so as to pass through all the calculated points [542]:

$$\begin{aligned}{}^3J_{COCH}\left(\varphi^{OC}\right) = {} & 3.58 - 2.74\cos\left(\varphi^{OC}\right) + 3.59\cos\left(2\varphi^{OC}\right) \\ & - 0.22\cos\left(3\varphi^{OC}\right) - 0.12\cos\left(4\varphi^{OC}\right) + 0.12\cos\left(5\phi^{OC}\right)\end{aligned} \tag{9.20}$$

The Karplus equation was used for conformational analysis of model compounds of poly(cyclohexene carbonate) (PCHC), (1*R*,2*R*)-*trans*-di(methoxycarbonyloxy)cyclohexane (*trans*-DMCC), and (1*R*,2*S*)-*cis*-di(methoxycarbonyloxy)cyclohexane (*cis*-DMCC) (see Figure 25.8). PCHC is the alternating copolymer of cyclohexene oxide and carbon dioxide.

The average coupling constant, $^3J_{CH}$, corresponding to the experimental value, was calculated from

$$^3J_{CH} = \frac{\sum_k {}^3J_{COCH}\left(\varphi_k^{OC}\right)\exp\left(-\Delta G_k/RT\right)}{\sum_k \exp\left(-\Delta G_k/RT\right)} \tag{9.21}$$

Table 9.3 Vicinal coupling constants of *trans*-DMCC and *cis*-DMCC: comparison between observed and calculated values.

		$^3J_{CH}$ (Hz)			
		trans-DMCC		*cis*-DMCC	
Medium	**Temp (°C)**	**Obs**	**Calc[a)]**	**Obs**	**Calc[a)]**
Chloroform-*d*	15	3.01	3.17	3.13	3.17
	25	3.02	3.17	3.14	3.17
	35	3.03	3.18	3.15	3.17
	45	3.03	3.19	3.15	3.17
	55	3.04	3.19	3.16	3.17
Dimethyl-d_6 sulfoxide	25	3.02	3.20	3.12	3.17
	35	3.02	3.21	3.12	3.17
	45	3.03	3.22	3.13	3.17
	55	3.04	3.22	3.13	3.17

a) From Eq. (9.21).
Source: [542], Table 3 (p. 9366)/Reproduced with permission of American Chemical Society.

where k indicates the conformation, φ_k^{OC} is its dihedral angle around the O—C bond, optimized at the B3LYP/6-311+G(2d,p) level, ΔG_k is its Gibbs free energy calculated from the electronic energy at the MP2/6-311+G(2d,p) level and the thermochemical term at the B3LYP/6-311+G(2d,p) level, and ${}^3J_{\mathrm{COCH}}\left(\varphi_k^{\mathrm{OC}}\right)$ was evaluated from Eq. (9.20). In Table 9.3, the ${}^3J_{\mathrm{CH}}$ values thus calculated are listed, being found to agree well with the observations.

As shown above, the quantum chemical calculations provide us with precise vicinal coupling constants enough for the conformational analysis.

10

Periodic Quantum Chemistry

A polymeric chain possesses a one-dimensional periodicity along the chain and forms a three-dimensional periodicity during crystallization. Ab initio molecular orbital (MO) and density functional theory (DFT) calculations under periodic boundary conditions have enabled us to determine crystal structures and evaluate or predict their properties [121]. All solid-state polymeric materials except the poly(methylene oxide) (PMO) whisker [213] include both crystalline and amorphous phases; therefore, the physical properties observed therefrom are affected by the amorphous phase. The properties of the perfect crystal of a given polymer, being the ultimate characteristic features that the polymer can exhibit but very difficult or impossible to evaluate experimentally, are expected to be predicted by the periodic quantum chemistry. Here, the principles of the periodic DFT methods are explained and their applications to polymer crystals are presented.

10.1 Direct Lattice and Reciprocal Lattice

All crystals are classified into one of triclinic, monoclinic, orthorhombic, tetragonal, trigonal, hexagonal, and cubic systems and, furthermore, one of 230 space groups [245, 273]. The atomic positions in the direct (real) lattice are represented with a position vector **r** by fractional coordinates (x, y, z) $(0 \leq x, y, z \leq 1)$:

$$\mathbf{r} = x\mathbf{a} + y\mathbf{b} + z\mathbf{c} \tag{10.1}$$

where **a**, **b**, and **c** are the lattice vectors in the a-, b-, and c-axis directions. The reciprocal vectors used in X-ray crystallography are defined as

$$\mathbf{a}^* = \frac{\mathbf{b} \times \mathbf{c}}{\mathbf{a} \cdot (\mathbf{b} \times \mathbf{c})} \quad \mathbf{b}^* = \frac{\mathbf{c} \times \mathbf{a}}{\mathbf{b} \cdot (\mathbf{c} \times \mathbf{a})} \quad \mathbf{c}^* = \frac{\mathbf{a} \times \mathbf{b}}{\mathbf{c} \cdot (\mathbf{a} \times \mathbf{b})} \tag{10.2}$$

where the denominators are equal to the unit-cell volume, and the $\mathbf{a}^*$, $\mathbf{b}^*$, and $\mathbf{c}^*$ vector are perpendicular to (**b** and **c**), (**c** and **a**), and (**a** and **b**), respectively.

Conformational Analysis of Polymers: Methods and Techniques for Structure-Property Relationships and Molecular Design, First Edition. Yuji Sasanuma.

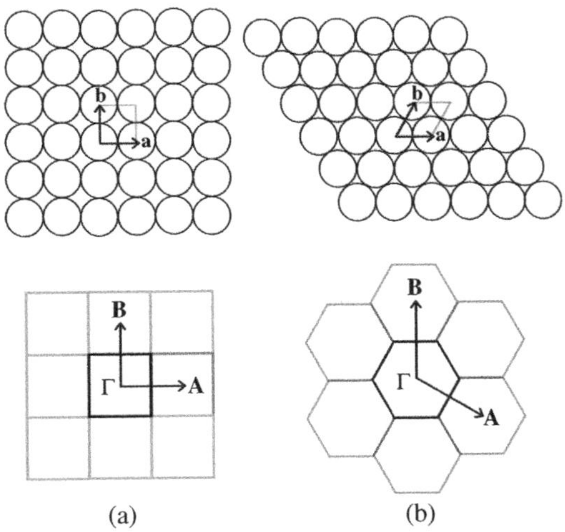

Figure 10.1 Direct (above) and reciprocal (below) lattices of (a) tetragonal and (b) hexagonal systems. The **c** (**C**)-axis is perpendicular to the **ab** (**AB**) plane. The thick (a) square and (b) hexagon of the pictures below represent the first Brillouin zones.

In sold-state physics, instead of $\mathbf{a}^*$, $\mathbf{b}^*$, and $\mathbf{c}^*$, the following reciprocal vectors are used:

$$\mathbf{A} = 2\pi\mathbf{a}^* \quad \mathbf{B} = 2\pi\mathbf{b}^* \quad \mathbf{C} = 2\pi\mathbf{c}^* \tag{10.3}$$

Therefore, between the direct and reciprocal vectors, the following relations hold:

$$\begin{array}{lll} \mathbf{a}\cdot\mathbf{A} = 2\pi & \mathbf{a}\cdot\mathbf{B} = 0 & \mathbf{a}\cdot\mathbf{C} = 0 \\ \mathbf{b}\cdot\mathbf{A} = 0 & \mathbf{b}\cdot\mathbf{B} = 2\pi & \mathbf{b}\cdot\mathbf{C} = 0 \\ \mathbf{c}\cdot\mathbf{A} = 0 & \mathbf{c}\cdot\mathbf{B} = 0 & \mathbf{c}\cdot\mathbf{C} = 2\pi \end{array} \tag{10.4}$$

The translation vector in the reciprocal lattice is defined as

$$\mathbf{K} = h\mathbf{A} + k\mathbf{B} + l\mathbf{C} \tag{10.5}$$

where h, k, and l are integers. The space surrounded by planes bisecting perpendicularly line segments that connect one reciprocal lattice point to all its nearest neighbors is termed the first Brillouin zone (see Figure 10.1).

10.2 Bloch Function

In a crystal, the potential function has a periodicity such as [120, 245, 273, 366]

$$V(\mathbf{r} - \mathbf{g}) = V(\mathbf{r}) \tag{10.6}$$

where $\mathbf{g}$ is the translation vector in the direct space

$$\mathbf{g} = u\mathbf{a} + v\mathbf{b} + w\mathbf{c} \tag{10.7}$$

with u, v, and w being integers. The Schrödinger equation of a crystal

$$\mathcal{H}(\mathbf{r})\Psi(\mathbf{r}) = E\Psi(\mathbf{r}) \tag{10.8}$$

satisfies the periodicity:

$$\mathcal{H}\left(\mathbf{r}-\mathbf{g}\right)\Psi\left(\mathbf{r}-\mathbf{g}\right) = E\Psi\left(\mathbf{r}-\mathbf{g}\right) \tag{10.9}$$

The eigenfunction of Eq. (10.8) is termed the Bloch function that is expressed by

$$\Phi\left(\mathbf{r}+\mathbf{g};\mathbf{k}\right) = e^{i\mathbf{k}\cdot\mathbf{g}}\Phi(\mathbf{r};\mathbf{k}) \tag{10.10}$$

where **k** is the wave vector

$$\mathbf{k} = \left(\frac{n_a}{N_a}\mathbf{A} + \frac{n_b}{N_b}\mathbf{B} + \frac{n_c}{N_c}\mathbf{C}\right) \tag{10.11}$$

with $0 \leq n_j(\text{integer}) \leq N_j$ ($j = a$, b, or c). If N_a, N_b, and N_c unit cells are arranged in the a- and b-, and c-axis directions, there are N cells in all:

$$N = N_a \times N_b \times N_c \tag{10.12}$$

As N_js increase, the number of **k** points increases. The Bloch function also possesses a periodicity in the reciprocal space:

$$\Phi\left(\mathbf{r}+\mathbf{g};\mathbf{k}+\mathbf{K}\right) = e^{i\mathbf{k}\cdot\mathbf{g}}\Phi(\mathbf{r};\mathbf{k}) \tag{10.13}$$

The nature of the Bloch function allows us to sample a finite number of **k** points only within the first Brillouin zone and formulate the Schrödinger equation at each **k** point in such a manner as

$$\mathcal{H}\Psi(\mathbf{r};\mathbf{k}) = E(\mathbf{k})\Psi(\mathbf{r};\mathbf{k}) \tag{10.14}$$

and solve it independently of those set up at different **k** points. In practice, on the basis of the symmetry of the reciprocal space, the **k** points for the calculations are sampled within the irreducible part of the Brillouin zone.

10.3 One-electron Crystal Orbital

Similar to the MO theory, the one-electron Hamiltonian is adopted for the periodic systems, and the crystal orbital is defined. The ith crystal orbital ($\phi_i(\mathbf{r};\mathbf{k})$) is expressed by a linear combination of Gaussian-type atomic orbitals [120, 122, 366, 488]:

$$\phi_i(\mathbf{r};\mathbf{k}) = \sum_{\mu} c_{\mu i}(\mathbf{k}) \sum_{\mathbf{g}} \chi_{\mu}\left(\mathbf{r}-\mathbf{r}_{\mu}-\mathbf{g}\right) e^{i\mathbf{k}\cdot\mathbf{g}} \tag{10.15}$$

where $c_{\mu i}$ is the crystal orbital coefficient, χ_{μ} is the atomic orbital given by Eq. (5.10), for example the 6-31G(d,p) basis set, and $\mathbf{r}_{\mu}$ is the μth atomic position.

Similar to the Hartree–Fock method, the Fock matrix **F**(**k**), the overlap matrix of the Block functions **S**(**k**), the diagonal energy matrix **E**(**k**), and the coefficient matrix **C**(**k**) are related by the Roothaan–Hall matrix equation:

$$\mathbf{F}(\mathbf{k})\mathbf{C}(\mathbf{k}) = \mathbf{S}(\mathbf{k})\mathbf{C}(\mathbf{k})\mathbf{E}(\mathbf{k}) \tag{10.16}$$

The matrix equation is solved self-consistently.

For DFT calculations, the electron density is computed from [488]

$$\rho(\mathbf{r}) = \int_{\mathrm{BZ}} \sum_{i}^{\mathrm{occupied}} |\phi_i(\mathbf{r};\mathbf{k})|^2 \theta\left(\epsilon_{\mathrm{F}} - \epsilon_i(\mathbf{k})\right) d\mathbf{k} \tag{10.17}$$

where the integration is carried out in the first Brillouin zone, the summation is performed only for occupied orbitals, θ is the Heaviside step function, ϵ_{F} is the Fermi energy, and $\epsilon_i(\mathbf{k})$ is the **k**-dependent eigenvalue of the ith crystal orbital.

10.4 Structural Optimization

The crystal structures, namely lattice constants and atomic fractional coordinates in the unit cell, are optimized under the given space group by minimizing the total energy of the crystal cell [122]. As the initial data, those determined by X-ray, electron, and neutron diffraction experiments are usually adopted. In the repeating computations for the optimization, if both the root-mean squares of the total-energy displacement and energy gradients between the previous and current cycles reach the given thresholds, the optimization will satisfactorily cease.

Here, the structural optimization of polyethylene (PE) crystal is exemplified [266]. The crystal structure of PE at room temperature was earlier reported by Bunn [54]: an orthorhombic lattice of space group P_{nam} with a = 7.40 Å, b = 4.93 Å, and c = 2.534 Å. The crystal structures at cryogenic temperatures have also been determined by neutron diffraction [30, 461], which may resolve the atomic positions of hydrogen as well as carbon atoms precisely. Table 10.1 shows the lattice constants and fractional coordinates of PE at ambient and low temperatures. The a and b lengths are seen to depend largely on temperature and increase with temperature because the lateral dimensions (a and b) are influenced by interchain interactions of weak van der Waals forces.

Inasmuch as the periodic DFT calculation corrected for the dispersion force (abbreviated as DFT-D) yields the crystal structure at 0 K, it should preferably be compared with those at low temperatures. As shown in Table 10.1, the optimized parameters satisfactorily reproduced the atomic coordinates at 10 and 4 K. The setting angle (45.5°), that is the angle made between the molecular plane and the b-axis (see Figure 10.2), agrees moderately with those of the neutron diffraction experiments.

Table 10.1 Lattice constants and atomic coordinates of PE.

	Lattice constant (Å)			Atomic position			
Temp (K)	*a*	*b*	*c*		*x*/*a*	*y*/*b*	$\phi^{a)}$ (°)
			Calcd				
0	7.004	4.849	2.567	C	0.0424	0.0605	45.5
				H	0.1967	0.0284	
				H	0.0217	0.2851	
			Exptl				
10[b)]	7.120	4.842		C	0.042	0.062	45.07
				H	0.193	0.023	
				H	0.016	0.284	
100[b)]	7.162	4.863		C	0.042	0.062	44.78
				H	0.192	0.024	
				H	0.015	0.283	
200[b)]	7.258	4.898		C	0.041	0.062	44.35
				H	0.189	0.026	
				H	0.013	0.282	
300[b)]	7.417	4.939		C	0.040	0.061	44.91
				H	0.185	0.023	
				H	0.015	0.278	
4[c)]	7.121	4.851	2.548	C	0.050	0.062	49
				H	0.198	0.034	
				H	0.019	0.283	
90[c)]	7.161	4.866	2.546	C	0.049	0.062	49
				H	0.197	0.030	
				H	0.018	0.278	
rt[d)]	7.40	4.93	2.534	C	0.038	0.065	41.2

a) The setting angle, that is the angle between the molecular plane and the *b*-axis (Figure 10.2).
b) By neutron diffraction for deuterated PE [461].
c) By neutron diffraction for deuterated PE [30].
d) By X-ray diffraction at room temperature [54].
Source: [266], Table 2 (p. 4826)/American Chemical Society/Licensed under CC BY 4.0.

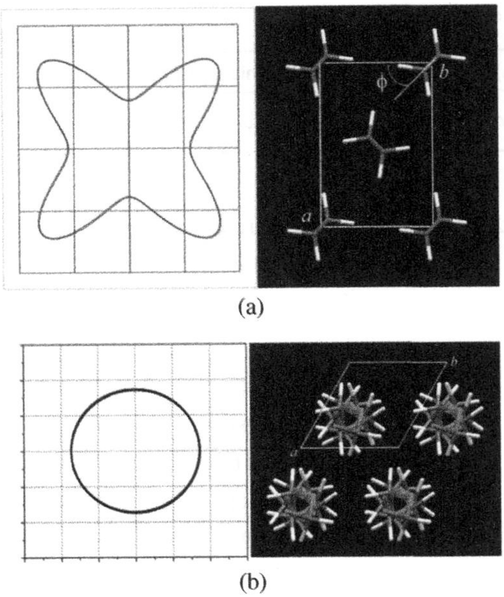

Figure 10.2 Young's modulus distributions on the plane perpendicular to the molecular axis of (a) polyethylene crystal and (b) trigonal (9/5 helical) poly(methylene oxide) crystal. The setting angle ϕ of polyethylene is defined as shown. The grid spacing corresponds to 10 GPa. Source: Adapted with permission from [266]/American Chemical Society/CC BY 4.0.

10.5 Crystal Elasticity

The stress tensor is defined in the orthogonal (x, y, z) coordinates as a symmetric matrix [342, 516]

$$\begin{bmatrix} \sigma_{xx} & \sigma_{xy} & \sigma_{xz} \\ \sigma_{yx} & \sigma_{yy} & \sigma_{yz} \\ \sigma_{zx} & \sigma_{zy} & \sigma_{zz} \end{bmatrix} \tag{10.18}$$

Only the six elements are irreducible and redefined according to Voigt's notation: $1 = xx$, $2 = yy$, $3 = zz$, $4 = yz$, $5 = xz$, and $6 = xy$.[1] Then, the stress σ and strain ϵ are expressed as 6×1 column vectors and related to each other via the stiffness tensor C:

$$\sigma = C\,\epsilon \tag{10.19}$$

where

$$C = \begin{bmatrix} c_{11} & c_{12} & c_{13} & c_{14} & c_{15} & c_{16} \\ & c_{22} & c_{23} & c_{24} & c_{25} & c_{26} \\ & & c_{33} & c_{34} & c_{35} & c_{36} \\ & & & c_{44} & c_{45} & c_{46} \\ & & & & c_{55} & c_{56} \\ & & & & & c_{66} \end{bmatrix} \tag{10.20}$$

1 Equation (10.18) has the same form as Eq. (9.4) (shielding tensor). Both the stress and shielding constant are conventionally represented by σ.

The C tensor is a 6×6 symmetric matrix with 21 independent elements. The crystal symmetry simplifies the C tensor. For example, the stiffness tensor of the cubic system has only three independent elements, and those of orthorhombic and monoclinic lattices, which have often been found for polymer crystals, include 9 and 13 independent elements, respectively, and the remainders are null. The compliance tensor S, the inverse matrix of C, is expressed as

$$S = \begin{bmatrix} s_{11} & s_{12} & s_{13} & s_{14} & s_{15} & s_{16} \\ & s_{22} & s_{23} & s_{24} & s_{25} & s_{26} \\ & & s_{33} & s_{34} & s_{35} & s_{36} \\ & & & s_{44} & s_{45} & s_{46} \\ & & & & s_{55} & s_{56} \\ & & & & & s_{66} \end{bmatrix} \tag{10.21}$$

Young's moduli in the x, y, and z directions (E_x, E_y, and E_z) are given by $E_x = s_{11}^{-1}$, $E_y = s_{22}^{-1}$, and $E_z = s_{33}^{-1}$, respectively. Young's modulus $E(l_1, l_2, l_3)$ in an arbitrary direction represented by the unit vector (l_1, l_2, l_3) can be calculated from

$$\frac{1}{E(l_1, l_2, l_3)} = \left\{ \begin{array}{llllll} l_1^4\, s_{11} & + 2l_1^2 l_2^2\, s_{12} & + 2l_1^2 l_3^2\, s_{13} & + 2l_1^2 l_2 l_3 s_{14} & + 2l_1^3 l_3\, s_{15} & + 2l_1^3 l_2 s_{16} \\ & + l_2^4\, s_{22} & - 2l_2^2 l_3^2\, s_{23} & + 2l_2^3 l_3 s_{24} & + 2l_1 l_2^2 l_3\, s_{25} & + 2l_1 l_2^3 s_{26} \\ & & - l_3^4\, s_{33} & + 2l_2 l_3^3 s_{34} & + 2l_1 l_3^3\, s_{35} & + 2l_1 l_2 l_3^2 s_{36} \\ & & & + l_2^2 l_3^2\, s_{44} & + 2l_1 l_2 l_3^2 s_{45} & + 2l_1 l_2^2 l_3\, s_{46} \\ & & & & + l_1^2 l_3^2\, s_{55} & + 2l_1^2 l_2 l_3 s_{56} \\ & & & & & + l_1^2 l_2^2\, s_{66} \end{array} \right. \tag{10.22}$$

The element c_{uv} (u and $v = 1$–6, Voigt's notation) of the stiffness tensor can be calculated from [121, 122, 126, 364]

$$c_{uv} = \frac{1}{V} \left[\frac{\partial^2 E_{\text{cryst}}}{\partial \epsilon_u \partial \epsilon_v} \right] \tag{10.23}$$

where ϵ is the rank-2 symmetric tensor of pure strain, and E_{cryst} and V are the total energy and volume per unit cell, respectively.

Young's modulus as a function of angles (θ, ϕ) in the polar coordinates is three-dimensionally averaged over the surface of the unit sphere to yield the surface-average Young's modulus (E_{av}):

$$\begin{aligned} E_{\text{av}} &= \left(\int_0^{2\pi} d\phi \int_0^{\pi} E(l_1, l_2, l_3) \sin\theta d\theta \right) \Big/ \left(\int_0^{2\pi} d\phi \int_0^{\pi} \sin\theta d\theta \right) \\ &= \frac{1}{4\pi} \int_0^{2\pi} d\phi \int_0^{\pi} E(l_1, l_2, l_3) \sin\theta d\theta \end{aligned} \tag{10.24}$$

where the unit vector (l_1, l_2, l_3) that begins at the origin and ends at a point on the surface of the sphere is expressed as

$$(l_1, l_2, l_3) = (\sin\theta\cos\phi, \sin\theta\sin\phi, \cos\theta) \tag{10.25}$$

By way of example, the stiffness and compliance tensors of the orthorhombic PE crystal at 0 K were obtained as [266]

$$C = \begin{bmatrix} 19.115 & 10.636 & 1.564 & 0.000 & 0.000 & 0.000 \\ & 13.754 & 4.734 & 0.000 & 0.000 & 0.000 \\ & & 334.578 & 0.000 & 0.000 & 0.000 \\ & & & 3.406 & 0.000 & 0.000 \\ & & & & 3.486 & 0.000 \\ & & & & & 9.490 \end{bmatrix} (\text{GPa}) \tag{10.26}$$

$$S = \begin{bmatrix} 91.947 & -71.307 & 0.579 & 0.000 & 0.000 & 0.000 \\ & 128.366 & -1.483 & 0.000 & 0.000 & 0.000 \\ & & 3.007 & 0.000 & 0.000 & 0.000 \\ & & & 293.599 & 0.000 & 0.000 \\ & & & & 286.881 & 0.000 \\ & & & & & 105.372 \end{bmatrix} (\text{TPa}^{-1}) \tag{10.27}$$

From the compliance tensor, the crystalline moduli in the a-, b-, and c-axis directions of the PE crystal were derived as shown in Table 10.2. The present DFT-D calculations ($E_a = 10.9$, $E_b = 7.8$, and $E_c = 333$ GPa) represent the ultimate mechanical properties of the perfect PE crystal of infinite size at 0 K, thus probably being superior to any experimental observations.

Neutron inelastic scattering (NIS) measurements provide phonon frequencies and wave vectors simultaneously. The slope of the frequency vs. wave vector plot yields the sound velocity v_x in the propagation direction (x) of the phonon [202]:

$$v_x = \left(\frac{E_x}{\rho}\right)^{1/2} \tag{10.28}$$

where E_x is Young's modulus in the x-direction, and ρ is density. This method has the advantage of informing us of the phonon frequency and wave vector in a specific direction. Therefore, the crystalline moduli along a, b, and c axes may be evaluated separately, while the Raman scattering yields the unidirectional longitudinal acoustic mode, from which only the $E_{\parallel}$ value (Young's modulus in the chain-axis direction) can be derived. In an early study on NIS of PE [129, 202], the $E_{\parallel}$ and $E_{\perp}$ (Young's modulus perpendicular to the chain axis) values at room temperature were determined to be 329 and 6 GPa, respectively. In a later study, the E_a and E_b values of PE at −196 °C were reported as 9 and 8 GPa, respectively [501].

Table 10.2 Crystalline moduli of some representative polymers, evaluated from periodic DFT-D calculations.[a)]

Polymer[b)]	Form	E_a	E_b	E_c	E_{av}	Conformation
PGA		26.5	29.3	**451**	33.2	all-trans
Nylon 4	α	53.6	**334**	16.8	15.1	all-trans
PE		10.9	7.8	**333**	15.9	all-trans
Nylon 6	α	44.5	**316**	19.4	13.7	all-trans
PCL		13.1	10.1	**252**	10.7	all-trans
PET		7.2	22.3	**182**	19.7	pseudo all-trans[c)]
Nylon 6	γ	25.4	**120**	38.1	24.8	partly distorted[d)]
PMO	9/5 helix	17.3	17.3	**115**	24.2	all-gauche
	2/1 helix	13.7	12.0	**82.9**	21.9	all-gauche
PBS	α	7.5	14.3	**67.2**	18.5	ttttttg$^+$tg$^-$t
PES		13.7	11.1	**48.6**	23.2	ttg$^-$tttg$^+$t
PBT	α	4.8	11.6	**20.8**	14.1	g$^+$g$^+$tg$^-$g$^-$ (bent)
PTT		6.9	18.4	**7.1**	13.2	tggt (bent)

a) In GPa. At 0 K. E_a, E_b, and E_c are Young's moduli in the *a*-, *b*-, and *c*-axis directions, respectively, and E_{av} is the three-dimensionally averaged Young's modulus. The boldfaced value represents the modulus parallel to or nearly parallel to the chain axis.
b) Abbreviations: PGA, poly(glycolic acid); Nylon 4, poly(2-pyrrolidone); PE, polyethylene; Nylon 6; poly(ϵ-caprolactam); PCL, poly(ϵ-caprolactone); PET, poly(ethylene terephthalate); PMO, poly(methylene oxide); PBS, poly(butylene succinate); PES, poly(ethylene succinate); PBT, poly(butylene terephthalate); PTT, poly(trimethylene terephthalate).
c) O—CH_2 = 141.9°.
d) HN—CH_2 = 111.8° and CH_2—C(O) = 117.5°.
Source: [238]/Reproduced with permission of Royal Society of Chemistry.

Figure 10.2 shows the $E_\perp$ distribution on the *ab*-plane of the PE crystal. The maxima can be found in the directions parallel to the molecular plane.

It is known that PMO whisker (of 9/5 helix), exceptionally, includes no amorphous phase [212]. Therefore, the PMO whisker is expected to give the true crystalline moduli directly; however, the whisker crystal itself is so small as to have been embedded in resin matrices and undergone the mechanical measurements, and Young's modulus ($E_\parallel$) of the whisker at room temperature was indirectly evaluated to be about 100 GPa [214]. The E_c value of the highly oriented PMO sample was calculated from its vibrational spectroscopic data to be 109 GPa [531]. At temperatures as low as −150 °C, the experimental E_c data stay within a narrow range of 100–105 GPa, being comparable to the theoretical value here (115 GPa).

Young's modulus distributions on the *ab*-plane, calculated from Eq. (10.22), are plotted in Figure 10.2. The distribution of PMO is seen to be a perfect circle, whose radius corresponds to the magnitude of E_a (= E_b). This is because the PMO chain strongly prefers the gauche conformation (the energy difference from the trans state is −1.5 to −2.5 kcal mol^{-1}) [423] and hence forms a uniform helical structure. The stiffness (C) and compliance (S) tensors of PMO crystal are [266]

$$C = \begin{bmatrix} 24.717 & 13.147 & 15.640 & 0.048 & 1.906 & 0.000 \\ & 24.717 & 15.640 & -0.048 & -1.906 & 0.000 \\ & & 126.813 & 0.000 & 0.000 & 0.000 \\ & & & 7.672 & 0.000 & -1.906 \\ & & & & 7.672 & 0.048 \\ & & & & & 5.785 \end{bmatrix} \text{(GPa)} \quad (10.29)$$

$$S = \begin{bmatrix} 61.778 & -32.371 & -3.627 & -0.591 & -23.393 & 0.000 \\ & 61.778 & -3.627 & 0.591 & 23.393 & 0.000 \\ & & 8.780 & 0.000 & 0.000 & 0.000 \\ & & & 141.973 & 0.000 & 46.787 \\ & & & & 141.973 & -1.182 \\ & & & & & 188.298 \end{bmatrix} (\text{TPa}^{-1}) \quad (10.30)$$

10.6 Vibrational Calculation

Vibrational calculations are carried out for the Γ point (the center of the first Brillouin zone, $\mathbf{k} = \mathbf{0}$, see Figure 10.1). The computational method is similar to that used for the molecular vibrations (Section 8.3). The Hessian matrix, whose elements are the second derivative of energy with respect to of the atomic coordinates, is calculated from [120, 121]

$$H_{\text{A}i,\text{B}j}(\mathbf{k} = \mathbf{0}) = \left(\frac{\partial^2 E}{\partial u_{\text{A}i} \partial u_{\text{B}j}} \right)_{\text{eq}} \quad (10.31)$$

Here, $u_{\text{A}i}$ is the displacement of the i coordinate of atom A from the equilibrium position, and $u_{\text{B}j}$ is defined similarly. The Hessian matrix is converted to the mass-weighted dynamical matrix:

$$W_{\text{A}i,\text{B}j}(\mathbf{k} = \mathbf{0}) = \frac{H_{\text{A}i,\text{B}j}}{\sqrt{M_\text{A} M_\text{B}}} \quad (10.32)$$

where M_A and M_B are masses of atoms A and B, respectively. The eigenvalues of the diagonalized W correspond to the squares of vibrational frequencies, and the eigenvectors represent the normal modes. Of them, three modes are assigned to translational motions at the Γ point, that is the acoustic modes, and the others correspond to vibrational frequencies observed by infrared (IR) and Raman spectroscopic methods.

The IR intensity A_i of the ith mode is given by

$$A_i \propto d_i \left| \frac{\partial \mu}{\partial Q_i} \right|^2 \tag{10.33}$$

where μ is the dipole moment, Q_i is the normal mode coordinate, and d_i is the degeneracy. The dipole moment is related to the Born effective charge tensor, which is the first derivative of the polarization per unit cell with respect to the atomic displacements.

Figure 10.3 shows the calculated IR spectra of trigonal (9/5 helix) and orthorhombic (2/1 helix) PMO crystals [166]. The DFT-D calculations were

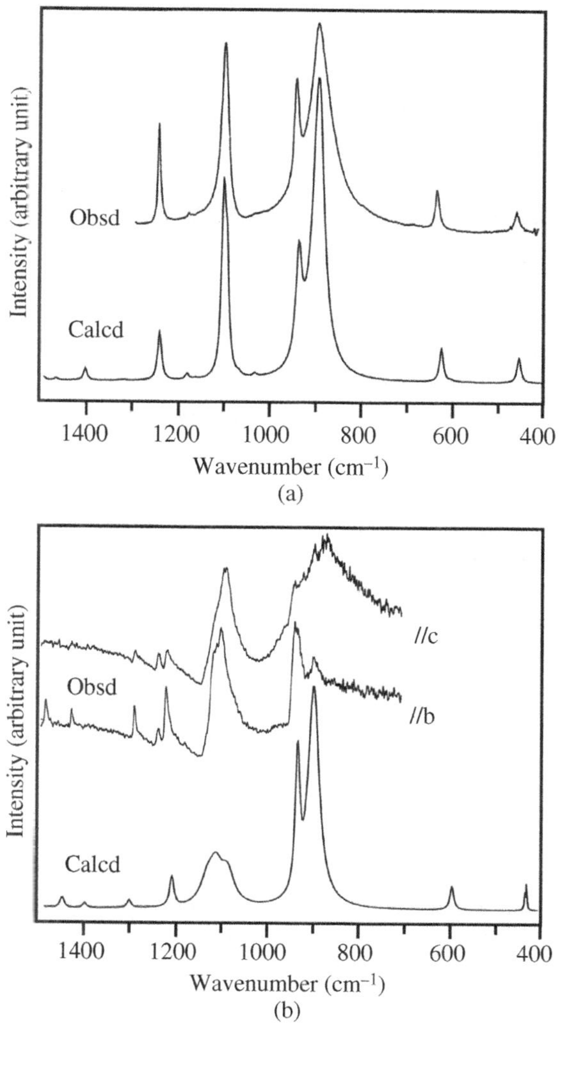

Figure 10.3 Infrared spectra of (a) trigonal and (b) orthorhombic PMO crystals. Each calculated spectral line was converted to a Lorentzian function whose position and integrated area correspond to the scaled frequency and intensity, respectively. The experimental spectra were observed from (a) fully extended chain crystals in needlelike single crystals [434] and (b) a micron-sized single crystal [249]. Source: [166], Figure 3 (p. 8675)/ Reproduced with permission of American Chemical Society.

carried out under three-dimensional periodic boundary conditions. As has often been pointed out, MO calculations overestimate vibrational frequencies, and hence, a scale factor for the frequency calibration has been proposed for each Hamiltonian and basis set [314]. Therefore, with respect to the most intense peak at 895 cm^{-1} of the trigonal PMO, the scale factor was adjusted so as to achieve the good agreement, and a value of 0.9846 was obtained, and the calibrated IR spectrum is shown in Figure 10.3a, where all the peaks exactly correspond to the observed bands. Surprisingly, even very small peaks around 1030–1070 and 1180 cm^{-1} were reproduced well.

The IR spectrum of orthorhombic PMO crystal was also calculated, the scale factor was similarly determined with respect to the most intense peak at 897 cm^{-1}, and the vibrational frequencies were calibrated with a scale factor of 0.9793. In Figure 10.3b, the calibrated spectrum is compared with polarized microfocus Fourier transform IR spectra observed from a micron-sized single PMO crystal [249]. In this case as well, the agreement between theory and experiment is generally satisfactory.

10.7 Thermal Chemistry

The thermodynamic quantities of crystals can also be calculated in the ways similar to those used for isolated molecules as described in Chapter 8; however, for crystals, the contributions of translational and rotational motions are considered to be unnecessary.

The Gibbs free energy, G, of a crystal may be calculated from [166]

$$G = E_{ee} + E_0 + E_T + PV - TS = H - TS \tag{10.34}$$

where the symbols are as follows: E_{ee}, electronic energy; E_0, zero-point energy; E_T, thermal contribution to the vibrational energy; P, pressure; V, volume; T, absolute temperature; S, entropy; and H, enthalpy. Here, all the thermodynamic quantities are defined per monomeric unit. The thermochemical data derived from periodic DFT calculations with atom-centered basis sets such as 6-31G(d) may suffer from the basis set superposition error (BSSE) [49, 450]. The BSSE-corrected free energy, G^{CP}, may be evaluated from

$$G^{CP} = (E_{ee} - \text{BSSE}) + E_0 + E_T + PV - TS = H^{CP} - TS = G - \text{BSSE} \tag{10.35}$$

Here, the BSSE-corrected enthalpy, H^{CP}, is given by H – BSSE.

Here, the phase transition from the orthorhombic to trigonal PMO crystals (Figure 10.4, abbreviated hereafter as o-PMO and t-PMO, respectively) is

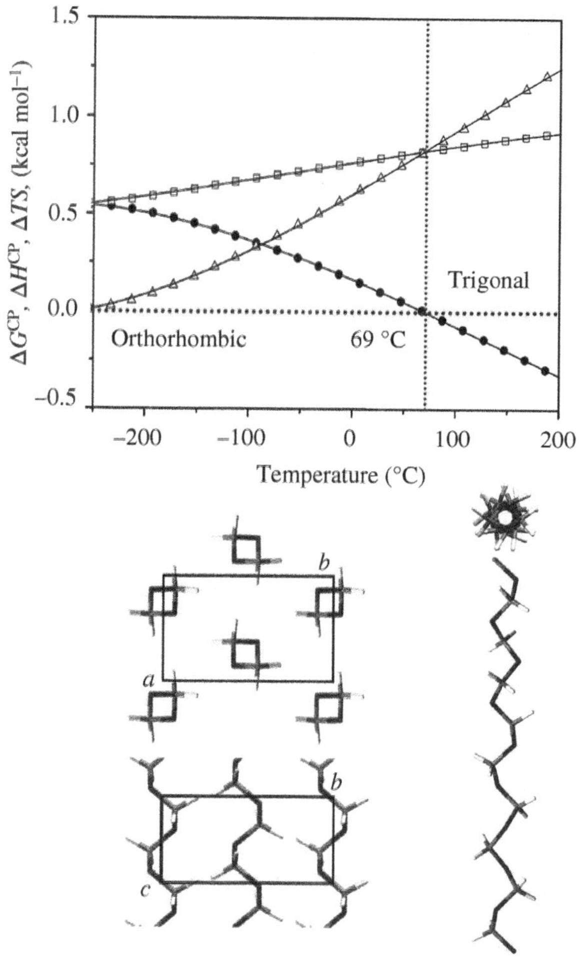

Figure 10.4 Differences in Gibbs free energy (ΔG^{CP}, filled circle), enthalpy (ΔH^{CP}, open square), and entropy ($T\Delta S$, open triangle) terms between trigonal 9/5 helical (lower right) and orthorhombic 2/1 helical (lower left) structures. Source: [166], Figures 4 (p. 8677) and 1 (p. 8674)/Reproduced with permission of American Chemical Society.

discussed. Let the difference in the thermodynamic function (A) between t-PMO and o-PMO be defined as

$$\Delta A = A_t - A_o \tag{10.36}$$

where $A = G^{CP}$, H^{CP}, or TS. In Figure10.4, the calculated ΔG^{CP} values are plotted against temperature, together with ΔH^{CP} and ΔTS.

In Figure 10.4, the solid line fitted to the data points at intervals of 20 K is expressed by a cubic function $\Delta G^{CP}(t) = at^3 + bt^2 + ct + d$, where t is temperature in °C, and a, b, c, and d are adjustable parameters. The cubic equation $\Delta G^{CP}(t) = 0$ yields $t = 69$ °C, at which the two crystalline phases are indicated to reach thermal equilibrium. Below 69 °C, $G_t^{CP} > G_o^{CP}$: this means that o-PMO is

Table 10.3 Interchain interaction energies (ΔE^{CP} (∞)) of some polymers.

Polymer	Form	Interchain interaction energy per		
		Monomer (kcal mol^{-1})	Bond (kcal bond^{-1})	Weight (cal g^{-1})
PGA[a)]		−15.47	−5.16	−267
Nylon 4	α	−17.90	−3.58	−210
Nylon 6	α	−21.25	−3.04	−188
	γ	−20.50	−2.93	−181
PMO	9/5 helix	−3.17	−1.59	−106
	2/1 helix	−3.38	−1.69	−113

a) See Figure 7.2.
Source: Adapted from [422].

thermodynamically more stable than t-PMO. Above 69 °C, $G_t^{CP} < G_o^{CP}$: t-PMO is preferred to o-PMO. Actually, Kobayashi et al. [251] observed the o → t transition at 69 °C as a sharp endothermic peak by DSC. Therefore, the BSSE-corrected DFT-D computations exactly reproduced the phase transition. Figure 10.4 also shows that o-PMO is enthalpically more stable and entropically less favored than t-PMO. This can be confirmed from the consequence that o-PMO is lower in ΔE^{CP} by 0.21 kcal mol^{-1} than t-PMO (Table 10.3). Because the ΔTS term increases with temperature, above 69 °C, t-PMO becomes thermodynamically more stable than o-PMO.

10.8 Cohesive (Interchain Interaction) Energy

The interaction energy (ΔE) between molecules A and B may be estimated from

$$\Delta E = E^{AB\{AB\}} - E^{A\{A\}} - E^{B\{B\}} \tag{10.37}$$

where $E^{AB\{AB\}}$ is the energy of the complex of molecules A and B, calculated with basis sets {AB} of the two molecules, and $E^{A\{A\}}$ and $E^{B\{B\}}$ are energies of molecules A and B, calculated with {A} and {B} basis sets, respectively. However, the ΔE value thus obtained includes the BSSE. To remove the BSSE, the counterpoise (CP) method has often been employed [48, 508]; the interaction energy (ΔE^{CP}) corrected for the BSSE can be evaluated from

$$\Delta E^{CP} = E^{AB\{AB\}} - E^{A\{AB\}} - E^{B\{AB\}} \tag{10.38}$$

Here, $E^{A\{AB\}}$ is the energy of molecule A, calculated with basis sets {AB}; molecule B is immaterial and supplies only its basis sets, thus being termed *ghost*, and $E^{B\{AB\}}$ is calculated similarly. Accordingly, the BSSE may be estimated from

$$\text{BSSE} = \left(E^{A\{AB\}} + E^{B\{AB\}}\right) - \left(E^{A\{A\}} + E^{B\{B\}}\right) \tag{10.39}$$

A CP method suitable for three-dimensional crystals in which polymeric chains of infinite length are packed was proposed (Figure 10.5). The BSSE per repeating unit may be defined as

$$\text{BSSE}\ (n_g) = E^{SC\{Domain\}}\ (n_g) - E^{SC\{SC\}} \tag{10.40}$$

where $E^{SC\{Domain\}}(n_g)$ is the energy of a single chain (SC) located in a domain filled with the n_g ghost chains, and $E^{SC\{SC\}}$ is the energy of the single isolated chain. Both $E^{SC\{Domain\}}(n_g)$ and $E^{SC\{SC\}}$ can be calculated under the one-dimensional periodic boundary condition (in the chain axis direction). On the basis of the above discussion, the interaction energy per repeating unit can be calculated from [165]

$$\Delta E^{CP}(n_g) = \left(\frac{E_{crystal}}{Z} - E^{SC\{SC\}+D}\right) - \text{BSSE}\ (n_g) \tag{10.41}$$

where $E_{crystal}$ is the total energy of the three-dimensional crystal including Z repeating units. The $E^{SC\{Domain\}}(n_g)$ and $E^{SC\{SC\}}$ terms are also evaluated per repeating unit. In principle, the $\Delta E^{CP}(n_g)$ value will be less affected by the

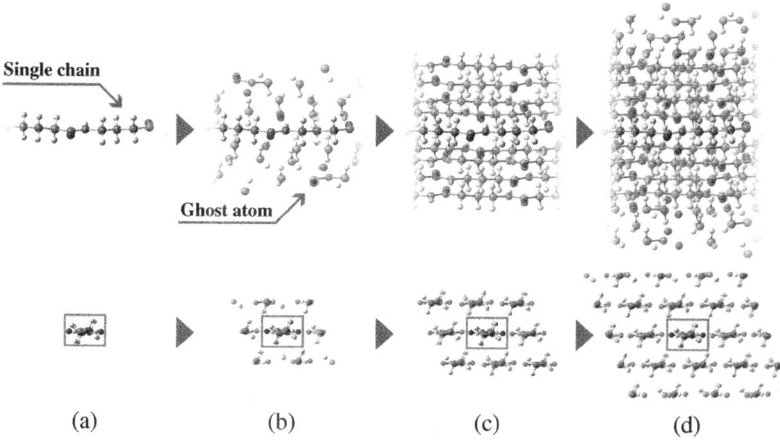

Figure 10.5 Schematic illustration of the CP method to correct the BSSE of polymer crystals: (above) side and (below) top views. (a) Target chain of infinite length. (b–d) Target chain surrounded by ghost chains of infinite length. As the number (n_g) of ghost chains increases (from b to d), the BSSE would approach a certain value, BSSE(∞) (see Figure 10.6). Source: [165], Figure 7 (p. 9550)/American Chemical Society/Licensed under CC BY 4.0.

BSSE with increasing domain size, namely the number (n_g) of ghost chains (see Figure 10.5). As described in Section 6.2, the E_{crystal} and $E^{\text{SC\{SC\}+D}}$ energies must undergo the dispersion correction of Eq. (6.24).

The most reliable BSSE would be

$$\text{BSSE}\ (\infty) = \lim_{n_g \to \infty} \text{BSSE}\ (n_g) \tag{10.42}$$

From BSSE (∞), we can obtain the ideal interaction energy, $\Delta E^{\text{CP}}(\infty)$. By way of example, a BSSE (n_g) vs. n_g plot of α form of nylon 4 is shown in Figure 10.6, yielding the BSSE (∞) = −6.06 kcal mol^{-1}. The ΔE^{CP} (∞) values of some polymer crystals, evaluated from Eq. (10.41), are listed in Table 10.3, where three kinds of units, kcal mol^{-1}, kcal bond^{-1}, and cal g^{-1} are used because the monomeric units of the polymers differ from each other in the number of bonds and formula weight. Of the polymers in Table 10.3, poly(glycolic acid) (PGA) exhibits the largest $|\Delta E^{\text{CP}}(\infty)|$ value in the unit of cal g^{-1}. This is because the PGA crystal has large dipole moments, between which strong intermolecular attractions are formed; therefore, the equilibrium melting point of PGA is as high as 231.4 °C although it is an aliphatic polyester. Nylons 4 and 6 form explicit intermolecular hydrogen bonds and hence have comparatively large $|\Delta E^{\text{CP}}(\infty)|$s.

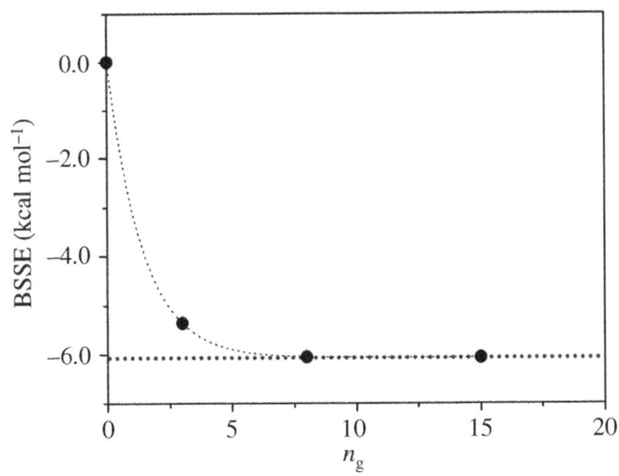

Figure 10.6 The basis set superposition error (BSSE) of the α form of nylon 4 as a function of the number (n_g) of ghost chains. As n_g increase, the BSSEs at B3LYP/6-31G(d,p) (filled circle) level approach −6.06 kcal mol^{-1} (horizontal dotted line). The calculated data were fitted to a function of $A + B\exp(-n_g/C)$ (dotted curve), where A, B, and C are adjustable parameters, and A corresponds to BSSE (∞). Source: [165], Figure 8 (p. 9551)/ American Chemical Society/Licensed under CC BY 4.0.

Part III

Statistical Mechanics of Chain Molecules: Rotational Isomeric State Scheme

Conformational Analysis of Polymers: Methods and Techniques for Structure-Property Relationships and Molecular Design, First Edition. Yuji Sasanuma.

Part II

Statistical [illegible]

11

Conventional RIS Scheme

In Chapter 2, a variety of models for polymeric chains are explained. As seen there, in general, as the model becomes more detailed, the characteristic ratio, $\langle r^2 \rangle / nl^2$, increases: the free jointed chain, the freely rotating chain, and the simple chain with rotational barrier show the $\langle r^2 \rangle / nl^2$ values of 1.0, 2.0, and 3.8, respectively. In fact, the experimental $\langle r^2 \rangle / nl^2$ values of polyethylene (PE) lying in the Θ state (in solutions and melts at 128–180 °C) range from 6.4 to 10.3 (see Table 29.4) [6, 132, 141, 294, 424]. The discrepancy between theory and experiment suggests the necessity of more exact models. As mentioned in Section 2.4, the repulsion occurring between methylene groups separated by four bonds, the so-called pentane effect, is required to be included in the polymer model.[1]

To include the pentane effect, the concept of the Markov chain [471] was introduced. According to the improved model (designated as the rotational isomeric state (RIS) scheme) [6, 141, 307, 383], the intramolecular interactions are divided into the first- and second-order interactions. The former and latter are assumed to act between atoms (atomic groups) separated by three and four bonds, respectively. The first-order interaction energy (usually expressed as E_σ) corresponds to the energy difference between conformations such as t, g$^+$, and g$^-$ around the centrally intervening bond. The first-order interaction of n-alkane (polyethylene) is represented by

$$—CH_2—CH_2—(t, g^+, \text{ or } g^-)—CH_2—CH_2—$$

The second-order interaction energy is an additional energy dependent on conformations around two in-between bonds. When two adjoining $CH_2—CH_2$ bonds

1 Only in this chapter describing the introductory RIS scheme, matrices are represented in bold to distinguish them from scalars. In other chapters, however, matrices are also written in italics unless special attention is required, because the variable types may be readily identified. Inasmuch as there are a large number of mathematical symbols, some of them are defined as different variables.

Conformational Analysis of Polymers: Methods and Techniques for Structure-Property Relationships and Molecular Design, First Edition. Yuji Sasanuma.

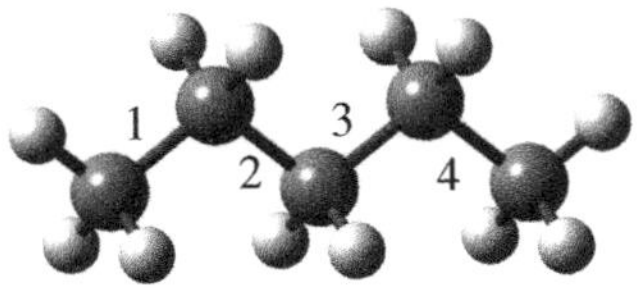

Figure 11.1 *n*-Pentane in the all-trans form. The bonds are numbered as shown.

of *n*-alkane lie in either g^+g^- or g^-g^+ pair, the steric repulsion (pentane effect) occurs between the outer (boldfaced) methylene groups separated by four bonds,

$$—\mathbf{CH_2}—CH_2—(g^{\pm})—CH_2—(g^{\mp})—CH_2—\mathbf{CH_2}—$$

the additional repulsive energy is 2–3 kcal mol^{-1} (expressed as E_ω). The second-order interaction appears in *n*-alkanes possessing at least five successive methylene (methyl) groups, thus being designated as the pentane effect. The statistical weight matrix of bond 3 of *n*-pentane (for the bond number, see Figure 11.1) is expressed as

$$\mathbf{U}_3 = \begin{array}{c} \\ t \\ g^+ \\ g^- \end{array} \begin{array}{c} \begin{array}{ccc} t & g^+ & g^- \end{array} \\ \begin{bmatrix} 1 & \sigma & \sigma \\ 1 & \sigma & \sigma\omega \\ 1 & \sigma\omega & \sigma \end{bmatrix} \end{array} \tag{11.1}$$

where the rows and columns of the matrix represent conformations (t, g^+, and g^-) of the current (3rd) and previous (2nd) bonds, respectively. The matrix expresses that, for example if the previous and current bonds, respectively, adopt the g^+ and g^- conformations, which corresponds to the (2,3) element of $\mathbf{U}_3$, then the statistical weight (the Boltzmann factor) is $\sigma\omega$, which is the product of two factors:

$$\sigma = \exp(-E_\sigma/RT) \tag{11.2}$$

and

$$\omega = \exp(-E_\omega/RT) \tag{11.3}$$

where E_σ is the gauche energy (0.4–0.7 kcal mol^{-1}), and E_ω is the additional energy for the pentane effect (2.0–3.0 kcal mol^{-1} [6, 141, 477, 497].

The statistical weight matrix of bond 2 of *n*-pentane is expressed by

$$\mathbf{U}_2 = \begin{array}{c} \begin{array}{ccc} t & g^+ & g^- \end{array} \\ \begin{bmatrix} 1 & \sigma & \sigma \\ 0 & 0 & 0 \\ 0 & 0 & 0 \end{bmatrix} \end{array} \tag{11.4}$$

Inasmuch as the first C—C bond has no previous bond and the fourth bond connected to the methyl terminal has the threefold (C_3) symmetry, the $\mathbf{U}_1$ and $\mathbf{U}_4$ matrices can be written as

$$\mathbf{U}_1 = \begin{bmatrix} 1 & 0 & 0 \\ 0 & 0 & 0 \\ 0 & 0 & 0 \end{bmatrix} \tag{11.5}$$

and

$$\mathbf{U}_4 = \begin{matrix} \mathrm{t} \\ \mathrm{g}^+ \\ \mathrm{g}^- \end{matrix} \begin{bmatrix} 1 & 0 & 0 \\ 0 & 1 & 0 \\ 0 & 0 & 1 \end{bmatrix} \tag{11.6}$$

The sequential product, $\mathbf{U}_1\mathbf{U}_2\mathbf{U}_3\mathbf{U}_4$, yields

$$\begin{bmatrix} 1+2\sigma & \sigma+\sigma^2+\sigma^2\omega & \sigma+\sigma^2+\sigma^2\omega \\ 0 & 0 & 0 \\ 0 & 0 & 0 \end{bmatrix} \tag{11.7}$$

The sum of all elements of $\mathbf{U}_1\mathbf{U}_2\mathbf{U}_3\mathbf{U}_4$ corresponds to the configurational partition function (Z): $1 + 4\sigma + 2\sigma^2 + 2\sigma^2\omega$, which can be extracted by

$$Z = \mathbf{J}^*\mathbf{U}_1\mathbf{U}_2\mathbf{U}_3\mathbf{U}_4\mathbf{J} \tag{11.8}$$

where

$$\mathbf{J}^* = \begin{bmatrix} 1 & 0 & 0 \end{bmatrix} \tag{11.9}$$

and

$$\mathbf{J} = \begin{bmatrix} 1 \\ 1 \\ 1 \end{bmatrix} \tag{11.10}$$

The Z function is the sum of statistical weights of the nine conformations formed in bonds 2 and 3: 1 (tt); σ (tg$^+$, tg$^-$, g$^+$t, and g$^-$t); σ^2 (g$^+$g$^+$ and g$^-$g$^-$); $\sigma^2\omega$ (g$^+$g$^-$ and g$^-$g$^+$).

If the correlations between up to two neighboring bonds are considered, the statistical weight matrix of bond 3 of n-pentane is applied to those of bonds 3 to $n - 1$ of PE, where n stands for the terminal C—C bond, and the partition function of PE is given by

$$Z = \mathbf{J}^* \left[\prod_{j=1}^{n} \mathbf{U}_j\right] \mathbf{J} = \mathbf{J}^* \left[\prod_{j=2}^{n-1} \mathbf{U}_j\right] \mathbf{J} \tag{11.11}$$

The bond conformation, that is the existing probability of conformation η (η = t, g$^+$, or g$^-$) in the jth bond of the polymeric chain, can be calculated from

$$p_{\eta;j} = Z^{-1}\mathbf{J}^* \left[\prod_{h=2}^{j-1} \mathbf{U}_h\right] \mathbf{U}'_{\eta;j} \left[\prod_{h=j+1}^{n-1} \mathbf{U}_h\right] \mathbf{J} \tag{11.12}$$

Here, in $\mathbf{U}'_{\eta;j}$, the column corresponding to η conformation is the same as that of $\mathbf{U}_j$, and the other columns are filled with zero. For example,

$$\mathbf{U}'_{\mathrm{g}^+;\,j} = \begin{bmatrix} 0 & \sigma & 0 \\ 0 & \sigma & 0 \\ 0 & \sigma\omega & 0 \end{bmatrix} \tag{11.13}$$

Problem 10 Express $p_{g+;3}$ of *n*-pentane in terms of σ and ω and numerically calculate the $p_{g+;3}$ value at 25 °C from $E_\sigma = 0.5$ kcal mol^{-1} and $E_\omega = 2.0$ kcal mol^{-1}.

The configurational (conformational) entropy per monomeric unit, S_{conf}, can be calculated from [8, 50, 293, 297, 484]

$$S_{conf} = \frac{R}{x}\left[\ln Z + T\frac{d(\ln Z)}{dT}\right] \tag{11.14}$$

where x is the degree of polymerization. The S_{conf} entropy is due to the conformational difference between in the Θ state and in the perfect regular form ($S_{conf} = 0$). The configurational internal energy is calculated from

$$U_{conf} = \frac{RT^2}{x}\frac{d(\ln Z)}{dT} \tag{11.15}$$

The U_{conf} value corresponds to the internal-energy difference between in the Θ state and in the reference state, where $Z = 1$ and $U_{conf} = 0$ are fulfilled (e.g. all-trans for PE).

The equilibrium melting point (T_m^0) of a polymer is related to the enthalpy (ΔH_u) and entropy (ΔS_u) of fusion by

$$T_m^0 = \frac{\Delta H_u}{\Delta S_u} \tag{11.16}$$

The entropy of fusion is divided into two terms:

$$\Delta S_u = \Delta S_v + (\Delta S_v)_V \tag{11.17}$$

where ΔS_v is the entropy change due to the latent volume change (ΔV_v) in melting, given by

$$\Delta S_v = \left(\frac{\partial S_v}{\partial V}\right)_T \Delta V_v = \left(\frac{\partial P}{\partial T}\right)_V \Delta V_v = \frac{\alpha}{\beta}\Delta V_v \tag{11.18}$$

Here, V is volume, P is pressure, α is the volume-temperature coefficient at constant pressure, and β is the volume-pressure coefficient at constant temperature. The entropy change ($(\Delta S_v)_V$) at constant volume corresponds to the configurational entropy:

$$(\Delta S_v)_V = S_{conf} \tag{11.19}$$

The contribution of S_{conf} to ΔS_u amounts to 60–90% [293, 421].

Problem 11 Prove that Eq. (11.14) is identical with the ordinary equation of entropy written in textbooks of physical chemistry

$$S_{conf} = \frac{1}{x}\left[R\ln Z + \frac{U(T) - U(0)}{T}\right] \tag{11.20}$$

except that the factor $1/x$ is put in front to set the unit of S_{conf} to cal K^{-1} (mole of monomer)$^{-1}$ or J K^{-1} (mole of monomer)$^{-1}$. Here, $U(T)$ and $U(0)$ are internal energies at T and 0 K, respectively. Note that

$$Z = \sum_{k}^{K} \exp(-E_k/RT) \tag{11.21}$$

where k stands for the chain conformational, E_k is its energy, and K is the total number of conformations.

11.1 Chain Dimension

The mean-square end-to-end distance of a polymer can be calculated from

$$\langle r^2 \rangle_0 = 2Z^{-1}\mathbf{J}^* \prod_{j=1}^{n} \mathbf{G}_j \mathbf{J}^{**} \tag{11.22}$$

and the partition function can be derived from

$$Z = \mathbf{J}^* \prod_{j=1}^{n} \mathbf{G}_j \mathbf{J}^{***} \tag{11.23}$$

where j is the bond number, $\mathbf{J}^* = [100]$, $\mathbf{J}^{**}$ is a column matrix whose first 12 elements are zero and last 3 elements are unity, $\mathbf{J}^{***}$ is a column matrix whose first 3 elements are unity and last 12 elements are zero, and $\mathbf{G}_j$ is a 15×15 matrix, termed *generator matrix*, expressed as

$$\mathbf{G}_j = \begin{bmatrix} \mathbf{U}_j & (\mathbf{U}_j \otimes \mathbf{l}_j^{\mathrm{T}})\|\mathbf{T}\|_j & (l_j^2/2)\mathbf{U}_j \\ \mathbf{0} & (\mathbf{U}_j \otimes \mathbf{I}_3)\|\mathbf{T}\|_j & \mathbf{U}_j \otimes \mathbf{l}_j \\ \mathbf{0} & \mathbf{0} & \mathbf{U}_j \end{bmatrix} \tag{11.24}$$

Here, $\otimes$ stands for the direct product, and $\mathbf{I}_3$ is the identity matrix of rank 3,

$$\mathbf{I}_3 = \begin{bmatrix} 1 & 0 & 0 \\ 0 & 1 & 0 \\ 0 & 0 & 1 \end{bmatrix} \tag{11.25}$$

If $\mathbf{U}_j$ is written as a 3×3 matrix such as

$$\mathbf{U}_j = \begin{bmatrix} u_{11} & u_{12} & u_{13} \\ u_{21} & u_{22} & u_{23} \\ u_{31} & u_{32} & u_{33} \end{bmatrix} \tag{11.26}$$

then it follows that

$$\mathbf{U}_j \otimes \mathbf{I}_3 = \begin{bmatrix} u_{11}\mathbf{I}_3 & u_{12}\mathbf{I}_3 & u_{13}\mathbf{I}_3 \\ u_{21}\mathbf{I}_3 & u_{22}\mathbf{I}_3 & u_{23}\mathbf{I}_3 \\ u_{31}\mathbf{I}_3 & u_{32}\mathbf{I}_3 & u_{33}\mathbf{I}_3 \end{bmatrix} \tag{11.27}$$

where, for example

$$u_{11}\mathbf{I}_3 = \begin{bmatrix} u_{11} & 0 & 0 \\ 0 & u_{11} & 0 \\ 0 & 0 & u_{11} \end{bmatrix} \quad (11.28)$$

The other factors in Eq. (11.24) are defined as

$$\mathbf{l}_j = l_j \begin{bmatrix} 1 \\ 0 \\ 0 \end{bmatrix} \quad (11.29)$$

and

$$\|\mathbf{T}\|_j = \begin{bmatrix} \mathbf{T}(\vartheta_j, \varphi_{\mathrm{t},j}) & \mathbf{0} & \mathbf{0} \\ \mathbf{0} & \mathbf{T}(\vartheta_j, \varphi_{\mathrm{g}+,j}) & \mathbf{0} \\ \mathbf{0} & & \mathbf{T}(\vartheta_j, \varphi_{\mathrm{g}-,j}) \end{bmatrix} \quad (11.30)$$

where l_j is the bond length of the jth bond, and

$$\mathbf{T}(\vartheta_j, \varphi_{\eta,j}) = \begin{bmatrix} \cos\vartheta_j & \sin\vartheta_j & 0 \\ \sin\vartheta_j \cos\varphi_{\eta,j} & -\cos\vartheta_j \cos\varphi_{\eta,j} & \sin\varphi_{\eta,j} \\ \sin\vartheta_j \sin\varphi_{\eta,j} & -\cos\vartheta_j \sin\varphi_{\eta,j} & -\cos\varphi_{\eta,j} \end{bmatrix} \quad (11.31)$$

with θ_j and $\varphi_{\eta,j}$ being the supplement of bond angle and the dihedral angle of the η conformation, respectively (Figure 11.2) [6, 139, 141, 151].

Instead of the bond length, the bond dipole moment (m_j),

$$\mathbf{m}_j = m_j \begin{bmatrix} 1 \\ 0 \\ 0 \end{bmatrix} \quad (11.32)$$

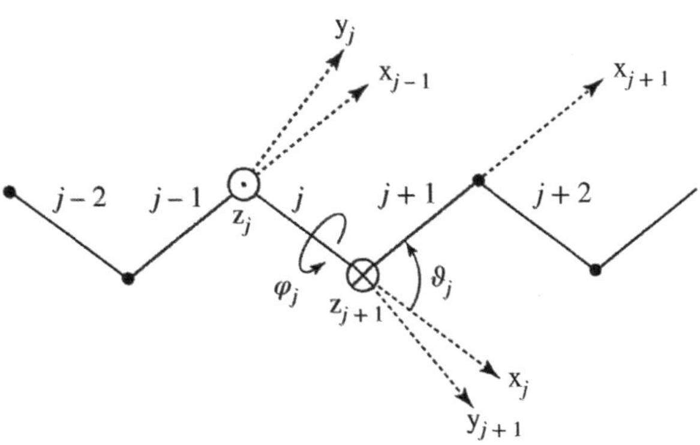

Figure 11.2 Local coordinate system defined for the jth bond. The x_j-axis is coincident with the jth bond, and the y_j-axis is located on the plane formed by the $(j-1)$th and jth bonds. The x, y, and z axes form a right-handed Cartesian coordinate system. ϑ_j and φ_j are the supplement of the bond angle and the dihedral angle of the jth bond, respectively.

is used; Eq. (11.22) yields the mean-square dipole moment, $\langle \mu^2 \rangle$, from which the dipole moment ratio $\langle \mu^2 \rangle / nm^2$ can be calculated.

So far, the method to calculate the mean square end-to-end distance $\langle r^2 \rangle_0$ and mean square dipole moment $\langle \mu^2 \rangle$ from the G_J matrix has been explained. Chain dimensions such as mean end-to-end vector $\langle \mathbf{r} \rangle_0$ and mean square radius of gyration $\langle S^2 \rangle_0$ can also be calculated from the successive multiplication of the respective generator matrices from one end to another along the polymeric chain [143, 144, 147, 191, 305, 307, 329].

12

Refined RIS Scheme

In the conventional rotational isomeric state (RIS) scheme, the product of the generator matrices G_js yields, for example the mean-square end-to-end distance and the partition function Z of a given polymeric chain [141, 307]. If the size of the statistical weight matrix U_j is 9×9, the intramolecular interaction (conformational) energies depend on the conformations of bonds $j-2, j-1$, and j. According to the additivity of conformational energies, it is allowed to formulate the U_j matrix with bonds $\geq j+1$ left out of consideration. On the other hand, the geometrical parameters of bond j depend significantly on the conformations of the neighboring bonds $j-1$ and $j+1$ as well as the current bond j, while the effects of distant bonds $\leq j-2$ and $\geq j+2$ may be negligible. In the *refined* RIS scheme here [402], the generator matrix $G_{\beta\gamma}$ for the 9×9 U_j matrix includes conformational energies of bonds $j-2, j-1$, and j and the geometrical parameters (bond length, bond angle, and dihedral angle) as a function of conformations of bonds $j-1$ (α), j (β), and $j+1$ (γ) (see Figure 12.1). Here, α, β, and γ, represent the rotamers of bonds $j-1$, j, and $j+1$, respectively. The $G_{\beta\gamma}$ matrices of all $\beta\gamma$ pairs are arranged in a larger Γ_j matrix (designated as *supergenerator* matrix), and the product of Γ_js yields the mean-square end-to-end distance, reflecting the conformational dependence of the geometrical parameters. Simultaneously, the partition function, Z, can also be derived from the product. If, in place of the bond length, the bond dipole moment is put in the $G_{\beta\gamma}$ matrix, the mean-square dipole moment will be derived. In principle, the refined RIS scheme will provide all such configurational properties and thermodynamic functions as the conventional RIS scheme gives. The formulation of the refined RIS is described below.

In the refined RIS scheme, the mean-square moment $\langle M^2 \rangle$ is given by

$$\langle M^2 \rangle = 2\, Z^{-1} \sum_{k=1}^{n_{\mathrm{RIS}}} \sum_{o=1}^{n_{\mathrm{RIS}}} \sum_{p=1}^{t_n} g_{5s_1(k-1)+1,\ 5t_n o-p+1} \tag{12.1}$$

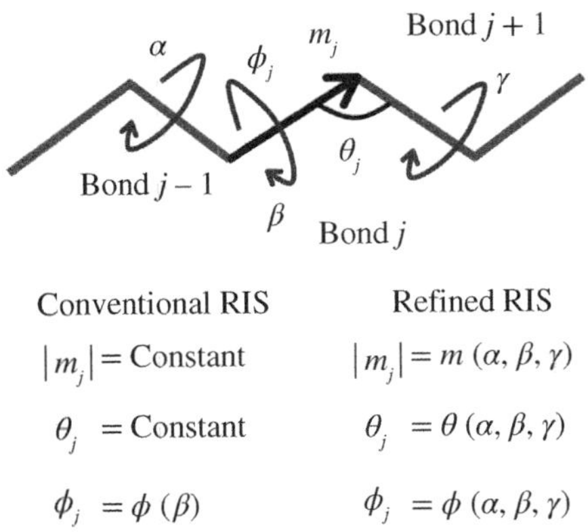

Figure 12.1 Schematic illustration of the refined rotational isomeric state (RIS) scheme, in which geometrical parameters as well as conformational energies are varied with conformations of the neighboring bonds to represent more realistic polymeric chains. For comparison, the conventional RIS scheme is also represented. m_j, θ_j, and ϕ_j are the bond length (or bond dipole moment), bond angle, and dihedral angle of the current bond j, and α, β, and γ stand for conformations of bonds, $j-1$, j, and $j+1$, respectively. Source: [402], Figure 4 (p.3491)/Reproduced with permission from American Chemical Society.

where n_{RIS} is the number of rotamers of each skeletal bond, and Z is the partition function of the polymeric chain, calculated from

$$Z = \sum_{k=1}^{n_{\mathrm{RIS}}}\sum_{o=1}^{n_{\mathrm{RIS}}}\sum_{p=1}^{t_n} g_{5s_1(k-1)+1,\ 5t_n(o-1)+p} \tag{12.2}$$

with $g_{q,r}$ being the (q, r) element of the product of all Γ_j matrices:

$$(g_{q,r}) = \Gamma_{\mathrm{all}} \equiv \prod_{j=1}^{n}\Gamma_j \tag{12.3}$$

The statistical weight matrices (U_1 and U_n) of the first and last skeletal bonds have sizes $s_1 \times t_1$ and $s_n \times t_n$, respectively, and the Γ_{all} matrix has a size of $15s_1 \times 15t_n$. The Γ_j matrix for bond j with three rotamers, t, g^+, and g^-, is composed of nine generator matrices:

$$\Gamma_j = \begin{bmatrix} G_{\mathrm{tt}} & G_{\mathrm{tg^+}} & G_{\mathrm{tg^-}} \\ G_{\mathrm{g^+t}} & G_{\mathrm{g^+g^+}} & G_{\mathrm{g^+g^-}} \\ G_{\mathrm{g^-t}} & G_{\mathrm{g^-g^+}} & G_{\mathrm{g^-g^-}} \end{bmatrix} \tag{12.4}$$

The rows and columns, respectively, correspond to the RISs of bonds j and $j+1$. In the multiplication between Γ_j and Γ_{j+1}, γ of every $G_{\beta\gamma}$ of the jth bond agrees with β of the counterpart of the $(j+1)$th. The sequential product, Γ_{all}, includes products of the G matrices corresponding to all possible conformations formed in the polymeric chain. The sums of the upper left $s \times t$ and upper right $s \times t$ parts of nine G blocks of Γ_{all} yield Z and $\langle M^2 \rangle/2$, respectively (see Eq. (12.15)).

The $G_{\beta\gamma}$ matrix for bond j is defined as

$$G_{\beta\gamma} = \begin{bmatrix} U_\beta & [(U_\beta\, M_\gamma) \otimes R_3]\, \|T\|_\gamma & (1/2)U_\beta\, (M_\gamma)^2 \\ 0 & (U_\beta \otimes I_3)\, \|T\|_\gamma & (U_\beta M_\gamma) \otimes C_3 \\ 0 & 0 & U_\beta \end{bmatrix} \tag{12.5}$$

where $\otimes$ stands for the direct product. In U_β, only the columns of the β state are equal to those of U_j, and the other columns are null; therefore,

$$U_j = U_{\text{t}} + U_{\text{g}^+} + U_{\text{g}^-} \tag{12.6}$$

The moment matrix, M_γ, is a diagonal matrix expressed as

$$M_\gamma = \text{diag}\left(m_{\text{t}\gamma}, m_{\text{g}^+\gamma}, m_{\text{g}^-\gamma}\right) \tag{12.7}$$

for 3×3 U_j matrices or

$$M_\gamma = \text{diag}\left(m_{\text{tt}\gamma}, m_{\text{tg}^+\gamma}, m_{\text{tg}^-\gamma}, m_{\text{g}^+\text{t}\gamma}, m_{\text{g}^+\text{g}^+\gamma}, m_{\text{g}^+\text{g}^-\gamma}, m_{\text{g}^-\text{t}\gamma}, m_{\text{g}^-\text{g}^+\gamma}, m_{\text{g}^-\text{g}^-\gamma}\right) \tag{12.8}$$

for 9×9 U_j matrices. Here, the moment $m_{\alpha\beta\gamma}$ is either bond length or bond dipole moment. In as much as $M_\gamma = \text{diag}(\ldots\, m_{\alpha\beta\gamma} \ldots)$, we have $(M_\gamma)^2 = \text{diag}(\ldots\, m^2_{\alpha\beta\gamma} \ldots)$. The R_3, C_3, and I_3 are row, column, and identity matrices of size 3, respectively:

$$R_3 = \begin{bmatrix} 1 & 0 & 0 \end{bmatrix} \tag{12.9}$$

$$C_3 = \begin{bmatrix} 1 \\ 0 \\ 0 \end{bmatrix} \tag{12.10}$$

and

$$I_3 = \begin{bmatrix} 1 & 0 & 0 \\ 0 & 1 & 0 \\ 0 & 0 & 1 \end{bmatrix} \tag{12.11}$$

The $\|T\|_\gamma$ matrix is defined as

$$\|T\|_\gamma = \text{diag}\left(T_{\text{t}\gamma}, T_{\text{g}^+\gamma}, T_{\text{g}^-\gamma}\right) \tag{12.12}$$

for 3×3 U_j matrices or

$$\|T\|_\gamma = \text{diag}\left(T_{\text{tt}\gamma}, T_{\text{tg}^+\gamma}, T_{\text{tg}^-\gamma}, T_{\text{g}^+\text{t}\gamma}, T_{\text{g}^+\text{g}^+\gamma}, T_{\text{g}^+\text{g}^-\gamma}, T_{\text{g}^-\text{t}\gamma}, T_{\text{g}^-\text{g}^+\gamma}, T_{\text{g}^-\text{g}^-\gamma}\right) \tag{12.13}$$

for 9×9 U_j matrices. The $T_{\alpha\beta\gamma}$ matrix transforms a vector from the jth to $(j-1)$th frame of reference. The $T_{\alpha\beta\gamma}$ matrix is expressed as

$$T_{\alpha\beta\gamma} = \begin{bmatrix} \cos\vartheta_{\alpha\beta\gamma} & \sin\vartheta_{\alpha\beta\gamma} & 0 \\ \sin\vartheta_{\alpha\beta\gamma}\cos\phi_{\alpha\beta\gamma} & -\cos\vartheta_{\alpha\beta\gamma}\cos\phi_{\alpha\beta\gamma} & \sin\phi_{\alpha\beta\gamma} \\ \sin\vartheta_{\alpha\beta\gamma}\sin\phi_{\alpha\beta\gamma} & -\cos\vartheta_{\alpha\beta\gamma}\sin\phi_{\alpha\beta\gamma} & -\cos\phi_{\alpha\beta\gamma} \end{bmatrix} \tag{12.14}$$

where $\vartheta_{\alpha\beta\gamma}$ is the supplement of the bonds angle, and $\phi_{\alpha\beta\gamma}$ is the dihedral angle for the β conformation of bond j, with bonds $j-1$ and $j+1$ being in the α and γ states, respectively. $T_{\beta\gamma}$ matrix is defined similarly. Note that the dihedral angle is defined here according to the convention of the RIS scheme: trans, $\sim 0°$; gauche$^{\pm}$, $\sim \pm120°$. The sizes of the block matrices of $G_{\beta\gamma}$ are

$$G_{\beta\gamma} : \begin{bmatrix} s \times t & s \times 3t & s \times t \\ 3s \times t & 3s \times 3t & 3s \times t \\ s \times t & s \times 3t & s \times t \end{bmatrix} \tag{12.15}$$

where (s, t) corresponds to the size of U_j.

If the Γ_j matrices of the repeating unit are combined into a matrix, H_i, Γ_{all} can also be obtained from

$$\Gamma_{\mathrm{all}} = \Gamma_1 H_1 \left(\prod_{i=2}^{x-1} H_i \right) \Gamma_n \tag{12.16}$$

where i stands for the repeating unit, and x is the degree of polymerization. The H_i matrices are given by

$$H_1 \equiv \Gamma_2\Gamma_3(\cdots) \tag{12.17}$$

and

$$H_i \equiv \Gamma_{\mathrm{a}}\Gamma_{\mathrm{b}}(\cdots) \tag{12.18}$$

where a, b... stand for the bonds in the repeating unit. This is because the first monomeric unit has U_js different in size and form from those of the subsequent units.

The characteristic ratio $\langle r^2\rangle_0/nl^2$ and dipole moment ratio $\langle \mu^2\rangle/nm^2$ can be calculated from

$$\langle r^2\rangle_0/nl^2 \text{ or } \langle \mu^2\rangle/nm^2 = \langle M^2\rangle \left[\sum_{j=1}^{n} \left(\overline{\lambda}_j\right)^2 \right]^{-1} \tag{12.19}$$

where $\overline{\lambda_j}$ represents the mean bond length or mean bond dipole moment, given by

$$\overline{\lambda}_j = \sum_{\alpha}\sum_{\beta}\sum_{\gamma} \lambda_{\alpha\beta\gamma j}\, p_{\alpha\beta\gamma j} \tag{12.20}$$

Here, $p_{\alpha\beta\gamma,j}$ is the probability of the $\alpha\beta\gamma$ conformation of bond j. The bond angle and dihedral angle are also averaged similarly:

$$\bar{\theta}_j = \sum_{\alpha}\sum_{\beta}\sum_{\gamma}\theta_{\alpha\beta\gamma,j}p_{\alpha\beta\gamma,j} \tag{12.21}$$

and

$$\bar{\phi}_{\beta,j} = \sum_{\alpha}\sum_{\gamma}\phi_{\alpha\beta\gamma,j}p_{\alpha\beta\gamma,j}\left(\sum_{\alpha}\sum_{\gamma}p_{\alpha\beta\gamma,j}\right)^{-1} \tag{12.22}$$

Note that these averaged geometrical parameters depend on the temperature.

The other physical properties that are related to the conventional RIS model may also be formulated under the refined RIS scheme. The configurational entropy, S_{conf}, can be calculated from

$$S_{\text{conf}} = \frac{R}{x}\left[\ln Z + T\frac{d(\ln Z)}{dT}\right] \tag{12.23}$$

The S_{conf} value corresponds to the entropy of the unperturbed states (for example melt, Θ solution, and amorphous state) relative to that of the perfect crystal ($S_{\text{cryst}} \equiv 0$). The configurational internal energy, U_{conf}, is calculated from

$$U_{\text{conf}} = \frac{RT^2}{x}\frac{d\ln Z}{dT} \tag{12.24}$$

If the energy of the crystal conformation is assumed to be null, the U_{conf} value corresponds to the difference in internal energy between the unperturbed and crystalline states.

12.1 RIS Scheme Including Middle-range Intramolecular Interactions

The RIS scheme described in Flory's book [141] uses 3×3 statistical weight matrices (U_js), which includes up to intramolecular interactions between atoms (atomic groups) separated by four bonds (dependent on the conformations of two successive bonds $j-1$ and j). If the molecule forms some specific interaction between atomic groups separated by five bonds (dependent on the conformations of bonds $j-2, j-1$, and j), and each bond has three rotamers, then the U_j matrices of size 9×9 are necessary to include the five-bond interaction. Similarly, as the distance of interactions increases by one bond, the number of conformations increases threefold: 6 bonds, 27×27 U_j (dependent on the conformations of bonds $j-3, j-2, j-1$, and j); 7 bonds, 81×81 U_j (bonds $j-4, j-3, j-2, j-1$, and j). The generator matrix, $G_{\beta\gamma}$, is five times as large as U_j, and the supergenerator matrix, Γ_j, is three times as large as $G_{\beta\gamma}$. Therefore, when the U_j matrix is defined as 81×81, the Γ_j matrix becomes 1215×1215 (1 476 225 elements!).

Aromatic polyesters of industrial importance, poly(ethylene terephthalate) (PET, $y = 2$), poly(trimethylene terephthalate) (PTT, $y = 3$), and poly(butylene terephthalate) (PBT, $y = 4$), have flexible O—$(CH_2)_y$—O bond sequences (designated herein as "spacers") connecting two benzene rings by ester linkages. The long spacers of PTT and PBT allow the neighboring benzene rings to occasionally come close to and interact with each other (intramolecular $\pi \cdots \pi$ interaction), which causes the fluorescent (dimer) emission, whereas PET with the shorter O—$(CH_2)_2$—O spacer exhibits no luminous behavior (see Figure 12.2) [310].

It is often found that the energy of each conformation is too intricately composed of various interaction energies to be divided into well-defined factors. For example, the tg^+g^- conformational energy of the O—CH_2—CH_2—O spacer of PET includes a CH_2—CH_2 gauche energy, a O—CH_2 gauche energy, a second-order g^+g^- energy for the CH_2—CH_2/CH_2—O pair, and, furthermore, some remaining energy. If the effect of the obscure interaction is significant, it is preferable that the energy of each conformation should be introduced as it is into the statistical weight matrix, and that the Gibbs free energy (ΔG_k, k: conformation) should be used as the conformational energy because the RIS scheme deals with the conformational equilibrium. The $U_a - U_e$ matrices (Eqs. (12.25)–(12.29)) of PET are filled with either unity or zero (for the bond symbols, see Figure 12.3): only the elements corresponding to the existent conformations are unity, and the others

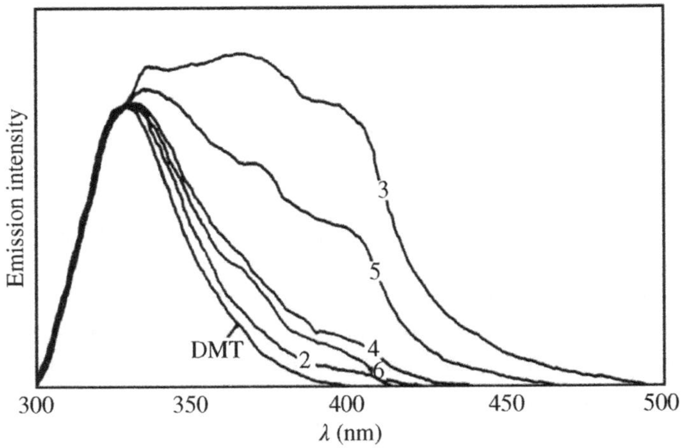

Figure 12.2 Emission spectra observed from dimethyl terephthalate (DMT) and the aromatic polyesters. The number in the figure corresponds to y: 2, PET; 3, PTT; 4, PBT. The solvent was dichloroethane, temperature was 25 °C, excitation was at 286 nm, and the emission spectra are normalized at 324 nm. Source: [310], Figure 2 (p.1671)/ Reproduced with permission of Elsevier.

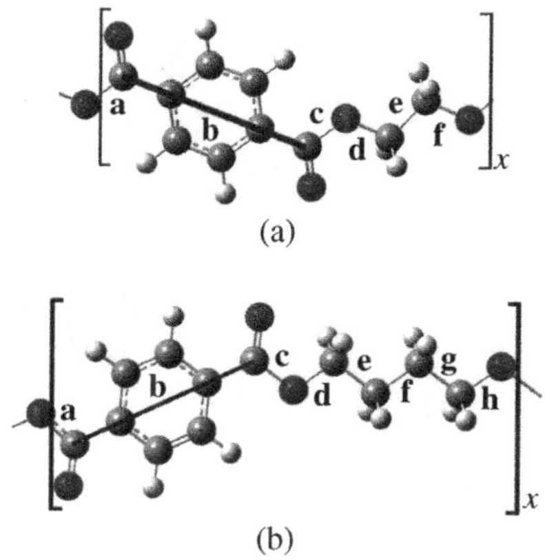

Figure 12.3 (a) PET and (b) PBT. The skeletal bonds are designated as shown, and the benzene ring is represented by the virtual bond b between the two carbonyl carbon atoms. Bonds a and c are not rotatable, and bond b adopts two states: trans and cis.

are null. The Boltzmann factors, $\exp(-\Delta G_k/RT)$s, are inserted in the U_f matrix (Eq. (12.30)) for the last bond in the repeating unit.

The conformational energies and geometrical parameters of these aromatic esters were obtained from accurate molecular orbital (MO) calculations and introduced into the refined RIS scheme using large statistical weight matrices (PET, 9×9 U_j; PTT, 27×27 U_j; and PBT, 81×81 U_j) to reveal the relationships between their conformational characteristics and luminous behaviors [400, 414, 419]. In Figure 12.4, the existing probabilities of the individual conformations of PET, PTT, and PBT are plotted as bar graphs against the intramolecular $C_g \cdots C_g$ distance, where C_g stands for the centroid of the benzene ring. The probabilities were evaluated from MO calculations on their model compounds, $C_6H_5COO(CH_2)_yOCOC_6H_5$ ($y = 2$, 3, and 4). The shortest distance (7.47 Å) of $y = 2$, assigned to the tg$^-$g$^-$ conformer, may be still too long as to cause the energy migration between the benzene rings, whereas the $C_g \cdots C_g$ distances of $y = 3$ are divided into two groups: 5.4–6.2 Å and 9.2–11.6 Å. The peaks of $y = 4$ are broadly distributed from 5.3 (tg$^+$g$^-$g$^+$t) to 13.3 (tttt) Å. It is obvious that the intensity of the dimer emission correlates to the probability of the short $C_g \cdots C_g$ distance, that is the frequency of the close $\pi \cdots \pi$ contact.

For example, the U_j matrices of bonds a–f of PET are expressed as

$$U_a = C_3 \otimes I_3 \otimes R_1 \tag{12.25}$$

$$U_b = C_3 \otimes S_1 \otimes R_\gamma \tag{12.26}$$

$$U_c = C_1 \otimes S_2 \otimes R_1 \tag{12.27}$$

$$U_d = C_2 \otimes S_1 \otimes R_3 \tag{12.28}$$

$$U_e = C_1 \otimes I_3 \otimes R_3 \tag{12.29}$$

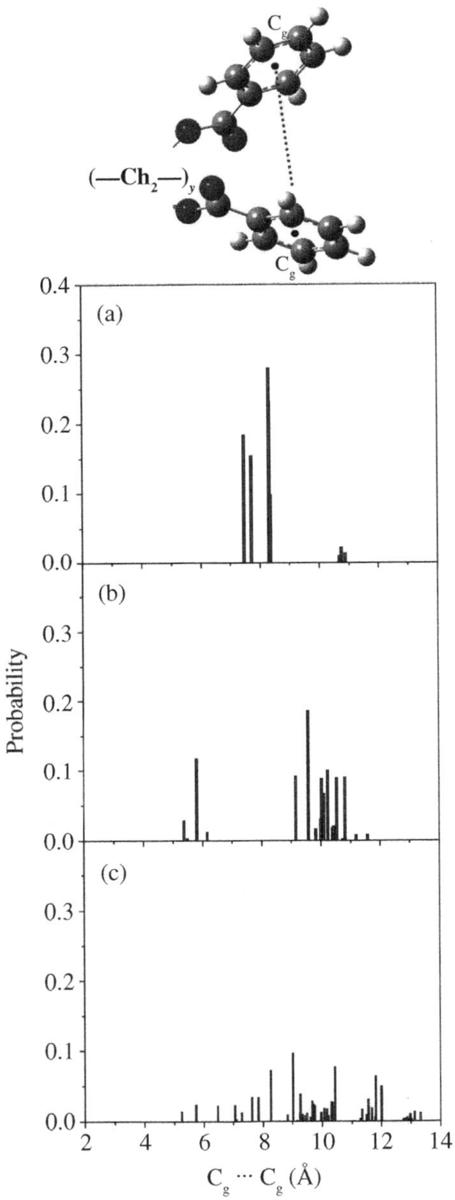

Figure 12.4 Existing probabilities of conformations of models of (a) PET, (b) PTT, and (c) PBT as a function of the distance ($C_g \cdots C_g$) between the centroids of the benzene rings. Source: [401], Figure 6 (p. 174)/Reproduced with permission of American Chemical Society.

and

$$U_{\mathrm{f}} = \begin{bmatrix} u_{\mathrm{ttt}} & u_{\mathrm{ttg^+}} & u_{\mathrm{ttg^-}} & 0 & 0 & 0 & 0 & 0 & 0 \\ 0 & 0 & 0 & u_{\mathrm{tg^+t}} & u_{\mathrm{tg^+g^+}} & u_{\mathrm{tg^+g^-}} & 0 & 0 & 0 \\ 0 & 0 & 0 & 0 & 0 & 0 & u_{\mathrm{tg^-t}} & u_{\mathrm{tg^-g^+}} & u_{\mathrm{tg^-g^-}} \\ u_{\mathrm{g^+tt}} & u_{\mathrm{g^+tg^+}} & u_{\mathrm{g^+tg^-}} & 0 & 0 & 0 & 0 & 0 & 0 \\ 0 & 0 & 0 & u_{\mathrm{g^+g^+t}} & u_{\mathrm{g^+g^+g^+}} & u_{\mathrm{g^+g^+g^-}} & 0 & 0 & 0 \\ 0 & 0 & 0 & 0 & 0 & 0 & u_{\mathrm{g^+g^-t}} & u_{\mathrm{g^+g^-g^+}} & u_{\mathrm{g^+g^-g^-}} \\ u_{\mathrm{g^-tt}} & u_{\mathrm{g^-tg^+}} & u_{\mathrm{g^-tg^-}} & 0 & 0 & 0 & 0 & 0 & 0 \\ 0 & 0 & 0 & u_{\mathrm{g^-g^+t}} & u_{\mathrm{g^-g^+g^+}} & u_{\mathrm{g^-g^+g^-}} & 0 & 0 & 0 \\ 0 & 0 & 0 & 0 & 0 & 0 & u_{\mathrm{g^-g^-t}} & u_{\mathrm{g^-g^-g^+}} & u_{\mathrm{g^-g^-g^-}} \end{bmatrix} \tag{12.30}$$

where I_3 is the identity matrix of order 3,

$$C_1 = \begin{bmatrix} 1 \\ 0 \\ 0 \end{bmatrix} \tag{12.31}$$

$$C_2 = \begin{bmatrix} 1 \\ 1 \\ 0 \end{bmatrix} \tag{12.32}$$

$$C_3 = \begin{bmatrix} 1 \\ 1 \\ 1 \end{bmatrix} \tag{12.33}$$

$$R_\gamma = \begin{bmatrix} 1 & \gamma & 0 \end{bmatrix} \tag{12.34}$$

$$R_1 = \begin{bmatrix} 1 & 0 & 0 \end{bmatrix} \tag{12.35}$$

$$R_3 = \begin{bmatrix} 1 & 1 & 1 \end{bmatrix} \tag{12.36}$$

$$S_1 = \begin{bmatrix} 1 & 0 & 0 \\ 0 & 0 & 0 \\ 0 & 0 & 0 \end{bmatrix} \tag{12.37}$$

and

$$S_2 = \begin{bmatrix} 1 & 0 & 0 \\ 0 & 1 & 0 \\ 0 & 0 & 0 \end{bmatrix} \tag{12.38}$$

The statistical weight, γ, is given by

$$\gamma = \exp(-E_\gamma/RT) \tag{12.39}$$

Table 12.1 Free energies (ΔG_ks) of conformations of PET, evaluated by ab initio MO calculations.

Conformation[b)]			Multiplicity	ΔG_k[a)], kcal mol^{-1} Gas 25 °C	Benzene 25 °C	Benzene 250 °C
t	t	t	1	0.00	0.00	0.00
t	t	g$^+$	4	0.50	0.55	0.80
t	g$^+$	t	2	−1.12	−1.24	−0.98
t	g$^+$	g$^+$	4	−0.96	−0.94	−0.82
t	g$^+$	g$^-$	4	−0.65	−0.69	−0.58
g$^+$	t	g$^+$	2	1.09	1.16	1.28
g$^+$	t	g$^-$	2	0.41	0.61	1.08
g$^+$	g$^+$	g$^+$	2	−0.84	−0.73	−0.19
g$^+$	g$^+$	g$^-$	4	−0.61	−0.59	−0.59
g$^+$	g$^-$	g$^+$	2		(absent)[c)]	
cis[d)](E_γ)				0.16	0.15	0.16

a) At the MP2/6-311+G(2d, p)//B3LYP/6-311+G(2d, p) level. At 25 °C. Relative to the all-trans conformation.
b) In the O–CH$_2$–CH$_2$–O bond sequence (spacer).
c) The local minimum of the potential was not found by the geometrical optimization at the B3LYP/6-311+G(2d, p) level.
d) The orientation between the two carbonyl groups attached to the benzene ring. Relative to the trans orientation.

Source: [400], Table 4 (p. 2859)/Reproduced with permission of American Chemical Society.

where E_γ is the cis energy of bond b (see Table 12.1). For example, the element $u_{\text{tg}^+\text{g}^-}$ of the U_f matrix is the Boltzmann factor of the corresponding conformer free energy: $u_{\text{tg}^+\text{g}^-} = \exp(-\Delta G_{\text{tg}^+\text{g}^-}/RT)$, where tg$^+g^-$ is the conformation in bonds d–f.

The U_j matrices of bonds a–h of PBT are given as follows:

$$U_\text{a} = C_3 \otimes I_3 \otimes I_3 \otimes I_3 \otimes R_1 \tag{12.40}$$

$$U_\text{b} = C_3 \otimes I_3 \otimes I_3 \otimes S_1 \otimes R_\gamma \tag{12.41}$$

$$U_\text{c} = C_3 \otimes I_3 \otimes S_1 \otimes S_2 \otimes R_1 \tag{12.42}$$

$$U_\text{d} = C_3 \otimes S_1 \otimes S_2 \otimes S_1 \otimes R_3 \tag{12.43}$$

$$U_\text{e} = C_1 \otimes S_2 \otimes S_1 \otimes I_3 \otimes R_3 \tag{12.44}$$

$$U_\text{f} = C_2 \otimes S_1 \otimes I_3 \otimes I_3 \otimes R_3 \tag{12.45}$$

Table 12.2 Configurational properties of PET, PTT, and PBT, evaluated from refined RIS calculations.[a)]

	PET		PTT		PBT	
	25 °C	280 °C[b)]	25 °C	237 °C[b)]	25 °C	230 °C[b)]
$\langle r^2 \rangle_0 / nl^2$[c)]	2.63	2.87	4.14	4.31	3.83	3.79
S_{conf} (cal K^{-1} mol^{-1})	7.12	7.51	8.93	9.18	10.80	11.30
U_{conf} (kcal mol^{-1})	−0.76	−0.43	0.49	0.31	0.10	0.27

a) Source: Adapted from [400, 414, 419].
b) The equilibrium melting point.
c) Calculated with the virtual bond.

$$U_g = C_1 \otimes I_3 \otimes I_3 \otimes I_3 \otimes R_3 \tag{12.46}$$

and

$$U_h = (u_{lm}) \; (1 \le l, m \le 81) \tag{12.47}$$

where

$$u_{lm} = \begin{cases} \exp(-\Delta G_{k'}/RT) & \text{when } l \text{ and } m \text{ satisfy the condition below} \\ 0 & \text{otherwise} \end{cases} \tag{12.48}$$

The U_h matrix is filled with $\exp(-\Delta G_{k'}/RT)$ only when l is equal to $\lfloor (k' - 1)/3 \rfloor + 1$ and m is equal to k' $(1 \le k' \le 81)$, $k' - 81$ $(82 \le k' \le 162)$, or $k' - 162$ $(163 \le k' \le 243)$, where $\lfloor \; \rfloor$ is the floor function, which gives the maximum integer fulfilling $\le (k' - 1)/3$, and ΔG_k is the free energy of conformation k' (k', 1 = tttt; 2 = tttg$^+$; 3 = tttg$^-$; 4 = ttg$^+$t; ⋯; 10 = ttg$^+$tt; ⋯; 28 = tg$^+$ttt; ⋯; 82 = g$^+$tttt; ⋯; 243 = g$^-$g$^-$g$^-$g$^-$g$^-$ in bonds d–h).

The configurational properties of PET, PTT, and PBT, calculated by the refined RIS scheme with the statistical weight matrices, are summarized in Table 12.2.

13

Inversional–Rotational Isomeric State (IRIS) Scheme

13.1 Pseudoasymmetry for Polyamines

Poly(ethylene imine) (PEI) has a hydrogen atom and a lone pair at the nitrogen site; accordingly, PEI is expected to form a variety of stereosequences, which may be represented in terms of the pseudoasymmetry proposed by Flory [140]. For example, *N*,*N*′-dimethylethylenediamine (di-MEDA), a monomeric model compound of PEI, in the all-trans form is put on paper as shown in Figure 13.1. When the hydrogen atom at the left-hand nitrogen site appears on this (that) side of the paper, the nitrogen site is considered to be the *d* (*l*) form. For other nitrogen sites, the *d* and *l* configurations can be defined similarly. The *dd* and *ll* diads are referred to as *meso*, and *dl* and *ld* as *racemo*.

13.2 Inversional–Rotational Isomerization

At ambient temperature, the nitrogen atom of amines can rapidly flip the orientation of the trigonal pyramid of three covalent bonds and a lone pair, which corresponds to a conversion between the *meso* and *racemo* forms. This phenomenon, nitrogen inversion, is illustrated for di-MEDA in Figure 1.6 and explained in Section 1.3. Figure 13.2 illustrates the inversional (solid line) and the rotational (dotted line) isomerizations of di-MEDA. The latter motion keeps the configuration, whereas the former always causes a *meso* ↔ *racemo* conversion. For polyamines undergoing the two isomerizations, the statistical mechanics, which was designated as inversional and rotational isomeric state (IRIS) scheme, was devised and formulated [403, 417].

Conformational Analysis of Polymers: Methods and Techniques for Structure-Property Relationships and Molecular Design, First Edition. Yuji Sasanuma.

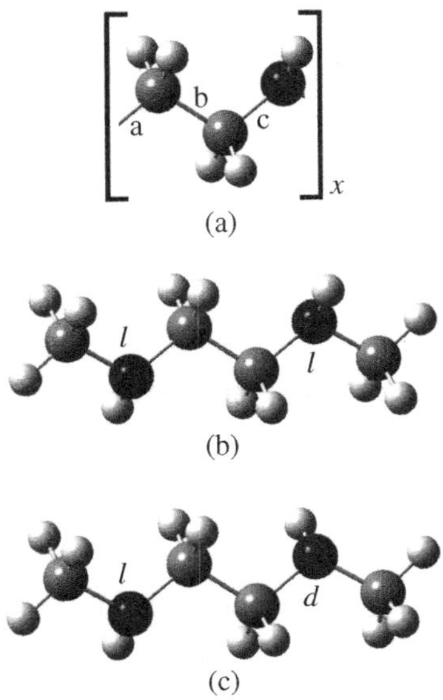

Figure 13.1 (a) All-trans forms of poly(ethylene imine) (PEI) and (b) *meso* (*ll*) *N,N′*- dimethylethylenediamine (di-MEDA) and (c) *racemo* (*ld*) di-MEDA.

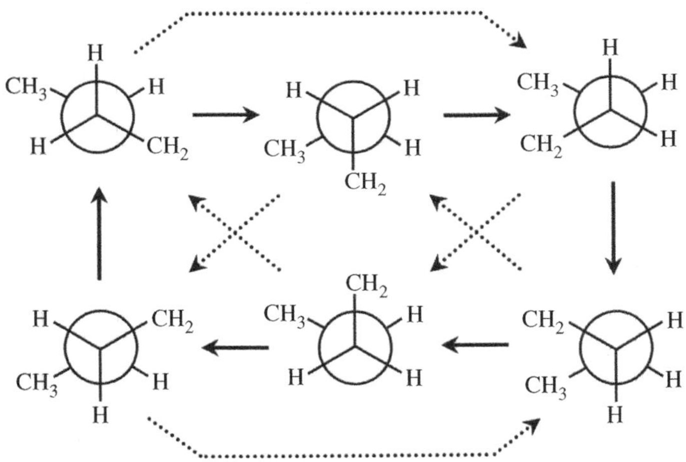

Figure 13.2 Inversional (solid line) and rotational (dotted line) isomerizations around the CH_2–NH bond of di-MEDA. Source: Sasanuma et al. [403], Figure 2 (p.9171)/ Reproduced with permission of American Chemical Society.

13.3 Statistical Weight Matrices of *Meso* and *Racemo* di-MEDA

The statistical weight matrix U_j of a given diad can be transformed to that of the mirror image by

$$U_j^{dd} = Q_k \, U_j^{ll} \, Q_k \tag{13.1}$$

and

$$U_j^{dl} = Q_k\, U_j^{ld}\, Q_k \tag{13.2}$$

and vice versa (dd to ll and dl to ld), where k represents the sizes of Q_k and U_j. For example,

$$Q_3 = \begin{bmatrix} 1 & 0 & 0 \\ 0 & 0 & 1 \\ 0 & 1 & 0 \end{bmatrix} \tag{13.3}$$

The higher Q_k matrix can be derived from

$$Q_k = \underbrace{Q_3 \otimes \cdots Q_3}_{k' \text{ times}} \tag{13.4}$$

where $k = 3^{(k'+1)}$, i.e. $k' = \log_3(k) - 1$, and $\otimes$ stands for the direct product.

13.4 Statistical Weight Matrices of PEI

Figure 13.3 illustrates the intramolecular interactions defined for PEI, and the corresponding conformational energies were evaluated from ab initio MO calculations as shown in Table 13.1. The interactions designated as η and ν, showing negative energies ($E_\eta = -1.54$ kcal mol^{-1} and $E_\nu = -0.58$ kcal mol^{-1}), can be regarded as N—H···N hydrogen bonds. The energy parameter clarifies the nature of each interaction: γ and δ, trans preferences of N—C bond; σ, flexibility of the C—C bond; ν', N···N repulsion; ω, N—H···H—C repulsion; ω', N···H—C, repulsion; and ω'', C—H···H—C repulsion. As an advantage of such definitions of interactions, it is possible that E_η and E_ν are selectively adjusted so as to investigate how the N—H···N hydrogen bonds affect the configurational properties of PEI. In Figure 13.4, the characteristic ratio is plotted as a function of the reciprocal degree of polymerization for each hydrogen bond strength (HBS). The two conformational energies (E_η and E_ν) representing the intramolecular N—H···N hydrogen bonds were treated as variables according to

$$E_\eta\ (\text{kcal mol}^{-1}) = -1.54 \times \text{HBS}(\%)/100 \tag{13.5}$$

and

$$E_\nu\ (\text{kcal mol}^{-1}) = -0.58 \times \text{HBS}\ (\%)/100 \tag{13.6}$$

When HBS = 100%, the hydrogen bond energies are as given by the MO calculations. Figure 13.4 shows that the $\langle r^2 \rangle_0/nl^2$ ratio increases with a decrease in HBS. When HBS = 100%, $\langle r^2 \rangle_0/nl^2 = 2.9$, whereas if the hydrogen bonds completely vanish (HBS = 0%), $\langle r^2 \rangle_0/nl^2$ will be 6.3 close to that of polyethylene [6].

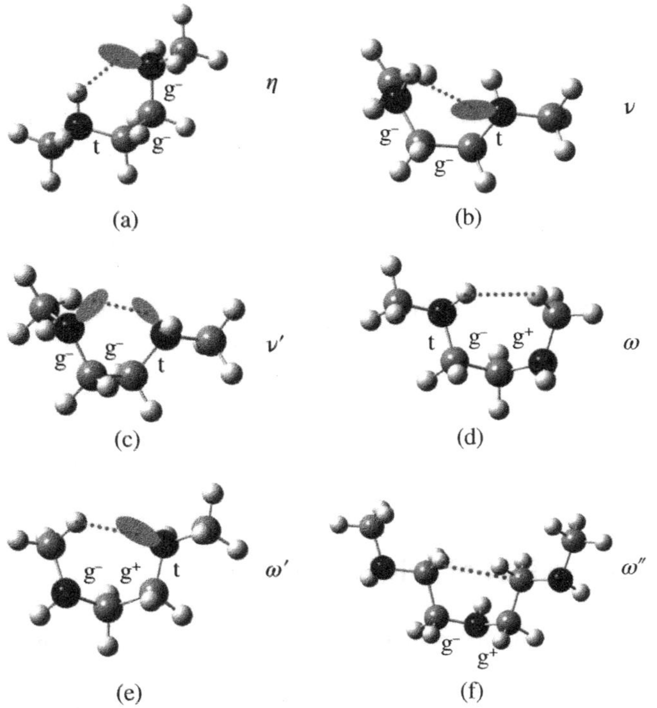

Figure 13.3 Second- and third-order intramolecular interactions formed in PEI, illustrated with its model compound, di-MEDA: (a) η, N—H· · ·N hydrogen bond; (b) ν, N—H· · ·N hydrogen bond; (c) ν', N· · ·N repulsion; (d) ω, N—H· · ·H—C repulsion; (e) ω', N· · ·H—C repulsion; and (f) ω'', C—H· · ·H—C repulsion. Source: Sasanuma et al. [403], Figure 11 (p.9177)/Reproduced with permission of American Chemical Society.

In the IRIS calculations on PEI, statistical weight matrices of bonds a–c were formulated as below. The weights are the Boltzmann factors corresponding to the conformational energies listed in Table 13.1. For example, $\eta = \exp(-E_\eta/RT)$.

$$U_{\mathrm{a}}^{ll} = U_{\mathrm{a}}^{ld} = \begin{bmatrix} 1 & \gamma & \delta & 0 & 0 & 0 & 0 & 0 & 0 \\ 0 & 0 & 0 & 1 & \gamma & \delta\omega'' & 0 & 0 & 0 \\ 0 & 0 & 0 & 0 & 0 & 0 & 1 & \gamma\omega'' & \delta \\ 1 & \gamma & \delta & 0 & 0 & 0 & 0 & 0 & 0 \\ 0 & 0 & 0 & 1 & \gamma & \delta\omega'' & 0 & 0 & 0 \\ 0 & 0 & 0 & 0 & 0 & 0 & 1 & 0 & \delta \\ 1 & \gamma & \delta & 0 & 0 & 0 & 0 & 0 & 0 \\ 0 & 0 & 0 & 1 & \gamma & 0 & 0 & 0 & 0 \\ 0 & 0 & 0 & 0 & 0 & 0 & 1 & \gamma\omega'' & \delta \end{bmatrix} \tag{13.7}$$

$$U_{\mathrm{a}}^{dd} = Q_9\, U_{\mathrm{a}}^{ll}\, Q_9 \tag{13.8}$$

Table 13.1 Conformational energies of di-MEDA and PEI, evaluated from ab initio MO calculations.[a]

	Conformational energy ($kcal\ mol^{-1}$)
First-order interaction	
E_γ	1.06
E_δ	0.54
E_σ	−0.09
Second- and third-order interactions[b]	
E_η	−1.54
E_ν	−0.58
$E_{\nu'}$	1.16
E_ω	0.97
$E_{\omega'}$	0.61
$E_{\omega''}$	0.94

a) At the MP2/6-311++G(3df,3pd)//HF/6-31G(d) level.
b) For the definitions, see Figure 13.3.
Source: Sasanuma et al. [403], Table 7 (p.9179)/Reproduced with permission of American Chemical Society.

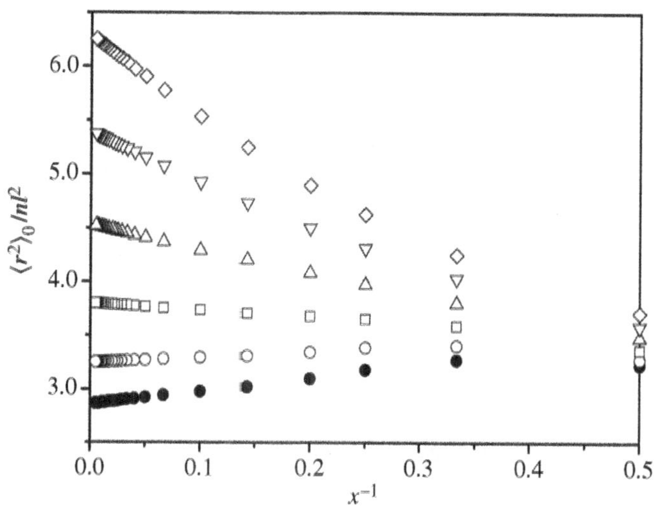

Figure 13.4 Characteristic ratios averaged over 512 chains of PEI with different hydrogen-bond strengths (HBSs) as a function of the reciprocal degree of polymerization (x^{-1}): HBS = 100% (filled circle), 80% (open circle), 60% (open square), 40% (open triangle), 20% (open inverted triangle), and 0% (open diamond). Source: Sasanuma et al. [403], Figure 13 (p.9179)/Reproduced with permission of American Chemical Society.

$$U_{\mathrm{a}}^{dl} = Q_9 \, U_{\mathrm{a}}^{ld} \, Q_9 \tag{13.9}$$

$$U_{\mathrm{b}}^{ll} = U_{\mathrm{b}}^{dd} = U_{\mathrm{b}}^{ld} = U_{\mathrm{b}}^{dl} = \begin{bmatrix} 1 & \sigma & \sigma & 0 & 0 & 0 & 0 & 0 & 0 \\ 0 & 0 & 0 & 1 & \sigma & \sigma & 0 & 0 & 0 \\ 0 & 0 & 0 & 0 & 0 & 0 & 1 & \sigma & \sigma \\ 1 & \sigma & \sigma & 0 & 0 & 0 & 0 & 0 & 0 \\ 0 & 0 & 0 & 1 & \sigma & \sigma & 0 & 0 & 0 \\ 0 & 0 & 0 & 0 & 0 & 0 & 1 & 0 & \sigma \\ 1 & \sigma & \sigma & 0 & 0 & 0 & 0 & 0 & 0 \\ 0 & 0 & 0 & 1 & \sigma & 0 & 0 & 0 & 0 \\ 0 & 0 & 0 & 0 & 0 & 0 & 1 & \sigma & \sigma \end{bmatrix} \tag{13.10}$$

$$U_{\mathrm{c}}^{ll} = \begin{bmatrix} 1 & \delta & \gamma & 0 & 0 & 0 & 0 & 0 & 0 \\ 0 & 0 & 0 & \eta & \delta\nu' & \gamma\omega' & 0 & 0 & 0 \\ 0 & 0 & 0 & 0 & 0 & 0 & \eta & \delta\omega & \gamma\nu \\ 1 & \delta & \gamma & 0 & 0 & 0 & 0 & 0 & 0 \\ 0 & 0 & 0 & \nu & \delta\eta & \gamma\omega & 0 & 0 & 0 \\ 0 & 0 & 0 & 0 & 0 & 0 & \omega' & 0 & \gamma\omega \\ 1 & \delta & \gamma & 0 & 0 & 0 & 0 & 0 & 0 \\ 0 & 0 & 0 & \omega & \delta\omega' & 0 & 0 & 0 & 0 \\ 0 & 0 & 0 & 0 & 0 & 0 & \nu' & \delta\omega' & \gamma\eta \end{bmatrix} \tag{13.11}$$

$$U_{\mathrm{c}}^{ld} = \begin{bmatrix} 1 & \gamma & \delta & 0 & 0 & 0 & 0 & 0 & 0 \\ 0 & 0 & 0 & \nu' & \gamma\eta & \delta\omega' & 0 & 0 & 0 \\ 0 & 0 & 0 & 0 & 0 & 0 & \nu & \gamma\omega & \delta\eta \\ 1 & \gamma & \delta & 0 & 0 & 0 & 0 & 0 & 0 \\ 0 & 0 & 0 & \eta & \gamma\nu & \delta\omega & 0 & 0 & 0 \\ 0 & 0 & 0 & 0 & 0 & 0 & \omega & 0 & \delta\omega' \\ 1 & \gamma & \delta & 0 & 0 & 0 & 0 & 0 & 0 \\ 0 & 0 & 0 & \omega' & \gamma\omega & 0 & 0 & 0 & 0 \\ 0 & 0 & 0 & 0 & 0 & 0 & \eta & \gamma\omega' & \delta\nu' \end{bmatrix} \tag{13.12}$$

$$U_{\mathrm{c}}^{dd} = Q_9 \, U_{\mathrm{c}}^{ll} \, Q_9 \tag{13.13}$$

and

$$U_{\mathrm{c}}^{dl} = Q_9 \, U_{\mathrm{c}}^{ld} \, Q_9 \tag{13.14}$$

13.5 Diad Probability and Bond Conformation

The *meso*-diad probability at the *i*th repeating unit, $P_{m;\,i}$, can be calculated from

$$P_{m;\,i} = Z^{-1} J^* \left(\prod_{h=1}^{i-1} W_h \right) W_i^m \left(\prod_{h=i+1}^{x} W_h \right) J \tag{13.15}$$

where x is the degree of polymerization, and Z is the partition function of the whole chain including all possible stereosequences:

$$Z = J^* \left(\prod_{i=1}^{x} W_i \right) J \tag{13.16}$$

Here, W_i is a combined statistical weight matrix of the ith unit

$$W_i = \begin{bmatrix} V_i^{ll} & V_i^{ld} \\ V_i^{dl} & V_i^{dd} \end{bmatrix} \tag{13.17}$$

W_i^m is the matrix for the *meso* form

$$W_i^m = \begin{bmatrix} V_i^{ll} & 0 \\ 0 & V_i^{dd} \end{bmatrix} \tag{13.18}$$

The V_i^α matrix ($\alpha = ll, dd, ld$, or dl) is defined as

$$V_i^\alpha = U_\mathrm{a}^\alpha\ U_\mathrm{b}^\alpha\ U_\mathrm{c}^\alpha\ (\cdots) \tag{13.19}$$

where a, b, c$\cdots$ are bond symbols in the repeating unit. The P_m value of the whole chain can be obtained from

$$P_m = x^{-1} \sum_{i=1}^{x} P_{m;\ i} \tag{13.20}$$

The *racemo*-diad probability P_r is

$$P_r = 1 - P_m \tag{13.21}$$

Bond conformations averaged over all possible stereosequences can be calculated as follows. For example, the trans fraction of bond a of the ith repeating unit is

$$p_{\mathrm{t;a};i} = Z^{-1} J^* \left(\prod_{h=1}^{i-1} W_h \right) W_{\mathrm{t;a}} \left(\prod_{h=i+1}^{x} W_h \right) J \tag{13.22}$$

where

$$W_{\mathrm{t;a}} = \begin{bmatrix} V_{\mathrm{t;a}}^{ll} & V_{\mathrm{t;a}}^{ld} \\ V_{\mathrm{t;a}}^{dl} & V_{\mathrm{t;a}}^{dd} \end{bmatrix} \tag{13.23}$$

with $V_{t;a}^\alpha$ being

$$V_{\mathrm{t;a}}^\alpha = U_{\mathrm{t;a}}^\alpha\ U_\mathrm{b}^\alpha\ U_\mathrm{c}^\alpha\ (\cdots) \tag{13.24}$$

In $U_{\mathrm{t;a}}^\alpha$, the columns corresponding to the trans state are equal to those of U_a^α, and the other elements are null. Bond conformations of the other bonds can be calculated similarly.

13.6 Characteristic Ratio

The mean-square end-to-end distance of each chain (mth chains) can be calculated from

$$\langle r^2 \rangle_{0,m} = 2z_m^{-1} J^* \left(\prod_{j=1}^{n} G_j \right)_m J^{**} \tag{13.25}$$

where J^{**} is the row matrix whose last k_n elements are unity, and the others are null (k_n depends on the size of G_n), and z_m is the partition function of the mth chain, being obtained from

$$z_m = J^* \left(\prod_{j=2}^{n-1} U_j^{\alpha} \right)_m J \tag{13.26}$$

and n is the number of skeletal bonds. The G_j matrix of the jth bond is defined as

$$G_j = \begin{bmatrix} U_j & (U_j \otimes \bar{l}_j^{T}) \|T\|_j & (l_j^2/2) U_j \\ 0 & (U_j \otimes I_3) \|T\|_j & U_j \otimes \bar{l}_j \\ 0 & 0 & U_j \end{bmatrix} \tag{13.27}$$

where $\bar{l}_j$ is the bond vector defined in Eq. (11.29). The $\|T\|_j$ matrix is defined as

$$\|T\|_j = \begin{cases} \tau_j & \text{for } k = 3 \\ I_3 \otimes \tau_j & \text{for } k = 9 \\ I_3 \otimes I_3 \otimes \tau_j & \text{for } k = 27 \end{cases} \tag{13.28}$$

where

$$\tau_j = \begin{bmatrix} T_{\mathrm{t};j} & 0 & 0 \\ 0 & T_{\mathrm{g}^+;j} & 0 \\ 0 & 0 & T_{\mathrm{g}^-;j} \end{bmatrix} \tag{13.29}$$

The T matrix transforms a vector from the jth to $(j-1)$th frame of reference. For example, $T_{t;j}$ is expressed as

$$T_{\mathrm{t};j} = \begin{bmatrix} \cos \vartheta_{\mathrm{t};j} & \sin \vartheta_{\mathrm{t};j} & 0 \\ \sin \vartheta_{\mathrm{t};j} \cos \phi_{\mathrm{t};j} & -\cos \vartheta_{\mathrm{t};j} \cos \phi_{\mathrm{t};j} & \sin \phi_{\mathrm{t};j} \\ \sin \vartheta_{\mathrm{t};j} \sin \phi_{\mathrm{t};j} & -\cos \vartheta_{\mathrm{t};j} \sin \phi_{\mathrm{t};j} & -\cos \phi_{\mathrm{t};j} \end{bmatrix} \tag{13.30}$$

where $\vartheta_{t;j}$ is the supplement of the bond angle, and $\phi_{t;j}$ is the dihedral angle for the trans state of the jth bond.

The configurational sequence of PEI can be chosen according to the algorithm of the Monte Carlo chain to satisfy the P_m (P_r) value obtained from Eq. (13.20). In the IRIS scheme, P_m cannot be selected arbitrarily but will be determined from the conformational energies. The $H_i^{\alpha_i}$ matrix of the ith unit is defined for each configuration α_i as

$$H_i^{\alpha_i} = \begin{cases} G_2^{\alpha_i}\ G_3^{\alpha_i}\ G_4^{\alpha_i}\ (\cdots) \\ G_a^{\alpha_i}\ G_b^{\alpha_i}\ G_c^{\alpha_i}\ (\cdots) \end{cases} \tag{13.31}$$

The mean-square end-to-end distance of the mth chain in the system can be calculated from

$$\langle r^2 \rangle_{0,m} = 2Z_m^{-1} J^* G_1 \left(\prod_{i=1}^{x} H_i^{\alpha_i} \right)_m G_n\, J^{**} \tag{13.32}$$

Here, the $H_i^{\alpha_i}$ matrices are arranged as determined by the Monte Carlo method, that is repeated Bernoulli trials [471], that will be explained in Chapter 14. The ensemble average of $\langle r^2 \rangle_{0,m}$s is given by (weight average)

$$\langle r^2 \rangle_0 = \frac{\sum_{m=1}^{n_c} \langle r^2 \rangle_{0,m} Z_m}{\sum_{m=1}^{n_c} Z_m} \tag{13.33}$$

where n_c is the total number of chains included in the system. When the sampling number ($x \times n_c$) is large enough, the value is satisfactorily approximated by (number average)

$$\langle r^2 \rangle_0 = \frac{\sum_{m=1}^{n_c} \langle r^2 \rangle_m}{n_c} \tag{13.34}$$

The characteristic ratio can be obtained from

$$\frac{\langle r^2 \rangle_0}{nl^2} = \frac{\sum_{m=1}^{n_c} \left[\frac{\langle r^2 \rangle_{0,m}}{\left(\sum_{j=1}^{n} l_j^2 \right)_m} Z_m \right]}{\sum_{m=1}^{n_c} Z_m} \approx \frac{\sum_{m=1}^{n_c} \left[\frac{\langle r^2 \rangle_{0,m}}{\left(\sum_{j=1}^{n} l_j^2 \right)_m} \right]}{n_c} \tag{13.35}$$

13.7 Orientational Correlation Between Bonds

In general, the characteristic ratio increases monotonously with increasing degree of polymerization (x); therefore, when the characteristic ratio is plotted against x^{-1}, the extrapolated curve will show a negative slope at $x^{-1} = 0$. However, exceptionally, PEI (HBS = 100%) exhibits a positive slope as shown in Figure 13.4.

Mattice et al. [306] have expressed the slope at $x^{-1} = 0$ as

$$\left(\frac{d\langle r^2 \rangle_0 / nl^2}{dx^{-1}} \right)_{x^{-1}=0} = -2 \sum_{k=1}^{\infty} k \langle \mathbf{u}_j \bullet \mathbf{u}_{j+k} \rangle_0 \tag{13.36}$$

where $\mathbf{u}_j$ is the unit vector along the jth bond. The scalar product between the unit vectors, $\mathbf{u}_j$ and $\mathbf{u}_{j+k}$, can be averaged over all possible conformations (e.g. n_1 = a − c and $n_b = 3$ for PEI) and all configurations of the n_c chains according to

$$\overline{\langle \mathbf{u}_j \bullet \mathbf{u}_{j+k} \rangle_0} = \frac{1}{n_b\, n_c} \sum_{n_1=a}^{c} \sum_{n_2=1}^{n_c} \langle T_j T_{j+1} \cdots T_{j+k-1} \rangle_{11}^{n_1,n_2} \tag{13.37}$$

where the overbar stands for averaging over bonds and configurations, the angular brackets represent the average for the conformation, and $\langle T_j T_{j+1} \cdots T_{j+k-1} \rangle_{11}^{n_1,n_2}$ corresponds to the (1,1) element of $\langle T_j T_{j+1} \cdots T_{j+k-1} \rangle^{n_1,n_2}$ calculated from [306]

$$\langle T_j T_{j+1} \cdots T_{j+k-1} \rangle^{n_1,n_2} = z_{n_2}^{-1} [(J^* U_1 \cdots U_{j-1}) \otimes I_3] \times$$
$$[(U_j \otimes I_3) \| T_j \| \cdots \times (U_{j+k-1} \otimes I_3) \| T_{j+k-1} \|] [(U_{j+k} \cdots U_n) \otimes I_3] \tag{13.38}$$

Here, T_j is the transformation matrix from the jth to $(j-1)$th frame of reference. The partition function of the n_2th chain, z_{n_2}, is given by

$$z_{n_2} = J^* \left(\prod_{j=2}^{n-1} U_j \right)_{n_2} J \tag{13.39}$$

where U_js are the statistical weight matrices arranged in accordance with the configurational sequence of the n_2th chain. In general, $\overline{\langle \mathbf{u}_j \bullet \mathbf{u}_{j+k} \rangle_0}$ decreases rapidly with increasing k; therefore, an amplification factor k has been introduced to detect up to the long-range correlation. Because $\mathbf{u}_j \bullet \mathbf{u}_{j+k} = \cos\theta_{j,\,j+k}$ ($\theta_{j,\,j+k}$: angle between $\mathbf{u}_j$ and $\mathbf{u}_{j+k}$), the term can also be expressed as

$$k\overline{\langle \mathbf{u}_j \bullet \mathbf{u}_{j+k} \rangle_0} = k\overline{\langle \cos\theta_{j,j+k} \rangle_0} \tag{13.40}$$

Accordingly, the $k\overline{\langle \mathbf{u}_j \bullet \mathbf{u}_{j+k} \rangle_0}$ term quantifies the orientation correlation between bonds j and $j + k$.

Figure 13.5 shows $k\overline{\langle \mathbf{u}_j \bullet \mathbf{u}_{j+k} \rangle_0}$ vs. k plots of PEI and an alternating copolymer, poly(ethylene imine-*alt*-ethylene oxide) (P(EI-EO)) [408, 409]. In Figure 13.5a, the ninth ($k = 9$) bond exhibits the minimum, and the 18th ($k = 18$) bond shows the second maximum, which suggests that helical paths of about 18 bonds per turn are formed owing to intramolecular N—H···N hydrogen bonds as illustrated. Figure 13.5b expresses helices of about 20 bonds per turn of P(EI-EO), and the peak around $k = 20$ of P(EI-EO) is higher than the second peak of PEI, which represents that the P(EI-EO) helix is more stable than the PEI one. This is because the N—H···O hydrogen bond of P(EI-EO) is stronger than N—H···N of PEI. The percentage in the figure expresses the hydrogen-bond strength (HBS): 100%, hydrogen-bond energy as given by the MO calculations; 0%, null hydrogen-bond energy [403]. As HBS decreases, the oscillation amplitude becomes small; that is, the bond correlation becomes weak. At HBS = 0%, the dotted curve decays monotonously with increasing k; the correlation completely disappears.

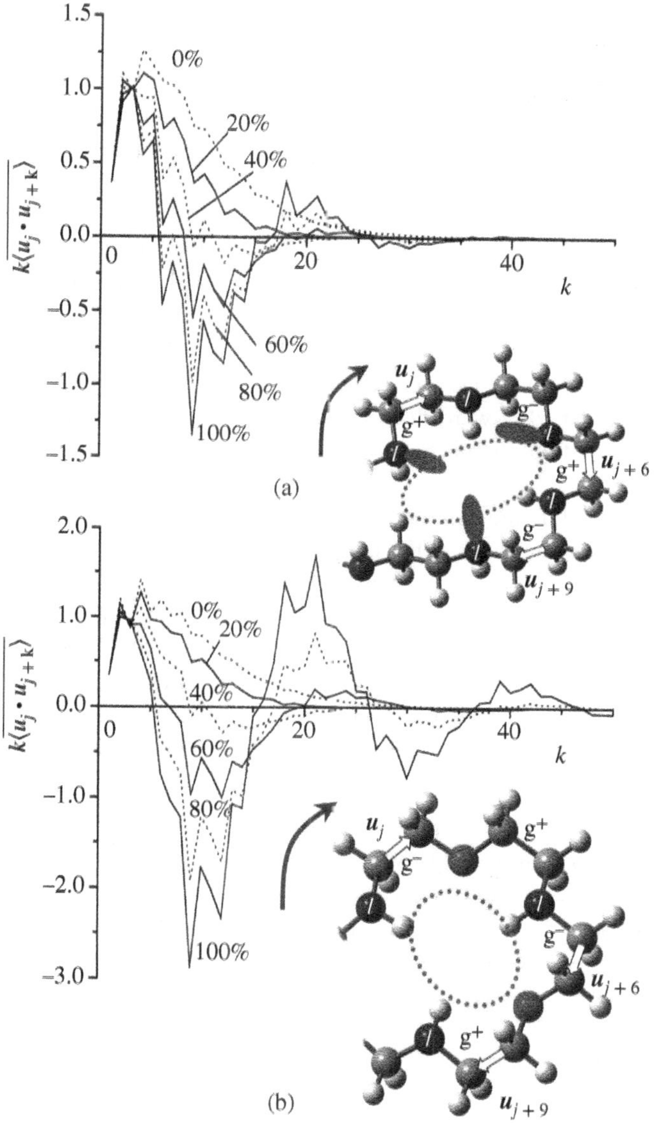

Figure 13.5 $k\overline{\langle \mathbf{u}_j \bullet \mathbf{u}_{j+k} \rangle_0}$ vs. k plots of (a) PEI and (b) P(EI-EO). The percentage represents the hydrogen bond strength (HBS). Source: Sasanuma et al. [409], Figures 10 & 11 (p.6759)/Reproduced with permission from American Chemical Society.

13.8 Solubility of Polyamines

PEI is soluble in hot water and acidic water but insoluble in cold water, whereas poly(*N*-methylethylene imine) (PMEI, $[—CH_2CH_2N(CH_3)—]_x$) is soluble even in cold water; thus, PMEI is more hydrophilic than PEI [164]. These experimental facts may be inconsistent with intuitive understanding because the methyl group seems to be more organic (hydrophobic) than the hydrogen atom. However, the solubilities of PEI and PMEI can be interpreted in terms of the configurational entropy (S_{conf}). Table 13.2 shows the S_{conf} values of PEI, PMEI, poly(ethylene oxide) (PEO) [412, 413], and isotactic poly(propylene oxide) (PPO) [398]. The partition functions of PEI and PMEI were calculated according to the IRIS scheme, while those of PEO and PPO were derived from the RIS calculations. Conformational energies of PEO were determined for the gas phase, organic solutions, and aqueous solution; therefore, three S_{conf} values of PEO are shown in Table 13.2.

The S_{conf} value corresponds to the difference in configurational entropy between the unperturbed and crystalline states of the polymeric chain. The polymer with a large S_{conf} value is easy to dissolve in solvents because $-S_{conf}\,T$ is a measure of the Gibbs free energy in dissolution (mixing free energy). PEI of HBS = 100% (4.5 cal K^{-1} mol^{-1}) is smaller in S_{conf} than that of HBS = 0% (6.3 cal K^{-1} mol^{-1}) because the intramolecular N—H···N hydrogen bond disturbs the dissolution. In place of the hydrogen atom, PMEI has a methyl group at the nitrogen site, and hence, its lone pair is free from the intramolecular hydrogen bond and can make contact with the surrounding solvents. When the $CH_2—CH_2$ bond of PEO

Table 13.2 Configurational entropies of PEI, PMEI, PEO, and isotactic PPO at 25 °C.[a)]

Polymer	Condition	S_{conf} (cal K^{-1} mol^{-1})	TS_{conf} (kcal mol^{-1})
PEI	HBS = 100%	4.54	1.35
	HBS = 0%	6.26	1.87
PMEI	As predicted by MO calc	5.81	1.73
PEO	Gas phase	5.00	1.49
	Organic solvent	4.80	1.43
	Water	4.25	1.27
Isotactic PPO	As predicted by MO calc	3.80	1.13

a) Abbreviations: HBS, hydrogen bond strength; MO, molecular orbital; PEI, poly(ethylene imine); PMEI, poly(*N*-methylethylene imine); PEO, poly(ethylene oxide); PPO, poly(propylene oxide)

Source: Adapted from [164].

adopts a gauche conformation, lone pairs of the oxygen atoms can interact with the solvents as found for crown ethers. The intramolecular C—H···O hydrogen bonds of polyethers are so weak as to switch to intermolecular interactions with solvents [412, 413]. Therefore, PEO is readily soluble in many organic solvents and even in water. PPO has the methyl side chain similar to PMEI; however, PPO is insoluble in water, and its methyl group plays the role of a hydrophobic group. The hydrophilicity and high solubility of PMEI are obviously due to the nitrogen inversion that augments the configurational entropy and adjusts the stereochemical arrangement to get in contact with solvents.

14

RIS Scheme Combined with Stochastic Process

Polymeric chains with an asymmetric carbon (*R* or *S*) or a pseudoasymmetric carbon (*d* or *l*) in the repeating unit may include a variety of stereosequences, thus being atactic or showing well-defined stereoregularities such as isotacticity and syndiotacticity. Furthermore, during polymerization, polymeric chains may form different regiosequences composed of head-to-head (H–H), head-to-tail (H–T), and tail-to-tail (T–T) linkages. Stereo- and regiosequences of a polymeric chain can be virtually generated via stochastic processes: an atactic chain is generated by repeating Bernoulli trials, and isotactic and syndiotactic chains can be arranged according to the Markov chain [471].

The atactic chain will be assembled via random-number generator of computer software. Some modules of computer languages (for example, *random_number*, a subroutine of FORTRAN) provide nearly uniform random numbers distributed between zero and unity. If the given random number, *ran*, is smaller than the predefined *R*-form probability, P_R, then the repeating unit will be set as an *R*-form; otherwise, it will be an *S*-form because the *S*-form probability, P_S, is equal to $1 - P_R$. This operation will be repeated x times to complete a single chain, where x is the degree of polymerization. If the system is composed of n_c polymeric chains, the above process will be repeated n_c times. The total number of trials, $x \times n_c$, must be large enough to satisfy $n_R/(n_R + n_S) \cong P_R$, where n_R and n_S are the *R*- and *S*-event numbers, respectively: $n_R + n_S = x \times n_c$. The configurational properties and thermodynamic quantities are calculated according to the rotational isomeric state (RIS) or inversional-rotational isomeric state (IRIS) scheme for the individual chains and averaged over the whole chains to yield the mean values.

On the other hand, because isotactic and syndiotactic chains have strong correlations between the neighboring repeating units (diad), the assumption of independent trial of the Bernoulli trial is invalid. Instead, therefore, the Markov chain is adopted to generate such regular stereosequences. The probability (P_R^m or P_S^m) at

Conformational Analysis of Polymers: Methods and Techniques for Structure-Property Relationships and Molecular Design, First Edition. Yuji Sasanuma.

which the mth unit will be the R- or S-form can be derived from

$$(P_R^m\ P_S^m) = (P_R^1\ P_S^1)\ \mathbf{P}^{m-1} \tag{14.1}$$

where

$$\mathbf{P} = \begin{bmatrix} P_{RR} & P_{RS} \\ P_{SR} & P_{SS} \end{bmatrix} \tag{14.2}$$

This **P** matrix is termed transition probability matrix, and P_{RR} and P_{RS} are probabilities that R and S forms will appear immediately after the R-form, respectively. By definition, $P_{RR}+ P_{RS} = 1$, $P_R^1 + P_S^1 = 1$, and $P_R^m+ P_S^m = 1$. Similarly, P_{SR} (S to R) and P_{SS} (S to S) are defined and fulfill $P_{SR} + P_{SS} = 1$. When m is fully large, the stationary distribution may be assumed:

$$(P_R\ P_S) = (P_R\ P_S)\ \mathbf{P} \tag{14.3}$$

and

$$P_R + P_S = 1 \tag{14.4}$$

Equations (14.3) and (14.4) lead to

$$P_R = \frac{P_{SR}}{P_{RS} + P_{SR}} \tag{14.5}$$

and

$$P_S = \frac{P_{RS}}{P_{RS} + P_{SR}} \tag{14.6}$$

Therefore, when the number of trials is fully large, the P_R and P_S values are independent of the initial P_R^1 and P_S^1 and depend only on the transition probabilities.

In the numerical calculations, the initial P_R^1 and P_S^1 are provided, and a random number, *ran*, is generated; if $ran \le P_R^1$, then the first unit will be R; otherwise, it will be S. For the subsequent units, the following operation will be repeated. When the preceding unit was R, and a random number *ran* is generated, if $ran \le P_{RR}$, then the current unit will be R, otherwise, S. When the preceding unit was S, if $ran \le P_{SS}$, the current unit will be S, otherwise, R. The operation will be repeated x times for a single chain, and the whole process will be repeated n_c times. The configurational properties and thermodynamic quantities are calculated for the individual chains including x repeating units and averaged over the n_c chains to yield the mean values.

The accuracy of averaging via the random-number generation was investigated as a function of the number of trials ($x \times n_c$), where the degree (x) of polymerization was chosen out of 5, 25, 75, 150, and 300, and the number (n_c) of chains was set to 1, 5, 25, 75, 150, or 300. The initial stochastic parameters were set as follows: $P_S = P_R = 1/2$ (Bernoulli); $P_{iso} = P_{SS} = P_{RR} = 1/2$, $p_{syn} = P_{SR} = P_{RS} = 1/2$, and $P_S^1 = P_R^1 = 1/2$ (Markov). As the trial is repeated, the Markov chain must approach the stationary distribution as predicted from Eqs. (14.5) and (14.6), i.e. $P_S = P_R = 1/2$. Here, a priori probabilities are expressed by P, and a posteriori data (results from the simulations) by f. In Figure 14.1, the averaged fractions of (S)-units (f_S) and *iso*-diads (f_{iso}) are plotted as the coordinates (f_S, f_{iso}) in the range of $0.40 \leq f_S$, $f_{iso} \leq 0.60$. When the $x \times n_c$ values are small, the markers are scattered, and some data are located out of the range. However, as the number of trials increases, f_S and f_{iso} approach 0.50, that is a priori P_S and P_{iso}, respectively. The markers of n_c = 75, 150, and 300 are close to or overlapped at the center (0.5, 0.5); it has been proven that the Bernoulli and Markov processes based on the random-number generation fulfill the theoretical prediction. When $x = n_c = 300$ in particular, the (f_S, f_{iso}) point is located exactly at the center of Figure 14.1a,b. In the RIS calculations, the degree of polymerization of 300 is large enough for us to consider the polymeric chain almost infinite in length. Probably, $x = n_c = 200$ would also be fully acceptable.

Figure 14.2 shows the characteristic ratio ($\langle r^2 \rangle_0/nl^2$) of poly(propylene carbonate) (PPC) calculated by the refined RIS scheme with the Markov stochastic process as a function of P_{meso} (*meso*-diad probability) and $P_{\text{H-T}}$ (head-to-tail probability) under $P_{meso} = P_{RR} = P_{SS}$ [415]. It can be seen that the characteristic ratio depends only a little on both stereo- and regioregularities. This is a characteristic feature of PPC as will be discussed in Chapter 25.

Figure 14.4a shows the characteristic ratio of atactic poly(2-hydoxybutyrate) (P2HB, Figure 14.3) as a function of the probability (P_S) of (S)-monomeric unit [422]. The curve is symmetric with respect to $P_S = 0.5$. Figure 14.4b shows the P_{meso} dependence of the characteristic ratio of P2HB. The isotactic chain ($P_{meso} = 1$) has a relatively large $\langle r^2 \rangle_0/nl^2$ value of 6.44, whereas that of syndiotactic P2HB ($P_{meso} = 0$) is close to unity. In the syndiotactic poly((S)-2HB-*alt*-(R)-2HB) chain, dihedral angles of the backbone of (R)-unit are exactly opposite in sign to those of the (S)-unit, and consequently, the chain dimension will be reduced considerably.

14.1 Polymeric Chains with Internally Rotatable Side Chains

Poly(2-hydroxybutyrate) has an ethyl side chain whose C*H—CH_2 bond (designated as bond d, see Figure 14.3) also has a rotational degree of freedom, and hence, its rotation interferes largely with conformational distributions of the main chain. The asterisk indicates the asymmetric carbon; therefore, the repeating unit of P2HB is either an R or an S enantiomer. Here, the S optical isomer is treated mainly as the model. Bond a is fixed at the trans form; bond b can adopt three rotamers, namely trans, gauche$^+$, and gauche$^-$ conformations; and bond c lies

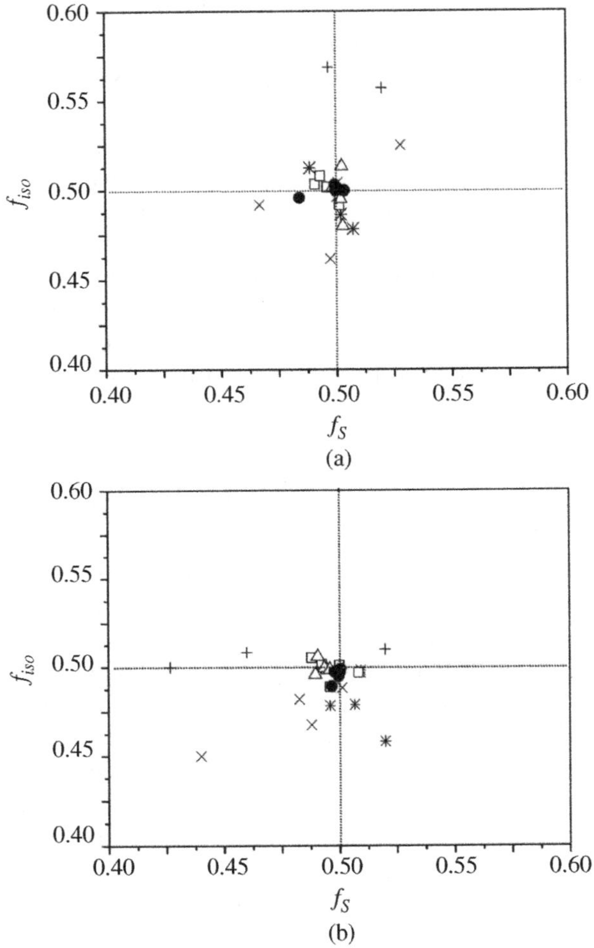

Figure 14.1 f_{iso} vs. f_S plots of (a) the Bernoulli and (b) Markov chains. The parameters were initially set as (a) $P_S = P_R = 1/2$ and (b) $P_{iso} = P_{SS} = P_{RR} = 1/2$, $P_{syn} = P_{SR} = P_{RS} = 1/2$, and $P_S^1 = P_R^1 = 1/2$, and the pseudorandom number was repeatedly generated. The degree (x) of polymerization was 5, 25, 75, 150, or 300, and the number (n_c) of chains was 1 (cross), 5 (saltire), 25 (asterisk), 75 (square), 150 (triangle), or 300 (filled circle). (a) If the Bernoulli chains are fully random or (b) if the infinite Markov-chain approximation ($x \times n_c \approx \infty$) is valid, then the marker is located at the center (0.50, 0.50). Source: Sasanuma and Touge [418], Figure 6 (p.1907)/Reproduced with permission of Elsevier.

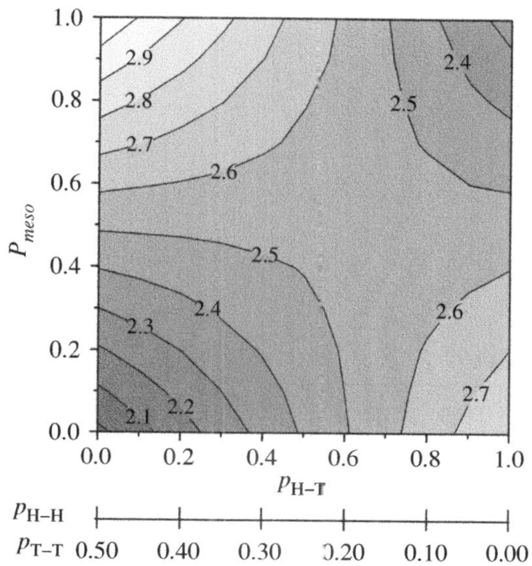

Figure 14.2 Contour map of the characteristic ratio of PPC with different regio- (P_{H-T}) and stereosequences (p_{meso}), calculated from the refined RIS scheme with the Markov stochastic process. $p_{meso} = 0$ and 1 correspond to syndio- and isotacticity, respectively. The numerical value written in the map represents the $\langle r^2 \rangle_0/nl^2$ ratio. Source: Sasanuma and Takahashi [415], Figure 8 (p.4815)/with permission of American Chemical Society.

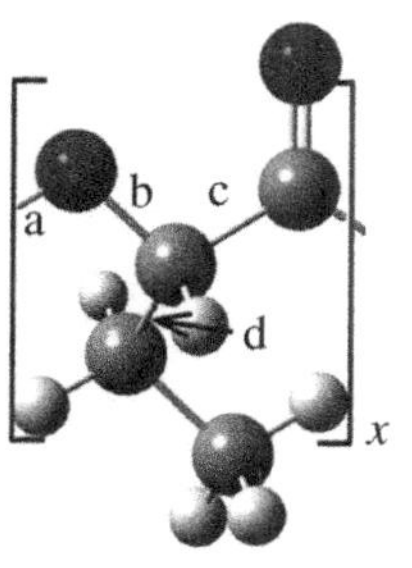

Figure 14.3 Poly(2-hydroxybutyrate) (P2HB). The bonds are designated as shown.

in one of trans, synperiplanar, and gauche$^-$ states [422]. Bond d of the ethyl side chain also forms trans, gauche$^+$, and gauche$^-$ conformations. The conformational energy of the repeating unit is also affected by the rotation of bond d. For example, the tg$^+$t(g$^+$) conformation, in which bonds a, b, c, and d adopt t, g$^+$, t, and g$^+$ (in the parenthesis) states, respectively, is more stable by about –0.8 kcal mol^{-1} than

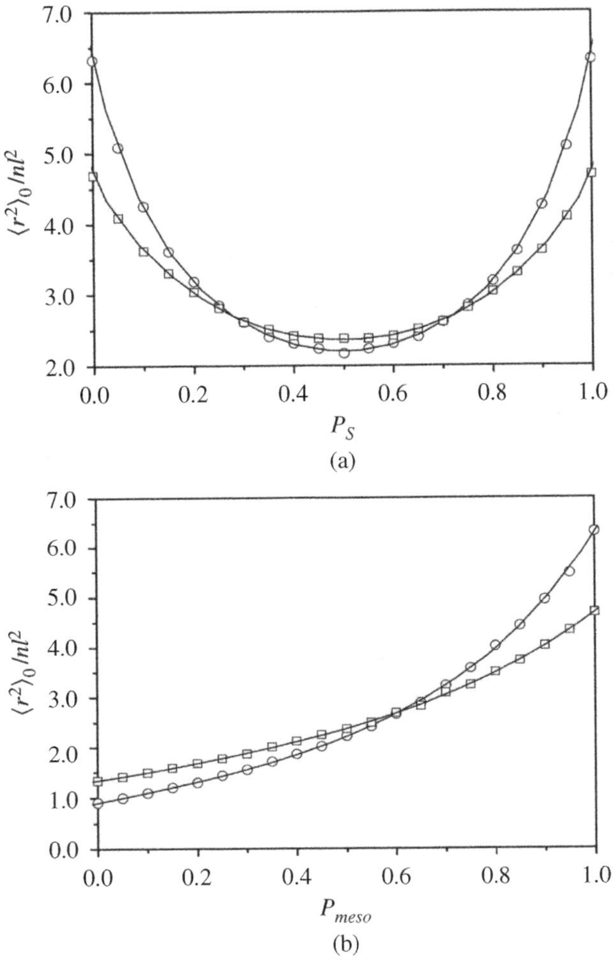

Figure 14.4 (a) Characteristic ratios of P2HB at 25 °C, calculated (a) as a function of the probability (P_S) of (*S*)-monomeric unit of atactic chains generated by Bernoulli trials or (b) as a function of the *meso*-diad probability (P_{meso}) based on the Markov stochastic process. $P_{meso} = 0$ and 1 correspond to syndio- and isotacticity, respectively. Two sets of energy parameters were used: (circle) without solvent; (square) in DMSO. Source: Sasanuma et al. [422], Figures 6 and 7 (p. 3738)/Reproduced with permission of American Chemical Society.

tg$^+$t(t). Accordingly, to accurately evaluate the configurational properties of P2HB, the side-chain effect must be integrated into the RIS calculations.

Statistical weight matrices of P2HB are formulated and explained here [422]. The U_{a1} matrix of the first repeating unit can be expressed in the 3 × 3 form:

$$U_{a1} = C_1 \otimes R_1 \tag{14.7}$$

and that of bond b1 is also a 3 × 3 matrix,

$$U_{b1} = C_1 \otimes R_3 \tag{14.8}$$

where ⊗ stands for the direct product,

$$C_1 = \begin{bmatrix} 1 \\ 0 \\ 0 \end{bmatrix} \tag{14.9}$$

$$R_1 = \begin{bmatrix} 1 & 0 & 0 \end{bmatrix} \tag{14.10}$$

and

$$R_3 = \begin{bmatrix} 1 & 1 & 1 \end{bmatrix} \tag{14.11}$$

The U_{c1} matrix (U_{c1}^S) of the (S)-form, including the weights of the conformations of bond d as well, must be extended to a 3 × 27 matrix containing the following elements:

(1,1)	(1,2)	(1,3)	(1,4)	(1,5)	(1,6)	(1,7)	(1,8)	(1,9)
$u_{ttt(t)}$	$u_{ttt(g^+)}$	$u_{ttt(g^-)}$	$u_{tts(t)}$	$u_{tts(g^+)}$	$u_{tts(g^-)}$	$u_{ttg^-(t)}$	$u_{ttg^-(g^+)}$	$u_{ttg^-(g^-)}$
(2,10)	(2,11)	(2,12)	(2,13)	(2,14)	(2,15)	(2,16)	(2,17)	(2,18)
$u_{tg^+t(t)}$	$u_{tg^+t(g^+)}$	$u_{tg^+t(g^-)}$	$u_{tg^+s(t)}$	$u_{tg^+s(g^+)}$	$u_{tg^+s(g^-)}$	$u_{tg^+g^-(t)}$	$u_{tg^+g^-(g^+)}$	$u_{tg^+g^-(g^-)}$
(3,19)	(3,20)	(3,21)	(3,22)	(3,23)	(3,24)	(3,25)	(3,26)	(3,27)
$u_{tg^-t(t)}$	$u_{tg^-t(g^+)}$	$u_{tg^-t(g^-)}$	$u_{tg^-s(t)}$	$u_{tg^-s(g^+)}$	$u_{tg^-s(g^-)}$	$u_{tg^-g^-(t)}$	$u_{tg^-g^-(g^+)}$	$u_{tg^-g^-(g^-)}$

where (l, m) (l = 1–3 and m = 1 – 27) represents the position in the matrix, and, for example

$$u_{tg^+s(g^-)} = \exp(-\Delta E_{tg^+s(g^-)}/RT) \tag{14.12}$$

is the statistical weight of the tg$^+$s(g$^-$) conformation, and $\Delta E_{tg^+s(g^-)}$ is its conformational energy. The other elements of U_{c1}^S are set equal to zero.

The U_{c1} matrix of the (R)-form, U_{c1}^R, can be derived from U_{c1}^S according to

$$U_{c1}^R = Q_3 U_{c1}^S (Q_3 \otimes I_3 \otimes Q_3) \tag{14.13}$$

where the transformation matrix Q_3 and the identity matrix I_3 are as defined previously. The statistical weight matrices of the subsequent repeating units can be formulated as follows:

$$U_a = C_3 \otimes I_3 \otimes I_3 \tag{14.14}$$

$$U_b = C_3 \otimes C_3 \otimes R_3 \tag{14.15}$$

$$U_c = U_{c1} \tag{14.16}$$

where

$$C_3 = \begin{bmatrix} 1 \\ 1 \\ 1 \end{bmatrix} \tag{14.17}$$

Table 14.1 Bond conformations of bonds a–d and c (d) of isotactic (*S*)-P2HB

Medium	Temp[a)] (°C)	Conformation t	g^+ (s)	g^-
		Bond b		
Gas	25.0	0.00	0.98	0.02
	130.3	0.02	0.94	0.04
DMSO	25.0	0.01	0.98	0.01
	130.3	0.01	0.95	0.04
		Bond c		
Gas	25.0	0.80	0.18	0.02
	130.3	0.71	0.24	0.05
DMSO	25.0	0.73	0.26	0.01
	130.3	0.67	0.30	0.03
		Bond d (side chain)		
Gas	25.0	0.14	0.46	0.40
	130.3	0.20	0.43	0.37
DMSO	25.0	0.18	0.45	0.37
	130.3	0.22	0.43	0.35

Medium	Temp (°C)	Conformations of bonds c and (d) t(t)	t(g^+)	t(g^-)	s(t)	s(g^+)	s(g^-)	g^-(t)	g^-(g^+)	g^-(g^-)
Gas	25.0	0.10	0.37	0.33	0.03	0.08	0.07	0.01	0.01	0.00
	130.3	0.12	0.31	0.28	0.05	0.10	0.09	0.03	0.02	0.00
DMSO	25.0	0.13	0.32	0.28	0.05	0.12	0.09	0.00	0.01	0.00
	130.3	0.14	0.28	0.25	0.07	0.13	0.10	0.01	0.02	0.00

a) The equilibrium melting point is 130.3 °C.
Source: Sasanuma et al. [422], Table 9 (p.3737)/Reproduced with permission of American Chemical Society.

The geometrical parameters of the (*R*)-form can be obtained from those of the (*S*)-form according to the correspondence relations shown below. Between the paired forms, the bond length and bond angle are identical, whereas the dihedral angles (ϕs) are the same in magnitude but opposite in sign: $\phi_R = -\phi_S$.

bond a									
(*S*)-form:	t								
(*R*)-form:	t								
bond b									
(*S*)-form:	t	g^+	g^-						
(*R*)-form:	t	g^-	g^+						
bonds c(d)									
(*S*)-form:	t(t)	t(g^+)	t(g^-)	s(t)	s(g^+)	s(g^-)	g^-(t)	g^-(g^+)	g^-(g^-)
(*R*)-form:	t(t)	t(g^-)	t(g^+)	s(t)	s(g^-)	s(g^+)	g^+(t)	g^+(g^-)	g^+(g^+)

Bond conformations can be calculated in the manner similar to that used in the conventional RIS scheme. The columns corresponding to the target conformation (ξ) of the statistical weight matrix (U_j) are kept as they are, the other columns are filled with zero to yield the $U'_{\xi;j}$ matrix, and the other statistical weight matrices are left as they are. The statistical weight matrices thus prepared are multiplied consecutively along the polymeric chain from one end to the other. The ξ probability at the *j*th bond, $p_{\xi;\,j}$, can be calculated from

$$p_{\xi;\,j} = Z^{-1}J^* \left(\prod_{h=2}^{j-1} U_h\right) U'_{\xi;j} \left(\prod_{h=j+1}^{n-1} U_h\right) J \tag{14.18}$$

where *n* is the number of bonds in the main chain, and *Z* is the partition function of the whole chain. Bond conformations of bond d of P2HB can also evaluated similarly: the columns for the target conformation (ξ) expressed in the parentheses are left as they are. For example, the trans fraction of bond d can be calculated from Eq. (14.18) with the (1,1), (1,4), (1,7), (2,10), (2,13), (2,16), (3,19), (3,22), and (3,25) elements of U_c left as they are, and the others set equal to zero. The bond conformations of bonds a–d and c (d) of isotactic (*S*)-P2HB, thus obtained, are shown in Table 14.1.

Part IV

Experimental Methods

Conformational Analysis of Polymers: Methods and Techniques for Structure-Property Relationships and Molecular Design, First Edition. Yuji Sasanuma.

15

Nuclear Magnetic Resonance (NMR)

In Chapter 9, the fundamentals of nuclear magnetic resonance (NMR) spectroscopy and quantum chemical calculations on NMR parameters were already described. Here, two examples of the application of NMR to conformational analysis of polymers are introduced: (1) Conformational analysis of isotactic poly(propylene oxide) (PPO), in which conformational characteristics and configurational properties of isotactic PPO were investigated via the dihedral-angle dependence of the vicinal ^1H–^1H coupling constants, ab initio molecular orbital calculations on its model compound, and rotational isomeric state (RIS) calculations on PPO [397, 398] and (2) Simulations of carbon-13 NMR chemical shifts of propylene oxide (PO) dimers, in which γ- and δ-substituent effects on ^{13}C NMR chemical shifts were used to determine their configurations and conformational energies [405].

15.1 Conformational Analysis of Isotactic Poly(propylene oxide)

In general, aliphatic ethers prefer a gauche conformation in the C—C bond adjacent to the O—C bond, and this phenomenon has been termed the attractive gauche effect [226, 227]. Poly(ethylene oxide) (PEO, $[—CH_2—CH_2—O—]_x$) may be a simple polyether including such a bond sequence. Conformational energies of PEO were determined so as to attain satisfactory agreement between the calculated and observed unperturbed characteristic ratios and dipole moment ratios [7, 300, 301]. The gauche state of the C—C bond was suggested to be more stable by 0.4–0.5 kcal mol^{-1} than the trans form. From the analysis of the NMR vicinal coupling constants observed from 1,2-dimethoxyethane (DME, $CH_3OCH_2CH_2OCH_3$), a monomeric model compound of PEO, it was suggested that the conformational energy (E_σ), that is, the gauche energy, depending largely

Conformational Analysis of Polymers: Methods and Techniques for Structure-Property Relationships and Molecular Design, First Edition. Yuji Sasanuma.

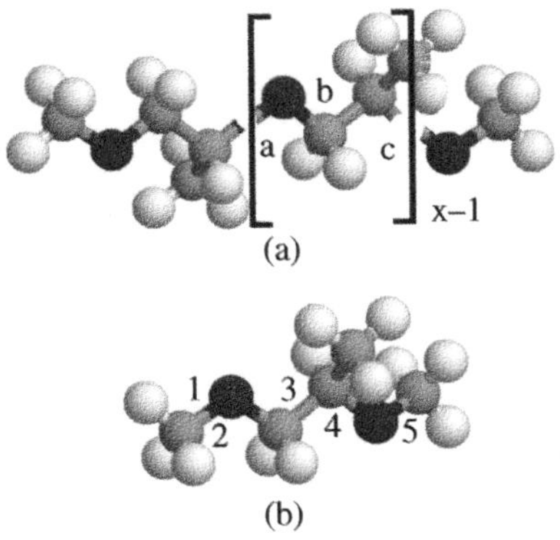

Figure 15.1 (a) Isotactic poly((*R*)-propylene oxide) dimethyl ether and (b) its monomeric model compound (*R*)-1,2,-dimethoxypropane (DMP) in the all-trans conformation. As indicated, the skeletal bonds are labeled. *x* stands for the degree of polymerization. Source: [399], Figure 1 (p.219)/Reproduced with permission of Elsevier.

on the medium, varies from −0.4 kcal mol^{-1} (in the gas phase) to −1.2 kcal mol^{-1} (in water) [216, 469].

The attractive gauche effect was also found for PPO [4, 198, 345], which has an asymmetric methine carbon atom (C*) in the repeating unit and hence two stereochemical arrangements, namely, (*R*)- and (*S*)-optical isomers. Herein, (*R*)-isomers are treated exclusively as models for isotactic PPO and its model compounds (Figure 15.1). This is because (*R*)- and (*S*)-isomers show identical NMR spectra, and descriptions about the (*R*)-isomer are also valid for the (*S*)-one. Owing to the chirality, two gauche states, g$^+$ and g$^-$, of the O—C, C—C*, and C*–O bonds of the repeating unit are nonequivalent; therefore, a number of energy parameters must be defined to represent the conformations of PPO.

15.1.1 ^{1}H NMR Vicinal Coupling Constant

Bond conformations, that is, fractions of conformations, of the C—C* bond of DMP were determined from the vicinal coupling constants between protons A and C ($^3J_{AC}$) and B and C ($^3J_{BC}$) (see Figure 15.2). According to Gutowsky's method [180], the observed $^3J_{AC}$ and $^3J_{BC}$ values are related to the bond conformations by

$$^3J_{AC} = J_T p_t + J_G p_{g^+} + J_G p_{g^-} \tag{15.1}$$

and

$$^3J_{BC} = J_G p_t + J_T p_{g^+} + J_G p_{g^-} \tag{15.2}$$

where ^3J_T and ^3J_G are, respectively, the vicinal coupling constants between the protons arranged in the trans (antiperiplanar) and gauche (synclinal) positions, and p_η is the fraction of conformation η (η = t, g$^+$, or g$^-$). By definition, it follows that

$$p_t + p_{g+} + p_{g-} = 1 \tag{15.3}$$

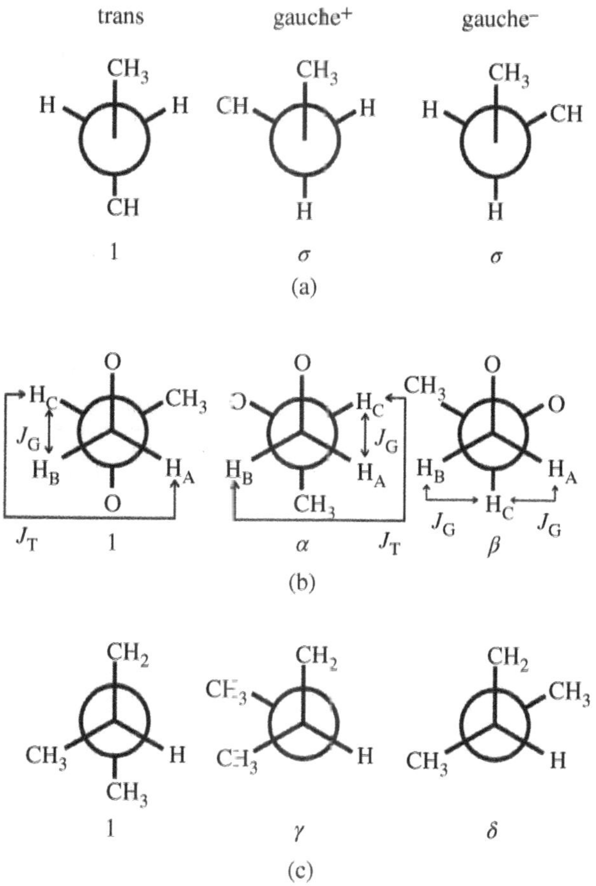

Figure 15.2 Newman projections of (a) O—C, (b) C—C*, and (c) C*—O bonds of DMP with definitions of vicinal coupling constants around the C—C* bond and the first-order intramolecular interactions that are represented by the corresponding statistical weights (Greek letters). Source: [399], Figure 2 (p.220)/Reproduced with permission of Elsevier.

Shown in Figure 15.3, parts a and b, are examples of the methine and methylene parts of 500 MHz ^{1}H NMR spectra of DMP-d_6 ($CD_3OCH_2CH(CH_3)OCD_3$). Since the methoxy signals of DMP partly overlap with those of the methine and methylene parts, the partially deuterated compound was prepared. The spin system is regarded as $ABCX_3$ or $ABMX_3$. An early NMR study on partially deuterated isotactic PPO, $[-CH_2CD(CH_3)O-]_x$ [198], assigned the upfield portion of the spectra to methylene A proton and the downfield to methylene B and methine C protons. As shown in Figure 15.3, the least-squares adjustment of calculated

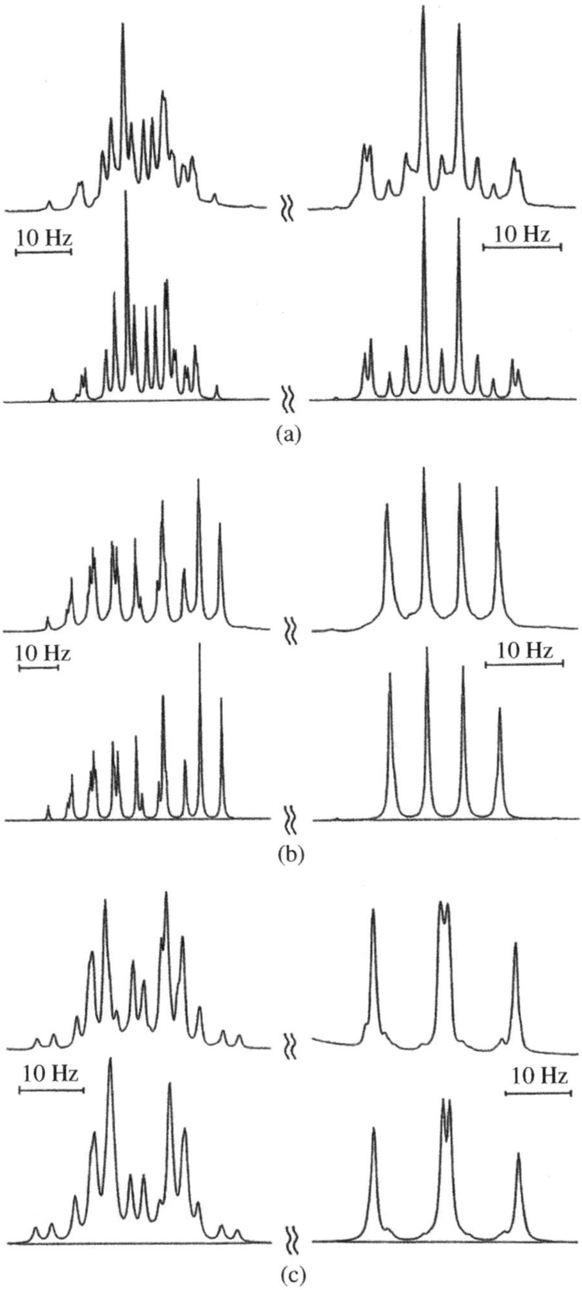

Figure 15.3 Methine and methylene parts of the observed (above) and calculated (below) ^{1}H NMR spectra of (a) $CD_3OCH_2C^{*}H(CH_3)OCD_3$ (DMP-d_6) in the gas phase at 180 °C, (b) DMP-d_6 in benzene-d_6 at 43 °C, and (c) *cis*-DMDO in dimethyl-d_6 sulfoxide at 80 °C. The spectra a and b were recorded at 500 MHz, and the spectrum c at 270 MHz. Source: [397], Figure 2 (p.13487)/Reproduced with permission of American Chemical Society.

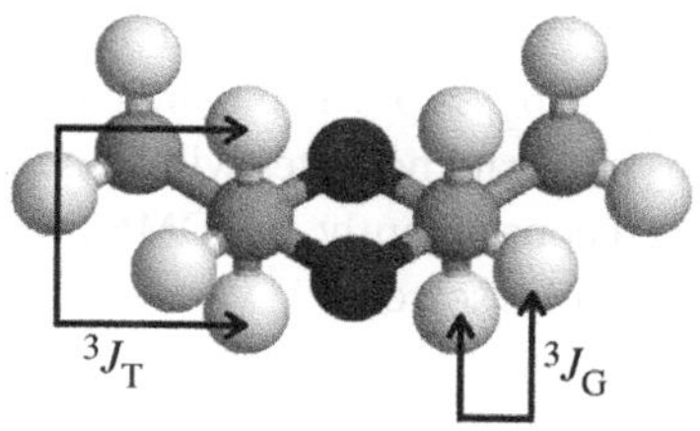

Figure 15.4 *cis*-2,6-Dimethyl-1,4-dioxane (*cis*-DMDO) with the definition of J_T and J_G. Source: [399], Figure 4 (p.222)/Reproduced with permission of Elsevier.

on observed nuclear magnetic resonance spectra (abbreviated as LAOCOON) III simulations [67] satisfactorily reproduced the experimental spectra. The J_T and J_G values were determined from a cyclic compound, *cis*-2,6,-dimethyl-1,4-dioxane (*cis*-DMDO), which has the same O—CH_2—C*H(CH_3)—O bond sequence as those of DMP and PPO (Figure 15.4). The two methyl groups of *cis*-DMDO always stay equatorial owing to the $CH_3 \cdots CH_3$ repulsion. Figure 15.3c shows a 270 MHz ^{1}H NMR spectrum of *cis*-DMDO dissolved in dimethyl sulfoxide-d_6 (DMSO-d_6). This spectrum was simulated as an $ABMX_3$ system and reproduced well. The chemical shifts varied slightly with temperature, whereas the coupling constants were invariant over the experimental temperatures. These facts indicate that the *cis*-DMDO ring ceases flip-flopping. The J_T and J_G values were determined as follows: $J_T = 9.80 \pm 0.04$ Hz and $J_G = 2.54 \pm 0.04$ Hz (cyclohexane-d_{12}); $J_T = 9.87 \pm 0.05$ Hz and $J_G = 2.54 \pm 0.05$ Hz (benzene-d_6); $J_T = 10.25 \pm 0.08$ Hz and $J_G = 2.52 \pm 0.05$ Hz (DMSO-d_6).

For example, from $^3J_{AC} = 5.24 \pm 0.02$ Hz and $^3J_{BC} = 5.47 \pm 0.01$ Hz of gaseous DMP at 180 °C, it follows that $p_t = 0.37$, $p_{g+} = 0.40$, and $p_{g-} = 0.23$, with the J_T and J_G values obtained from the cyclohexane-d_{12} solution. In Table 15.1, the bond conformations for individual media are listed. The bond conformations

Table 15.1 Bond conformations of the C—C* bond of DMP.

Medium	Dielectric constant of medium	Temp (°C)	p_t	p_{g+}	p_{g-}
	^{1}H NMR				
Gas	1.00	145	0.38	0.41	0.21
Cyclohexane-d_{12}	2.02	26	0.41	0.41	0.18
Benzene-d_6	2.28	26	0.34	0.44	0.22
Dimethyl-d_6 sulfoxide	45.0	25	0.26	0.43	0.31
	Ab initio MO calculations				
Gas	1.00	145	0.37	0.46	0.17

Source: [399], Table 1 (p.222)/Reproduced with permission of Elsevier.

were evaluated in the order of $p_{g+} > p_t > p_{g-}$ except for the polar DMSO solution ($p_{g+} > p_{g-} > p_t$). The bond conformations explicitly depend on the solvent polarity; the gauche states will be more stable with increasing polarity of the medium, and consequently, the DMP molecule renders itself more polar to be suitable for the polar solvent.

15.1.2 Ab initio MO Calculation

Ab initio MO calculations were carried out for DMP using the Gaussian program [159]. For all the possible conformers, the Gibbs free energies were calculated at the MP2/6-31+G*//HF/6-31G* level. The solvent effects on the conformer free energies were estimated by the self-consistent reaction field calculations [526].

In Table 15.2, the conformer free energies ΔG_ks (k: conformer number) obtained from the MO calculations are listed. In Table 15.1, the bond conformations, evaluated from the ΔG_k values according to the Boltzmann distribution, are compared with those from the NMR experiments. The conformational stability of the order $g^+ > t > g^-$, indicated by the NMR experiments, is well reproduced by the MO calculations.

According to the RIS scheme including up to the third-order intramolecular interactions [529, 532], the statistical weight matrices of DMP may be expressed as

$$U_2 = \begin{array}{c} \\ t \\ g^+ \\ g^- \end{array} \begin{array}{c} \begin{array}{ccc} t & g^+ & g^- \end{array} \\ \begin{bmatrix} 1 & \sigma & \sigma \\ 0 & 0 & 0 \\ 0 & 0 & 0 \end{bmatrix} \end{array} \tag{15.4}$$

$$U_3 = \begin{array}{c} \\ t \\ g^+ \\ g^- \end{array} \begin{array}{c} \begin{array}{ccccccccc} t & g^+ & g^- & t & g^+ & g^- & t & g^+ & g^- \end{array} \\ \begin{bmatrix} 1 & \alpha & \beta & 0 & 0 & 0 & 0 & 0 & 0 \\ 0 & 0 & 0 & 0 & \alpha & \beta\omega_2 & 0 & 0 & 0 \\ 0 & 0 & 0 & 0 & 0 & 0 & 1 & \alpha\omega_1 & 0 \end{bmatrix} \end{array} \tag{15.5}$$

$$U_4 = \begin{array}{c} \\ tt \\ tg^+ \\ tg^- \\ g^+t \\ g^+g^+ \\ g^+g^- \\ g^-t \\ g^-g^+ \\ g^-g^- \end{array} \begin{array}{c} \begin{array}{ccccccccc} t & g^+ & g^- & t & g^+ & g^- & t & g^+ & g^- \end{array} \\ \begin{bmatrix} 1 & \gamma & \delta & 0 & 0 & 0 & 0 & 0 & 0 \\ 0 & 0 & 0 & 1 & \gamma & \delta\omega_1 & 0 & 0 & 0 \\ 0 & 0 & 0 & 0 & 0 & 0 & 1 & \gamma\omega_2 & \delta \\ 1 & \gamma & \delta & 0 & 0 & 0 & 0 & 0 & 0 \\ 0 & 0 & 0 & 1 & \gamma\chi & \delta\omega_1 & 0 & 0 & 0 \\ 0 & 0 & 0 & 0 & 0 & 0 & 1 & 0 & \delta \\ 1 & \gamma & \delta & 0 & 0 & 0 & 0 & 0 & 0 \\ 0 & 0 & 0 & 1 & \gamma & 0 & 0 & 0 & 0 \\ 0 & 0 & 0 & 0 & 0 & 0 & 1 & \gamma\omega_2 & \delta \end{bmatrix} \end{array} \tag{15.6}$$

Table 15.2 Free energies (ΔG_k) of DMP, evaluated by ab initio molecular orbital calculations.

k	Conformation[a)]	Statistical weight[b)]	ΔG_k[c)] (kcal mol^{-1})
1	t t t	1	0.00
2	t t g^+	γ	3.20
3	t t g^-	δ	0.43
4	t g^+ t	α	0.62
5	t g^+ g^+	$\alpha\gamma$	3.80
6	t g^+ g^-	$\alpha\delta\omega_1$	−0.20
7	t g^- t	β	1.57
8	t g^- g^+	$\beta\gamma\omega_2$	2.18
9	t g^- g^-	$\beta\delta$	1.38
10	g^+ g^+ t	$\sigma\alpha$	2.30
11	g^+ g^+ g^+	$\sigma\alpha\gamma\chi$	4.02
12	g^+ g^+ g^-	$\sigma\alpha\delta\omega_1$	1.34
13	g^+ g^- t	$\sigma\beta\omega_2$	0.99
14	g^+ g^- g^-	$\sigma\beta\delta\omega_2$	1.51
15	g^- t t	σ	1.25
16	g^- t g^+	$\sigma\gamma$	4.46
17	g^- t g^-	$\sigma\delta$	1.87
18	g^- g^+ t	$\sigma\alpha\omega_1$	0.80
19	g^- g^+ g^+	$\sigma\alpha\gamma\omega_1$	4.28

a) For example, the conformation tg^+g^- represents that bonds 2, 3, and 4 adopt t, g^+, and g^- states, respectively.
b) For the definition of statistical weights, see Figures 15.2 and 15.5.
c) At the MP2/6-31+G*//HF/6-31G* level. Relative to the ΔG_k value of the all-trans conformation. At 25 °C and 1 atm.
Source: [399], Table 2 (p.223)/Reproduced with permission of Elsevier.

Here, the rows and columns of these matrices are indexed to the rotational states for the preceding and current bonds. The first-, second-, and third-order interactions are defined between atoms (atomic groups) separated by three, four, and five bonds, respectively. The first-order α, β, γ, δ, and σ interactions are defined in Figure 15.2, and the second-order ω_1 and ω_2 and the third-order χ interactions are illustrated in Figure 15.5. The intramolecular interactions are represented by the corresponding statistical weights, which are related to the conformational energies through the Boltzmann factor; for example, $\alpha = \exp(-E_\alpha/RT)$. The statistical

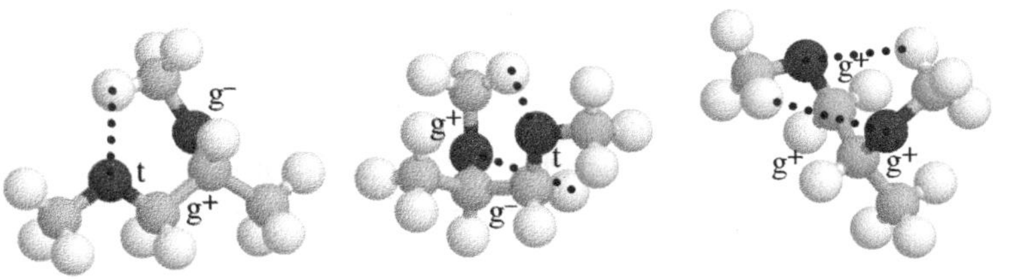

ω_1 ω_2 χ

Figure 15.5 Second-order (ω_1 and ω_2) and third-order (χ) intramolecular interactions defined for DMP and PPO: ω_1, in the tg^+g^- conformation of (*R*)-DMP; ω_2, in tg^-g^+; χ, $g^+g^+g^+$. The dotted lines represent the C—H···O close contacts (weak hydrogen bonds). Source: [399], Figure 5 (p.225)/Reproduced with permission of Elsevier.

weight matrices were formulated partly on the basis of a previous study [4], in addition, via inspection of the molecular model.

In the conventional RIS scheme [141, 307], the conformer free energy ΔG_k is assumed to be represented as the sum of conformational energies. For example, the $g^+g^+g^+$ conformation of DMP has a weight of $\sigma\alpha\gamma\chi$; therefore, its ΔG_k value may be approximated by $E_\sigma + E_\alpha + E_\gamma + E_\chi$. The E_ξ values ($\xi = \alpha, \beta, \gamma, \delta, \sigma, \omega_1, \omega_2$, and χ) were determined by minimizing the standard deviation between the ΔG_ks and the sums of E_ξs of the individual conformers. The temperature T was set to 298.15 K. The conformational energies thus derived are shown in Table 15.3. All the first-order interaction energies, E_α, E_β, E_γ, E_δ, and E_σ, were obtained to be positive, whereas the second-order ($E_{\omega 1}$ and $E_{\omega 2}$) and third-order (E_χ) interaction energies are negative, and hence, the corresponding interactions are suggested to be attractive. Figure 15.5 illustrates the tg^+g^-, tg^-g^+, and $g^+g^+g^+$ conformers with the ω_1, ω_2, and χ interactions, respectively. Here, for example the conformer tg^+g^- represents that bonds 2, 3, and 4 adopt t, g^+, and g^- states, respectively. The O···H distances of the tg^+g^-, tg^-g^+, and $g^+g^+g^+$ conformations, predicted by the MO calculations, are, respectively, 2.506 Å, (2.590 Å and 2.495 Å), and 2.631 Å, thus being smaller than the sum (2.70 Å) of the van der Waals radii of oxygen and hydrogen atoms [45]. When bond 3 lies in the g^- conformation, a hydrogen atom of the pendant methyl group also comes in contact with its nearest oxygen atom; thus, the $E_{\omega 2}$ value is close to the double of $E_{\omega 1}$. The $g^+g^+g^+$ conformer is seen to have two O···H close contacts; however, one of them does not satisfy the above criterion (the O···H distance is 2.880 Å), and consequently, the E_χ value is comparable to $E_{\omega 1}$. From the above discussion, it seems reasonable to conclude that the ω_1, ω_2, and χ interactions correspond to a sort of intramolecular hydrogen bond (hereafter referred to as the C—H···O attraction). Such weak hydrogen bonds have been found in molecular crystals and supramolecules [108, 223]. However, the nonbonded C—H···O attractions strongly affect the conformational characteristics of DMP and PPO undergoing rapid molecular motions.

Table 15.3 Conformational energies of DMP and isotactic PPO and polypropylene oxide (PO) dimers.[a)]

	DMP and isotactic PPO						PO dimers[b)]	
	Gas phase		Cyclohexane		Benzene			Benzene
	MO[c)]	^{1}H NMR[d)]	MO[c)]	^{1}H NMR[d)]	MO[c)]	^{1}H NMR[d)]		^{13}C NMR
				First order				
E_α	0.71	0.77 ± 0.04	0.53	0.71 ± 0.00	0.50	0.54 ± 0.03	E_{A}	0.36
E_β	1.30	1.2 ± 0.1	1.17	1.1 ± 0.0	1.15	0.83 ± 0.04	E_{B}	1.00
E_γ	3.06		2.98		2.97		E_Γ	2.86
E_δ	0.35		0.24		0.22		E_Δ	0.44
E_σ	1.52		1.42		1.41		E_Σ	1.40
				Second order				
E_{ω_1}	−1.24		−1.06		−1.04		E_{Ω_1}	−0.88
E_{ω_2}	−1.88		−1.77		−1.75		E_{Ω_2}	−1.53
				Third order				
E_χ	−1.27		−0.95		−0.91		E_{X}	−0.93

a) For the interactions, see Figures 15.2 and 15.5.
b) The interactions of the dimers are represented with the capital Greek letters. Determined by simulations for ^{13}C NMR chemical shifts of six PO dimers.
c) Determined from ab initio MO calculations for DMP at the MP2/6-31+G*//HF/6-31G* level including the solvent effects.
d) By RIS analysis of ^{1}H NMR vicinal coupling constants observed from DMP.
Source: [399], Table 3 (p.225)/Reproduced with permission of Elsevier.

15.1.3 RIS Analysis of Bond Conformations

In the RIS scheme, the bond conformation of the C—C* bond of DMP, p_η, is given by

$$p_\eta = \frac{J^*[U_2 U_3'(\eta) U_4]J}{J^* \left[\prod_{i=2}^{4} U_i\right] J} \tag{15.7}$$

where $J^* = [100]$, and J is the 9×1 column matrix of which all elements are unity. The U_3' matrix is obtained as follows: the columns of U_3 corresponding to the η state are left as they are, and the other columns are filled with zero. Equation (15.7) leads to the bond conformations:

$$\frac{p_{\mathrm{g}^+}}{p_{\mathrm{t}}} = \frac{1 + \sigma + \sigma\omega_1 + \gamma(1 + \sigma\chi + \sigma\omega_1) + \delta\omega_1(1 + \sigma)}{(1 + \sigma)(1 + \gamma + \delta)}\alpha \tag{15.8}$$

$$\frac{p_{\mathrm{g}^-}}{p_{\mathrm{t}}} = \frac{1 + \sigma\omega_2 + \gamma\omega_2 + \delta(1 + \sigma\omega_2)}{(1 + \sigma)(1 + \gamma + \delta)}\beta \tag{15.9}$$

Substitution of the p_t, p_{g^+}, and p_{g^-} values derived from the NMR experiments and the E_γ, E_δ, E_σ, $E_{\omega 1}$, $E_{\omega 2}$, and E_χ energies given by the ab initio MO calculations into Eqs. (15.8) and (15.9) yields the E_α and E_β values (in the "NMR" columns of Table 15.3), which are seen to be consistent with the MO calculations.

15.1.4 Configuration-dependent Properties

The characteristic ratio ($\langle r^2\rangle_0/nl^2$) and dipole moment ratio ($\langle \mu^2\rangle/nm^2$) of isotactic PPO in the Θ state were calculated with the conformational energies determined as above. The following geometrical parameters were used [4]: bond lengths, $l_{\mathrm{C-C^*}} = 1.53$ Å and $l_{\mathrm{C-O}} = 1.43$ Å; bond angles, ∠COC = ∠CCO = 111.5°; and dihedral angles, $\phi_t^a = 0°$, $\phi_{g\pm}^a = \pm 100°$, $\phi_t^b = 0°$, $\phi_{g\pm}^b = \pm 120°$, $\phi_t^c = -20°$, $\phi_{g^+}^c = 120°$, and $\phi_{g^-}^c = -100°$, where the superscripts represent the bond symbols (Figure 15.1). The bond dipole moments, $m_{\mathrm{C-O}}$ ($m_{\mathrm{C^*-O}}$) and $m_{\mathrm{C-C^*}}$, were assumed to be 1.17 and 0.00 D, respectively [271]. The dihedral angles are defined according to the convention of polymer chemistry [141].

Inspection of the molecular model led to statistical weight matrices for bonds a, b, and c of the repeating unit of isotactic PPO as follows:

$$U_a = \begin{array}{c} \\ \mathrm{tt} \\ \mathrm{tg^+} \\ \mathrm{tg^-} \\ \mathrm{g^+t} \\ \mathrm{g^+g^+} \\ \mathrm{g^+g^-} \\ \mathrm{g^-t} \\ \mathrm{g^-g^+} \\ \mathrm{g^-g^-} \end{array} \begin{array}{c} \begin{array}{ccccccccc} \mathrm{t} & \mathrm{g^+} & \mathrm{g^-} & \mathrm{t} & \mathrm{g^+} & \mathrm{g^-} & \mathrm{t} & \mathrm{g^+} & \mathrm{g^-} \end{array} \\ \begin{bmatrix} 1 & \sigma & 0 & 0 & 0 & 0 & 0 & 0 & 0 \\ 0 & 0 & 0 & 1 & 0 & 0 & 0 & 0 & 0 \\ 0 & 0 & 0 & 0 & 0 & 0 & 1 & 0 & \sigma \\ 1 & \sigma & 0 & 0 & 0 & 0 & 0 & 0 & 0 \\ 0 & 0 & 0 & 1 & 0 & 0 & 0 & 0 & 0 \\ 0 & 0 & 0 & 0 & 0 & 0 & 1 & 0 & \sigma \\ 1 & \sigma & 0 & 0 & 0 & 0 & 0 & 0 & 0 \\ 0 & 0 & 0 & 1 & 0 & 0 & 0 & 0 & 0 \\ 0 & 0 & 0 & 0 & 0 & 0 & 1 & 0 & \sigma \end{bmatrix} \end{array} \tag{15.10}$$

$$U_b = \begin{array}{c} \\ \mathrm{tt} \\ \mathrm{tg^+} \\ \mathrm{tg^-} \\ \mathrm{g^+t} \\ \mathrm{g^+g^+} \\ \mathrm{g^+g^-} \\ \mathrm{g^-t} \\ \mathrm{g^-g^+} \\ \mathrm{g^-g^-} \end{array} \begin{array}{c} \begin{array}{ccccccccc} \mathrm{t} & \mathrm{g^+} & \mathrm{g^-} & \mathrm{t} & \mathrm{g^+} & \mathrm{g^-} & \mathrm{t} & \mathrm{g^+} & \mathrm{g^-} \end{array} \\ \begin{bmatrix} 1 & \alpha & \beta & 0 & 0 & 0 & 0 & 0 & 0 \\ 0 & 0 & 0 & 0 & \alpha & \beta\omega_2 & 0 & 0 & 0 \\ 0 & 0 & 0 & 0 & 0 & 0 & 1 & 0 & 0 \\ 1 & \alpha & \beta & 0 & 0 & 0 & 0 & 0 & 0 \\ 0 & 0 & 0 & 0 & \alpha & 0 & 0 & 0 & 0 \\ 0 & 0 & 0 & 0 & 0 & 0 & 1 & 0 & 0 \\ 1 & \alpha & \beta & 0 & 0 & 0 & 0 & 0 & 0 \\ 0 & 0 & 0 & 0 & \alpha & 0 & 0 & 0 & 0 \\ 0 & 0 & 0 & 0 & 0 & 0 & 1 & \alpha\omega_1 & 0 \end{bmatrix} \end{array} \tag{15.11}$$

and

$$U_c = U_4 \tag{15.12}$$

The $\langle r_0^2 \rangle / nl^2$ and $\langle \mu^2 \rangle / nm^2$ values of isotactic 100mer PPO were calculated to be 5.91 (6.01) at 50 °C and 0.43 (0.43) at 25 °C, respectively, where the values in the parentheses are the corresponding experimental data. In the RIS calculations, both ratios of a 100mer are almost equal to those of the infinite-length chain. The experimental $\langle r^2 \rangle_0 / nl^2$ value was estimated from the viscosity measurements under the Θ condition (isooctane at 50 °C) [22], and the dipole moment was observed from the benzene solution at 25 °C [197]. The $\langle r^2 \rangle_0 / nl^2$ and $\langle \mu^2 \rangle / nm^2$ values were calculated from the conformational energy sets for the cyclohexane and benzene solutions, respectively. The good agreement between theory and experiment suggests that the intramolecular hydrogen bonds are probably formed. By analogy with the "pentane effect," [141] which gives rise to severe steric repulsion between methylene (methyl) groups in *n*-alkanes, it had been believed that the ω_1 and ω_2 interactions of PPO, appearing to be similar to the pentane effect, should also be repulsive. Accordingly, we attempted to offer further evidence of the C—H···O attractions of PPO as described in the following section.

15.2 Carbon-13 NMR Chemical Shifts of Dimeric Propylene Oxides

PPO is prepared by ring-opening polymerization of PO. If both C—O bonds of the monomer are cleaved, three types of linkages are formed between the monomeric units: head-to-tail (H—T); head-to-head (H—H); and tail-to-tail (T—T), where H and T stand for the methine and methylene ends, respectively. In addition, propylene oxide, possessing a chiral methine carbon, exists in either (*R*)- or (*S*)-isomer. Accordingly, the ^{13}C NMR spectra observed from atactic PPO are complicated [189, 428].

In order to assign ^{13}C NMR signals of polymers to the regio- and stereosequences, substituent effects on the chemical shifts have often been utilized [47, 485]. In particular, the γ substituent, which is three bonds distant from the observed carbon atom, tends to shield or deshield the observed nucleus from the applied magnetic field; the magnitude of the shielding (deshielding) effect depends on the distance between the two nuclei, thus being sensitive to the conformation of the intervening bond [47, 365, 485]. By this method, the ^{13}C NMR chemical shifts of a variety of polymers have been related to their microstructures.

This section describes an attempt to determine the conformational energies of six dimeric model compounds (Figure 15.6) of PO from ^{13}C NMR chemical shifts by a combined use of the substituent effects and the RIS scheme. The dimers may be the simplest compounds having different regiosequences (H—T, H—H, and T—T) and stereosequences (*RR*, *RS*, *SR*, and *SS*). Hereafter, the compounds (and molecules) with the H—T, H—H, and T—T linkages are, respectively, simply

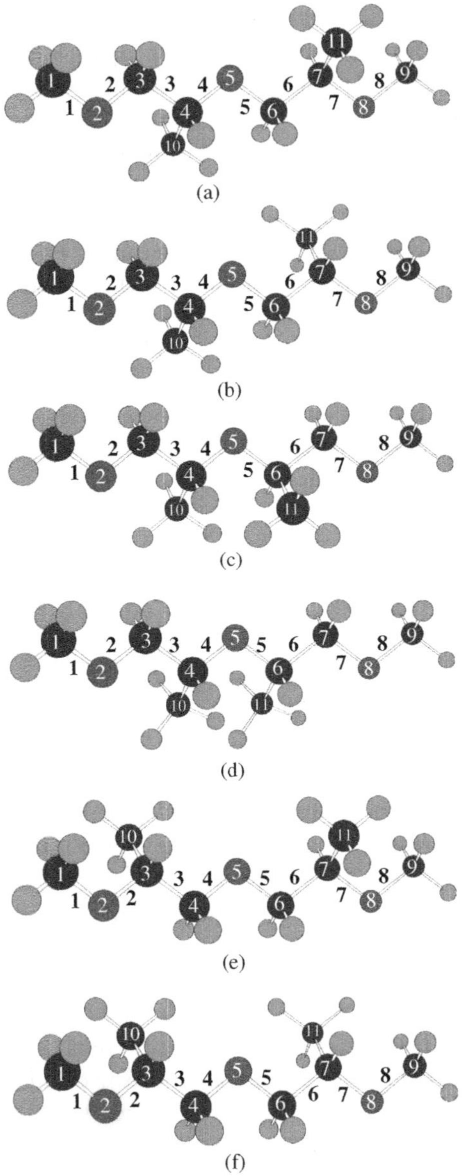

Figure 15.6 Dimeric model compounds of propylene oxide: (a) H–T (*RR*), (b) H–T (*RS*), (c) H–H (*RR*), (d) H–H (*RS*), (e) T–T (*RR*), and (f) T–T (*RS*) lying in the all-trans conformation. As indicated, the atoms and bonds are numbered. Source: [405], Figure 2 (p.3278)/Reproduced with permission of American Chemical Society.

referred to as H—T, H—H, and T—T. For each linkage, four optical isomers *RR*, *RS*, *SR*, and *SS*, possibly exist. However, the *RR* and *SS* (*RS* and *SR*) isomers are mirror images of each other and hence indistinguishable by NMR; therefore, the four isomers are represented by only two, *RR* and *RS*.

15.2.1 Theoretical Basis

On the basis of the empirical additivity relationship [34, 365], the chemical shift of carbon i_C is given by

$$\delta_{i_\mathrm{C}} = \sum_{n_{j,\alpha}} \Delta\delta_{\alpha,j} + \sum_{n_{j,\beta}} \Delta\delta_{\beta,j} + \sum_{\eta}\sum_{n_{j,\gamma}} p_\eta \Delta\delta^{\eta}_{\gamma,j} + \sum_{\eta\eta'}\sum_{n_{j,\delta}} p_{\eta\eta'} \Delta\delta^{\eta\eta'}_{\delta,j} + S \tag{15.13}$$

where $n_{j,\alpha}$, $n_{j,\beta}$, $n_{j,\gamma}$, and $n_{j,\delta}$ are the numbers of nonhydrogen atoms j at the α, β, γ, and δ positions (separated by one, two, three, and four bonds from the observed carbon i_C, respectively), $\Delta\delta_{\alpha,j}$ and $\Delta\delta_{\beta,j}$ are the chemical shift increments due to the atoms at the α and β positions, and $\Delta\delta^{\eta}_{\gamma,j}$ and $\Delta\delta^{\eta\eta'}_{\delta,j}$ are those due to the γ and δ atoms with the intervening bond(s) being in the η (= t, g$^+$, or g$^-$) and $\eta\eta'$ (= tt, tg$^+$… or g$^-$g$^-$) conformations, respectively. Thus, the third and fourth terms related to the γ- and δ-effects include the conformational probabilities, p_η and $p_{\eta\eta'}$, respectively. The α- and β-substituents induce downfield shifts (i.e. $\Delta\delta_{\alpha,j} > 0$ and $\Delta\delta_{\beta,j} > 0$), whereas the γ-substituent yields an upfield shift ($\Delta\delta^{\eta}_{\gamma,j} < 0$). The sign of the δ effect ($\Delta\delta^{\eta\eta'}_{\delta,j}$) is changeable with structure and conformation. The steric factor S depends on the extent of branching at the carbon i_C and its adjacent carbons.

If the constant terms of Eq. (15.13) are unified, Eq. (15.13) can be simplified to

$$\delta_{i_\mathrm{C}} = \Delta\delta_{0,i_\mathrm{C}} + \sum_{\eta}\sum_{n_{j,\gamma}} p_\eta \Delta\delta^{\eta}_{\gamma,j} + \sum_{\eta\eta'}\sum_{n_{j,\delta}} p_{\eta\eta'} \Delta\delta^{\eta\eta'}_{\delta,j} \tag{15.14}$$

where $\Delta\delta_{0,i_\mathrm{C}}$ is the summation of the first, second, and fifth terms of Eq. (15.13). The $\Delta\delta_{0,i_\mathrm{C}}$ term, being independent of conformation, may be given from the molecule that has the same atoms at the α and β positions and only hydrogen atoms at the γ positions. Such a molecule is designated as the *parent* compound. Parent compounds for the methine, methylene, pendant methyl, and terminal methoxy carbons of the six model compounds are 2-methoxy-1-propanol ($CH_3OCH(CH_3)CH_2OH$), 1-methoxy-2-propanol ($CH_3OCH_2CH(CH_3)OH$), 2-propanol ($CH_3CH(CH_3)OH$), and dimethyl ether (CH_3OCH_3), respectively. The individual carbon atoms have the same numbers and kinds of α and β atoms (CH, 2 α—C, 1 α—O, 1 β—C, and 1 β—O; CH_2, 1 α—C, 1 α—O, 2 β—C, and 1 β—O; CH_3, 1 α—C, 1 β—C, and 1 β—O; and CH_3O, 1 α—O, and 1 β—C), irrespective of the linkage type and chirality. However, the β substituent groups are not always common, for example for the methine carbons, ^{4}CH of H—T, 1 β—CH_2, and 1 β—O; ^{7}CH of H—T, 1 β—CH_3, and 1 β—O; ^{4}CH and ^{6}CH of H—H, 1 β—CH, and 1 β—O; and ^{3}CH and ^{7}CH of T—T, 1 β—CH_3, and 1 β—O. Here, the superscripts correspond to the carbon numbers in Figure 15.6. For this reason, it is preferable that the $\Delta\delta_{0,\zeta}$ (ζ = CH, CH_2, CH_3, or CH_3O) value should be defined for each β-substituent group.

The γ-effect ($\Delta\delta^{t}_{\gamma,\mathrm{C}}$) of carbon in the trans position, i.e. the γ-anti effect, is negligibly small, while that ($\Delta\delta^{t}_{\gamma,\mathrm{O}}$) of oxygen is, in general, comparatively large (−2 to −3 ppm) [34, 365]. In the ^{13}C NMR studies on alkanes and their oxygenated derivatives, the γ-gauche effects of carbon and oxygen ($\Delta\delta^{g\pm}_{\gamma,\mathrm{C}}$ and $\Delta\delta^{g\pm}_{\gamma,\mathrm{O}}$) have been found within ranges of −4 to −6 ppm and −6 to −8 ppm, respectively [47, 428, 485]. The δ-effects, which are typically much smaller in magnitude ($\leq$0.5 ppm) than those of α-, β-, and γ-effects, have not been taken into account, except for the $\mathrm{g}^{\pm}\mathrm{g}^{\mp}$ conformations. In hydrocarbon polymers, the $\mathrm{g}^{\pm}\mathrm{g}^{\mp}$ conformations lead to a severe steric interaction called the pentane effect. Thus, the δ-effect contribution ($\Delta\delta^{g\pm g\mp}_{\delta,\mathrm{C}}$) to the chemical shift may be negligible for alkanes. As shown in the previous section, however, the $\mathrm{g}^{\pm}\mathrm{g}^{\mp}$ conformations for the C—O/C—C* bond pairs of PPO and its model compounds are expected to be stabilized by the intramolecular C—H···O hydrogen bonds. The $\Delta\delta^{g\pm g\mp}_{\delta,\mathrm{O}}$ values, estimated as 2–3 ppm [34, 365], have been included here, whereas the $\Delta\delta^{g\pm g\mp}_{\delta,\mathrm{C}}$ effect may be negligible because of the pentane effect. On the other hand, the δ-effects related to other conformations were assumed to be null.

According to the RIS scheme [141, 307], for example, the fraction $f_{\mathrm{g^+tg^-}\ldots}$ of conformation $\mathrm{g^+tg^-}\ldots$ can be calculated from statistical weight matrices U_is according to

$$f_{\mathrm{g^+tg^-}\ldots} = \frac{J^*[U_2'(\mathrm{g}^+)U_3'(\mathrm{g^+t})U_4'(\mathrm{g^+tg^-})\ldots]J}{J^*\left[\prod_{i=2}^{n-1} U_i\right]J} \tag{15.15}$$

where n is the number of skeletal bonds. The $U_2'(\mathrm{g}^+)$ matrix can be obtained by filling columns of U_2 other than that of the g^+ state with zero, the $U_3'(\mathrm{g^+t})$ matrix is obtained by filling elements of U_3 other than that of the $\mathrm{g^+t}$ state with zero, $U_4'(\mathrm{g^+tg^-})$ includes only $\mathrm{g^+tg^-}$ element of U_4 as it is with the others filled with zero, etc. To facilitate the analysis, conformational fractions of the six dimers were assumed to be expressed as a function of the same eight conformational energies. To distinguish the energy parameters here from those established for isotactic PPO, statistical weights and interactions on the dimers are represented by the corresponding capital letters: A, B, Γ, Δ, Σ, Ω_1, Ω_2, and X. The bond conformation p_η of the ith bond is given as the sum of fractions of conformers with the η state in the ith bond, and $p_{\eta\eta'}$ for the ith and $(i+1)$th bond pair is the sum of fractions of conformers with the $\eta\eta'$ state in the two bonds. The chemical shift δ_{i_C} can be calculated from $\Delta\delta_0$s, $\Delta\delta^{t}_{\gamma,\mathrm{O}}$, $\Delta\delta^{g\pm}_{\gamma,\mathrm{C}}$, $\Delta\delta^{g\pm}_{\gamma,\mathrm{O}}$, $\Delta\delta^{g\pm g\mp}_{\delta,\mathrm{O}}$, p_ηs, and $p_{\eta\eta'}$s.

15.2.2 ^{13}C NMR Spectra and Assignment

PO was reacted with methanol to yield a mixture of 2-methoxy-1-propanol and 1-methoxy-2-propanol, and these products were further reacted with PO to produce the dimerized alcohols, which were further treated with sodium hydride and iodomethane to yield a mixture of the model compounds. The mixture was fractionated by supercritical fluid chromatography [538] into T—T, H—T, and H—H dimers.

Figure 15.7 shows the ^{13}C NMR spectra with broadband ^{1}H decoupling observed from benzene solutions of T—T, H—T, and H—H at 25 °C. The peaks at 17–18 ppm can be assigned to the pendant methyl carbons, and those around 56–59 ppm to terminal methoxy carbons. The signals around 74–78 ppm were assigned to methine and methylene carbons by reference to the distortionless enhancement by polarization transfer (DEPT) spectra [51, 88, 241].

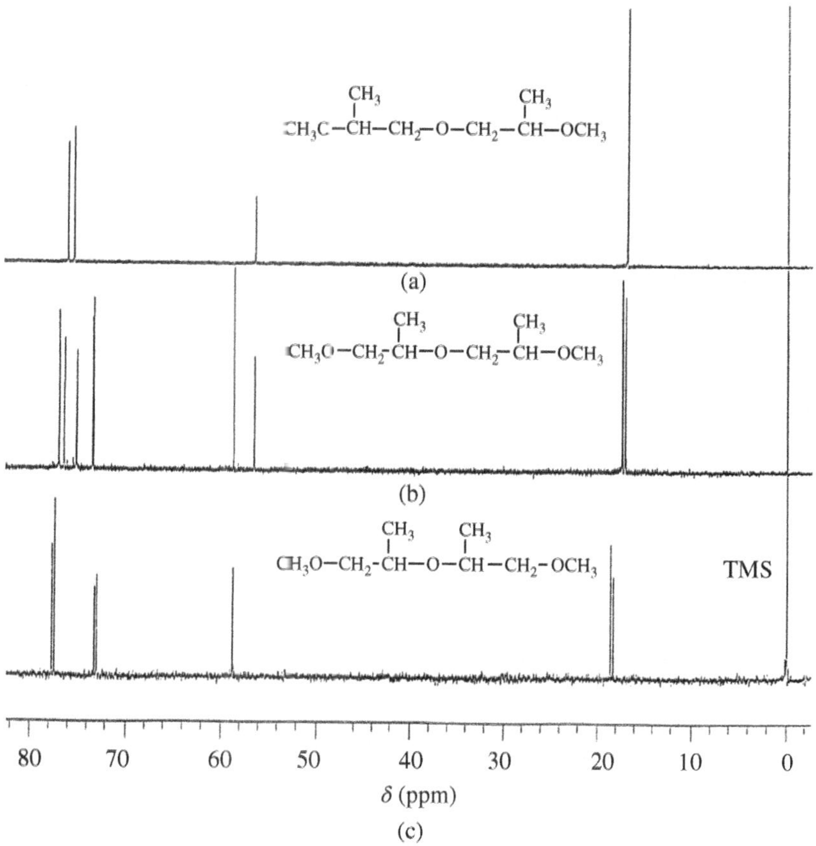

Figure 15.7 ^{13}C NMR spectra observed from benzene-d_6 solutions at 25 °C with broadband ^{1}H decoupling: (a) T−T, (b) H−T, and (c) H−H. The internal standard was tetramethylsilane (TMS, $\delta = 0$ ppm). Source: [405], Figure 4 (p.3280)/Reproduced with permission of American Chemical Society.

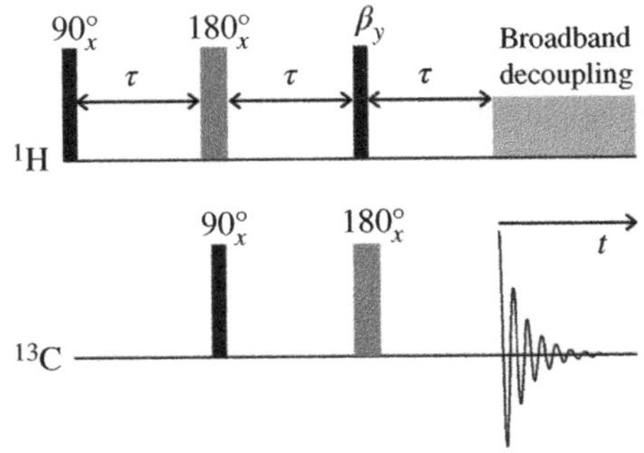

Figure 15.8 DEPT (distortionless enhancement by polarization transfer) pulse sequence. The third ^{1}H pulse has a flip angle β. The optimum value of delay τ is $1/(2J)$ (J, coupling constant), and t is the acquisition period.

Figure 15.8 illustrates the DEPT pulse sequence. The third ^{1}H-spin pulse has a flip angle β that determines the intensity of the observed ^{13}C signal, depending on the carbon type. In Table 15.4, the ^{13}C signal intensities of CH, CH_2, and CH_3 groups are expressed as a function of β. Figure 15.9 shows the methine and methylene parts of DEPT(90) and DEPT(135) spectra of H—T in benzene at 25 °C, where the β values are written in the parentheses. The methylene signals are removed by DEPT(90) and inverted by DEPT(135). In H—T, only two methine and two methylene groups exist; however, four doublets are observed: one of the doublets arises from *RR* (*SS*), and the other from *RS* (*SR*). The narrow doublet spacing, corresponding to the chemical shift difference between *RR* and *RS*, ranges from 0.00 to 0.14 ppm for H—T. The number of peaks of H—H or T—T is smaller than that of H—T because of the high structural symmetry. The doublets of the H—H mixture show comparatively large spacings; CH and CH_2, 0.27 ppm; CH_3, 0.29 ppm; and CH_3O, 0.02 ppm. For T—T, the doublet was observed only from the methylene carbons at 75.61 and 75.66 ppm. This is because the two chiral centers are apart from each other. The similarity in structure between H—T and H—H in atoms 1–5 and that between H—T and T—T in atoms 5–9 enables us to assign all peaks of H—T.

Table 15.4 DEPT intensities of CH, CH_2, and CH_3 groups as a function of the β angle.

Spin system	Intensity
CH	$\sin(\beta)$
CH_2	$2\cos(\beta)\sin(\beta)$
CH_3	$3\cos^2(\beta)\sin(\beta)$

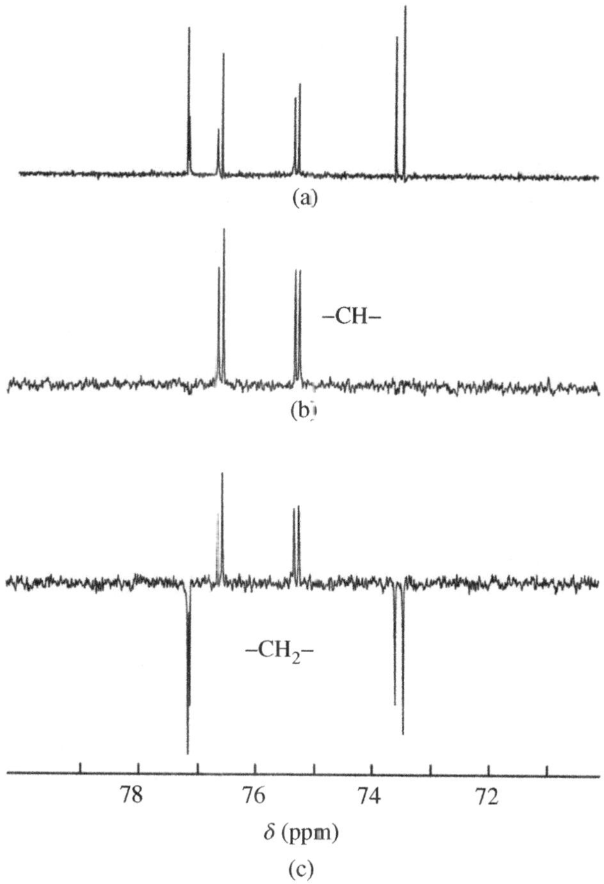

Figure 15.9 Methine and methylene parts of the ^{13}C NMR spectra of H–T in benzene-d_6 at 25 °C: (a) standard broadband ^{1}H decoupling; (b) DEPT(90); and (c) DEPT(135). The internal standard was tetramethylsilane. Source: [405], Figure 5 (p.3281)/Reproduced with permission of American Chemical Society.

15.2.3 Calculation of Chemical Shift by RIS Scheme

Statistical weight matrices for the six dimeric models were formulated by inspection of the molecular models and their structural symmetries [405]. The simulation based on Eq. (15.14) was carried out by the simplex method [338], and the average δ_{i_C} values of *RR* and *RS* were set as the targets. As a consequence, a good agreement between calculated and observed δ_{i_C}s was achieved.

Table 15.5 Observed and calculated chemical shifts of dimeric model compounds of PPO in benzene at 25 °C.

	Carbon number[a)] i_C	δ_{i_C} (ppm)				
		obsd		calcd		
		RR + *RS*	Average	*RR*	*RS*	Average
H-T	1	58.82	58.82	58.90	58.89	58.90
	3	77.11, 77.13	77.12	77.54	77.42	77.48
	4	75.27, 75.34	75.31	75.28	75.33	75.31
	6	73.50, 73.64	73.57	73.54	73.62	73.58
	7	76.55, 76.62	76.59	76.59	76.65	76.62
	9	56.65, 56.66	56.66	56.83	56.81	56.82
	10, 11[d]	17.17, 17.25	17.38	17.40	17.53	17.53
		17.52, 17.56		17.58	17.59	
$RMSE_{H-T}$[b)]			0.16			
H—H	1	58.82, 58.84	58.83	58.83	58.74	58.79
	3	77.48, 77.75	77.62	77.55	77.29	77.42
	4	73.06, 73.33	73.20	73.21	73.13	73.17
	6	73.06, 73.33	73.20	73.21	73.13	73.17
	7	77.48, 77.75	77.62	77.55	77.29	77.42
	9	58.82, 58.84	58.83	58.83	58.74	58.79
	10	18.35, 18.64	18.50	18.24	18.50	18.37
	11	18.35, 18.64	18.50	18.24	18.50	18.37
$RMSE_{H-H}$[b)]			0.12			
T—T	1	56.64	56.64	56.57	56.57	56.57
	3	76.27	76.27	76.24	76.27	76.26
	4	75.61, 75.66	75.64	75.61	75.64	75.63
	6	75.61, 75.66	75.64	75.61	75.64	75.63
	7	76.27	76.27	76.24	76.27	76.26
	9	56.64	56.64	56.57	56.57	56.57
	10	17.07	17.07	17.03	17.08	17.06
	11	17.07	17.07	17.03	17.08	17.06
$RMSE_{T-T}$[b)]			0.04			
$RMSE_{all}$[c)]			0.12			

a) See Figure 15.6.
b) The root-mean-square error is defined as RMSE (ppm) = $[\sum_{i_C=1}^{I_C}(\delta^{av}_{i_C,calc} - \delta^{av}_{i_C,obsd})^2/I_C]^{1/2}$, where I_C is the number of data.
c) The RMSE for all the data.
Source: [405], Table 2 (p.3282)/Reproduced with permission of American Chemical Society.

In Table 15.5, the calculated δ_{i_C} values are compared with the experimental data. The overall root-mean-square error ($RMSE_{all}$) was minimized to 0.12 ppm. The optimized conformational energies (E_A to E_X) (Table 15.3) slightly differ from those of isotactic PPO and DMP (E_α to E_χ) (Table 15.3).

From the $E_A - E_X$ values, the t, g^+, and g^- fractions of the central C—C bond of DMP were, respectively, evaluated as 0.37 (0.34), 0.47 (0.44), and 0.16 (0.22), where the values in the parentheses were determined from ^{1}H–^{1}H vicinal coupling constants of DMP in benzene at 26 °C [397]. The γ- and δ-effect parameters were also optimized as follows: $\Delta\delta^{t}_{\gamma,O} = -2.6$ ppm (-2 to -3 ppm) [34, 365], $\Delta\delta^{g\pm}_{\gamma,C} = -4.9$ ppm (-4 to -6 ppm) [47, 428, 485], $\Delta\delta^{g\pm}_{\gamma,O} = -7.9$ ppm (-6 to -8 ppm) [47, 428, 485], and $\Delta\delta^{g\pm g\mp}_{\delta,O} = 2.1$ ppm (2–3 ppm) [34, 365]. All these parameters stay within the literature ranges written in the parentheses.

In the previous section, conformational energies of isotactic PPO were determined from ab initio molecular orbital calculations and NMR vicinal coupling constants of DMP and the RIS analysis of the characteristic ratio and dipole moment ratio of isotactic PPO. The gauche stability of the C—C* bond in the main chain was indicated to be due to the weak C—H···O hydrogen bonds. Although the six dimers treated here differ significantly from each other in regio- and stereosequence, a minor modification of the conformational energy set of isotactic PPO gave a good agreement between calculated and observed δ_{i_C}s of all the dimers. These facts undoubtedly show the formation of the intramolecular C—H···O attractions in PPO.

15.3 Model Compound of Poly(ethylene terephthalate)

Poly(ethylene terephthalate) (PET, Figure 15.10a) is the most common polyester. Here, as a model compound of PET, ethylene glycol dibenzoate (EGDB) is adopted. For NMR experiments, two compounds were prepared: EGDB and EGDB-$^{13}C_1$ (Figure 15.10b) [400]. The Newman projections in Figure 15.11 illustrate how the vicinal ^{13}C–^{1}H coupling constant ($^3J_{CH}$) is expressed as a function of trans (p_t) and gauche (p_g) fractions of the O—CH_2 bond:

$$^3J_{CH} = J_G\, p_t + \frac{J'_T + J'_G}{2}\, p_g \tag{15.16}$$

where J'_T, J_G, and J'_G are defined in Figure 15.11. By definition, the bond conformations fulfill

$$p_t + p_g = 1 \tag{15.17}$$

For methylene protons of EGDB, two vicinal ^{1}H–^{1}H coupling constants determined experimentally, $^3J_{HH}$ ($= J_{AB} = J_{A'B'}$) and $^3J'_{HH}$ ($= {}^3J_{AB'} = {}^3J_{A'B}$), are related

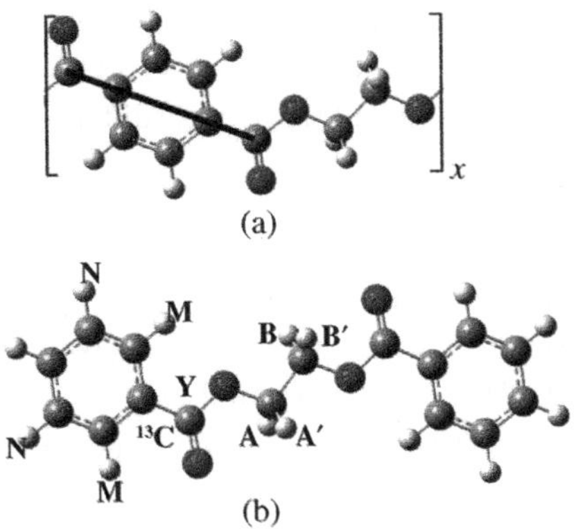

Figure 15.10 All-trans forms of (a) poly(ethylene terephthalate) (PET) and (b) ethylene glycol dibenzoate (EGDB). The hydrogen and carbon atoms of EGDB are partly designated to represent the spin systems of the ^{1}H and ^{13}C NMR spectra shown in Figure 15.12. The thick solid line segment expresses a virtual bond for the benzene ring, being used in the RIS calculations; the bond lengths of PET are $l_{O-C(=O)}$ = 1.357 Å, $l_{C(=O)-C(=O)}$ = 5.778 Å (virtual bond), l_{O-CH_2} = 1.439 Å, and $l_{CH_2-CH_2}$ = 1.515 Å.

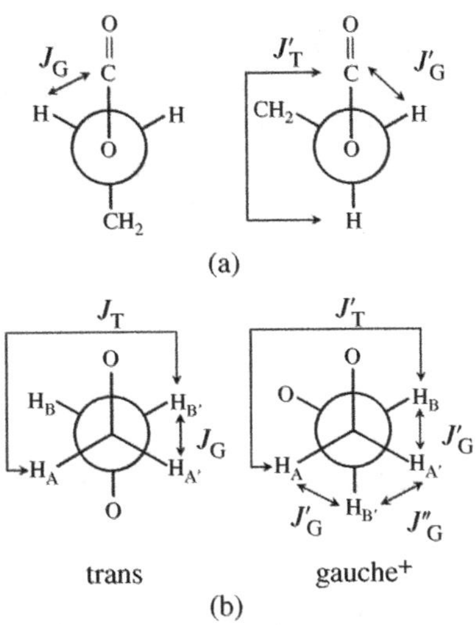

Figure 15.11 Rotational isomeric states around (a) $O-CH_2$ and (b) CH_2-CH_2 bonds of the spacers of EGDB and PET with the definition of J_Ts and J_Gs. Source: [400], Figure 2 (p.2855)/Reproduced with permission of American Chemical Society.

to the trans and gauche fractions of the CH_2—CH_2 bond:

$$^3J_{HH} = J_G\, p_t + \frac{J'_T + J''_G}{2}\, p_g \tag{15.18}$$

and

$$^3J'_{HH} = J_T\, p_t + J'_G\, p_g \tag{15.19}$$

where J_Ts and J_Gs are illustrated in Figure 15.11. To derive the p_t and p_g values of the O—CH_2 and CH_2 – CH_2 bonds from the above equations, the ^{1}H and ^{13}C NMR spectra of EGDB and EGDB-$^{13}C_1$ were measured and analyzed.

Figure 15.12c shows one of two ^{1}H NMR satellite bands of EGDB. It appears at a remove of $^1J_{CH}/2$ (73 Hz) from the intense methylene singlet (Figure 15.12a)

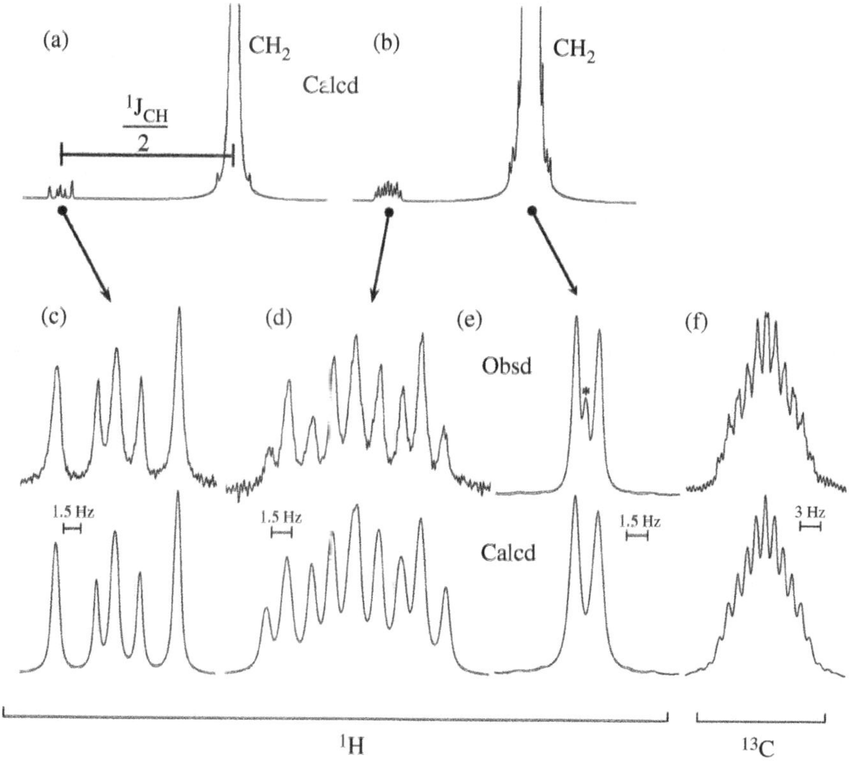

Figure 15.12 Observed (above) and calculated (below) ^{1}H (a–e) and ^{13}C (f) NMR spectra of EGDB (a and c) and EGDB-$^{13}C_1$ (b, d–f) in benzene-d_6 at 35 °C. The asterisk in part e indicates a peak from an impurity, unlabeled EGDB. Source: [400], Figure 3 (p.2855)/ Reproduced with permission of American Chemical Society

and provides both $^{3}J_{HH}$ and $^{3}J'_{HH}$ values. The satellite bands are due to the natural abundance $^{13}CH_2$ (1.1%), which disturbs the magnetic equivalence of the four methylene protons and changes the spin system from A_4 to AA′BB′. Figure 15.12 also shows an ^{1}H NMR spectrum of methylene groups of EGDB-$^{13}C_1$. The main signal (Figure 15.12e) is a doublet, and its satellite band (Figure 15.12d) includes more peaks than that of EGDB because the ^{13}C-labeled carbonyl carbon as well as the four methylene protons form the spin system. Figure 15.12f shows an observed ^{13}C NMR spectrum of the ^{13}C-labeled carbonyl carbon; it also exhibits fine structures due to the ^{13}C–^{1}H couplings with the methylene and phenyl protons. The spin systems of spectra in parts c, d, e, and f of Figure 15.12 may be expressed as AA′BB′(X), AA′BB′(X)Y, and AA′BB′M_2N_2Y, respectively (A, A′, B, and B′: methylene protons, M and N: phenyl protons, (X): natural abundance methylene carbon, and Y: ^{13}C-labeled carbonyl carbon. For the assignment, see Figure 15.10b).

It is reasonable to assume that the four spectra (Figures 15.12c—f), measured with the same solvent at the same temperature, can be reproduced by a set of ^1H–^1H and ^{13}C–^1H coupling constants; accordingly, the following procedure was employed.

First, spectrum c was simulated with the gNMR program [53] to yield the $^3J_{HH}$ and $^3J'_{HH}$ values. With these coupling constants, spectra d and e were analyzed to derive the $^3J_{CH}$ value. Spectrum f was confirmed to be satisfactorily reproduced from the ^3J data, and furthermore, both chemical shifts and coupling constants were adjusted until a best agreement was attained for all the spectra. For the sake of comparison, the calculated spectra are also shown in Figure 15.12, and the coupling constants were optimized to be $^3J_{HH} = 6.50$, $^3J'_{HH} = 3.10$, $^3J_{CH} = 3.32$, $^1J_{CH} = 145.85$, $^3J_{YM} = 4.10$, $^4J_{YN} = 1.10$, and $^4J_{YB} = -0.30$ Hz. The $^3J_{HH}$, $^3J'_{HH}$, and $^3J_{CH}$ values for all the solutions are listed in Table 15.6.

The bond conformations of the CH_2—CH_2 bond were evaluated from the $^3J_{HH}$ and $^3J'_{HH}$ values. Then, the coefficients (J_Ts and J_Gs) in Eqs. (15.18) and (15.19) were taken from (set A) those optimized for DME and PEO [469], (set B) those of *cis*-2,6-dimethyl-1,4-dioxane, a cyclic model compound for PPO [397], and (set C) those obtained from density functional MO calculations at the B3LYP/6-311++G(3df,3pd) level for EGDB itself. For sets A and B, $^3J_T = {}^3J'_T$ and $^3J_G = {}^3J'_G = {}^3J''_G$ were assumed. For the individual ^3J_T and ^3J_G values, see the footnote of Table 15.7. The p_t and p_g values derived from Eqs. (15.18) and (15.19) were divided by their sum to satisfy Eq. (15.17). The trans fractions thus obtained are listed in Table 15.7.

Table 15.6 Observed vicinal ^1H–^1H and ^{13}C–^1H coupling constants of EGDB and EGDB-$^{13}C_1$.[a)]

Solvent	Dielectric constant	Temp (°C)	$^3J_{HH}$	$^3J'_{HH}$	$^3J_{CH}$
Cyclohexane-d_{12}	2.0	25	6.43	3.47	3.38
Benzene-d_6	2.3	15	6.55	3.00	3.25
		25	6.50	3.05	3.30
		35	6.50	3.10	3.32
		45	6.45	3.15	3.35
		55	6.45	3.25	3.36
Methanol-d_4	32.7	25	6.50	2.85	3.45
Dimethyl-d_6 sulfoxide	46.7	25	6.45	2.70	3.25

a) In Hz.
Source: [400], Table 2 (p.2857)/Reproduced with permission of American Chemical Society.

Table 15.7 Trans fractions (p_ts) in O—CH_2—CH_2—O bond sequence of EGDB.

		p_t				
		CH_2—CH_2			O—CH_2	
Medium	**Temp (°C)**	**Set A[a)]**	**Set B[b)]**	**Set C[c)]**	**Set a[d)]**	**Set b[e)]**
	NMR expt					
Cyclohexane-d_{12}	25	0.12	0.09	0.19	0.43	0.42
Benzene-d_6	15	0.08	0.03	0.15	0.48	0.45
	25	0.08	0.04	0.15	0.46	0.44
	35	0.09	0.05	0.16	0.45	0.43
	45	0.09	0.05	0.16	0.44	0.42
	55	0.10	0.06	0.17	0.44	0.42
Methanol-d_4	25	0.06	0.03	0.13	0.41	0.40
Dimethyl-d_6 sulfoxide	25	0.05	0.02	0.12	0.48	0.45
	MO calc					
Gas phase	25	0.06			0.45	
Benzene	15	0.05			0.49	
	25	0.05			0.49	
	35	0.06			0.49	
	45	0.06			0.48	
	55	0.06			0.48	

a) Optimized values for poly(ethylene oxide): $^3J_T = 11.4$ and $^3J_G = 2.3$ Hz [469].
b) From *cis*-2,6-dimethyl-1,4-dioxane, a model compound of PPO: in cyclohexane, $^3J_T = 9.80$ and $^3J_G = 2.54$ Hz; in benzene, $^3J_T = 9.87$ and $^3J_G = 2.54$ Hz; in methanol and DMSO, $^3J_T = 10.25$ and $^3J_G = 2.52$ Hz [397].
c) MO calculations for EGDB at the B3LYP/6-311++G(3df,3pd) level: $^3J_T = 11.50$, $^3J_G = 4.90$, $^3J'_T = 10.96$, $^3J'_G = 1.48$, and $^3J''_G = 2.35$ Hz [400].
d) From the Karplus equation: $^3J_G = 1.7$ Hz ($\phi = 120.5°$), $^3J'_T = 5.4$ Hz ($\phi = 28.8°$), and $^3J'_G = 4.0$ Hz ($\phi = 147.3°$) [500].
e) MO calculations for EGDB at the B3LYP/6-311++G(3df,3pd) level: $^3J_G = 1.17$, $^3J'_T = 6.31$, and $^3J'_G = 3.59$ Hz [400].

Source: [400], Table 3 (p.2857)/Reproduced with permission of American Chemical Society.

In the analysis for $^3J_{CH}$ with Eqs. (15.16) and (15.17), the three coefficients, J_T, J_G, and J'_G, were derived from (set a) the Karplus equation established for the C—O—C—H bond sequence of carbohydrates: $^3J_{COCH} = 5.7\cos^2\phi - 0.6\cos\phi + 0.5$ (ϕ: dihedral angle between ^{13}C and ^{1}H according to the convention of polymer science) [500]. Then, the ϕ values were set equal to those optimized at the

B3LYP/6-311+G(2d, p) level. In addition, the three J coefficients obtained from the MO calculations for EGDB at the B3LYP/6-311++G(3df,3pd) level were used (see the footnote of Table 15.7).

As shown in Table 15.7, in the bond conformations, the NMR experiments and MO calculations are fully consistent. For the CH_2—CH_2 bond of EGDB, the p_t values, depending on the used J coefficients, are somewhat smaller than those ($p_t \sim 0.05$ [in water] to 0.2 [in cyclohexane]) of PEO and its model compounds [7, 9, 300, 412, 469]. On the other hand, the trans preference ($p_t \sim 0.4$–0.5) of the O—CH_2 bond is not so strong as found for PEO ($p_t \geq 0.65$). The conformational preference of EGDB somewhat differs from that of PEO.

In order to derive bond conformations of compounds as precisely as possible from NMR vicinal coupling constants, it is desirable to obtain the ^3J values, if possible, to two decimal places. For that purpose, it is preferable that the spectrum should be highly resolved by a proper zero filling and undergo the Fourier transform, if necessary, with a weak window function to keep the fine structures well defined.

As demonstrated here and in Section 9.2, it is of particular significance to prepare compounds suitable for the experimental purposes. For example, compounds replaced partly with deuterium get rid of unnecessary signals, and those labeled with carbon-13 enhance the signal-to-noise ratio (S/N) of the ^{13}C NMR spectra and yield additional information although ^{13}C-labeled chemicals in particular are expensive. It is known that the S/N will be improved in proportional to (number of accumulation)$^{1/2}$; however, ^{13}C-unlabeled EGDB here merely exhibited a broad carbonyl ^{13}C peak lacking in fine structures even after a large number of scans.

16

Scattering Methods

The static light scattering (SLS) experiment provides us with information on molecular properties of polymers in equilibrium in solutions [83, 429, 472]. The Zimm plot enables simultaneous determination of the weight-average molecular weight (M_w), radius of gyration (R_g), and second virial coefficient (A_2) of a given polymer [550, 551]. When the polymeric chain is unperturbed, i.e. $A_2 = 0$, the SLS data can be directly compared with, for example, the characteristic ratio that the rotational isomeric state (RIS) scheme yields [141]:

$$R_g = \langle S^2 \rangle_0^{1/2} = \sqrt{\frac{1}{6}} \langle r^2 \rangle_0^{1/2} \tag{16.1}$$

where

$$\langle S^2 \rangle = \frac{1}{2N^2} \sum_{j=1}^{N} \sum_{i=1}^{N} \left\langle \left| \mathbf{R}_j - \mathbf{R}_i \right|^2 \right\rangle = \frac{1}{N} \sum_{j=1}^{N} \langle S_{G,j}^2 \rangle \tag{16.2}$$

Here, N is the total number of scatterers, $\mathbf{R}_j$ is the position vector of the jth scatterer, and $S_{G,j}$ is the distance between the center of mass and the jth scatterer. The dynamic light scattering (DLS) experiment observes the photon autocorrelation function and yields the hydrodynamic radius via the diffusion coefficient of polymer particles in Brownian motion. Both SLS and DLS are established techniques, and the equipment are commercially available. The small-angle neutron scattering (SANS) [196, 390] and small-angle X-ray scattering (SAXS) [172] are based on theories similar to that of light scattering (LS).

16.1 Static Light Scattering (SLS)

Light enters a particle, which will be polarized in the direction of the electric field of the incident light. The polarization is oscillated in synchronization with the electronic field, which emits the electromagnetic wave of the same frequency

Conformational Analysis of Polymers: Methods and Techniques for Structure-Property Relationships and Molecular Design, First Edition. Yuji Sasanuma.

and wave length as those of the incident light. When, in vacuo (permittivity ϵ_0), a particle of polarizability α is irradiated with an incident light of wave length λ_0 and strength I_0, a light will be emitted therefrom and observed at the position separated by a distance r from the particle, and the observed intensity I is expressed as

$$\frac{I}{I_0} = \left(\frac{\alpha}{4\pi\epsilon_0}\right)^2 \left(\frac{2\pi}{\lambda_0}\right)^4 \frac{1}{r^2} = \frac{4\pi^2\alpha^2}{\epsilon_0^2\lambda_0^4 r^2} \tag{16.3}$$

This phenomenon is the Rayleigh scattering, which is characterized by $I \propto \lambda_0^{-4}$. Accordingly, the shorter the wave length is, the more strongly the light will be scattered. This is the reason why the canopy looks blue.

The Rayleigh ratio, R_θ, is adopted in LS experiments:

$$R_\theta = \frac{Ir^2}{I_0 V} \tag{16.4}$$

Since it is difficult to evaluate the R_θ value directly from the sample, the observed intensity (I) will be reduced by that (I_{solv}) of the solvent and scaled by the intensity (I_{ref}) and the Rayleigh ratio (R_{ref}) of a reference according to

$$R_\theta = \frac{I - I_{\text{solv}}}{I_{\text{ref}}} \left(\frac{n}{n_{\text{ref}}}\right)^2 R_{\text{ref}} \tag{16.5}$$

where n and n_{ref} are the refractive indexes of the solution and reference, respectively. As a reference, benzene or toluene has been used because their absolute R_{ref} values are established. The Rayleigh ratio is related to the particle (solute) structure by

$$R_\theta = KcMP(q) \tag{16.6}$$

where c and M are the concentration and molecular weight of the solute, respectively. The optical constant, K, is given by

$$K = \frac{4\pi^2 n_{\text{solv}}^2}{N_A \lambda_0^4} \left(\frac{\partial n}{\partial c}\right)^2 \tag{16.7}$$

The particle scattering factor, $P(q)$, depends on the size and shape of the particle, and q is the magnitude of the scattering vector

$$q = |\mathbf{q}| = \frac{4\pi n_{\text{solv}}}{\lambda_0} \sin\left(\frac{\theta}{2}\right) \tag{16.8}$$

where n_{solv} is the refractive index of the solvent, and θ is the scattering angle, namely the angle between the incident and scattering beams. For a spherical particle, $P(q)$ is expressed as

$$P(q) = \left[\left(\frac{3}{x^3}\right)(\sin x - x\cos x)\right]^2, \qquad \text{with} \quad x \equiv qR \tag{16.9}$$

and the $P(q)$ function of the Gaussian chain is known as the Debye function:

$$P(q) = \frac{2}{x^2}\left(e^{-x} - 1 + x\right) \qquad x \equiv q^2\langle S^2\rangle \tag{16.10}$$

The general expression of $P(q)$, being valid irrespective of the particle shape, is expressed as

$$P(q) = 1 - \frac{1}{3}\langle S^2\rangle q^2 + O\left(q^4\right) \tag{16.11}$$

where $O\left(q^4\right)$ is negligibly small as compared to the first and second terms.

At $\theta = 0$, the Rayleigh ratio is related to the second virial coefficient (A_2) via the osmotic pressure (Π):

$$\frac{Kc}{R_\theta} = \frac{\partial}{\partial c}\left(\frac{\Pi}{RT}\right) = \frac{1}{M_W} + 2A_2c + O\left(c^2\right) \tag{16.12}$$

For a fully diluted solution, it follows that

$$\frac{Kc}{R_\theta} = \frac{1}{M_W P(q)} = \frac{1}{M_W}\left[1 + \frac{1}{3}\langle S^2\rangle q^2 - O\left(q^4\right)\right] \tag{16.13}$$

The scattering intensities of solutions of different concentrations are recorded with θ being changed, for example, between 20° and 150° at intervals of 5°, and the excess Rayleigh ratios are calibrated according to Eq. (16.5). The Kc/R_θ values are plotted against $\sin^2(\theta/2) + kc$, where k is a scale factor, if necessary, chosen so as to shift the Kc/R_θ vs. $\sin^2(\theta/2)$ line properly. From the plots, the data at infinite dilution ($c = 0$) may be extrapolated to be expressed by a line, whose slope and intercept will yield the $\langle S^2\rangle$ and M_W^{-1} values, respectively. The line corresponding to $\theta = 0$ may also be simulated similarly, and its slope and intercept give the A_2 and M_W^{-1} values, respectively. Therefore, ideally, both lines are expected to cross each other on the vertical axis. This analytical procedure is termed the Zimm plot [550, 551]. After the Zimm plot, the $(Kc/R_\theta)^{1/2}$ vs. $\sin^2(\theta/2)$ plot, which yields a straighter line than that of the original, was proposed (the Berry plot) [42]:

$$\left(\frac{Kc}{R_\theta}\right)^{1/2} = \frac{1}{M_W^{1/2}}\left(1 + \frac{1}{6}\langle S^2\rangle q^2 + \cdots\right) \tag{16.14}$$

In SAXS experiments, the I vs. q^2 plot (the Guinier plot) [172, 179], has often been used. Similarly, the $\ln(R_\theta/Kc)$ vs q^2 plot has also been adopted in SLS experiments:

$$\ln\left(\frac{R_\theta}{Kc}\right) = \ln M_W - \frac{1}{3}\langle S^2\rangle q^2 + \cdots \tag{16.15}$$

16.1.1 Instrumentation and Sample Preparation for SLS

Figure 16.1 is a photograph of commercially available instruments of LS, and Figure 16.2 shows the optical setup placed in a clean bench that collects floating

Figure 16.1 Light scattering instruments, DLS-8000, commercially available from Otsuka Electronics.

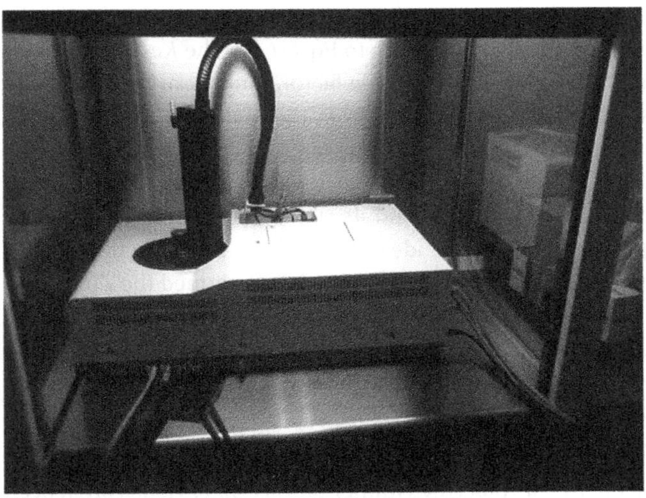

Figure 16.2 Light scattering optical setup installed in a clean bench.

dust and prepares a clean space inside. Figure 16.3 schematically illustrates the optical setup. The light is generated from a He–Ne laser, reflected by two mirrors, and weakened properly by neutral-density filters, and reaches the sample cell, which is cylindrical, made of quartz, filled with the sample solution up to an appropriate height, and immersed in silicone oil. The scattered light making an angle θ with the incident beam is reflected orthogonally by a total reflecting prism, collimated through pinholes, and focused on a photomultiplier.

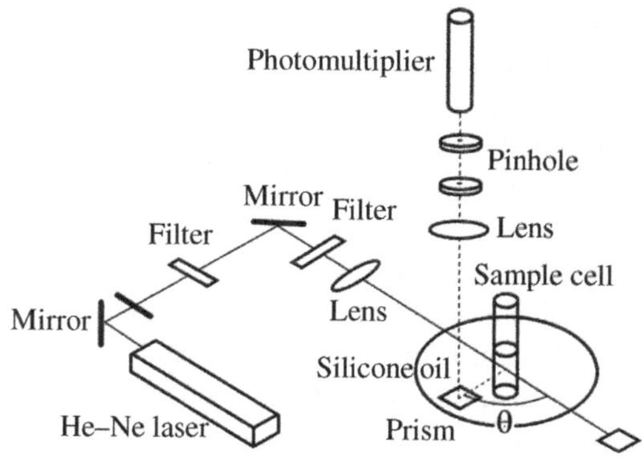

Figure 16.3 Schematic illustration of the light scattering optical setup.

The concentration of the solution is gravimetrically determined; therefore, one should pay particular attention to the accuracy of the chemical balance and significant figures of the weight. For cleaning, the glassware, syringe, and cell are exposed to acetone vapor for a few hours before use.

Before the measurement, the silicone oil in which the sample cell will be immersed is circulated to be filtrated to get rid of dirt and dust. First, the reference, benzene or toluene, is slowly injected through a fine membrane filter to the bottom of the cell and subjected to a θ scan to record I_{ref} of Eq. (16.5). After the reference is removed, the solvent is similarly poured into the sample cell and undergoes a measurement to record I_{solv}. The cell ought not to be moved and exchanged for another during all the measurements, so that both the incident and the scattering lights always pass through the same path to keep r and V of Eq. (16.4) constant. Following the solvent, the sample solutions undergo measurements in order of increasing concentration. Whenever the sample is changed, the cell should preferably be washed with the new sample, if possible, a few times.

16.1.2 Application of SLS: Chain Dimensions of Polysilanes in the Θ State

The backbones of polysilanes are composed of the Si—Si bond, which forms the bonding σ and antibonding σ^* localized orbitals; therefore, polysilanes show near-UV absorptions, whose wave lengths vary depending on the chemical species of the side chain [321]. This is because steric repulsions between the side chains affect the backbone conformation. In the crystalline state, poly(dimethylsilane) (PDMS) lies in the all-trans form [287], poly(di-*n*-butylsilane) (PDBS) adopts a 7/3 helical structure [426], and poly(di-*n*-hexylsilane) (PDHS) also exits in the all-trans structure [288]. The solution and fresh solid film of PDHS exhibit the UV absorption around 310–320 nm; however, after the film stands for a while at room temperature, the absorption moves to 370–375 nm, and simultaneously, the former intensity decreases, and the latter increases [320, 376]. The dihedral angle of the 7/3 helix and the absorption of PDBS, respectively, deviate by 30° and 60 nm from those of the all-trans PDHS [426]. Kato et al. earnestly conducted SLS experiments and determined the Θ conditions and unperturbed chain dimensions of PDHS, PDBS, and poly(methyl-*n*-propylsilane) (PMPrS) to reveal relationships between backbone conformations and solution and optical properties of these polysilanes [235].

It was suggested that a mixed solvent of *n*-hexane and 2-propanol would be a Θ solvent for PDHS [92]. The refractive indexes of *n*-hexane and 2-propanol at 25 °C are, respectively, 1.3675 and 1.3704, thus being close to each other. In Figure 16.4, the second virial coefficients (A_2) of the PDHS solutions at 25 °C are plotted as a function of the weight fraction (w_p) of 2-propanol. The A_2 vs. w_p curve crosses the horizontal line of $A_2 = 0$ at $w_p = 0.418$. Therefore, the Θ solvent of PDHS is a mixed solvent of *n*-hexane (58.2%) and 2-propanol (41.8%) at 25 °C. Figure 16.5 shows the Zimm plot under the Θ condition. The open circles represent the observed data, and the filled circles are extrapolated points, and the extrapolated line corresponding to $\sin^2(\theta/2) = 0$ is almost horizontal and yields a fully small A_2 value of -8.09×10^{-6} ml mol g^{-2}. The other extrapolation to $c = 0$ leads to $\langle S^2 \rangle_0^{1/2} = 324$ Å. Both extrapolated lines cross each other on the ordinate axis, and the reciprocal of the intercept yields $M_w = 5.21 \times 10^5$. With the Si—Si bond length of 2.35 Å [321], the characteristic ratio can be evaluated as

$$\langle r^2 \rangle_0 / nl^2 = \frac{6\langle S^2 \rangle_0}{(M_w/M_0)\, l_{\text{Si-Si}}^2} = \frac{6 \times 324^2}{(5.21 \times 10^5/198.4) \times 2.35^2} = 43.4 \qquad (16.16)$$

Surprisingly enough, the unperturbed PDHS chain exhibits a large $\langle r^2 \rangle_0 / nl^2$ value of 43.4. The M_w values of four PDHS samples range from 5.21×10^5 to 2.21×10^6, and the radius of gyration can be expressed as a function of the molecular weight:

$$\langle S^2 \rangle_0^{1/2} = 0.519 M_w^{0.49} \qquad (16.17)$$

The exponent of M_w is very close to that (1/2) of the ideal chain.

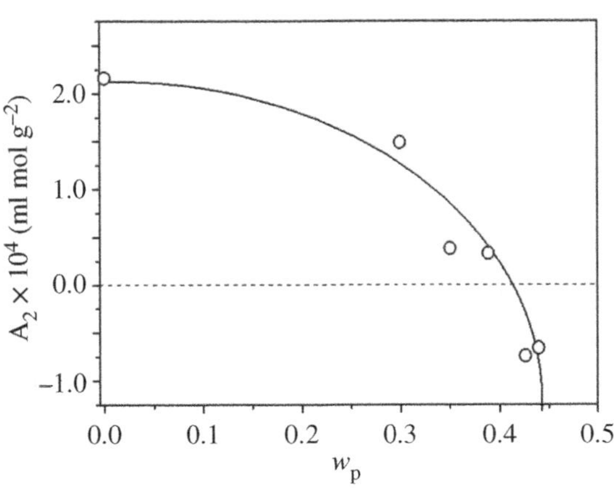

Figure 16.4 Second virial coefficient of PDHS and a mixed solvent of *n*-hexane and 2-propanol at 25 °C as a function of weight fraction of 2-propanol. Source: Reproduced with permission from reference [235]. Copyright 2001 American Chemical Society.

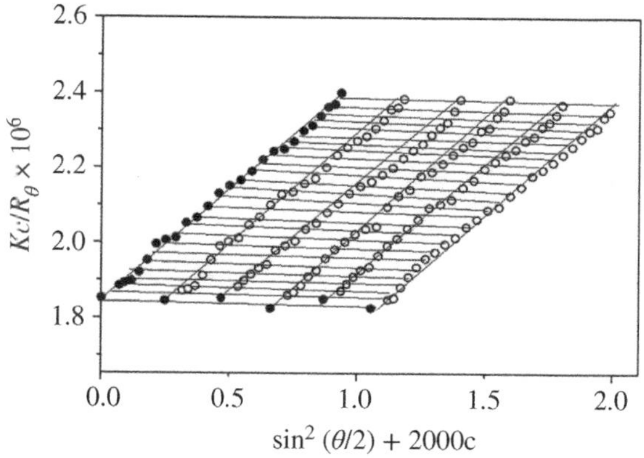

Figure 16.5 Zimm plot for PDHS in the Θ state dissolved in *n*-hexane (58.2%) and 2-propanol (41.8%) at 25 °C. Source: Reproduced with permission from reference [235]. Copyright 2001 American Chemical Society.

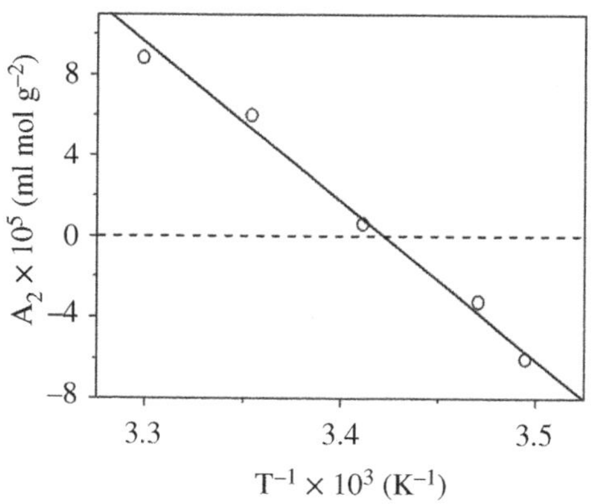

Figure 16.6 Second virial coefficient of PDBS and *n*-hexane as a function of reciprocal temperature. Source: Reproduced with permission from reference [235]. Copyright 2001 American Chemical Society.

Figure 16.6 shows the temperature dependence of the second virial coefficient of *n*-hexane solutions of PDBS. The A_2 vs. T^{-1} line crosses the horizontal dashed line of $A_2 = 0$ at 19.1 °C, which is the Θ point. The average $\langle r^2 \rangle_0 / nl^2$ value of four samples of $M_w = 3.83 \times 10^5$–2.40×10^6 is 42.3, close to that of PDHS. This fact suggests the similarity in backbone conformation between unperturbed PDHS and PDBS chains. The radius of gyration of PDBS can be related to M_w as

$$\langle S^2 \rangle_0^{1/2} = 0.613 M_w^{0.49} \tag{16.18}$$

The Θ solvent of PMPrS is also a mixture of *n*-hexane (62.6%) and 2-propanol (37.4%) at 25 °C. The average $\langle r^2 \rangle_0 / nl^2$ value of four samples ($M_w = 1.913 \times 10^5$ – 7.44×10^5) is 19.1, and the relation between $\langle S^2 \rangle_0$ and M_w was obtained as

$$\langle S^2 \rangle_0^{1/2} = 0.446 M_w^{0.50} \tag{16.19}$$

Figure 16.7 shows the UV absorption spectra observed from the three unperturbed polysilanes. The absorption maxima are found at 306 nm (PMPrS) and 314 nm (PDBS and PDHS), which supports the above suggestion that the

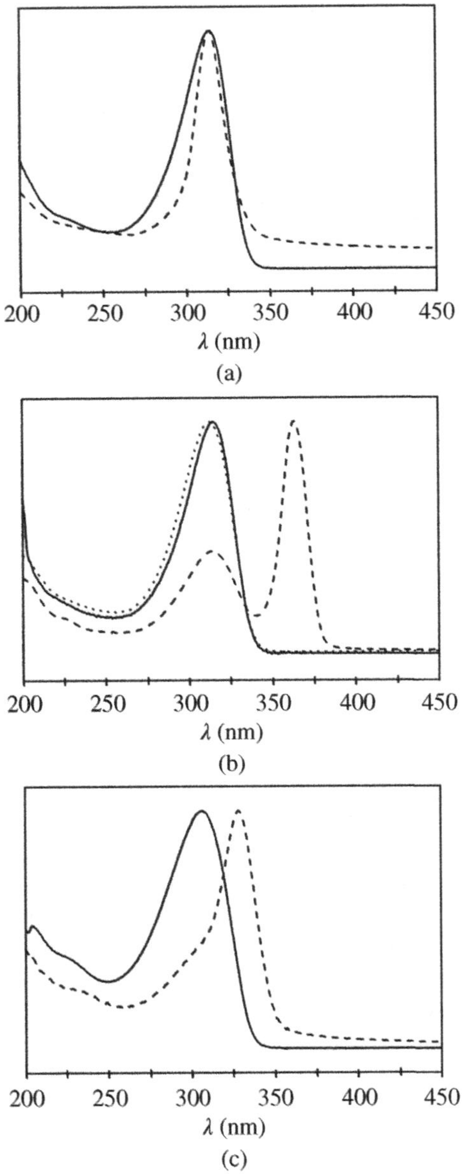

Figure 16.7 Ultraviolet absorption spectra of (a) PDBS, (b) PDHS, and (c) PMPrS: solid line, the θ solution; dotted line, fresh film; dashed line, film after standing at room temperature for a day. Source: Reprinted with permission from reference [235]. Copyright 2001 American Chemical Society.

unperturbed PDBS and PDHS adopt almost the same backbone conformation. Crystallization of PDHS induces an absorption shift from 314 to 329 nm, whereas that of PDBS stays at 314 nm. The PDBS chain may keep the helical conformation in the Θ solution and crystal, while PDHS changes its conformation from the helical (314 nm) to the all-trans (363 nm) form. After crystallization, the absorption of PMPrS is separated into two (306 and 329 nm). Because the asymmetric side chains of PMPrS cause various configurations, its crystal conformation would probably be ill-defined.

Molecular dynamics (MD) simulations on small model compounds of the three polysilanes provided conformational energies of the polymeric chains. With the MD energies, the RIS calculations were carried out for the three polysilanes and yielded characteristic ratios of the unperturbed chains: 42.0 (PDBS); 54.0 (PDHS); 12.3 (atactic PMPrS of 50% *meso* and 50% *racemo*) [406]. These $\langle r^2 \rangle_0 / nl^2$ values are sufficiently close to the experimental observations. By the MD/RIS simulations, the dihedral angles (φs) around the Si—Si bond of PDBS, PDHS, and PMPrS were averaged to be 150.6°, 151.2°, and 139.6°, respectively. In conclusion, the average φ angle is correlated with the absorption wavelength irrespective of the chemical species of the side chain.

16.2 Dynamic Light Scattering (DLS)

Diffusion due to the Brownian motion of polymeric particles in solutions can be traced by DLS [83, 429, 472]. The time autocorrelation of scattering intensity is given as the average of products between scattering intensities at two times separated by the delay τ:

$$\langle I(t)I(t+\tau)\rangle = \lim_{T\to\infty} \frac{1}{T} \int_0^T I(t)I(t+\tau)dt \tag{16.20}$$

where T is the period. The autocorrelation function corresponds to $\langle I^2 \rangle$ at $\tau = 0$ and approaches the baseline equal to $\langle I \rangle^2$ with increasing τ (Figure 16.8).

The autocorrelation function, being normalized as

$$g_2(\tau) = \frac{\langle I(t)I(t+\tau)\rangle}{\langle I(t)\rangle^2} \tag{16.21}$$

Figure 16.8 Light scattering intensity $I(t)$ as a function of time t (above) and the intensity autocorrelation function $\langle I(t)I(t+\tau)\rangle$ as a function of the delay time τ (below).

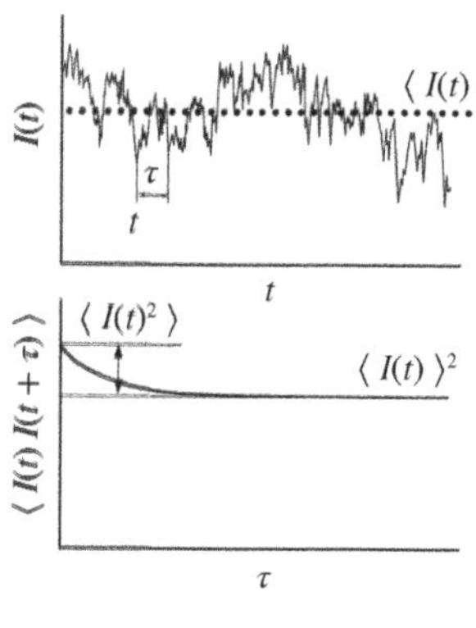

can be related to the electric-field autocorrelation function

$$g_1(\tau) = \frac{\langle E^*(t)E(t+\tau)\rangle}{\langle E^*(t)E(t)\rangle} \tag{16.22}$$

according to

$$g_2(\tau) = 1 + |g_1(\tau)|^2 \tag{16.23}$$

Here, $E(t)$ represents the electric field. Equation (16.23) is called the Siegert relation. In fact, Eq. (16.23) is modified by introducing the coherence factor β to be

$$g_2(\tau) = 1 + \beta|g_1(\tau)|^2 \tag{16.24}$$

Because the incident and scattering lights are not perfectly punctate but somewhat broadened, β becomes smaller than unity. Ideally, if both beams are very thin, then $\beta \approx 1$.

The $g_1(\tau)$ function is related to the diffusion coefficient D as

$$g_1(\tau) = \exp\left(-Dq^2\tau\right) \tag{16.25}$$

where q is the magnitude of the scattering vector defined in Eq. (16.8). The diffusion coefficient is expressed by the Stokes–Einstein equation:

$$D = \frac{kT}{6\pi\eta R_S} \tag{16.26}$$

where k is the Boltzmann constant, η is the viscosity of the solvent, and R_S is the Stokes radius. For polymers, the hydrodynamic radius (R_H) based on Eq. (16.26) is defined as

$$R_H = \frac{kT}{6\pi\eta D} \tag{16.27}$$

Double the R_H value is considered to be the particle size. If a parameter Γ is defined by

$$\Gamma = Dq^2 \tag{16.28}$$

then Eq. (16.25) is rewritten as

$$g_1(\tau) = \exp(-\Gamma\tau) \tag{16.29}$$

In DLS measurements, the θ angle is fixed at, for example, 90°, and hence, q is constant. If the particles are uniform in size, the $\ln g_1(\tau)$ vs. τ plot forms a straight line, and its slope yields the Γ value from which, via the D value, the particle size can be obtained with Eq. (16.27).

However, the particles usually have a size distribution. Then, Eq. (16.29) is replaced by

$$g_1(\tau) = \int_0^\infty G(\Gamma)\exp(-\Gamma\tau)\,d\Gamma \tag{16.30}$$

where $G(\Gamma)$ represents the size distribution. Equation (16.30) is none other than the Laplace transform of $G(\Gamma)$.

One solving method for Eq. (16.30) is the cumulant expansion:

$$\ln|g_1(\tau)| = \sum_{m=1} \frac{K_m}{m!}(-\tau)^m = -K_1\tau + \frac{1}{2}K_2\tau^2 - \frac{1}{6}K_3\tau^3 + \cdots \tag{16.31}$$

where

$$K_1 = \langle\Gamma\rangle \quad K_2 = \langle\Gamma^2\rangle - \langle\Gamma\rangle^2 \tag{16.32}$$

Equation (16.31) is fitted to the experimental $\ln g_1(\tau)$ values to obtain the coefficients K_ms. As in Eq. (16.32), for example, the K_1 and K_2 values give the average particle size and size fluctuation, respectively. Another solving method is the inverse Laplace transform, for which the established software termed constrained regularization method for inverting data (CONTIN) is available [373, 374].

16.2.1 Application of DLS: Size Distribution of Polystyrene Latex Particles

Figure 16.9 shows the $g_2(\tau)$ function of polystyrene (PS) latex particles dispersed in water at 26 °C, and the $\ln g_1(\tau)$ vs. τ curve is depicted in Figure 16.10. The particle size distribution is displayed in Figure 16.11. The average size is 115.0 nm.

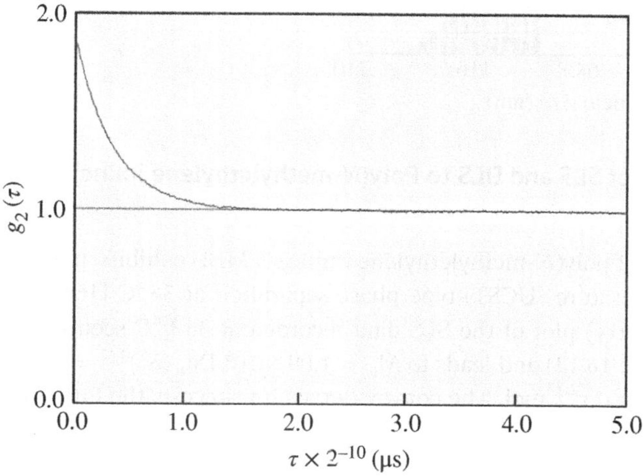

Figure 16.9 $g_2(\tau)$ vs. τ plot of polystyrene latex particles at 22.6 °C.

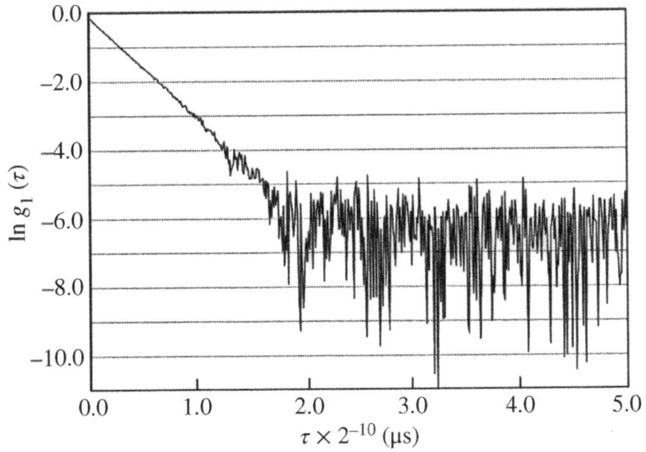

Figure 16.10 ln $g_1(\tau)$ vs. τ plot of the polystyrene latex particles at 22.6 °C.

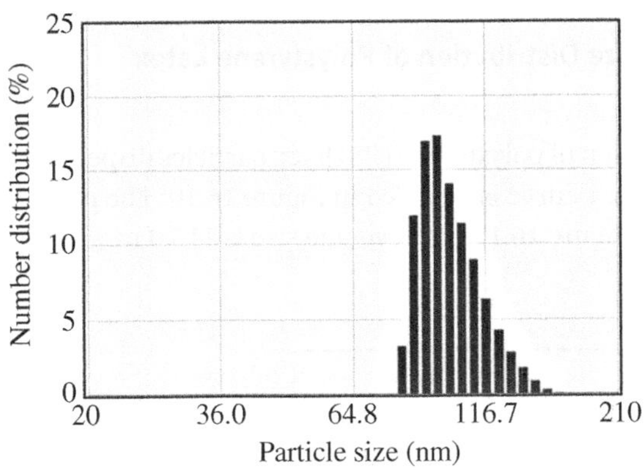

Figure 16.11 Particle size distribution of the polystyrene latex. The average size is 115 nm.

16.2.2 Application of SLS and DLS to Poly(*N*-methylethylene imine) Solutions

The toluene solution of poly(*N*-methylethylene imine) (PMEI) exhibits an upper critical solution temperature (UCST)-type phase separation at 36 °C [164]. The square-root Zimm (Berry) plot of the SLS data recorded at 36.1 °C seems to be convex upward (Figure 16.12) and leads to $M_w = 1.09 \times 10^6$ Da, $\langle S^2 \rangle^{1/2} = 41$ nm, and $A_2 = -3.4 \times 10^{-4} \text{cm}^3 \text{ g}^{-2}$ mol. The convex Berry plot suggests the formation of extraordinarily large particles [18].

The apparent molecular weight of the self-aggregates of PMEI is larger than one million, whereas a size-exclusion chromatographic measurement estimated M_w of the isolated PMEI chain to be 2.6×10^4 Da. For huge polymers, Fujita derived an

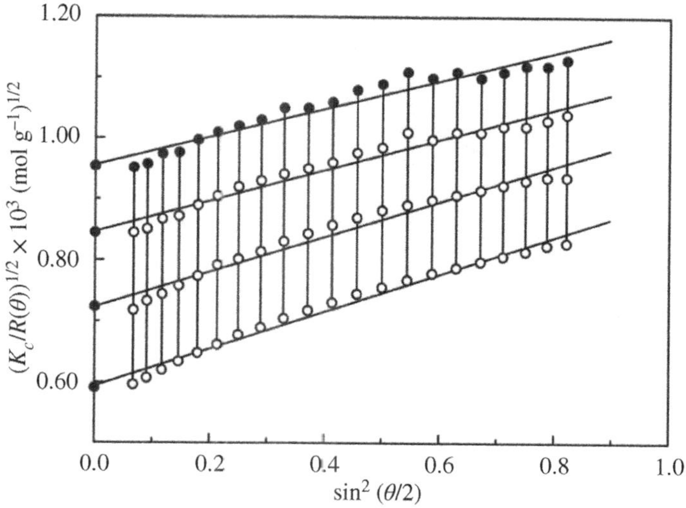

Figure 16.12 Square-root Zimm (Berry) plot of PMEI dissolved in toluene at 36.1 °C. The PMEI concentrations are 0.3171, 0.6342, and 1.022 mg mL. Source: Material from reference [164], published 2016, Springer Nature.

analytical method [162]:

$$\frac{1}{y(u)} = \frac{1}{M_w} + bZ(u) + \text{higher terms in } Z \tag{16.33}$$

where

$$y(u) = \lim_{c \to 0} \frac{R(\theta)}{Kc} \tag{16.34}$$

$$b = \frac{32\pi^2}{3\lambda^2} \frac{\langle S^2 \rangle}{M_w} \tag{16.35}$$

$$Z(u) = \left[y(u)u^{4/3}\right]^{-1} \int_0^u y(u)u \, du \tag{16.36}$$

and

$$u = \sin^3 \frac{\theta}{2} \tag{16.37}$$

The $Z(u)$ values were numerically calculated from the observed data. To exclude the effect of the higher terms in Eq. (16.33), a straight line was fitted to the data in the comparatively small Z range (Figure 16.13); the intercept and slope of the fitted line yielded, respectively, M_w^{-1} and b, from which $M_w = 1.17 \times 10^6$ Da and $\langle S^2 \rangle^{1/2} = 71$ nm were determined.

It is well known that poly(ethylene imine) forms complexes with DNA [98]. To examine whether PMEI forms aggregates with negatively charged molecules, if

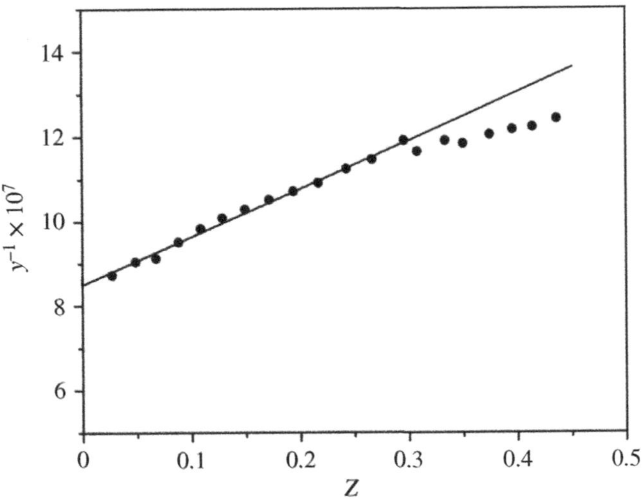

Figure 16.13 Fujita plot to analyze huge PMEI aggregates formed in toluene at 36.1 °C. Source: Material from reference [164], published 2016, Springer Nature.

that happens, to investigate the aggregation behaviors, PMEI was mixed in a PS latex aqueous solution. The particle size, namely, hydrodynamic diameter (D_h), and zeta potential (ζ) of the pure latex particle were determined to be 119 ± 2 nm and −53.8 mV, respectively.

Figure 16.14 shows the D_h and ζ values plotted as a function of PMEI concentration (c_{PMEI}). In the range of $0 \leq c_{PMEI} \lesssim 10\ \mu\text{mol l}^{-1}$ (MR, monomer ratio of PS to PMEI, $\gtrsim 290$), $D_h = 121$ nm and $\zeta = -47.3$ mV. At $c_{PMEI} = 38.5\ \mu\text{mol l}^{-1}$ (MR = 75), the ζ value was raised to −6.6 mV, and the D_h value of the PMEI and PS latex (PMEI-PS) aggregates reached as large as 2.48×10^4 nm, increased still further with time, and, finally, the aggregates were deposited. At $c_{PMEI} = 57.8\ \mu\text{mol l}^{-1}$ (MR = 50), the D_h value dropped down to 133 nm and stayed constant independently of time, and the ζ potential was positive (+20.1 mV). At $c_{PMEI} = 96.3\ \mu\text{mol l}^{-1}$ (MR = 30), $D_h = 121$ nm (equal to the initial value) and $\zeta = +33.5$ mV.

The experimental results can be interpreted as illustrated in Figure 16.14.

- Range I: $c_{PMEI} \leq 10\ \mu\text{mol l}^{-1}$, the PS latex particles absorb the PMEI chains on the surface but still stay charged negatively and repulse each other.
- Range II: $10 \leq c_{PMEI} \leq 60\ \mu\text{mol l}^{-1}$, the PMEI-PS particles, being close to (or equal to) neutrality in charge, combine with each other and form huge aggregates.
- Range III: $c_{PMEI} \geq 60\ \mu\text{mol l}^{-1}$, the PMEI-PS particles are charged positively and dispersed owing to the electrostatic repulsions.

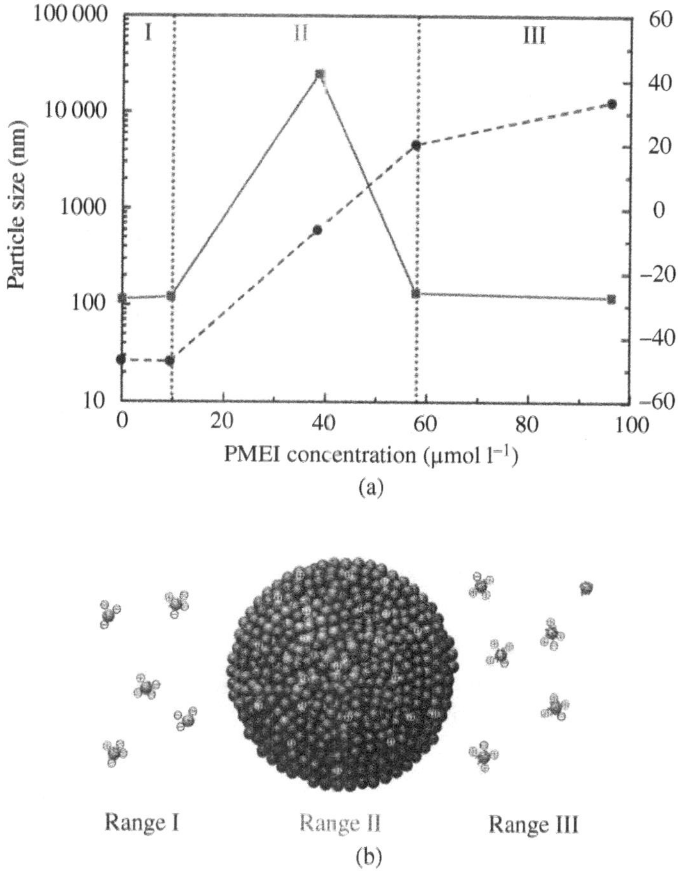

Figure 16.14 Aggregation behaviors of PMEI and PS latex particles in water: (a) particle size (hydrodynamic diameter) and zeta potential as a function of the PMEI concentration and (b) schematic illustration of the aggregations formed in the three concentration ranges. Source: Material from reference [164], published 2016, Springer Nature.

In ranges II and III, PMEI clearly behaves as a flocculant and a dispersant, respectively.

16.3 Small-angle Neutron Scattering (SANS)

SANS can be used to investigate shapes and dimensions of polymeric chains placed in solutions and in condensed phases. LS and SAXS originate from interactions of the electromagnetic waves with electrons surrounding molecules, while SANS stems from those of neutrons with inner nuclei; LS, SAXS, and SANS detect differences in refractive index, electron density, and scattering length, respectively [83, 172, 196, 390]. Except for the source and instrumentation, the data treatments are, in principle, common to the three scattering techniques. Compared with LS and SAXS, SANS has an advantage to differentiate between hydrogen and deuterium because the two elements have significantly different neutron scattering lengths [390, 521]. Hydrogen is mostly included in most polymers, and deuterium labeling may not disturb the static chemical properties. SANS is divided into elastic and inelastic scatterings. The former has been used to investigate conformational and configurational properties of polymers, and hence, its theoretical basis and analytical methods are exclusively dealt with herein. In addition, its application to amorphous poly(ethylene terephthalate) (PET) [171] is exemplified for the sake of comparison with molecular orbital (MO) and RIS calculations on unperturbed PET [400].

Small angle means that the scattering angle 2θ is small. It should be noted that the scattering angle, the angle between the incident and the scattering beams, is defined as 2θ and different from that (θ) of LS. According to Bragg's law, $2d \sin\theta = \lambda$, the lattice spacing d in the real space is calculated from

$$d = \frac{\lambda}{2\sin\theta} \tag{16.38}$$

where λ is the wavelength. Therefore, small-angle scattering and diffraction provide us with information on comparatively large structures. The magnitude of the scattering vector $\mathbf{q}$ is calculated from

$$q = |\mathbf{q}| = \frac{4\pi}{\lambda}\sin\theta \tag{16.39}$$

Accordingly, d is related to q by

$$d = \frac{2\pi}{q} \tag{16.40}$$

For example, the SANS of a spherical particle of radius R is expressed with Eq. (16.9) as

$$I(q) = \rho_0^2\, v^2 \left[\frac{3}{x^3}(\sin x - x\cos x)\right]^2 \qquad x \equiv qR \tag{16.41}$$

where ρ_0 is the scattering length density, and v is the particle volume. The SANS intensity of the Gaussian chain is given by

$$I(q) = \rho_0^2\, v^2 D(x) \tag{16.42}$$

where $D(q)$ is the Debye function (Eq. (16.10)). In a small q range (the Guinier region), the Debye function is approximated as

$$D(q) = 1 - \frac{q^2\langle S^2\rangle}{3} + \cdots \tag{16.43}$$

while, in a large q range, $D(q)$ varies depending on q^{-2}; thus, an $I(q)q^2$ vs. q plot (designated as the Kratky plot) exhibits a plateau, whose formation shows the applicability of the Gaussian coil to the polymer. For polydisperse polymers, the Debye function is modified to [174]

$$D(x) = \frac{2\left[(1+Ux)^{-\frac{1}{U}} + x - 1\right]}{(1+U)x^2} \qquad x \equiv q^2 \frac{\langle S^2 \rangle_Z}{1+2U} \tag{16.44}$$

where U is the polydispersity defined as

$$U = \frac{M_w}{M_n} - 1 \tag{16.45}$$

and $\langle S^2 \rangle_Z$, the mean-square Z-average radius of gyration, is related to the weight-average one ($\langle S^2 \rangle_W$) by

$$\langle S^2 \rangle_W = \frac{1+U}{1+2U} \langle S^2 \rangle_Z \tag{16.46}$$

The fitting of the original or modified Debye function to the SANS intensity may lead to the $\langle S^2 \rangle$ value.

Another method to determine the radius of gyration is the Guinier plot ($\ln I(q)$ vs. q^2), which is based on the Guinier law [179] represented as

$$I(q) = \rho_0^2\, v^2 \left(1 - \frac{\langle S^2 \rangle}{3} q^2 + \cdots \right) = \rho_0^2\, v^2 \exp\left(-\frac{\langle S^2 \rangle}{3} q^2\right) \tag{16.47}$$

The Guinier plot exhibits a straight line in a very small q range, and its slope corresponds to the $-\langle S^2 \rangle/3$ value.

Similar to LS, the Zimm plot is applicable to the SANS data and yields $\langle S^2 \rangle$ as the slope and M_w as the intercept. In LS, a number of solutions of different polymer concentrations are required to be prepared and subjected to the SLS measurements, and the extrapolation to null concentration leads to the $\langle S^2 \rangle$ value. In SANS, however, it is possible to derive the $\langle S^2 \rangle$ and M_w values from only a single sample containing both deuterium-labeled and unlabeled polymers [16, 522, 523]. As stated above, neutron scattering lengths of hydrogen and deuterium are, respectively, $b_H = -0.374 \times 10^{-12}$ cm and $b_D = 0.667 \times 10^{-12}$ cm, being greatly different from each other. The scattering intensity observed from a mixed sample of deuterium-labeled and unlabeled polymers is expressed as [522]

$$I(q) = \left(b_H - b_D\right)^2 \phi_H\, \phi_D\, N\, Z^2\, P(q) \tag{16.48}$$

where ϕ_H and ϕ_D are the volume fractions of unlabeled and labeled species ($\phi_H + \phi_D = 1$), N is the number of molecules, Z is the weight-average degree of polymerization, and $P(q)$ is the form factor of a single molecule. The scattering intensity may be maximized at $\phi_H = \phi_D = 0.5$. Equation (16.48) indicates that the form factor $P(q)$ can be determined from only a single sample of an appropriate ϕ_D. If the normalized scattering intensity is defined as

$$S(q) = \frac{I(q)}{ZN\phi_H\phi_D\left(b_H - b_D\right)^2} \tag{16.49}$$

the slope of the Zimm plot ($S(q)^{-1}$ vs. q^2) yields the $\langle S^2\rangle_Z$ value because the form factor is expressed as

$$P(q)^{-1} = 1 + \frac{\langle S^2\rangle_Z}{3}q^2 + \cdots \tag{16.50}$$

16.3.1 Application of SANS to Amorphous PET

Deuterium-labeled PET of $M_w = 4.6 \times 10^4$ Da and unlabeled PET of $M_w = 4.0 \times 10^4$ Da were mixed at given ϕ_D fractions, melt-pressed at 250 °C for 20 s to remove voids that would exhibit intense scattering, and quenched in ice water to freeze the molten state [171]; therefore, the quenched sample was expected to keep the amorphous state of molten PET at 250 °C.

Figure 16.15 shows the Zimm plots of $\phi_D = 0.5$ (square) and 0.7 (circle). Both data are seen to lie on the same line, from which $M_w = 2.73 \times 10^4$ Da and $\langle S^2\rangle_z = 65$ Å were derived. With $U = 0.96$ obtained from the Kratky plot in Figure 16.16, the $(\langle S^2\rangle_w/M_w)^{1/2}$ value was determined to be 0.32 Å.

The reduction in M_w was ascribed to the melt-pressing at 250 °C [261]. Figure 16.16 represents the Kratky plot for the same samples as in Figure 16.15. As explained above, the horizontal region is formed, indicating that the PET chain can be properly approximated by the Gaussian coil. The solid line expressing Eq. (16.44) was fitted to the observed data, and as a consequence, the polydispersity and $(\langle S^2\rangle_w/M_w)^{1/2}$ were obtained as 0.96 and 0.34 Å, respectively. The dotted line

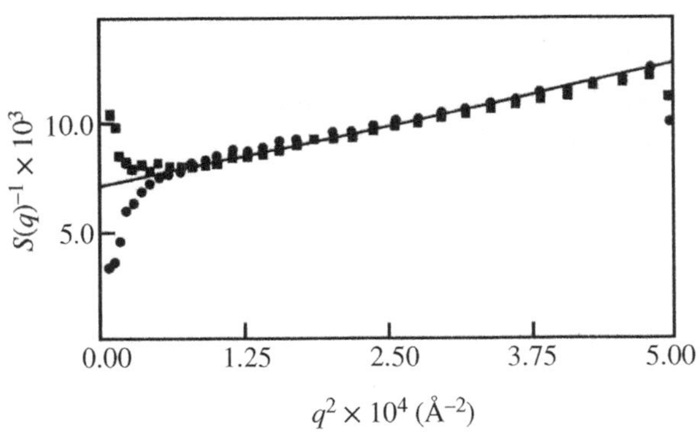

Figure 16.15 Zimm plots of amorphous PET blends (melt-pressed at 250 °C for 20 s and quenched in ice water): $\phi_D = 50\%$ (square) and 70% (circle). Source: Reproduced with permission from reference [171]., Copyright 1986, with permission from Elsevier.

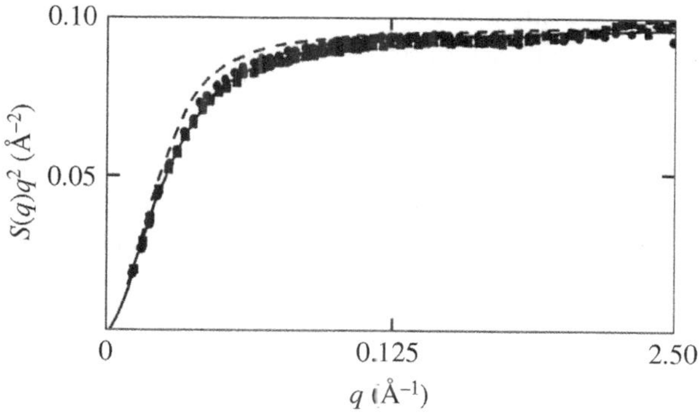

Figure 16.16 Kratky plots of the amorphous PET blends: $\phi_D = 50\%$ (square) and 70% (circle). The solid and dashed lines were calculated from Eq. (16.44), with $U = 0.96$ and 0.00, respectively. Source: Reproduced with permission from reference [171], Copyright 1986, with permission from Elsevier.

in Figure 16.16 represents Eq. (16.44) using $U = 0$, deviating somewhat from the data points.

The $(\langle S^2\rangle_w/M_w)^{1/2}$ values of the Zimm (0.32 Å) and Kratky (0.34 Å) plots lead to the characteristic ratios $(\langle r^2\rangle_0/nl^2)$ of 2.7 and 3.1, respectively; the bond lengths used are $l_{O-C(=O)} = 1.357$ Å, $l_{C(=O)-C(=O)} = 5.778$ Å, $l_{O-CH_2} = 1.439$ Å, and $l_{CH_2-CH_2} = 1.515$ Å. A virtual bond (5.778 Å) between two *para* carbonyl carbon atoms attached to the benzene ring was assumed as illustrated in Figure 15.10. From the $(\langle S^2\rangle_w/M_w)^{1/2}$ value, the $\langle r^2\rangle_0/nl^2$ ratio can be calculated as follows:

$$\frac{\langle r^2\rangle_0}{nl^2} = \frac{6\langle S^2\rangle_0}{(M/M_0)\,l_0^2} = 6\left(\frac{\langle S^2\rangle_0}{M}\right)\frac{M_0}{l_0^2} = 6(0.32)^2 \times \frac{192}{43.51} = 2.7 \qquad (16.51)$$

where M_0 and l_0^2 are the formula mass and the sum of square bond lengths of the repeating unit, respectively.

The conformational energies of PET at 25 and 250 °C, obtained from the model compound of PET, ethylene glycol dibenzoate (EGDB, see Figure 15.10b), are listed in Table 12.1. From the energy parameter, the $\langle r^2\rangle_0/nl^2$ values of unperturbed PET chains at 25 and 250 °C were calculated by the refined RIS method to be 2.63 and 2.84, respectively [400]. The latter value falls within the experimental results of 2.7 (Zimm)–3.1 (Kratky), being in good agreement with the experiment.

On the basis of the discussion regarding Eq. (3.20) described earlier [145], Flory predicted that molten polymers must lie in the Θ state. The studies cited here support his statement [146].

One may observe that statistical mechanical treatments of systems of polymer chains invariably lead to a partition function that is factorable into intra- and intermolecular parts, each independent of the other. Only the latter factor depends on the composition; the former factor is unaffected by dilution (in a Θ-solvent). It follows at once that the configuration of the chain should be independent of the polymer concentration, and hence, that its spatial configuration should be unperturbed by the requirement that it must coexist in the amorphous state with other chains at high density.

Part V

Applications: Conformational Analysis and Elucidation of Structure–Property Relationships of Polymers

Figure V.1 shows a portion of the periodic table, which displays the elements contained in the main chains of most synthetic and natural polymers: group 14, carbon and silicon; group15, nitrogen and phosphorus; and group 16, oxygen, sulfur, and selenium. Carbon, nitrogen, and oxygen form four, three, and two covalent bonds, respectively. Nitrogen and oxygen have one and two lone pairs, respectively. The atoms except carbon and hydrogen are designated as heteroatoms. We have so far studied the polymers listed in Table V.1 mainly in terms of conformation and found that conformational characteristics and physical properties of the polymers are influenced by intramolecular interactions due to the heteroatoms (the representative examples are depicted in Figure V.2). In Table V.1, the conformational characteristic, namely, intramolecular interactions, stable conformations, and characteristic ratios of the polymers, are summarized. In the following chapters, the study results are individually described in detail.

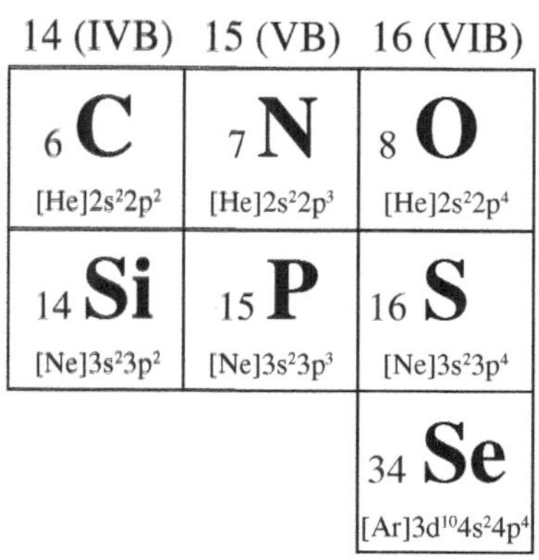

Figure V.1 Elements forming the main chains of most synthetic and natural polymers. Source: Reproduced from reference [411] with permission from the PCCP Owner Societies.

Conformational Analysis of Polymers: Methods and Techniques for Structure-Property Relationships and Molecular Design, First Edition. Yuji Sasanuma.

Table V.1 Conformational characteristics of polymers treated herein: intramolecular interactions, stable conformations, and characteristic ratios.

Polymer	Intramolecular interaction (kcal mol^{-1})		
	Stable conformation(s)	Characteristic ratio	Reference
	Polyether (C, H, O)		
Poly(methylene oxide)	Dipole–dipole, $n_O \rightarrow \sigma^*_{C-O}$, and $E_\sigma = -1.5$ to -2.0		[399, 423]
[—O—CH_2—]	gg	8.0–9.2	
Poly(ethylene oxide)	C—H· · ·O, $E_\sigma = -1.4$ to $+0.1$, and $E_\omega = -1.2$ to $+0.4$		[399, 412, 413]
[—O—CH_2—CH_2—]	ttt, tgt, and $tg^{\pm}g^{\mp}$	4.1–5.1	
Poly(trimethylene oxide)	$E_\sigma = -0.7$ to -0.5		[420]
[—O—CH_2—CH_2—CH_2—]	tggt	3.7	
Poly(propylene oxide)	C—H· · ·O and $E_\omega = -1.9$ to -0.9		[397–399, 405]
[—O—CH_2—$CH(CH_3)$—]	tg^+g^- and ttt [(*R*)-form]	5.9 (isotactic)	
Poly(tetramethylene oxide)	C—H· · ·O, $E_{\sigma 1} = -0.6$, and $E_\omega = -0.4$		[270, 271, 399]
[—O—CH_2—CH_2—CH_2—CH_2—]	tgtgt	5.7	
	Polyamine (C, H, N)		
Poly(ethylene imine)	N—H· · ·N ($E_\eta = -1.54$ and $E_\nu = -0.58$)		[403, 408]
[—NH—CH_2—CH_2—]	*meso* (*ll*) tgt	2.9 (HBS=100%) and 6.3 (HBS=0%)	
Poly(*N*-methylethylene imine)	C—H· · ·N ($E_{\nu'} = -0.27$) and $E_\sigma = -0.42$		[164, 411]
[—$N(CH_3)$—CH_2—CH_2—]	*meso* (*ll*) tg^+g^+ and *racemo* (*ld*) tg^+t		
Poly(trimethylene imine)	N—H· · ·N ($E_{\psi''} = -0.83$) and C—H· · ·N ($E_\eta = -0.15$)		[417]
[—NH—CH_2—CH_2—CH_2—]	*racemo* (*ld*) $tg^{\pm}g^{\mp}t$	3.5	
Poly(*N*-methyltrimethylene imine)	C—H· · ·N ($E_\eta = -0.40$)		[417]
[—$N(CH_3)$—CH_2—CH_2—CH_2—]	*racemo* (*ld*) $tg^+g^+g^+$	4.2	

(Continued)

Table V.1 (Continued)

Polymer	Intramolecular interaction (kcal mol^{-1})		
	Stable conformation(s)	Characteristic ratio	Reference
	Polyphosphine (C, H, P)		
Poly(1-methylphosphirane)	No specific interaction		[411]
[—P(CH_3)—CH_2—CH_2—]	*meso* (*ll*) ttt	7.4 (isotactic) and 6.7 (syndiotactic)	
Poly(1-phenylphosphirane)	$\pi - \pi$ ($G_{racemo\ ttt} - G_{meso\ ttt} = -0.89$)		[411]
[—P(C_6H_5)—CH_2—CH_2—]	*racemo* (*ld*) ttt	7.3 (isotactic) and 38 (syndiotactic)	
	Polysulfide (C, H, S)		
Poly(methylene sulfide)	Dipole–dipole, $n_S \rightarrow \sigma^*_{C-S}$, and $E_\sigma \approx -1.0$ to -1.5		[423]
[—S—CH_2—]	gg	7.6	[399]
Poly(ethylene sulfide)	Dipole–dipole, $E_\rho \approx -0.4$, and S···S repulsion		[412]
[—S—CH_2—CH_2—]	$g^\pm tg^\mp$	3.1	
Poly(trimethylene sulfide)	$E_\rho \approx -0.2$ to -0.3 and $E_\sigma \approx -0.3$ to -0.6		[420]
[—S—CH_2—CH_2—CH_2—]	gggg	3.6	
Poly(propylene sulfide)	$E_\delta \approx -0.4$, $E_\sigma \approx -0.5$, and C—H···S repulsion		[404]
[—S—CH_2—CH(CH_3)—]	g^-tg^- [(*R*) form]	4.0 (isotactic) and 3.9 (atactic)	[399]
	Polyselenide (C, H, Se)		
Poly(methylene selenide)	Dipole–dipole, $n_{Se} \rightarrow \sigma^*_{C-Se}$, and $E_\sigma \approx -1.0$		[421]
[—Se—CH_2—]	gg	7.3	
Poly(ethylene selenide)	$E_\rho \approx -0.3$, $E_\sigma \approx 1$, and $E_\kappa \approx -0.3$		[421]
[—Se—CH_2—CH_2—]	$g^\pm tg^\mp$	5.4	
Poly(trimethylene selenide)	$E_\rho \approx -0.4$ and $E_\sigma \approx -0.1$		[421]
[—Se—CH_2—CH_2—CH_2—]	$g^\pm ttg^\mp$	5.3	

	Copolymer (C, H, N, O, S)		
Poly(ethylene imine-*alt*-ethylene oxide)	N—H· · ·O, C—H· · ·N, and C—H· · ·O		[409]
[—NH—CH_2—CH_2—OCH_2CH_2—]	tg^-t [*l* form]	1.3 (HBS = 100%) and 6.9 (HBS = 0%)	[408]
Poly(ethylene imine-*alt*-ethylene sulfide)	N—H· · ·S and $E_{\rho(C-S)} = -0.4$		[203, 408]
[—NH—CH_2—CH_2—SCH_2CH_2—]	tg^-g^- [*l* form]	5.2 (HBS = 100%) and 6.4 (HBS = 0%)	
Poly(ethylene oxide-*alt*-ethylene sulfide)	C—H· · ·O, $E_{\rho(C-S)} = -0.4$, and $E_{\kappa} = -0.3$ (dipole–dipole)		[402]
[—O—CH_2—CH_2—SCH_2CH_2—]	$ttg^{\pm}$	6.0	
	Aromatic polyester (C, H, O)		
Poly(ethylene terephthalate)	C=O· · ·H—C, —O· · ·H—C, dipole–dipole, and $E_{\sigma} \approx -1.2$		[400]
[—COC_6H_4COO—$(CH_2)_2$—O—]	*trans*-tgt	2.6	
Poly(trimethylene terephthalate)	$\pi - \pi$, C=O· · ·H—C, —O· · ·H—C, dipole–dipole, and $E_{\sigma} \approx -0.6$		[414]
[—COC_6H_4COO—$(CH_2)_3$—O—]	*trans*-tggt, tggg, and tgtg	4.1	[419]
Poly(butylene terephthalate)	$\pi - \pi$, C=O· · ·H—C, —O· · ·H—C, dipole–dipole, and $E_{\sigma} \approx -0.3$		[419]
[—COC_6H_4COO—$(CH_2)_4$—O—]	*trans*-tgtgt, tgtgg, and $tg^{\pm}tg^{\mp}t$	3.8	
Poly(ethylene naphthalate)	C=O· · ·H—C, —O· · ·H—C, dipole–dipole, and $E_{\sigma} \approx -1.2$		[400]
[—$COC_{10}H_6COO$—$(CH_2)_2$—O—]	tgt	2.4	
	Aliphatic Polyester (C, H, O)		
Poly(glycolic acid)	C(=O)· · ·O repulsion, electrostatic, and dipole–dipole		[422]
[—CO—O—CH_2—]	tgt	5.3	
Poly(L-lactide)	C(=O)· · ·O repulsion		[418, 422]
[—CO—O—$CH(CH_3)$—]	tg^+t	9.0 (MP2) and 11.8 (B3LYP)	
Poly((*S*)-2-hydroxybutyrate)	C(=O)· · ·O repulsion		[422]
[—CO—O—CH(CH_2—CH_3)—]	$tg^+t(g^+)$	4.7–6.4	
Poly((*R*)-3-hydroxybutyrate)	C=O· · ·H—C and —O· · ·C—H		[407]
[—CO—O—$CH(CH_3)$—CH_2—]	ttg^+g^+ and ttg^+t	5.6	

(*Continued*)

Table V.1 (Continued)

Polymer	Intramolecular interaction (kcal mol^{-1}) Stable conformation(s)	Characteristic ratio	Reference
Poly(ε-caprolactone)	C=O· · ·H—C and —O· · ·C—H		[238]
[—CO—O—$(CH_2)_5$—]	$tg^{\pm}ttg^{\pm}t$ and $tg^{\pm}ttg^{\mp}t$	4.1–4.5	
Poly(ethylene succinate)	$E_{\sigma} \approx -1.3$ and $E_{\rho} \approx 0.3 - 0.5$		[410]
[—CO—$(CH_2)_2$—CO—O— $(CH_2)_2$—O—]	ttttttgt	6.2	
Poly(butylene succinate)	$E_{\sigma} \approx -0.1$ and $E_{\rho} \approx 0.5 - 0.6$		[410]
[—CO—$(CH_2)_2$—CO—O— $(CH_2)_4$—O—]	$tttttg^{\pm}tg^{\mp}t$	6.7–7.1	
	Polycarbonate (C, H, O)		
Poly(ethylene carbonate)	C=O· · ·H—C and electrostatic ⊖· · ·⊖ repulsion		[415]
[—OCO—CH_2—CH_2—]	tgt, tgg, and $g^{\pm}g^{\pm}g^{\mp}$	2.4–2.5	
Poly(propylene carbonate)	C=O· · ·H—C and electrostatic ⊖· · ·⊖ repulsion		[415]
[—OCO—CH_2—$CH(CH_3)$—]	$g^{+}g^{+}t$ and $tg^{+}t$ [(*R*)-form]	2.3–2.4 (isotactic)	
Poly(cyclohexene carbonate)	C=O· · ·H—C		[542, 543]
[—OCO—C*H—C*H—]	$tg^{+}t$ [(*R*,*R*)-form]	[*trans*-form] 21–37 (iso) and 0.7–1.7 (syn)	
*$CH_2CH_2CH_2$*CH_2 (* connection point)	$g^{-}g^{+}g^{+}$ and $g^{-}g^{-}g^{+}$ [(*R*,*S*)-form]	[*cis*-form] 0.5–0.65 (iso) and 16–60 (syn)	
	Aliphatic Polyamide (C, H, N, O)		
Nylon 4	N—H· · ·O=C		[165]
[—NH—CH_2—CH_2—CH_2—C(=O)—]	$g^{\pm}g^{\mp}g^{\mp}g^{\pm}$, $tg^{\pm}g^{\mp}t$, and $g^{\pm}g^{\mp}g^{\pm}t$		
	Aromatic Polyamide (C, H, N, O)		
Poly(ethylene terephthalamide)	N—H· · ·O and C—H· · ·O		[331]
[—COC_6H_4CONH—$(CH_2)_2$—NH—]	*cis*-*trans*-$tg^{\pm}g^{\mp}$	3.2	
Poly(trimethylene terephthalamide)	N—H· · ·O=C, C—H· · ·π, and $\pi - \pi$		[331]
[—COC_6H_4CONH—$(CH_2)_3$—NH—]	*cis*-*trans*-tttg	3.7	

	Aromatic Polythioamide (C, H, N, S)		
Poly(ethylene terephthalthioamide)	N—H· · ·S and C—H· · ·S		[331, 332]
[—CSC_6H_4CSNH—$(CH_2)_2$—NH—]	*cis-trans*-ggg	3.6	
Poly(trimethylene terephthalthioamide)	N—H· · ·S=C, C—H· · ·π, and $\pi - \pi$		[331, 333]
[—CSC_6H_4CSNH—$(CH_2)_3$—NH—]	*cis-trans*-tttt	5.7	
	Polythionoester (C, H, O, S)		
Poly(ethylene thionoterephthalate)	C=S· · ·H—C, —O· · ·H—C, and $E_\sigma \approx -2.1$		[416, 468]
[—CSC_6H_4CSO—$(CH_2)_2$—O—]	tgt	2.1	
	Polythioester (C, H, O, S)		
Poly(ethylene dithioterephthalate)	Dipole–dipole, C=O· · ·H—C, and S· · ·S repulsion		[12, 331]
[—COC_6H_4COS—$(CH_2)_2$—S—]	$g^{\pm}tg^{\mp}$	16.7	
Poly(trimethylene dithioterephthalate)	Dipole–dipole, C=O· · ·H—C, and S· · ·S repulsion		[10, 11, 331]
[—COC_6H_4COS—$(CH_2)_3$—S—]	gttg	10.5	
	Polydithioester (C, H, S)		
Poly(ethylene tetrathioterephthalate)	Dipole–dipole, C=S· · ·H—C, and S· · ·S repulsion		[12, 13, 331]
[—CSC_6H_4CSS—$(CH_2)_2$—S—]	*trans-cis*-$g^{\pm}tg^{\mp}$	5.2	
Poly(trimethylene tetrathioterephthalate)	Dipole–dipole, C=S· · ·H—C, and S· · ·S repulsion		[10, 11, 331]
[—CSC_6H_4CSS—$(CH_2)_3$—S—]	*trans-cis*-gttg	6.1	
	Polysilane (C, H, Si)		
Poly(methyl-*n*-propylsilane)	Repulsion between side chains		[235, 406]
[—$Si(CH_3)(C_3H_7)$—]	$T_{\pm}T_{\pm}$ and $T_{\pm}T_{\mp}$ (isotactic)	19.9 (expt) and 12.3 (calcd, $P_{meso} = 0.5$)	
Poly(di-*n*-butylsilane)	Repulsion between side chains		[235, 406]
[—$Si(C_4H_9)_2$—]	$D_{\pm}$	42.3 (expt) and 42.0 (calcd)	
Poly(di-*n*-hexylsilane)	Repulsion between side chains		[235, 406]
[—$Si(C_6H_{13})_2$—]	$D_{\pm}$	42.5 (expt) and 54.0 (calcd)	
	Polyolefin (C, H)		
Poly(ethylene)	Steric repulsion and dispersion force		
[—CH_2—CH_2—]	tt	7.98	

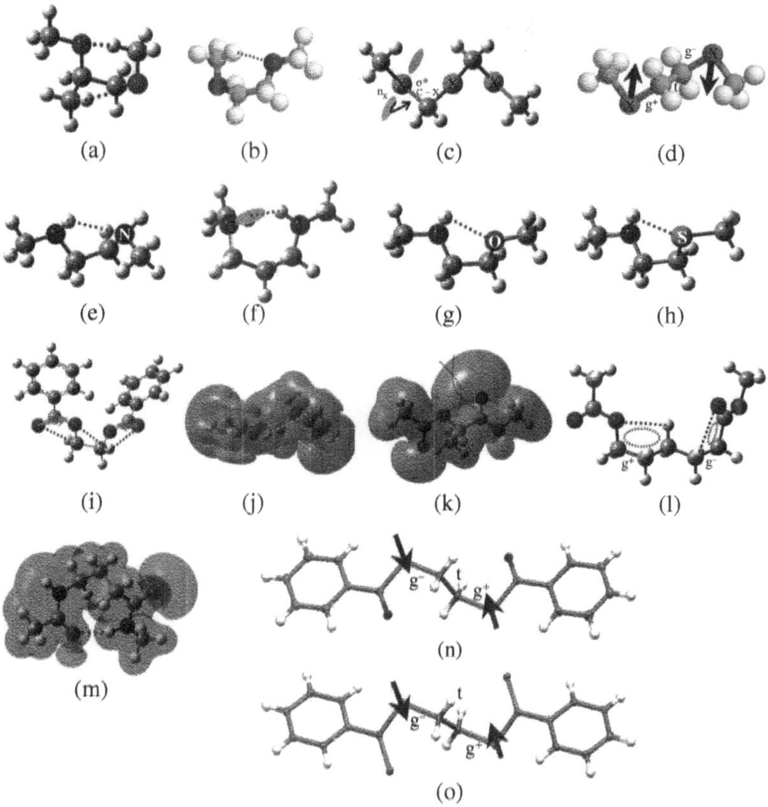

Figure V.2 Examples of intramolecular interactions characteristic of polymers including heteroatoms in the backbone, illustrated for model compounds: (a) C−H···O of poly(propylene oxide); (b) C−H···O of poly(ethylene oxide); (c) X→ σ^*_{C-X} (X = O, S, and Se) of poly(methylene oxide), poly(methylene sulfide), and poly(methylene selenide); (d) dipole–dipole interaction of poly(ethylene sulfide); (e) N−H···N of poly(ethylene imine); (f) N−H···N of poly(trimethylene imine); (g) N−H···O of poly(ethylene imine-*alt*-ethylene oxide); (h) N−H···S of poly(ethylene imine-*alt*-ethylene sulfide); (i) C=O···H−C and dipole–dipole interactions of poly(ethylene terephthalate); (j) π–π interaction of poly(trimethylene terephthalate); (k) C(=O)···O repulsion of poly(L-lactide); (l) C=O···H−C and −O···C−H of poly(ε-caprolactone); (m) N−H···O=C of nylon 4; (n) dipole–dipole interaction of poly(ethylene dithioterephthalate); (o) dipole–dipole interaction of poly(ethylene tetrathioterephthalate). All the interactions except O···O of poly(L-lactide) are attractive. Source: [401]/Reproduced with permission of American Chemical Society. [165, 401]/Reproduced from reference [12, 238] with permission from the Royal Society of Chemistry.

17

Polyethers

17.1 Poly(methylene oxide) (PMO)

In the community of organic stereochemistry, it is well known that the alkoxy substituent at the C-2 site of pyranose ring strongly prefers the axial (*gauche*) to the equatorial (*trans*) form (Figure 17.1). This phenomenon was designated as the anomeric effect [227, 475]. Some explanations for the anomeric effect have been suggested (see Figure 17.1): (1) Dipole–dipole interaction between the C—O bonds of the pyranose ring and alkoxy substituent (the *gauche* state with smaller resultant dipole moment is more stabilized); (2) Delocalization of lone pair electrons (n_O) of the oxygen atom by hyperconjugation ($n_O \rightarrow \sigma^*_{C\text{-}O}$) with the antibonding orbital ($\sigma^*_{C\text{-}O}$) (the *gauche* conformation, in which the lone pair is *antiperiplanar* to the acceptor C—O bond, is stabilized).

The simplest polymer containing the O—C—O—C bond sequence is PMO, which has been widely used as engineering thermoplastics because of its high crystallinity, superior lubricity properties, and good chemical resistance [290]. As described in section 10.7, PMO crystallizes in either 2/1 or 9/5 helical structure of the all-gauche conformation. The conformational energy (E_σ) representing the gauche stability relative to the trans form in the C—O bond was evaluated to be −1.4 [7], −1.5 [141, 153], −1.74 [502], −2.5 [438, 439], −2.5 [5], or −3.08 to −3.38 [323] kcal mol^{-1}.

The density functional theory (DFT) calculations at the B3LYP/6-311+G(2d,p) level evaluated the gauche stability of bond 2 of PMO-2 as −2.15 kcal mol^{-1} [423] (for the model compounds, see Figure 17.2). The E_σ energies for bonds 2 (tg$^+$g$^+$g$^+$) and 3 (g$^+$tg$^+$g$^+$ and g$^+$g$^+$tg$^-$) of PMO-3 were obtained as −2.08, −1.71, and −1.57 kcal mol^{-1}, respectively. In addition, Gibbs free energies of the tg$^+$g$^+$g$^+$g$^+$g$^+$, g$^+$tg$^+$g$^+$g$^+$g$^+$, and g$^+$g$^+$tg$^+$g$^+$g$^+$ conformations of PMO-4 were also calculated, and consequently, the E_σ value was found to show the position dependence; bonds 2, 3, and 4 have E_σs of −1.98, −1.61 and −1.66 kcal mol^{-1}, respectively. On this ground, it seems reasonable to accept Flory's (−1.5 kcal mol^{-1}) and Mark and

Conformational Analysis of Polymers: Methods and Techniques for Structure-Property Relationships and Molecular Design, First Edition. Yuji Sasanuma.

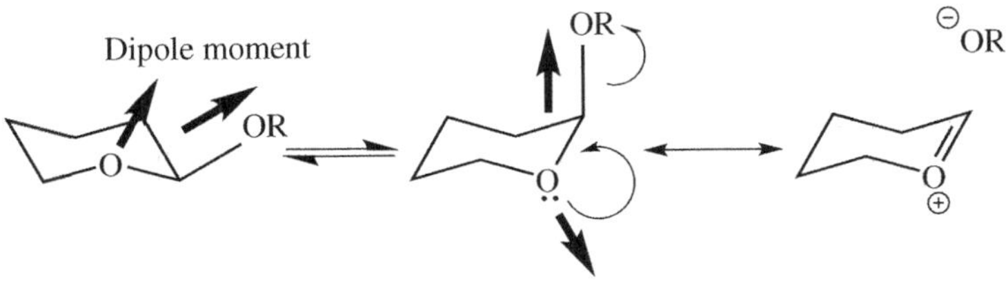

Figure 17.1 Explanations for the anomeric effect: dipole–dipole interaction and lone pair (n_O) → antibonding orbital (σ^*_{C-O}) hyperconjugation. The thick arrow and R stand for the C—O bond dipole moment and an alkyl group, respectively. Source: [423]/Reproduced with permission of American Chemical Society.

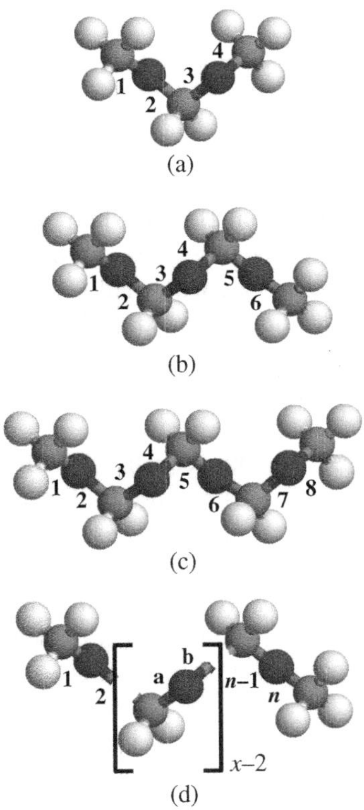

Figure 17.2 Schematic representation of PMO and its oligomeric model compounds: (a) dimer, dimethoxymethane (PMO-2); (b) trimer, 1,3-dimethoxy-dimethylether (PMO-3); (c) tetramer, bis [(methoxy)methoxy]methane (PMO-4); (d) polymer, PMO. The bonds are numbered as indicated, and *x* is the degree of polymerization. Source: [423]/Reproduced with permission of American Chemical Society.

Abe's (−1.4 kcal mol^{-1}) values for long PMO chains with the common statistical weight matrix

$$U_j = \begin{bmatrix} 1 & \sigma & \sigma \\ 1 & \sigma & \sigma\omega \\ 1 & \sigma\omega & \sigma \end{bmatrix} \tag{17.1}$$

where their E_ω value of 1.5 kcal mol^{-1} [7] was adopted. For PMO of $x = 300$, the conformational energies and geometrical parameters offered by Flory and Mark [153] and Abe and Mark [7] yield $\langle r^2 \rangle_0 / nl^2$s of 8.0 and 9.2 at 25 °C, respectively. The experimental $\langle r^2 \rangle_0 / nl^2$ values of PMO in the Θ state were estimated to be 7.5 at 90 °C and 10.5 ± 1.5 at 25 °C [141]. The Abe–Mark parameters gave the dipole moment ratio of 0.092 at 25 °C and configurational entropy of 3.1 cal mol^{-1} K^{-1} at 180 °C. The experimental entropy of fusion (ΔS_u) was reported as 3.5 cal mol^{-1} K^{-1} at 183 °C [487]; therefore, the $S_{\mathrm{conf}} / \Delta S_u$ ratio is 0.89.

17.2 Poly(ethylene oxide) (PEO)

Poly(ethylene oxide) (PEO) is a unique polymer; PEO is soluble in a variety of solvents, for example common organic solvents, water, and aqueous solutions of inorganic salts, and harmless to organisms, thus being widely used for medical, chemical, biological, and industrial applications. As in crown ethers, the O—C—C—O bond sequence acts as an effective electron donor. These characteristics due to the lone pairs and electronegativity of oxygen enable us to apply PEO for ion conductors [545].

Since the 1960s, conformational characteristics of PEO have been investigated extensively, and in the 1990s, refocused to be a subject of controversy. From the experimental characteristic ratio and the dipole moment ratio of PEO, Mark and Flory determined the conformational energies, E_σ and E_ρ, for the gauche states of the C—C and C—O bonds as −0.43±0.07 and 0.90±0.07 kcal mol^{-1}, respectively [141, 300, 301]. The conformational energy E_ω for the second-order interaction (referred hereafter as the C—H· · ·O interaction) between the methylene (methyl) group and oxygen atom occurring in the g$^\pm$g$^\mp$ conformations for the C—O/C—C bond pairs was evaluated to be 0.35±0.20 kcal mol^{-1} (repulsive). The intramolecular interactions are illustrated in Figure 17.3. Abe and Mark [7] proposed a minor modification of the energy parameters: $E_\sigma = -0.5$, $E_\rho = 0.9$, and $E_\omega = 0.4$ kcal mol^{-1}. These data mean that the gauche stability in the C—C bond of PEO is due to the first-order σ interaction; the C—C bond itself has an intrinsic gauche preference. However, ab initio molecular orbital (MO) calculations on

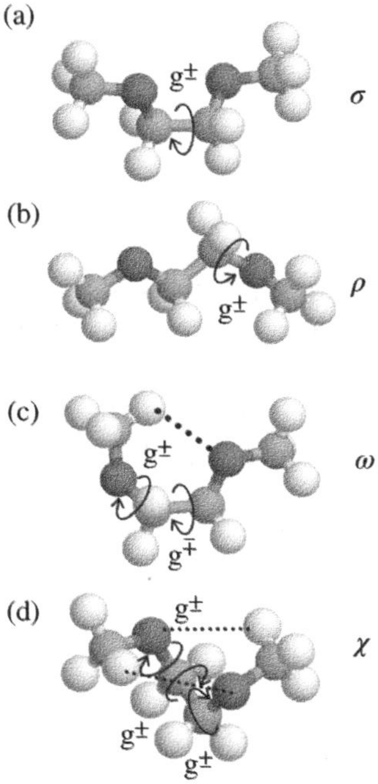

Figure 17.3 Intramolecular interactions defined for the ethylene oxides: (a) σ and (b) ρ: the first-order interactions around the C–C and C–O bonds, respectively; (c) ω: the second-order interaction occurring in $g^{\pm}g^{\mp}$ conformations for the C–O/C–C bond pair; (d) χ: the third-order interaction formed in $g^{\pm}g^{\pm}g^{\pm}$ conformations of the O–C–C–O bond sequence. The ω and χ interactions correspond to intramolecular C–H· · ·O hydrogen bonds. The model is DME. Source: Material from reference [413], published 2006, Springer Nature.

1,2-dimethoxyethane (DME) yielded the E_σ and E_ω values of 0.1–0.5 kcal mol^{-1} and −1.2 to −1.4 kcal mol^{-1}, respectively [221, 498]; the attractive C—H· · ·O contact was suggested to be the indirect source of the gauche stability in the C—C bond. The MO calculations fairly well reproduced the experimental bond conformations of gaseous DME [27, 544].

As a consequence of deliberation [3, 33, 412, 437], we reached an idea of the competitive balance between intra- and intermolecular attractions of ethylene oxides. As the MO calculations on DME predicted, the isolated (gaseous) ethylene-oxide chains form the intramolecular hydrogen bonds, which cause an apparent gauche stability of the C—C bond, whereas in polar solvents, the O—C—C—O segment tends to prefer the tgt conformation because of attractive interactions with solvent molecules. The phenomena may be observed as continuous variations of two conformational energies, E_σ and E_ω, which, depending on the environment, are significantly shifted in the opposite directions: $E_\sigma = +0.32$ and $E_\omega = -1.12$ kcal mol^{-1} for DME in the gas phase [412]; $E_\sigma = -0.25$ and $E_\omega = -0.79$ kcal mol^{-1} for PEO in weakly polar solvents such as 1,4-dioxane and benzene [412]; $E_\sigma = -0.5$ and $E_\omega = +0.4$ kcal mol^{-1} for PEO in the Θ solutions, for example 0.45 M K_2SO_4 at 34.5 °C [7, 141, 300]. In water, a good solvent, PEO shows as small an E_σ value as −1.2 kcal mol^{-1} [469].

The natural bond orbital (NBO) analysis [382] on DME indicated that the attractive gauche effect stems mainly from the C—H bond → C—O antibond

($\sigma_{C—H} \rightarrow \sigma^*_{C—O}$) delocalizations of electrons [412]. As a consequence, the tgt state becomes more stable than ttt. To our best knowledge, however, all ab initio MO calculations carried out so far for DME have suggested that the ttt conformation is more stable than tgt. In the crystalline state, PEO can adopt either tgt or ttt state in the O—C—C—O bond sequence. The former and latter conformations form a distorted (7/2) helix [456, 464] and a planar zigzag structure [463], respectively. The PEO chain is suggested to slightly prefer tgt to ttt because the ttt conformation appears only under stretching [463]. The four (7/2) helical chains form a $P2_1/a$ monoclinic cell [464] in which no clue to the specific interactions such as O· · ·H close contacts to stabilize the tgt form can be found.

As found for PMO–x, the gauche energy (E_σ) of the central C—C bond of ethylene oxides may show the chain-position dependence. Accordingly, accurate MO calculations at the MP2/6–311++G(3df, 3pd) level were carried out for triglyme as well as DME [413]. In addition, ^{1}H NMR of triglyme was measured and analyzed to evaluate its bond conformations, compare them with the MO calculations, and furthermore, investigate whether the E_σ value depends on both the environment (medium) and the chain position, and whether the attractive gauche effect exists independently of the C—H· · ·O hydrogen bonds as suggested by the NBO analysis.

Conformational free energies were calculated at the MP2/6–311++G(3df, 3pd) level for 10 and 11 conformers of DME and triglyme, respectively, and divided into three kinds of conformational energies, $E_{\rho j}$, $E_{\sigma j}$, and $E_{\omega j}$ (j: bond number) by the least-squares method, and the energy parameters were determined as shown

Table 17.1 Conformational free energies of DME and triglyme.[a]

	DME	Triglyme
E_{ρ_2}	1.30	1.25
E_{ρ_4}		1.09
E_{ρ_5}		1.05
E_{σ_3}	0.08	0.06
E_{σ_6}		−0.08
E_{ω_3}	−1.02	−1.15
E_{ω_4}		−1.18
E_{ω_6}		−1.14
E_{χ_4}	−0.42	−0.44
E_{χ_7}		−0.37

a) In kcal mol^{-1}. The subscript of the statistical weight corresponds to the bond number (Figure 17.4).

Source: Material from reference [413], published 2006, Springer Nature.

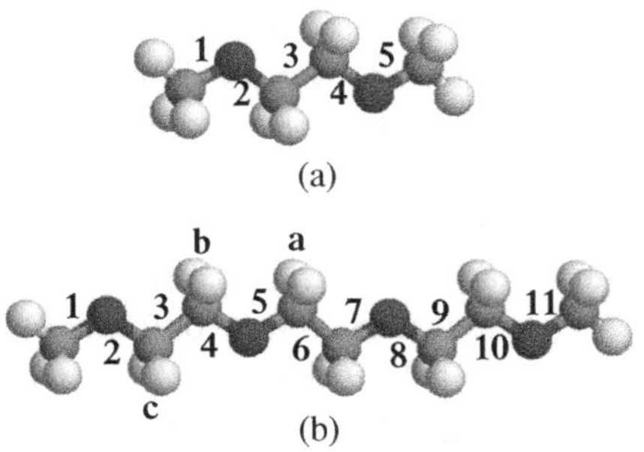

Figure 17.4 (a) Monomeric (1,2-dimethoxyethane: DME) and (b) trimeric (triglyme) model compounds of poly(ethylene oxide) (PEO). As indicated, the skeletal bonds are numbered. Source: Material from reference [413], published 2006, Springer Nature.

in Table 17.1. The results can be summarized as follows. (1) A narrow range (−1.02 to −1.18 kcal mol^{-1}) of E_ωs were obtained from the two model compounds; the strength of the C—H· · ·O attraction may be essentially independent of the chain length and bond position. (2) The E_{σ_3} value is slightly positive, whereas E_{σ_6} is slightly negative; the terminal monomeric unit has an E_σ value larger than the inner unit. This suggests the possibility that the C—C bond of PEO would adopt the gauche conformation even without the C—H· · ·O interaction. This is consistent with the fact that, in the crystal, the PEO chain prefers the helical structure rather than the planar zigzag form. The densities of the monoclinic (all-gauche helix) and triclinic (all-trans zigzag) lattices are 1.228 and 1.197 g cm^{-3}, respectively; thus, the helical chains are more densely packed. The small magnitude of E_{σ_6} suggests that PEO may switch the C—C bond between gauche and trans even in the crystal.

Figure 17.5 shows an ^{1}H NMR spectrum observed from methylene protons of triglyme. The spectrum simulations with the gNMR program [53] yielded vicinal coupling constants from which the trans and gauche fractions (p_t and p_g) of bonds 3 and 6 were derived in the manner similar to that explained in Section 15.1.1. In Table 17.2, the p_t and p_g values are compared with those evaluated from the MO calculations. It can be seen that p_t decreases with increasing permittivity of the medium. In Figure 17.6, the p_t values, calculated from the conformational energies except E_{σ_6} of triglyme (Table 17.1), are plotted as a function of E_{σ_6} (curve A). The curve intersects with the horizontal line for the cyclohexane solution around E_{σ_6} = −0.1 kcal mol^{-1}. This value probably corresponds to E_{σ_6} of trigylme in nonpolar cyclohexane (dielectric constant ϵ = 2.0) and close to that in the gas phase (ϵ = 1.0). From bond conformations of PEO dissolved in 1,4-dioxane [469] and dipole moment ratios for the benzene solution [32, 244, 391], we determined E_ρ = 1.17 and E_ω = −0.79 kcal mol^{-1} [412]. These energy parameters are considered to represent the ethylene oxides in weakly polar solvents (curve B). Similarly, curve C was obtained from E_ρs = 0.9 and

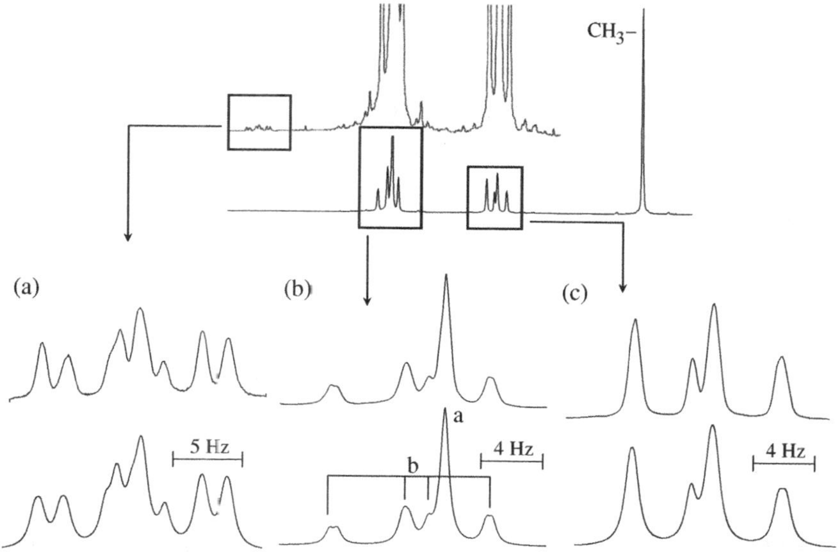

Figure 17.5 Observed (above) and calculated (below) ^{1}H NMR spectra of triglyme dissolved in cyclohexane-d_{12} at 25 °C: (a) methylene group a (satellite peaks); (b) methylene groups a and b; (c) methylene group c. For the designation of protons, see Figure 17.4. Source: Material from reference [413], published 2006, Springer Nature.

Table 17.2 Bond conformations of triglyme at 25 °C.

		Bond 3		Bond 6	
Environment	**Permittivity**	p_t	p_g	p_t	p_g
	MO calc[a)]				
Gas	1.0	0.21	0.79	0.17	0.83
	NMR exptl				
Cyclohexane-d_{12}	2.0	0.21	0.79	0.18	0.82
Chloroform-d_1	4.8	0.11	0.89	0.12	0.88
Methanol-d_4	32.7	0.11	0.89	0.11	0.89
Dimethyl sulfoxide-d_6	46.7	0.13	0.87	0.13	0.87
Deuterium oxide	78.5	0.05	0.95	0.05	0.95

a) Evaluated from the conformational energies shown in Table 17.1.
Source: Material from reference [413], published 2006, Springer Nature.

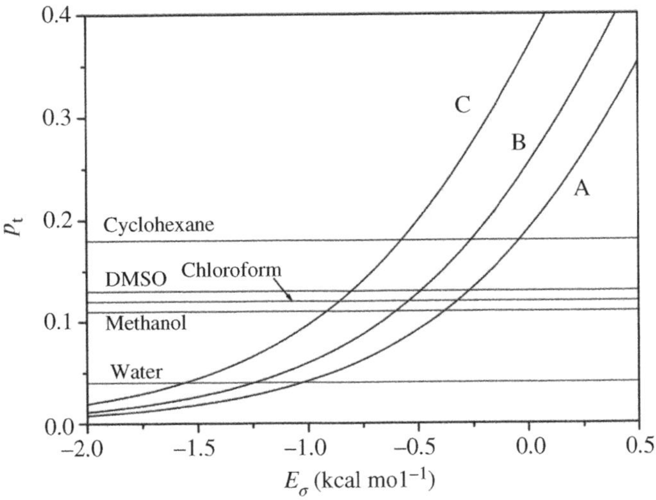

Figure 17.6 Trans fraction (p_t) of bond 6 of triglyme as a function of E_{σ_6}. The horizontal lines represent the p_t values determined from the NMR experiments using the solvents indicated. Curves A, B, and C were obtained from the following energy parameters and $E_{\sigma_3} = E_{\sigma_6}$: curve A, conformational energies of triglyme in Table 17.1, except for E_{σ_6}; curve B, E_{ρ_i}s = 1.17 and E_{ω_i}s = −0.79 kcal mol^{-1}; curve C, E_{ρ_i}s = 0.9 and E_{ω_i}s = 0.4 kcal mol^{-1}. Source: Material from reference [413], published 2006, Springer Nature.

E_ωs = 0.4 kcal mol^{-1} (Abe–Mark) [7]. The horizontal lines for chloroform, dimethyl sulfoxide, and methanol intersect with curve B at E_{σ_6} = −0.5 kcal mol^{-1} and with curve C around E_{σ_6} = −0.8 to −0.9 kcal mol^{-1}. Therefore, the E_{σ_6} values of triglyme dissolved in these solvents probably stay within the range. From the intersection between curve C and the horizontal line of water, we can estimate the E_{σ_6} value of triglyme in water to be as small as ≤ −1.6 kcal mol^{-1}.

Both MO calculations and NMR experiments suggest that E_{σ_6} of triglyme is slightly negative even in the gas phase. This means that the C—C bond adjacent to O—C has the inherent gauche stability; in other words, the attractive gauche effect is not merely due to the C—H· · ·O attraction. In addition, the small differences in the energy parameters between DME and triglyme (Table 17.1) fairly justify using monomeric model compounds in conformational analysis of polymers, unless very strict discussion is required.

For PEO, a wide range of $\langle r^2 \rangle_0/nl^2$ values (4.1–9.7) and various Θ conditions have been reported [21, 41, 46, 173, 237, 260, 300, 440, 511], which is obviously due to the competitive balance between E_σ and E_ω; the two energy parameters vary to a large extent, depending on the environment (medium). The temperature coefficient, $10^3 d \ln \langle r^2 \rangle_0/dT$, ranges from positive and negative values: 0.23±0.2 K^{-1}

(amorphous network at 60 °C) [300], 0.2±0.2 K^{-1} (extrapolated from benzene solution at 35 °C) [44], −1.5±0.5 K^{-1} (1.24 M KOH at 25 °C) [173], and −0.3 K^{-1} (melt at 130 °C) [440]. In contrast, experimental dipole moment ratios ($\langle \mu^2 \rangle / nm^2$) of benzene solutions at ambient temperature stay within a narrow range of 0.40–0.42 [32, 244, 391]. This is probably because the dipole moment ratio is always free from the excluded-volume effect [111, 330], thus being the truly unperturbed value. Here, the $\langle \mu^2 \rangle / nm^2$ values are calibrated with the bond dipole moment of the C-O bond (m_{C-O} = 1.18 D) determined from the least-squares fitting for dipole moments of different conformers of DME. The temperature coefficient, $10^3 d \ln\langle \mu^2 \rangle / dT$, was estimated experimentally to be 2.6 K^{-1} [32]. The $\langle r^2 \rangle_0 / nl^2$, $10^3 d \ln \langle r^2 \rangle_0 / dT$, and $10^3 d \ln\langle \mu^2 \rangle / dT$ values (Table 17.3) were calculated from two sets of energy parameters: curve B (in Figure 17.6), $E_\sigma = -0.25$, $E_\rho = 1.17$, and $E_\omega = -0.79$ kcal mol^{-1}; Abe–Mark, $E_\sigma = -0.5$, $E_\rho = 0.9$, and $E_\omega = 0.4$ kcal mol^{-1}. The $\langle r^2 \rangle_0 / nl^2$ (4.1) and 10^3it ln $\langle r^2 \rangle_0 / dT$ (0.23 K^{-1}) values based on the curve B data agree with those (4.1 and 0.2 – 0.23 K^{-1}) estimated experimentally [44, 300], and the $10^3 d \ln\langle \mu^2 \rangle / dT$ value (1.9 K^{-1}) is comparable to the experiment (2.6 K^{-1}) [32]. The characteristic ratio (5.1) calculated from the Abe–Mark energies agrees with that (5.2±0.1) determined by light scattering for a 0.45 M K_2SO_4 aqueous solution at 34.5 °C [237]. The entropy of fusion at T_m^0 (80 °C) was reported as 5.9 cal K^{-1} mol^{-1} [293], and the S_{conf} value was calculated to be 5.0–5.1 cal K^{-1} mol^{-1}.

Table 17.3 Configurational properties of PEO, evaluated from RIS calculations at 25 °C.

	Curve B[a)]	Abe–Mark[b)]
E_ρ	1.17	0.9
E_σ	−0.25	−0.5
E_ω	−0.79	0.4
E_χ	0.00	0.00
$\langle r^2 \rangle_0 / nl^2$ [c)]	4.1	5.1
$10^3 d \ln \langle r^2 \rangle_0 / dT$ [c)] (K^{-1})	0.23	0.36
$\langle \mu^2 \rangle / nm^2$ [c)]	0.41	0.49
$10^3 d \ln\langle \mu^2 \rangle / dT$ [c)] (K^{-1})	1.9	2.9
S_{conf} [d)] (cal K^{-1} mol^{-1})	5.1	5.0

a) The conformational energies were determined from NMR vicinal coupling constants of the 1,4-dioxane solution and the dipole moment ratio of the benzene solution [412].
b) Reference [7].
c) At 25 °C.
d) At T_m^0.
Source: Adapted from [412].

As shown above, the characteristic features of PEO, such as high solubilities in various solvents, stem mainly from the competitive balance between intra- and intermolecular attractions, namely, the capability to adjust itself to the environment.

It was first reported that PEO crystallizes in a (7/2) helical structure in which seven monomeric units form two turns. The four PEO chains lying in a tgt conformation in the O—CH_2—CH_2—O sequence are included in a monoclinic unit cell of $a = 8.16$ Å, $b = 12.99$ Å, $c = 19.30$ Å, and $\beta = 126.5°$ [456]. Afterward, the crystal structure was refined as follows: monoclinic; space group $P2_1/a$; lattice constants, $a = 8.05$ Å, $b = 13.04$ Å, $c = 19.48$ Å, and $\beta = 125.4°$ [464]. The chain structure slightly deviates from the regular (7/2) helix.

When a PEO sample is extended twofold at room temperature, a different X-ray diffraction pattern appeared and was assigned to the following structure: triclinic; space group, $P\bar{1}$; lattice constants, $a = 4.71$ Å, $b = 4.44$ Å, $c = 7.12$ Å, $\alpha = 62.8°$, $\beta = 93.2°$, and $\gamma = 111.4°$ [463]. The chain adopts the all-trans conformation, which seems to be unstable, because, as soon as the tension is removed, the chain returns to the helix.

With each of the experimental crystal structures set initially, the structural optimization was carried out by the periodic DFT calculations at the B3LYP-D/6-31G(d) level with the dispersion-force correction [541], and the lattice constants were obtained as shown in Table 17.4. The difference between theory and experiment in lattice constants is small for the helical structure ($\Delta_{LC} = 0.88\%$) but moderate for the all-trans structure ($\Delta_{LC} = 3.16\%$), where Δ_{LC} is defined as

$$\Delta_{LC}\ (\%) = \frac{1}{N_{LC}}\left[\sum_{i=1}^{N_{LC}}\left(\frac{x_{i,\ calc} - x_{i,\ obs}}{x_{i,\ obs}}\right)^2\right]^{1/2} \times 100 \tag{17.2}$$

where N_{LC} is the number of lattice constants, and $x_{i,calc}$ and $x_{i,obs}$ are the calculated and observed lattice constants, respectively. The reason for the moderate reproduction ($\Delta_{LC} = 3.16\%$) of the all-trans structure may be explained as follows. In the experiment, the sample was extended twofold, and the all-trans structure was not necessarily optimal, whereas the DFT calculation reached the minimum electronic energy of the crystal lattice. The helical structure thus optimized is depicted in Figure 17.7.

In Table 17.4, the electronic energies at 0 K and Gibbs free energies at room temperature (273.0 K) are compared between the helical and all-trans crystals. The electronic energies suggest that the all-trans structure is preferable by −0.30 kcal mol^{-1}, whereas the Gibbs energies show that the helical structure is more stable by −0.59 kcal mol^{-1} at room temperature, where the unit "mol" means mole of the repeating unit. The free-energy difference is due to that in vibrational entropy:

Figure 17.7 (a) Crystal structure of the helical PEO chain, optimized by the periodic DFT-D calculations at the B3LYP-D/6-31G(d) level. Crystallographic data: monoclinic; space group $P2_1/a$; lattice constants, $a = 7.93$ Å, $b = 12.63$ Å, $c = 19.48$ Å, and $\beta = 126.2$. (Above) The $a'b$-plane perpendicular to the c (helical)-axis. (Below) The bc'-plane perpendicular to the a-axis. (b) Young's modulus distribution on the $a'b$-plane. The grid spacing corresponds to 10 GPa.

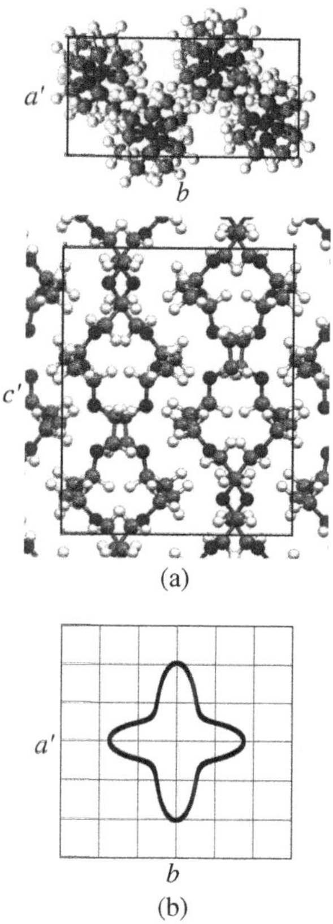

helical, 13.0 cal K^{-1} mol^{-1}; all-trans, 7.6 cal K^{-1} mol^{-1}. The contribution of $-(\Delta S)T$ to the free energy renders the helix more stable than the extended form.

Young's moduli (E_a, E_b, and E_c) in the a-, b-, and c-axis directions of both crystals, evaluated from the periodic DFT-D calculations, are also given in Table 17.4. Interestingly, the helical structure exhibits rather isotropic crystalline moduli: E_a (19.6 GPa) $\approx$ E_b (20.3 GPa) $\approx$ E_c (helical axis) (23.3 GPa), whereas the all-trans structure shows very anisotropic distribution: E_a = 9.0 GPa, E_b = 14.9 GPa, and E_c (molecular axis) = 412 GPa. In general, all-trans polymeric chains exhibit very large Young's modulus along the molecular axis. The E_c value of the all-trans PEO exceeds those of polyethylene (333 GPa), nylon 4 (334 GPa), and nylon 6 (316 GPa) but still falls short of that (451 GPa) of poly(glycolic acid) (see Table 10.2). It should be noted, however, that the all-trans structure is formed only under a tension and hence essentially unstable at room temperature.

Table 17.4 Optimized and experimental crystal structures and thermodynamic and mechanical properties of helical and all-trans PEO chains.

	Helical[a)]			All-trans[b)]		
	Lattice constant (Å, °)					
	Optimized by DFT-D method[c)]					
	a	b	c	a	b	c
	7.93	12.63	19.48	4.87	4.29	7.10
		β		α	β	γ
		126.2		57.4	107.1	118.5
	X-ray experiment[d)]					
	a	b	c	a	b	c
	8.05	13.04	19.48	4.71	4.44	7.12
		β		α	β	γ
		125.4		62.8	93.2	111.4
		$\Delta_{LC} = 0.88\%$[e)]			$\Delta_{LC} = 3.16\%$[e)]	
	Thermodynamics					
ΔE_{EL}[f)] (kcal mol^{-1})		0.0			−0.30	
ΔG[g)] (kcal mol^{-1})		0.0			0.59	
S (cal K^{-1} mol^{-1})		13.0			7.6	
	Young's modulus[h)] (GPa)					
	E_a	E_b	E_c	E_a	E_b	E_c
	19.6	20.3	23.3	9.0	14.9	412

a) Monoclinic, space group $P2_1/a$.
b) Triclinic, space group $P\bar{1}$.
c) Reported by Yoshida [541].
d) Reported by Takahashi and Tadokoro [464] and Takahashi et al. [463].
e) Defined in Eq. (17.2).
f) Electronic energy at 0 K. Relative to the helical structure.
g) Gibbs free energy at room temperature. Relative to the helical structure.
h) Young's moduli in the a-, b-, and c-axis directions of the crystal.
Source: Adapted from [464, 467, 544].

17.3 Poly(propylene oxide) (PPO)

Conformational analysis of poly(propylene oxide) (PPO) via NMR experiments and MO calculations on a model compound, 1,2-dimethoxypropane, is already described in detail (Section 15.1). The crystal structure of isotactic PPO was optimized by the periodic DFT with the dispersion-force correction at the B3LYP-D/6-31G(d) level, and the crystalline moduli and the interchain

Table 17.5 Optimized and experimental crystal structures of isotactic PPO.[a)]

	Optimized[b)]			Experimental[c)]		
			Lattice constant (Å)			
	a	*b*	*c*	*a*	*b*	*c*
	9.97	4.45	7.05	10.46	4.66	7.03
			$\Delta_{LC} = 2.2\%$[d)]			
			Fractional coordinates			
	x/a	y/b	z/c	x/a	y/b	z/c
O	0.259	0.918	0.331	0.260	0.909	0.330
C	0.265	0.094	0.163	0.275	0.079	0.166
C	0.293	0.086	0.498	0.275	0.079	0.494
C	0.444	0.138	0.516	0.419	0.148	0.522
H	0.209	0.304	0.183	0.223	0.275	0.185
H	0.369	0.155	0.128	0.375	0.127	0.147
H	0.241	0.304	0.492	0.223	0.275	0.475
H	0.466	0.273	0.640	0.430	0.280	0.646
H	0.485	0.251	0.391	0.455	0.263	0.402
H	0.497	0.924	0.534	0.471	0.951	0.541
			$\Delta_{CHO} = 0.029$[e)]			

a) Orthorhombic, space group $P2_12_12_1$.
b) At 0 K. Reported by Ishida [220].
c) Reported by Cesari et al. [70]
d) Defined in Eq. (17.2)
e) $\Delta_{CHO} = \frac{1}{N_{atom}} \sum_{atom} \left\{ \left[\left(\frac{x}{a}\right)_{calc} - \left(\frac{x}{a}\right)_{expt} \right]^2 + \left[\left(\frac{y}{b}\right)_{calc} - \left(\frac{y}{b}\right)_{expt} \right]^2 + \left[\left(\frac{z}{c}\right)_{calc} - \left(\frac{z}{c}\right)_{expt} \right]^2 \right\}^{1/2}$
where N_{atom} is the number of atoms used in the calculation.
Source: Adapted from [71, 221].

interaction energy were evaluated [220]. The crystal structure determined by X-ray diffraction (orthorhombic and space group $P2_12_12_1$) [70] was set as the initial structure. In Table 17.5, the initial and optimized structures are compared. The differences in lattice constants and atomic coordinates are $\Delta_{LC} = 2.2\%$ and $\Delta_{CHO} = 0.029$, where Δ_{LC} and Δ_{CHO} are defined in Eq. (17.2) and the footnote of Table 17.5, respectively. Young's moduli in the *a*-, *b*-, and *c*-axis directions were evaluated to be 12.4, 11.4, and 208 GPa, respectively, and the Young's modulus distribution on the *cb*-plane perpendicular to the molecular axis is depicted in Figure 17.8b. The basis set superposition error (BSSE)-corrected interchain interaction energy is -5.23 kcal mol^{-1} per repeating unit, i.e. -90 cal g^{-1}.

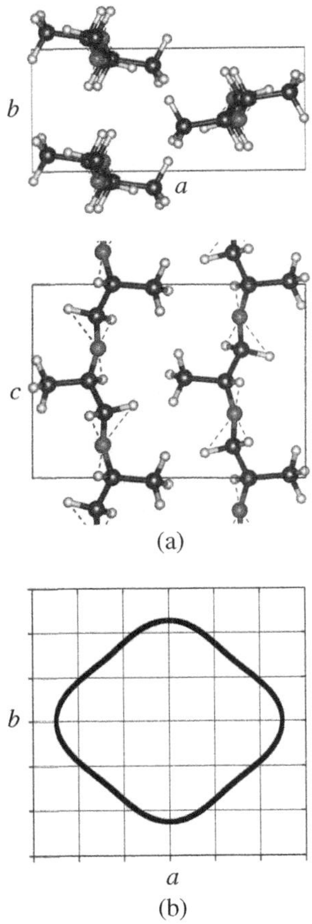

Figure 17.8 (a) Isotactic PPO crystal optimized by the periodic DFT calculations at the B3LYP-D/6-31G(d) level. Crystallographic data: orthorhombic, space group $P2_12_12_1$, $a = 9.97$ Å, $b = 4.45$ Å, and 7.05 Å. (b) Young's modulus distribution on the *ab*-plane perpendicular to the molecular axis of PPO. The grid spacing corresponds to 5 GPa. The horizontal and vertical lines are parallel to the *a* and *b* axes, respectively.

17.4 Poly(trimethylene oxide) (PTrMO)

Because of the chain structure, poly(trimethylene oxide) (PTrMO, Figure 17.9) does not exhibit the C—H···O close contact. Ab initio MO calculations on a monomeric model compound of PTrMO, 1,3-dimethoxypropane ($CH_3OCH_2CH_2$-CH_2OCH_3), suggested that all second-order interactions formed between atomic groups separated by four bonds are repulsive (see Figure 17.10 and Table 17.6). The O—C bond prefers the trans conformation (E_ρ), and the $CH_2 - CH_2$ bond adjacent to O—C shows a gauche preference (E_σ). The three energy parameters, E_ρ, E_σ, and $E_{\omega'}$, were optimized so as to simultaneously reproduce all the experimental values of the characteristic ratio, dipole moment ratio, and their temperature coefficients of PTrMO and bond conformations of 1,3-dimethoxypropane. The results are shown in the column "optimized" under "calcd" of Table 17.6. The E_σ value thus obtained is still negative (−0.47 kcal mol^{-1}). This means that the gauche stability is the nature of the $CH_2 - CH_2$ bond itself. The calculated $\langle r^2\rangle_0/nl^2$, $d\ln\langle r^2\rangle_0/dT$, $\langle\mu^2\rangle/nm^2$, $d\ln\langle\mu^2\rangle/dT$, p_t^{C-O}, and p_t^{C-C} values are in agreement with the experiments except for the $d\ln\langle r^2\rangle_0/dT$ value.

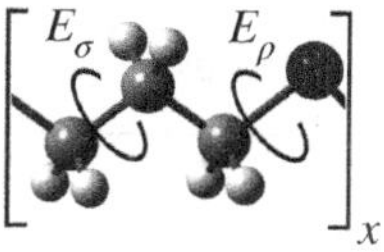

Figure 17.9 Poly(trimethylene oxide) (PTrMO). The first-order interaction energies, E_ρ and E_σ, correspond to the gauche energies of the O−C and C−C bonds, respectively. x is the degree of polymerization.

It was found that PTrMO shows three crystal modifications (designated as I, II, and III): (I) monoclinic, $a = 12.3$ Å, $b = 7.27$ Å, c (fiber axis) = 4.8C Å, and $\beta = 91°$; (II) trigonal, $a = 14.13$ Å and c (fiber axis) = 8.41 Å; (III) orthorhombic, $a = 9.23$ Å, $b = 4.82$ Å, c (fiber axis) = 7.21 Å[457]. In modifications I, II, and III, the PTrMO chain adopts tttt, tgtt, and tggt conformations in the O—CH_2—CH_2—CH_2—O bond sequence, respectively. In modification II, the tgtt sequence repeats in such a manner as (tg$^+$tt)(tg$^-$tt). Modification I appears only in the presence of water. Modification II is formed only in oriented samples. Modification III seems to be the most stable. As the number of the gauche conformation in the C—C bond increases, the crystalline state becomes more stable. This tendency is due to the conformational characteristics of the PTrMO chain.

17.5 Poly(tetramethylene oxide) (PTetMO)

Poly(tetramethylene oxide) (PTetMO, [—$CH_2CH_2CH_2$-CH_2O—]$_x$, Figure 17.11a), also being termed polytetrahydrofuran, can form the C—H···O close contact (see Figure 17.11). The conformational energy (E_ω) expressing the second-order C—H···O interaction was evaluated from MO calculations for a model compound, 1,4-dimethoxybutane (DMB), at the MP2/6-31+G(d)//HF/6-31G(d) level to be −0.43 kcal mol^{-1} [270]. The negative sign of E_ω shows the C—H···O contact to be attractive, but its magnitude is smaller than those (approximately −1 kcal mol^{-1}) of PEO and PPO. The C—C bond adjacent to C—O also has a negative gauche energy: E_σ = −0.65 kcal mol^{-1}. In PTetMO, both the first- (E_σ) and second-order (E_ω) interactions are attractive.

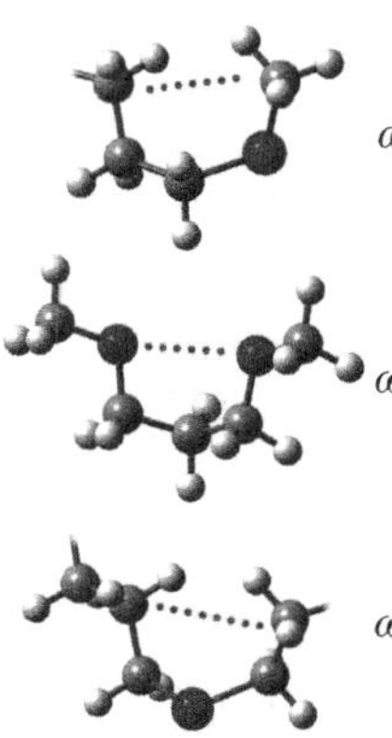

Figure 17.10 Second-order interactions defined for PTrMO. Source: [420]/ Reproduced with permission of American Chemical Society.

To reveal the nature of the C—H···O interaction, the conformational energies were evaluated for α,ω-dimethoxyalkanes with different numbers (n = 4–8)

Table 17.6 Conformational energies and calculated and observed configurational properties of PTrMO.

	Calcd		Obsd
	ΔG_k [a]	Optimized [b]	
First-order interaction			
E_ρ	1.13	0.97	
E_σ	−0.71	−0.47	
Second-order interaction [c]			
E_ω	∞	∞	
$E_{\omega'}$	2.41	0.39	
$E_{\omega''}$	∞	∞	
$\langle r^2 \rangle_0/nl^2$		3.7	3.9±0.3 [d]
$10^3 d \ln \langle r^2 \rangle_0/dT$ (K^{-1})		0.6	0.08 [e]
$\langle \mu^2 \rangle/nm^2$		0.32	0.35 [f]
$10^3 d \ln \langle \mu^2 \rangle/dT$ (K^{-1})		1.9	1.8 [g]
S_{conf} (cal K^{-1} mol^{-1})		6.3	
ΔS_u (cal K^{-1} mol^{-1})			6.5 [h]
T_m^0 (°C)			50.0 [h]
$p_t^{C—O}$		0.81	0.82 [i]
$p_t^{C—C}$		0.25	0.24 [i]

a) As obtained from the MO calculations.
b) Optimized so as to reproduce the experimental data.
c) See Figure 17.10.
d) Determined from the Θ solution at 27 °C (solvent: C_6H_{12}) [82].
e) Obtained from elasticities of the amorphous network at 60 °C [462].
f) Obtained from a CCl_4 solution at 20 °C [299, 388].
g) Obtained from a CCl_4 solution at 35 °C [299, 388].
h) Quoted from reference [293].
i) Trans fractions of the C—O ($p_t^{C—O}$) and C—C ($p_t^{C—C}$) bonds, determined from the NMR of the model compound, 1,3-dimethoxypropane, in C_6H_{12} at 25 °C.

Source: Adapted from [420].

of methylene units between two oxygen atoms (Figure 17.11). Of them, only DMB (n = 4) shows a negative E_ω, whereas the longer ethers (n = 5, 6, 7, and 8) have positive E_ωs. In DMB, the methylene unit relating to the C—H···O interaction adjoins the electron-withdrawing oxygen atom. If the oxygen atom is replaced by fluorine of still larger electronegativity, then the compound will be 1-fluoro-4-methoxybutane (FMB) and shows an E_ω value of −0.84 kcal mol^{-1}; that is, the C—H···O attraction becomes stronger. The Mulliken population

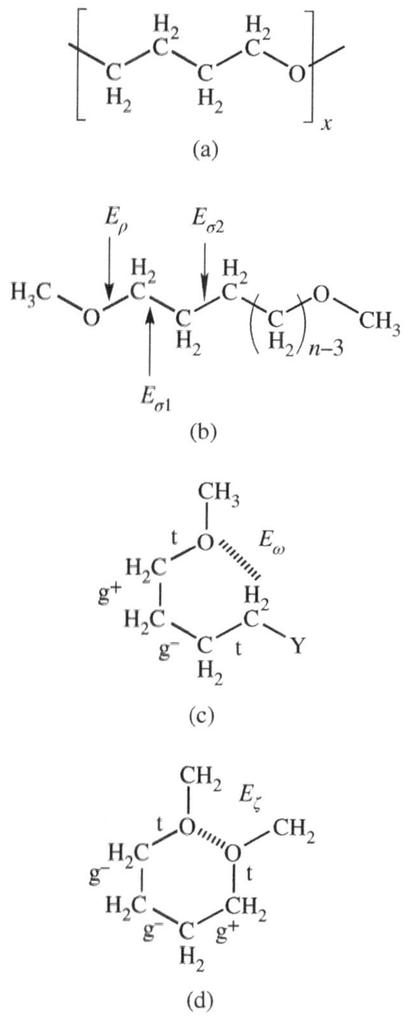

Figure 17.11 (a) Poly(tetramethylene oxide) (PTetMO). (b) First-order interaction energies, that is gauche energies around the C—O (E_ρ), OCH_2—CH_2 ($E_{\sigma1}$), and CH_2CH_2—CH_2 ($E_{\sigma2}$) bonds. n is the number of methylene units between oxygen atoms. (c) The second-order C—H···O interaction (dashed line), and its energy is E_ω. Y is $(CH_2)_{n-4}OCH_3$ (α,ω-dimethoxyalkanes) or F (1-fluoro-4-methoxybutane, FMB). See Table 17.7. (d) The third-order O···O interaction (dashed line), and its energy is E_ζ. Source: [271]/Reproduced with permission of American Chemical Society.

analysis yielded the partial charges at the methylene groups bonded to oxygen and fluorine as 0.19 e (DMB) and 0.34 e (FMB), respectively; the latter CH_2 group is charged more positively than that of the former. The above facts suggest that the C—H···O attraction represented by the negative E_ωs may be interpreted, in the classical sense, as an electrostatic interaction.

The RIS calculations using the MO energies yielded configurational properties of PTetMO (Table 17.8) [271]. A new energy parameter, E_ζ (= 0.74 kcal mol^{-1}), representing a third-order interaction formed between groups separated by five bonds in PTetMO (Figure 17.11), was added to the RIS calculations. The characteristic

Table 17.7 Conformational energies of $CH_3O-(CH_2)_n-OCH_3$ (n = 4–8) and 1-fluoro-4-methoxybutane (FMB)[a]

n	E_ρ	$E_{\sigma1}$	$E_{\sigma2}$	E_ω
4	0.83	−0.65	0.83	−0.43
5	1.14	−0.57	0.65	0.66
6	1.21	−0.41	0.78	0.24
7	1.38	−0.61	0.69	0.42
8	1.41	−0.55	0.74	0.55
FMB	1.31	−0.42	0.90	−0.84

a) In kcal mol^{-1}. For the definition of the conformational energies, see Figure 17.11.
Source: [270]/Reproduced with permission of the Royal Society of Chemistry.

Table 17.8 Configurational properties of PTetMO, evaluated from RIS calculations

	Calcd[a]	Obsd			
$\langle r^2\rangle_0/nl^2$	5.68[b]	5.40[c]	5.03[d]	5.81[e]	6.89[f]
		5.13[g]	6.18[h]	5.61[i]	5.81[j]
$10^3 d\ln\langle r^2\rangle_0/dT$ (K^{-1})	−0.72[k]	−1.33[l]			
$\langle\mu^2\rangle/nm^2$	0.42[m]	0.42[n]	0.41[o]		
$10^3 d\ln\langle\mu^2\rangle/dT$ (K^{-1})	2.4[p]	2.7[q]	1.8[r]		

a) With $E_\rho = 0.83$, $E_{\sigma1} = -0.65$, $E_{\sigma2} = 0.83$, $E_\omega = -0.43$, and $E_\zeta = 0.74$ kcal mol^{-1}.
b) At 30 °C.
c) In ethyl acetate (22.7%) and n-hexane (77.3%) at 31.8 °C [264].
d) In ethyl acetate at 30 °C [264].
e) In cyclohexane at 30 °C [264].
f) In benzene at 30 °C [264].
g) In different solvents [124].
h) In ethyl acetate (22.7%) and n-hexane (77.3%) at 30.4 °C [128].
i) In 2-propanol at 44.6 °C [128].
j) In diethyl malonate at 33.5 °C [128].
k) At 60 °C.
l) In cross-linked network at 60 °C [32].
m) At 20 °C.
n) In benzene at 20 °C. Reestimated with $m_{C-O} = 1.18$ D and $m_{C-C} = 0.0$ D [32].
o) In benzene at 20 °C. Reestimated with $m_{C-O} = 1.18$ D and $m_{C-C} = 0.0$ D [384].
p) At 20 °C.
q) In benzene [32].
r) In benzene [384].
Source: Adapted from [271].

ratio at 30 °C was calculated to be 5.68, while those evaluated experimentally with a variety of solvents at different temperatures range widely from 5.03 to 6.89: the average is 5.73, and the median is 5.71. The calculated temperature coefficient of $\langle r^2 \rangle_0/nl^2$ is -0.72×10^{-3} K^{-1}, and the experimental value is $-1.33 \times 10^{-3-1}$; both are negative. The calculated dipole moment ratio is 0.42, in exact agreement with the experimental values of 0.42 and 0.41. In the $\langle \mu^2 \rangle / nm^2$ calculations, the C—O bond dipole moment ($m_{C—O}$) was set to 1.18 D, which was determined for PPO [398], and the C—C bond dipole moment ($m_{C—C}$) was assumed to be null. The temperature coefficient of the dipole moment ratio was calculated to be 2.4, being close to the experimental values of 2.7 and 1.8 (the average 2.3). As above, the RIS calculations on PTetMO satisfactorily reproduced all the experimental data.

18

Polyamines

Poly(ethylene imine) (PEI), a representative polyamine, has been extensively investigated in expectation of its potential use as a solid electrolyte [548], a transfection vector [17], and a carbon dioxide absorbent [279] owing to its cationic basicity. Commercially available PEIs have been produced by ring-opening polymerization of aziridine, being highly branched at the nitrogen site, while linear PEI has been synthesized from hydrolysis of poly(2-oxazoline)s (see Figure 18.1) [363, 392].

As pointed out in Section 1 3, polyamines exhibit a unique structural characteristic, nitrogen inversion: at room temperature, the substituent and lone pair connected to the nitrogen atom rapidly exchange their positions, and the activation free energies are as small as approximately 4 kcal mol^{-1} [403]. The facts indicate that it is impossible for us to control configurations of the polyamines in polymerization. The polyamines will reach the equilibrium ratios of conformations (for example, between trans and gauche states) and configurations (between *meso* and *racemo* diads) according to the Boltzmann distribution under given conditions (temperature and environment). For such polyamines, we have formulated the inversional and rotational isomeric state (IRIS) scheme and applied it to PEI. In addition, we carried out conformational analysis of poly(trimethylene imine) (PTMI) [417]. The repeating unit of PTMI is often included in hyperbranched polymers known as dendrimers [26, 546]. Furthermore, we treated their *N*-methyl-substituted polymers: poly(*N*-methylethylene imine) (PMEI) and poly(*N*-methyltrimethylene imine) (PMTMI). As described in Section 16.2.2, PMEI shows interesting solution properties, which stem partly from a comparatively large configurational entropy (S_{conf}) due to the inversional–rotational isomerization [164]. The electronic structure of nitrogen generates the nitrogen inversion and some attractive interactions and leads to the physical properties unique to the polyamines.

Conformational Analysis of Polymers: Methods and Techniques for Structure-Property Relationships and Molecular Design, First Edition. Yuji Sasanuma.

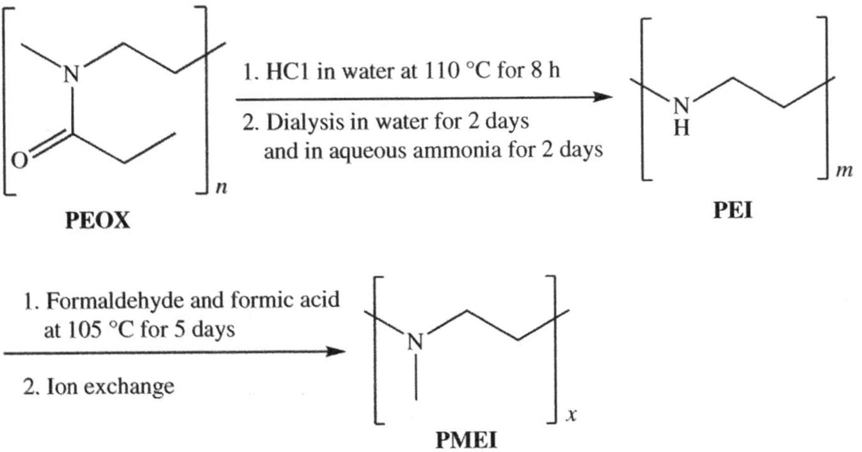

Figure 18.1 Synthesis of linear poly(ethylene imine) (PEI, $M_w \sim 32\,000$) and poly(*N*-methylethylene imine) (PMEI, $M_w \sim 1100$) from poly(2-ethyl-2-oxazoline) (PEOX, $M_w \sim 610\,000$). Source: Material from reference [164], published 2016, Springer Nature.

18.1 Poly(ethylene imine) (PEI)

PEI forms a variety of intramolecular interactions as illustrated in Figure 13.3. Of them, the η and ν close contacts are attractive and considered to be N—H···N hydrogen bonds. The MO calculations at the MP2/6-311++G(3df,3pd) level led to the energies as $E_\eta = -1.54$ and $E_\nu = -0.58$ kcal mol^{-1} (Table 13.1) [403]. Owing to these attractive interactions, the C—C bond is rich in the gauche conformation, the C—N bond appears to prefer the trans state (Table 18.1), and the PEI chain occasionally forms helical segments even in solutions and melts as shown by the Mattice analysis (see Section 13.7), which is suggestive of a helix of approximately 18 bonds per turn [408, 409]. The helical segments become rare with decreasing hydrogen bond strength (HBS) and finally vanish at HBS = 0%. The structural change can also be seen in characteristic ratio ($\langle r^2 \rangle_0/nl^2$) and *meso* probability (P_m): At HBS = 100%, 80%, 60%, 40%, 20%, and 0%, the $\langle r^2 \rangle_0/nl^2$ value will be 2.9, 3.3, 3.8, 4.5, 5.4, and 6.3, respectively, and simultaneously, the P_m value will be changed from 0.63 (HBS = 100%) to 0.61, 0.59, 0.56, 0.54, and 0.52 (0%), and the chain configuration approaches randomness ($P_m = 1/2$). This is because the η hydrogen bond stabilizes the *meso* tgt conformation. In anhydrous PEI crystal, the molecular chain adopts a 5/1 helix (3 bonds × 5 units = 15 bonds per turn) of *meso* tgt conformation (Figure 1.12 in Section 1.6) [71]. The helix in the crystal is tighter than the helical segment in solutions. The conformation of PEI is characterized mainly by the nitrogen inversion and the N—H···N hydrogen bonds.

Table 18.1 Trans fractions of *N,N′*-dimethylethylenediamine,[a] a model compound of PEI, at 25 °C.

Environment	Permittivity	p_t^{CC}	p_t^{CN}
	MO calc[b]		
Gas	1.0	0.08	0.73
	NMR exptl		
Cyclohexane-d_{12}	2.0	0.07	0.78
Chloroform-*d*	4.8	0.15	0.80
Methanol-d_4	33	0.29	0.77
Dimethyl-d_6 sulfoxide	47	0.19	0.73
Deuterium oxide	79	0.29	0.72

a) $CH_3NHCH_2CH_2NHCH_3$.
b) At the MP2/6-311++G(3df,3pd) level.

18.2 Poly(*N*-methylethylene imine) (PMEI)

PMEI does not form any N—H···N hydrogen bond because it has a methyl group at the nitrogen site in place of a hydrogen atom. Therefore, the lone pair is free from the intramolecular attractions and directed to the surroundings such as solvents. In this respect, PMEI is similar to PEO; however, in contrast with PEO, PMEI performs the nitrogen inversion because even bulky substituents such as *tert*-butyl group cannot suppress the inversional isomerization, which enhances the configurational entropy (S_{conf}). PEI has S_{conf} values of 4.54 cal K^{-1} mol^{-1} at HBS = 100% and 6.26 cal K^{-1} mol^{-1} at HBS = 0%, while that of PMEI is 5.81 cal K^{-1} mol^{-1}. PEI is soluble in hot water but insoluble in cold water, whereas PMEI is soluble in a variety of organic solvents and cold water like PEO. The pKa value of PMEI is 7.28, suggesting that 43% nitrogen atoms of PMEI will be protonated to be cationic in human blood (pH 7.4). We studied solution properties of PMEI in detail mainly via light scattering (see Section 16.2.2) [164]. Toluene solutions of PMEI exhibit the upper critical solution temperature (UCST)-type phase separation, while aqueous solutions of PMEI show the lower critical solution temperature (LCST)-type one. The latter phase behavior was investigated with the aid of the salting-out effect. Figure 18.2 shows the cloud point vs. salt concentration plots of the aqueous solutions including different inorganic salts (PMEI concentration was 0.50 wt%). The extrapolation lines of the plots seem to converge to a single point (116 °C) on the ordinate axis; therefore, the LCST of the pure aqueous solution of PMEI may be determined to be 116 °C.

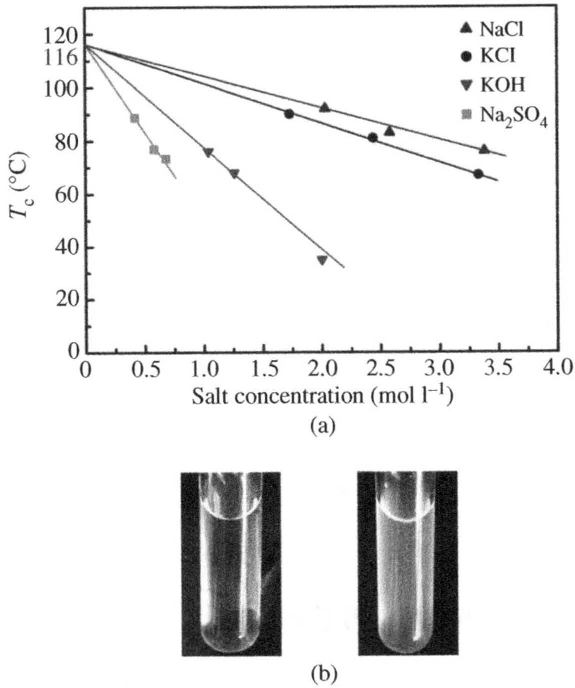

Figure 18.2 LCST-type phase separation of PMEI in water. (a) Cloud point vs. salt concentration plots of aqueous solutions including different salts. The extrapolation lines converge at a single point (116 °C) on the ordinate axis. (b) Visual observation of phase separation behavior of an aqueous solution containing KCL (3.33 mol l^{-1}) at temperatures that were lower (left) and higher (right) by 2 °C than the cloud point. The PMEI concentration was 0.50 wt%. Material from: reference [164], published 2016, Springer Nature.

18.3 Poly(trimethylene imine) (PTMI) and Poly(*N*-methyltrimethylene imine) (PMTMI)

PTMI with an NH group and three methylene units in the backbone also undergoes the nitrogen inversion. As seen from Table 18.2, the C—C and C—N bonds have gauche and trans preferences, respectively. Figure 18.3 illustrates intramolecular interactions formed in PTMI and PMTMI. Of them, the attractive η and ψ'' interactions are regarded as weak C—H$\cdots$N and N—H$\cdots$N hydrogen bonds, respectively. The interaction energies of PTMI were evaluated at the MP2/6-311++G(3df,3pd) level as $E_\eta = -0.15$ and $E_{\psi''} = -0.83$ kcal mol^{-1}. The gauche energy (E_σ) of the C—C bond is essentially null. PMTMI also forms a C—H$\cdots$N hydrogen bond (η interaction in Figure 18.3), whose energy was evaluated to be −0.40 kcal mol^{-1} [417].

The $\langle r^2 \rangle_0/nl^2$ values of the infinite chains ($x \to \infty$) of PTMI and PMTMI are 3.5 and 4.2, respectively. The *meso* probabilities are 0.44 (PTMI) and 0.48 (PMTMI). Figure 18.4 shows the characteristic ratio, trans fractions of the C—C and C—N bonds, and *meso* probability as a function of the interaction energies: E_σ and $E_{\psi''}$ for PTMI, and E_σ and E_η for PMTMI. The ordinate quantities except p_t^{CN} vary significantly with the abscissa-axis energies. As the energies are moved along the abscissa axis to the positive (right) direction, p_t^{CC}, $\langle r^2 \rangle_0/nl^2$, and P_m increase

Table 18.2 Trans fractions of *N,N′*-dimethyl-1,3-propanediamine,[a] a model compound of PTMI, at 25 °C.

Environment	Permittivity	p_t^{CC}	p_t^{CN}
	MO calc[b]		
Gas	1.0	0.25	0.73
	NMR exptl		
Cyclohexane-d_{12}	2.0	0.36	0.86
Chloroform-d	4.8	0.35	0.86
Methanol-d_4	33	0.48	0.67
Dimethyl-d_6 sulfoxide	47	0.35	0.88
Deuterium oxide	79	0.46	0.77

a) $CH_3NHCH_2CH_2CH_2NHCH_3$.
b) At the MP2/6-311++G(3df,3pd) level.

simultaneously: the molecular chain is extended, and the chain configuration approaches randomness ($P_m \rightarrow 1/2$).

In summary, the polyamines show the nitrogen inversion and the N—H· · ·N and C—H· · ·N hydrogen bonds. The attractions, even though being weak, strongly influence conformational characteristics and configurational properties of the polyamines. The substitution of *N*-hydrogen with an alkyl group renders the lone pair free from the N—H· · ·N interactions, and consequently, the molecular chain will be expanded, and the chain configuration becomes random.

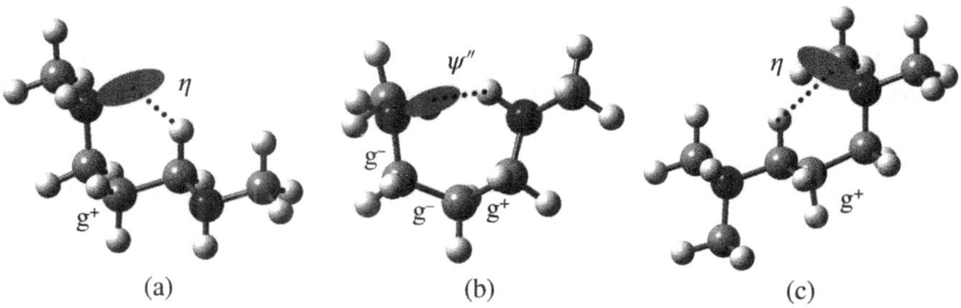

Figure 18.3 Intramolecular attractions formed in PTMI and PMTMI: (a) C—H· · ·N (*η*) and (b) N—H· · ·N (*ψ″*) of PTMI and (c) C—H· · ·N (*η*) of PMTMI. Source: Reproduced with permission from reference [417]. Copyright 2005 American Chemical Society.

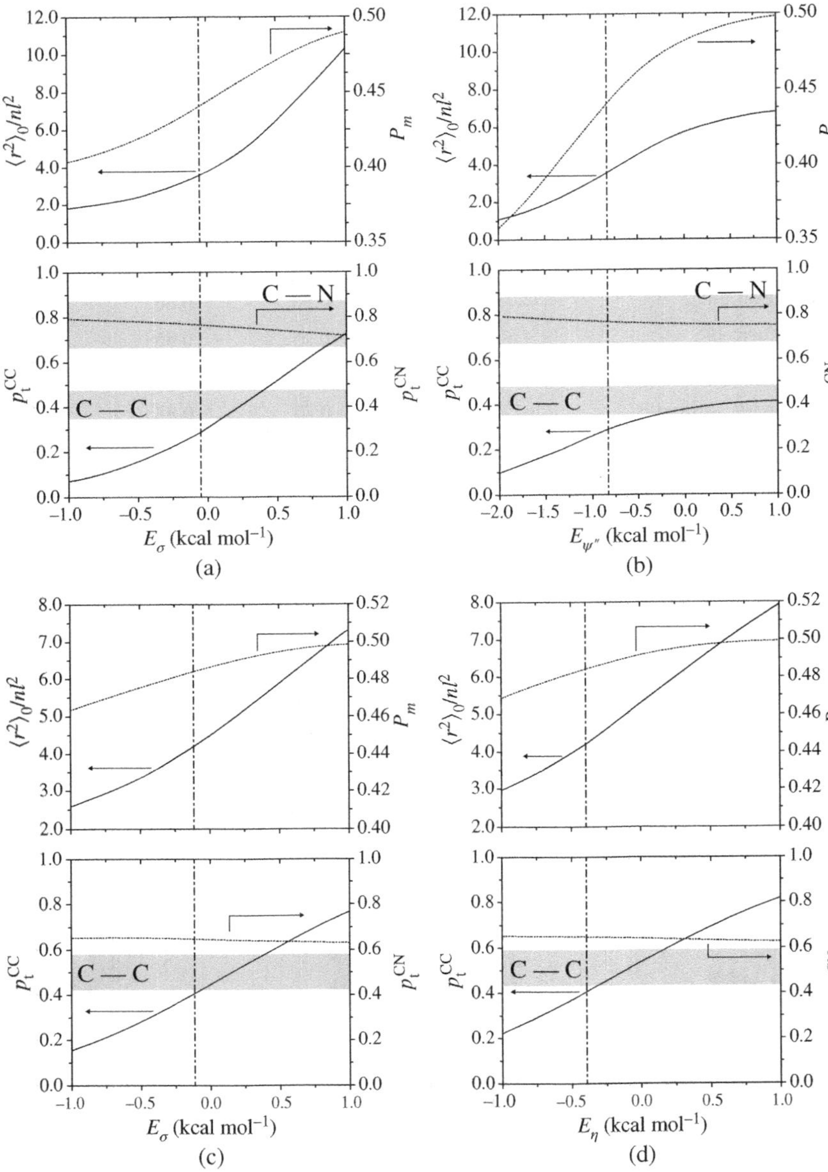

Figure 18.4 Characteristic ratio ($\langle r^2\rangle_0/nl^2$), *meso* probability ($P_m$), and trans fractions of the C–C (p_t^{CC}) and C–N (p^{CN}_t) bonds as a function of (a) E_σ (gauche energy) or (b) $E_{\psi''}$ (N–H···N interaction energy) of PTMI and (c) E_σ (gauche energy) or (d) E_η (C–H···N interaction energy) of PMTMI. The vertical dash-dotted line corresponds to the energy value obtained from the MO calculations. The horizontal shaded regions represent the p_t^{CC} and p_t^{CN} ranges derived from the NMR experiments. Source: Reproduced with permission from reference [417]. Copyright 2005 American Chemical Society.

19

Polyphosphines

Polyphosphines (PPPs) include trivalent phosphorus atoms in the backbone. Phosphorus is the same pnictogen (group 15) element as nitrogen. PPPs are prepared by the reduction of poly(phosphine oxide)s obtained from deoxophostones [252] or by ring-opening polymerization of phosphiranes. [250] Inasmuch as the trivalent phosphorus atom has a lone pair of electrons like nitrogen, PPPs are expected to exhibit structures and properties similar to those of polyamines. As described above, the most characteristic features of polyamines are nitrogen inversion and intramolecular N—H· · ·N and C—H· · ·N attractions. In order to investigate whether PPPs will show similar phenomena, the electronic structures of PPPs were investigated [411].

The PPPs treated here are poly(1-methylphosphirane) (PMePP), a simple polymer similar to poly(ethylene imine) (PEI), and poly(1-phenylphosphirane) (PPhPP) with phenyl groups at the phosphorus sites (Figure 19.1). As their model compounds, 1,2-bis(dimethylphosphino)ethane (BDMePE) and 1,2-bis (methylphenylphosphino)ethane (BMePhPE) have been adopted. According to the pseudoasymmetry [140], each phosphorus site of the PPPs and models are assigned to either *l*- or *d*-form, and the *ll* and *dd* diads are designated as *meso*, and the *ld* and *dl* diads as *racemo*.

19.1 Possibility of Phosphorus Inversion

To examine the possibility of phosphorus inversion, a molecule simpler than BDMePE and BMePhPE, 1,2-bis(methylphosphino)ethane (BMePE) with a hydrogen atom at the phosphorus site, was adopted, and its transition state between *meso* ttt and *racemo* ttt conformations was searched for by the synchronous transit-guided quasi-Newton (QST2) method [360, 361] combined with density functional theory (DFT) calculations at the B3LYP/6-31G(d) level. In the transition state thus determined, the lone pair of the P atom is shaped

Conformational Analysis of Polymers: Methods and Techniques for Structure-Property Relationships and Molecular Design, First Edition. Yuji Sasanuma.

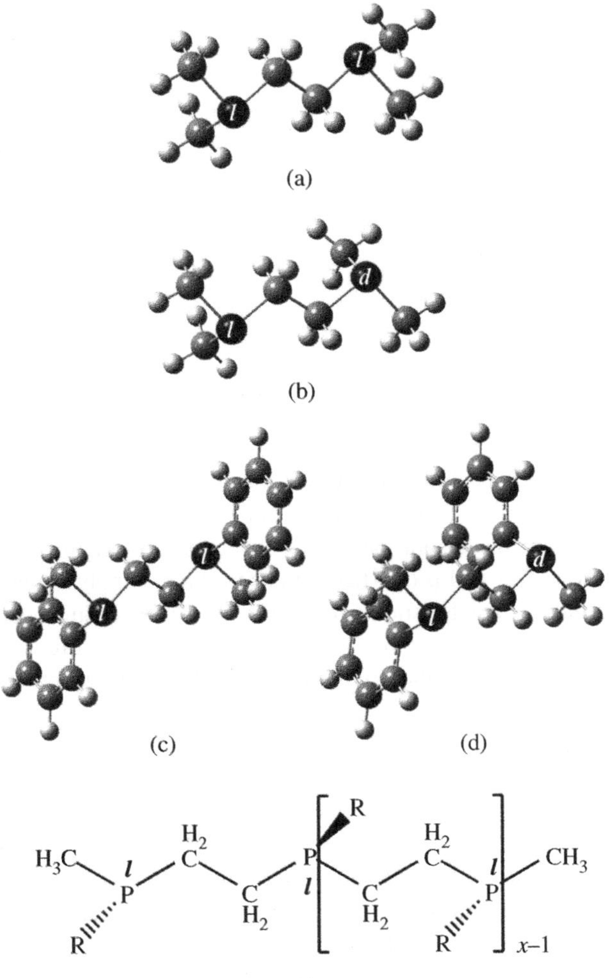

Figure 19.1 All-*trans* states of (a) *ll*-1,2-bis(dimethylphosphino)ethane (*ll*-BDMePE), (b) *ld*-BDMePE, (c) *ll*-1,2-bis(methylphenylphosphino)ethane (*ll*-BMePhPE), (d) *ld*-BMePhPE, and (e) isotactic poly(1-methylphosphirane) (R = CH_3, PMePP) or isotactic poly(1-phenylphosphirane) (R = C_6H_5, PPhPP). According to Flory's concept of pseudoasymmetry, the *l* and *d* configurations are defined. For details, see Section 1.1. Source: [411]/Reproduced with permission of the PCCP Owner Societies.

like a p orbital, and the hydrogen atom bonded to the P atom lies on the C—P—C plane. The transitional state has virtually the same form as that of *N*,*N*′-dimethylethylenediamine (di-MEDA), the model for PEI. However, the free energy of the transition state is as high as 35.3 (35.0) kcal mol^{-1} relative to that of the *meso* (*racemo*) ttt state, and the Boltzmann factor at room temperature is essentially null (2×10^{-26}); therefore, the phosphorus inversion ought never to occur, and PPPs will keep the configurations (tacticities) as polymerized.

19.2 Intramolecular Interactions Related to Phosphorus

The zero-point energies of the conformers of BDMePE and BMePhPE were evaluated from ab initio MO calculations at the MP2/6-311++G(3df.3pd) level. The most stable state of BDMePE was determined to be *meso* ttt. For the sake of comparison, the zero-point energies of *N*,*N*,*N*′,*N*′-tetramethylethylenediamine (tetra-MEDA), a model compound of poly(*N*-methylethylene imine) (PMEI), were also calculated, and its lowest energy states are *meso* tg^+g^+, *meso* g^-g^-t, *racemo* tg^+t, and *racemo* $g^-g^-g^-$ and have a zero-point energy of −0.69 kcal mol^{-1} relative to *meso* ttt. The negative energy suggests that intramolecular C—H···N attractions are formed in the conformations, where the H···N distances are 2.41 and 2.72 Å. In contrast, C—H···P contacts of BDMePE can be interpreted as repulsions for the following reasons. The natural bond orbital (NBO) analysis [382] expressed the lone pair of BDMePE as a hybrid orbital of $sp^{1.02}$; therefore, the s character is 50%, whereas that of tetra-MEDA was given as $sp^{5.99}$, and the s character is 14%. It is known that the proton (hydrogen) affinity of a lone pair, that is basicity, decreases with increasing s character. Therefore, tetra-MEDA (PMEI) is a strong base, while BDMePE (PMePP) is a weak base. According to the hard and soft acids and bases (HSAB) theory [184, 358], a strong (weak) base has an affinity for a strong (weak) acid. The strong base, PMEI, will interact with a strong acid, hydrogen, and the weak base, PMePP, attracts weak acids such as transition, noble, and heavy metals. In addition, the NBO analysis allotted a positive charge of +0.85 to the phosphorus atom, which will repulse positively charged hydrogen atoms. In conclusion, the C—H···P contact must be repulsive.

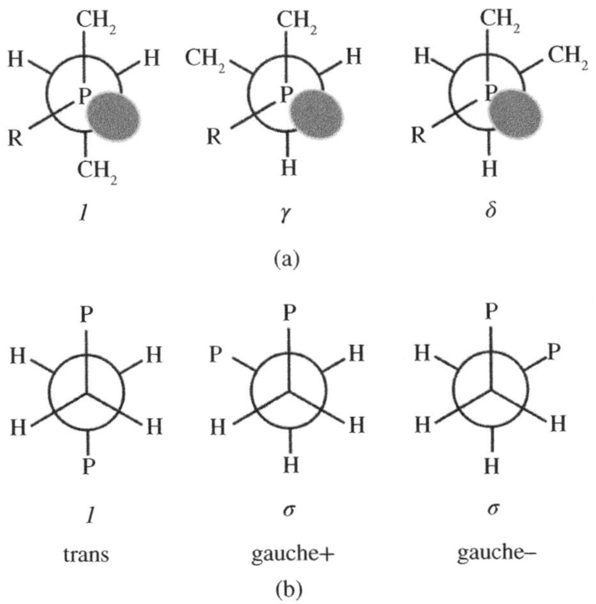

Figure 19.2 Rotational isomeric states around (a) P–CH_2 and (b) CH_2–CH_2 bonds of poly(1-R substituted phosphirane) (R=CH_3, PMePP or R=C_6H_5, PPhPP). Here, the phosphorus sites adopt the *l* configuration. The Greek letters, γ, δ, and σ, represent first-order interactions. Source: [411]/Reproduced with permission of the PCCP Owner Societies.

19.3 RIS Calculation

Gibbs free energies of the conformers of BDMePE were evaluated from ab initio MO calculations at the MP2/6-311++G(3df.3pd)//B3LYP/6-311G(3df,2p) level and divided into well-defined interaction (conformational) energies (see Table 19.1 and Figures 19.2 and 19.3). For the sake of comparison, the conformational energies of tetra-MEDA (PMEI) are also listed there. The statistical

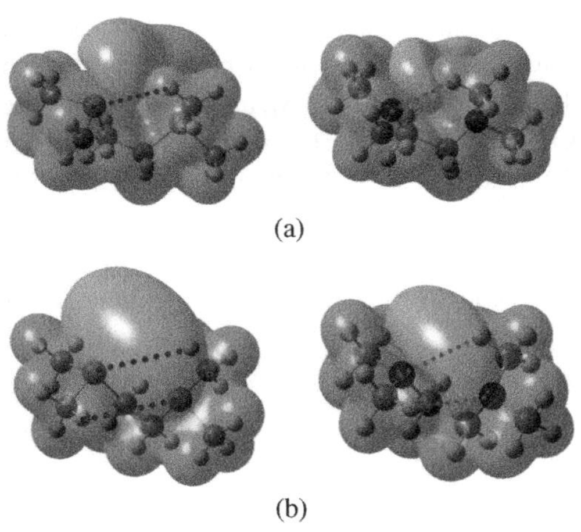

Figure 19.3 Higher order interactions defined for BDMePE and PMePP (left), compared in part with those of tetra-MEDA (right, a and b): (a) η; (b) ν'; (c) ν; (d) ω; and (e) ω'. Source: [411]/Reproduced with permission of the PCCP Owner Societies.

Figure 19.3 (Continued)

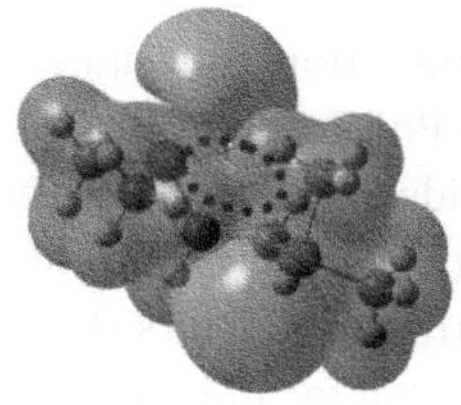
(c)

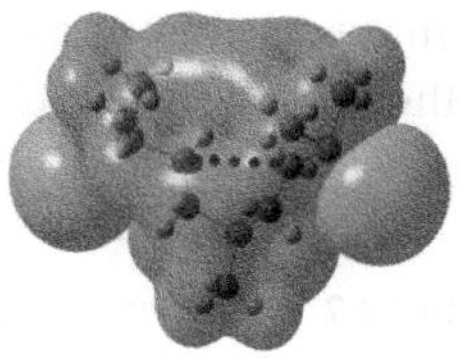
(d)

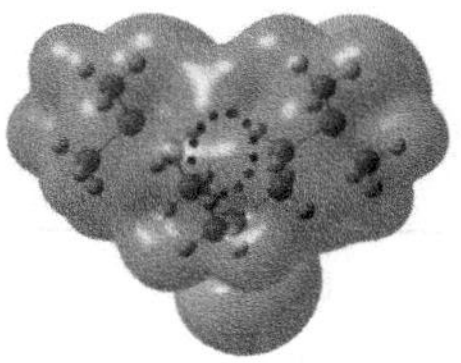
(e)

Table 19.1 Conformational energies (E_ξ) of PMePP and PMEI.

Interaction[a)]	E_ξ (kcal mol^{-1})	
	PMePP	PMEI
First order		
γ	0.81	1.27
δ	0.29	0.04
σ	0.91	−0.42
Higher order		
η	0.29	0.34
ν	1.50	∞
ν'	−0.01	−0.27
ω	0.60	1.69
ω'	3.23	3.09

a) For the definitions, see Figures 19.2 and 19.3.
b) From MO calculations for BDMePE and tetra-MEDA.

weight matrices (U_js) were formulated with the conformational energies. This formulation is designated as model I. However, the intramolecular interactions of PPhPP are so complicated that the conformer free energies could not be simply divided into a small number of conformational energies (Table 19.2); thus, the statistical weights were directly evaluated from the Gibbs energies and arranged in the U_j matrices (this method is named model II). Both models I and II were adopted for PMePP to compare the two results. The geometrical parameters were chosen from those of the model compounds. Each pseudochiral center of the polymeric chain was assigned to either *l*- or *d*-form by the so-called Monte Carlo method based on the random number generation [403, 418] so as to match the given *meso* ratio (P_m). The number (n_c) of polymeric chains in the ensemble

Table 19.2 Conformer free energies (ΔG_ks) of PPhPP, evaluated by ab initio MO calculations on BMePhPE.

	meso (*ll*)			*racemo* (*ld*)		
k	Conformation		ΔG_k a)	Conformation		ΔG_k a)
1	ttt		0.00	ttt		−0.89
2	ttg$^+$	g$^-$tt	1.26	ttg$^+$	g$^+$tt	1.07
3	ttg$^-$	g$^+$tt	−0.45	ttg$^-$	g$^-$tt	1.45
4	tg$^+$t	tg$^-$t	0.81	tg$^+$t		0.73
5	tg$^+$g$^+$	g$^-$g$^-$t	1.88	tg$^+$g$^+$	g$^+$g$^+$t	2.50
6	tg$^+$g$^-$	g$^+$g$^-$t	3.51	tg$^+$g$^-$	g$^-$g$^+$t	2.74
7	tg$^-$g$^+$	g$^-$g$^+$t	3.48	tg$^-$t		1.10
8	tg$^-$g$^-$	g$^+$g$^+$t	(absent)b)	tg$^-$g$^+$	g$^+$g$^-$t	3.70
9	g$^+$tg$^+$	g$^-$tg$^-$	2.16	tg$^-$g$^-$	g$^-$g$^-$t	2.59
10	g$^+$tg$^-$		1.41	g$^+$tg$^+$		0.55
11	g$^+$g$^+$g$^+$	g$^-$g$^-$g$^-$	4.13	g$^+$tg$^-$	g$^-$tg$^+$	1.75
12	g$^+$g$^+$g$^-$	g$^+$g$^-$g$^-$	(absent)b)	g$^+$g$^+$g$^+$		(absent)b)
13	g$^+$g$^-$g$^+$	g$^-$g$^+$g$^-$	(absent)b)	g$^+$g$^+$g$^-$	g$^-$g$^+$g$^+$	4.72
14	g$^-$tg$^+$		2.28	g$^+$g$^-$g$^+$		8.54
15	g$^-$g$^+$g$^+$	g$^-$g$^-$g$^+$	4.10	g$^+$g$^-$g$^-$	g$^-$g$^-$g$^+$	4.34
16				g$^-$tg$^-$		2.60
17				g$^-$g$^+$g$^-$		5.26
18				g$^-$g$^-$g$^-$		2.97

a) In kcal mol^{-1}. From MO calculations for BMePhPE.
b) The local minimum of the potential was not found by the geometrical optimization.
Source: [411]/Reproduced with permission of the PCCP Owner Societies.

must be large enough to reproduce the given P_m even for small degrees (xs) of polymerization. In the previous studies [403, 418], the n_c value of 512 (2^9) was proved to satisfy the requirement and also used here.

Figure 19.4 shows the characteristic ratios of isotactic ($P_m = 1$), atactic ($P_m = 0.5$), and syndiotactic ($P_m = 0$) chains of PMePP as a function of the reciprocal degree

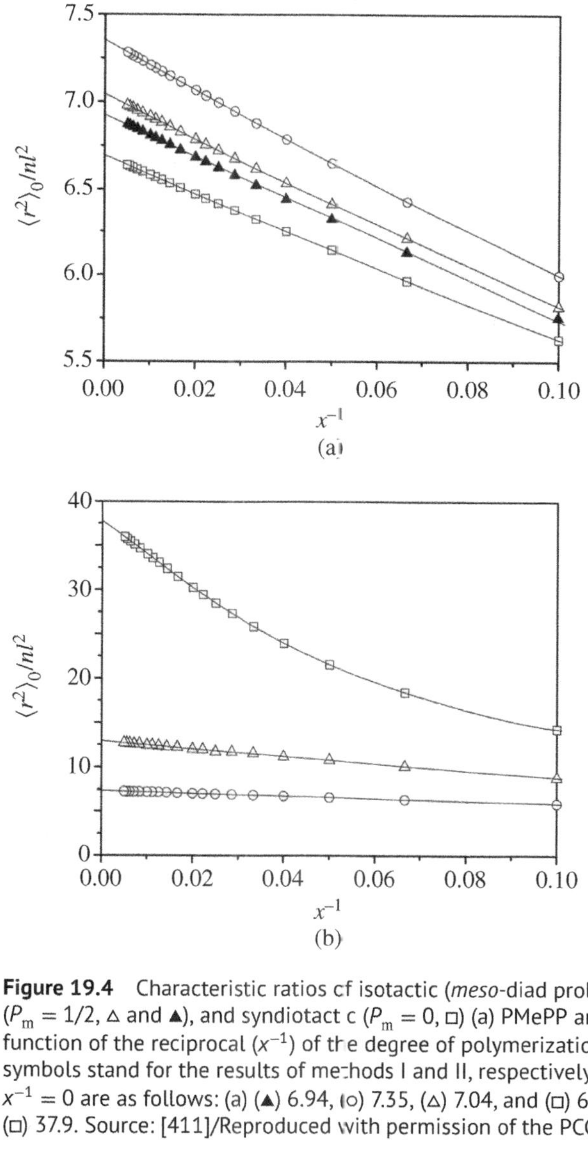

Figure 19.4 Characteristic ratios of isotactic (*meso*-diad probability $P_m = 1$, ○), atactic ($P_m = 1/2$, △ and ▲), and syndiotactic ($P_m = 0$, □) (a) PMePP and (b) PPhPP at 25 °C as a function of the reciprocal (x^{-1}) of the degree of polymerization. The filled and open symbols stand for the results of methods I and II, respectively. The $\langle r^2 \rangle_0 / nl^2$ values at $x^{-1} = 0$ are as follows: (a) (▲) 6.94, (○) 7.35, (△) 7.04, and (□) 6.70; (b) (○) 7.33 (△) 13.0 and (□) 37.9. Source: [411]/Reproduced with permission of the PCCP Owner Societies.

of polymerization. The intercept at $x^{-1} = 0$ gives the characteristic ratio of the infinite-length chain. The $\langle r^2 \rangle_0/nl^2$ values of the atactic chain, derived from models I and II, are 6.94 and 7.04, respectively, being very close to each other; therefore, both models are acceptable for, at least, PMePP. The isotactic and syndiotactic chains yielded the $\langle r^2 \rangle_0/nl^2$ values of 7.35 and 6.70, respectively; thus, the chain dimension depends a little on tacticity. In contrast, the characteristic ratio of PPhPP shows large tacticity dependence: isotactic, 7.33; atactic, 13.0; and syndiotactic 37.9.

The *meso* segment of PMePP (PPhPP) has trans fractions of 0.60 (0.64) and 0.82 (0.88) in the P—C and C—C bonds, respectively: both *meso* segments adopt similar conformations. This is because, for example, the *meso* ttt conformation sticks out the two side groups in the opposite directions, and hence, the side groups seldom interact with each other and, consequently, give only slight effects on the backbone conformation. The *racemo* segment of PPhPP has trans fractions close to 0.9 in all the bonds, thus being much extended, and the syndiotactic PPhPP chain shows as large a characteristic ratio as 37.9. When the *racemo* part adopts the all-trans conformation, the neighboring phenyl groups face each other and form $\pi - \pi$ attraction, which stabilizes the extended form. The *racemo* ttt conformation is lower in free energy by 0.89 kcal mol^{-1} than *meso* ttt. The spatial configuration of PPhPP depends largely on tacticity in contrast to that of PMePP.

19.4 Functions and Stability

The soft bases, PPPs, have affinities for soft acids, such as transition, noble, and heavy metals and hence act as ligands for these metals. However, PPPs may be readily oxidized by oxygen in the atmosphere and dissolved in solvents to be pentavalent. If the oxidation of BDMePE is suppose

$$(CH_3)_2P(CH_2)_2P(CH_3)_2 + \frac{1}{2}O_2 \rightarrow (CH_3)_2P(=O)(CH_2)_2P(CH_3)_2$$

then the reaction enthalpy calculated at the MP2/6-311++G(3df,3pd) level would be −93.1 kcal mol^{-1}. This result suggests that, as soon as PPPs are synthesized, they will be oxidized rapidly or gradually. Therefore, it is preferable that PPPs will be used under the environments without oxygen and/or for usages where the oxidation will be advantageous.

20

Polysulfides

Polysulfides, aka polythioethers, include bivalent sulfur atoms in the backbone. Sulfur is a chalcogen element, one of the oxygen family. In this section, conformational characteristics and configurational properties of poly(methylene sulfide) (PMS), poly(ethylene sulfide) (PES), poly(propylene sulfide) (PPS), and poly(trimethylene sulfide) (PTrMS), which are the analogs of poly(methylene oxide) (PMO), poly(ethylene oxide) (PEO), poly(propylene oxide) (PPO), and poly(trimethylene oxide) (PTrMO), respectively, are treated and compared with those of the corresponding polyethers.

20.1 Poly(methylene sulfide) (PMS)

As explained in Section 17.1, PMO shows a strong gauche stability. Such phenomena are designated as the anomeric effect. Similarly, the S—CH_2—S bond sequence of 2-alkylthiotetrahydrothiopyrans is known to show a weak gauche preference ($\Delta G \sim -0.4$ kcal mol^{-1}) [227]. As will be discussed in Section 21.1, the Se–CH_2–Se sequence of poly(methylene selenide) also tends to be gauche ($\Delta G \sim -1.0$ kcal mol^{-1}) [421]. PMS and PMO are insoluble in common organic solvents at room temperature. The crystalline PMS chain forms the all-gauche 17/9 helix [61] and melts at high temperatures of 220–245 °C [269].

As a model compound of PMS, bis(methylthio)methane (abbreviated as PMS-2) was employed and subjected to density functional theory (DFT) calculations and nuclear magnetic resonance (NMR) experiments [399, 423]. Here, PMS-x corresponds to PMO-x (see Figure 17.2). For the NMR measurements, PMS-2 labeled with carbon-13 (PMS-2-^{13}C, $^{13}CH_3SCH_2SCH_3$) was prepared. From the vicinal coupling constant ($^3J_{CH}$) between the methyl carbon-13 and methylene

Conformational Analysis of Polymers: Methods and Techniques for Structure-Property Relationships and Molecular Design, First Edition. Yuji Sasanuma.

protons, the trans fraction p_t of the S—CH$_2$ bond can be derived according to

$$p_t = \frac{J_T + J_G - 2\,{}^3J_{CH}}{J_T - J_G} \tag{20.1}$$

where J_T and J_G are, respectively, the vicinal $^{13}C-^1H$ coupling constants in the antiperiplanar and synclinal positions and were experimentally determined from 2-methyl-1,3,5-trithiane (MTT), a cyclic compound with three S–CH$_2$–S sequences. For example, $J_T = 7.13$ Hz and $J_G = 2.62$ Hz were obtained from the C_6D_6 solution. The p_t values were calculated from Eq. (20.1) with $^3J_{CH}$, J_T, and J_G of the individual solutions, being given in Table 20.1.

The statistical weight matrices of bonds 2 and 3 (for the bond numbers, see Figure 17.2) may be formulated with the weight of the gauche state set unity:

$$U_2 = \begin{bmatrix} \sigma^{-1} & 1 & 1 \\ 0 & 0 & 0 \\ 0 & 0 & 0 \end{bmatrix} \tag{20.2}$$

and

$$U_3 = \begin{bmatrix} \sigma^{-1} & 1 & 1 \\ \sigma^{-1} & 1 & 0 \\ \sigma^{-1} & 0 & 1 \end{bmatrix} \tag{20.3}$$

The elements of U_3 corresponding to the $g^{\pm}g^{\mp}$ conformations are set null because of strong S· · ·S steric repulsion occurring there. The p_t value can be related to σ as

$$p_t = \frac{2\sigma + 1}{2\sigma^2 + 4\sigma + 1} \tag{20.4}$$

Table 20.1 Observed vicinal $^{13}C-^1H$ coupling constants and S–CH$_2$ bond conformations of PMS-2-^{13}C.

Medium[a)]	$^3J_{CH}$ (Hz)	p_t	E_σ (kcal mol^{-1})
Gas	4.52	0.16	−1.43
Cyclohexane-d_{12}	4.63	0.11	−1.21
Benzene-d_6	4.57	0.14	−1.05
Chloroform-d	4.51	0.15	−0.98
Methanol-d_4	4.55	0.15	−0.98
Dimethyl-d_6 sulfoxide	4.49	0.15	−0.96

a) At 180 °C for the gas phase, and at 25 °C for the solutions.

The gauche energy (E_σ) of the S—CH_2 bond can be obtained from $^3J_{CH}$ via Eqs. (20.1) and (20.4) because of

$$E_\sigma = -RT \ln \sigma \tag{20.5}$$

The E_σ value is given for each medium in Table 20.1. Gibbs free energies of PMS-2 at 25 °C, evaluated from DFT calculations at the B3LYP/6-311+G(2d,p)//B3LYP/6-31G(d) level, are listed for the individual conformers (Table 20.2), and the E_σ value of −1.54 kcal mol^{-1} was obtained from ΔG_k of gaseous PMS-2 (cf. E_σ = −2.15 kcal mol^{-1} of PMO-2). The DFT calculations also yield E_σ = −1.38 kcal mol^{-1} at 180 °C, which agrees with the NMR experiment (−1.43 kcal mol^{-1}) for gaseous PMS-2-^{13}C at 180 °C.

The statistical weight matrix of bond 4 of PMS-3 can be formulated as

$$U_4 = \begin{bmatrix} \sigma^{-1} & 1 & 1 \\ \sigma^{-1} & 1 & \omega \\ \sigma^{-1} & \omega & 1 \end{bmatrix} \tag{20.6}$$

For PMS-3, the DFT calculations indicated the formation of the $g^\pm g^\mp$ states, and hence, the second-order ω interaction is introduced into U_4. For example, the tg^+g^-t state has a weight of $\sigma^{-2}\omega$, and hence, its Gibbs energy may be approximated as $-2E_\sigma + E_\omega$. The E_σ and E_ω values were optimized by the least-squares method so as to reproduce the Gibbs free energies of all conformer of PMS-3. The results were E_σ = −1.54 kcal mol^{-1} and E_ω = 2.09 kcal mol^{-1} (cf. E_σ = −1.72 kcal mol^{-1} and E_ω = 1.98 kcal mol^{-1} of PMO-3). The gauche energy (E_σ) of the S—CH_2 bond is somewhat smaller in magnitude than that of the O—CH_2 bond.

The bond dipole moment of the S—CH_2 bond, $m_{S\text{-}C}$, was determined from the conformer dipole moments (μ_ks) given by the DFT computations: by the least-squares method, the difference between μ_k and the magnitude of $\mathbf{m}_{S\text{-}C}$ sum of each conformer was minimized, and as a consequence, $m_{S\text{-}C}$ = 1.23 D was obtained. The geometrical parameters used for the RIS calculations were chosen from the model compounds: bond length, $l_{S\text{-}C}$ = 1.833 Å; bond angles, ∠CSC = 100.06° and ∠SCS = 117.30°; and dihedral angles, $\phi_t = 0.00°$ and $\phi_{g\pm} = \pm 113.07°$. The E_σ value was set to that (−1.05 kcal mol^{-1}) derived from the NMR experiment for the benzene solution of PMS-2-^{13}C, and E_ω was as above, 2.09 kcal mol^{-1}. The statistical weight matrices of the repeating unit were

$$U_a = U_3 \tag{20.7}$$

and

$$U_b = U_4 \tag{20.8}$$

For the bond symbols, see Figure 17.2. The characteristic ratio and dipole moment ratio were calculated by the RIS scheme. For a fully long PMS chain of x = 300

Table 20.2 Conformer free energies (ΔG_k) of PMS-2 and PMS-3, evaluated by DFT calculations at the B3LYP/6-311+G(2d,p)//B3LYP/6-31G(d) level.

k	Conformation	M_k [b)]	Statistical weight(s)	ΔG_k [a)] (kcal mol^{-1})
PMS-2				
1	t t	1	σ^{-2}	5.12 (5.88)
2	t g$^+$	4	σ^{-1}	1.54 (1.38)
3	g$^+$g$^+$	2	1	0.00 (0.00)
PMS-3 [c)]				
1	t t t t	1	σ^{-4}	9.66
2	t t t g$^+$	4	σ^{-3}	— [d)]
3	t t g$^+$ t	4	σ^{-3}	— [d)]
4	t g$^+$ t g$^+$	4	σ^{-2}	2.76
5	t g$^+$ t g$^-$	4	σ^{-2}	2.94
6	t g$^+$ g$^+$ t	2	σ^{-2}	2.78
7	t g$^+$ g$^+$ g$^+$	4	σ^{-1}	1.50
8	t g$^+$ g$^-$ t	2	$\sigma^{-2}\omega$	6.02
9	t g$^+$ g$^-$ g$^-$	4	$\sigma^{-1}\omega$	3.28
10	g$^+$ t t g$^+$	2	σ^{-2}	2.85
11	g$^+$ t t g$^-$	2	σ^{-2}	3.16
12	g$^+$ t g$^+$ g$^+$	4	σ^{-1}	1.61
13	g$^+$ g$^+$ t g$^-$	4	σ^{-1}	1.69
14	g$^+$ g$^+$ g$^+$ g$^+$	2	1	0.00
15	g$^+$ g$^+$ g$^-$ g$^-$	2	ω	3.07

a) Relative to the all-gauche conformation. At 25 °C. For comparison with the ^{13}C NMR experiment for gaseous PMS-2-^{13}C, the ΔG_k values for PMS-2 at 180 °C are shown in the parentheses.
b) Multiplicity.
c) By the least-squares method, the E_σ and E_ω values were determined to be −1.54 and 2.09 kcal mol^{-1}, respectively (PMS-3).
d) The local minimum of the potential was not found by the geometrical optimization.
Source: [423]/with permission of American Chemical Society.

at 25 °C, $\langle r^2 \rangle_0 / nl^2 = 7.6$ and $\langle m^2 \rangle / nm^2 = 0.17$ were obtained; however, the experimental values of the configurational properties are unrevealed owing to its poor solubility. The configurational entropy at 25 °C was calculated to be 2.7 cal K^{-1} mol^{-1}, being larger than that (2.0–2.3 cal K^{-1} mol^{-1}) of PMO. The S_{conf}

difference between PMS and FMO is due to that in E_σ and expresses that PMS is somewhat more flexible than PMO.

20.1.1 Crystal Structure of PMS

The PMS chain crystallizes to form the all-gauche (17/9) helix and packed in a hexagonal cell (space group, P_1) (abbreviated as h-PMS). The h-PMS structure was optimized by DFT calculations at the B3LYP-D/6-31G(d,p) level, and the crystalline moduli and interchain interactions were also calculated [228]. The optimized structure is compared with the experiment in Table 20.3 and illustrated in Figure 20.1. The stiffness tensor is

$$C = \begin{bmatrix} 16.483 & 6.302 & 7.203 & -0.055 & -0.029 & 0.027 \\ & 15.424 & 7.255 & 0.007 & 0.017 & 0.015 \\ & & 46.450 & 0.082 & -0.080 & -0.009 \\ & & & 4.826 & -0.210 & -0.005 \\ & & & & 4.876 & 0.037 \\ & & & & & 5.001 \end{bmatrix} \text{(GPa)} \qquad (20.9)$$

Crystalline Young's moduli in the a-, b-, and c-axis directions, calculated from the compliance tensor (the C^{-1} tensor), are as follows (in GPa): $E_a = 13.6$, $E_b = 13.4$, and $E_c = 41.9$. Figure 20.1 shows the distribution of Young's modulus on the plane perpendicular to the fiber axis. The circular shape is due to the uniform helical structure.

The interchain interaction energy was also calculated via the basis set superposition error (BSSE) correction and compared with those of the trigonal (t-PMO) and orthorhombic (o-PMO) PMO crystals: h-PMS (-4.51 kcal mol^{-1}), t-PMO (-3.17 kcal mol^{-1}), and o-PMO (-3.38 kcal mol^{-1}) [166, 228].

20.2 Poly(ethylene sulfide) (PES)

PES is insoluble in most solvents and poorly soluble in only a few solvents at temperatures higher than 140 °C [303]. The equilibrium melting pint was reported as 216 °C [1, 102], which is much higher than that (80 °C) of PEO [293]. Conformer Gibbs free energies of 1,2-bis(methylthio)ethane (BMTE), a model compound of PES, were evaluated from molecular orbital (MO) calculations at the MP2/6-311+G(3df,2p)//HF/6-31G(d) level [412]. The conformers of the lowest ΔG are g$^\pm$tg$^\mp$, where the two S—CH$_2$ bonds adopt gauche states of opposite signs, and the central CH$_2$—CH$_2$ bond is trans.

Figure 20.2 shows the ^{1}H and ^{13}C NMR spectra of BMTE. The satellite signals appearing around the CH$_2$ main peak gave the $^3J_{HH}$ and $^3J'_{HH}$ values

Table 20.3 Optimized and experimental crystal structures of PMS[a].

	Optimized[b]			Experimental[c]		
			Lattice constant (Å)			
	a	*b*	*c*	*a*	*b*	*c*
	5.127	5.127	38.21	5.07	5.07	36.52
			$\Delta_{LC} = 1.63\%$			
			Fractional coordinates[d]			
	x/*a*	*y*/*b*	*z*/*c*	*x*/*a*	*y*/*b*	*z*/*c*
S	0.2270	0.0428	0.0013	0.1953	0.0000	0.0000
C	0.0442	0.1860	0.0304	0.0942	0.2245	0.0294
S	−0.2430	−0.0855	0.0600	−0.2127	−0.0414	0.0588
C	−0.0158	−0.1780	0.0888	−0.0550	−0.2169	0.0882
S	0.2500	0.1180	0.1180	0.2228	0.0815	0.1176
C	−0.0063	0.1650	0.1470	0.0139	0.2018	0.1471
S	−0.2450	−0.1500	0.1760	−0.2254	−0.1187	0.1765
C	0.0355	−0.1510	0.2050	0.0277	−0.1799	0.2059
S	0.2370	0.1740	0.2340	0.2203	0.1519	0.2353
C	−0.0631	0.1210	0.2640	−0.0684	0.1519	0.2647
S	−0.2030	−0.2000	0.2940	−0.2076	−0.1799	0.2941
C	0.1150	−0.0704	0.3240	0.1067	−0.1187	0.3235
S	0.1780	0.2330	0.3540	0.1880	0.2018	0.3529
C	−0.1480	0.0382	0.3830	−0.1414	0.0815	0.3824
S	−0.1530	−0.2450	0.4130	−0.1619	−0.2169	0.4118
C	0.1630	−0.0101	0.4410	0.1712	−0.0414	0.4412
S	0.1210	0.2500	0.4700	0.1303	0.2245	0.4706
C	−0.1780	−0.0138	0.4990	−0.1953	0.0000	0.5000
S	−0.0902	−0.2450	0.5280	−0.0942	−0.2245	0.5294
C	0.1840	0.0409	0.5580	0.2127	0.0414	0.5588
S	0.0458	0.2280	0.5870	0.0550	0.2169	0.5882
C	−0.1870	−0.0781	0.6170	−0.2228	−0.0815	0.6176
S	0.0120	−0.1960	0.6470	−0.0139	−0.2018	0.6471
C	0.1830	0.1260	0.6770	0.2254	0.1187	0.6765
S	−0.0710	0.1690	0.7070	−0.0277	0.1799	0.7059
C	−0.1850	−0.1550	0.7360	−0.2203	−0.1519	0.7353
S	0.1020	−0.1450	0.7650	0.0684	−0.1519	0.7647

(Continued)

Table 20.3 (Continued)

	Optimized[b)]			Experimental[c)]		
C	0.1660	0.1650	0.7940	0.2076	0.1799	0.7941
S	−0.1420	0.1050	0.8230	−0.1067	0.1187	0.8235
C	−0.1570	−0.1840	0.8520	−0.1880	−0.2018	0.8529
S	0.1670	−0.0735	0.8810	0.1414	−0.0815	0.8824
C	0.1290	0.1840	0.9111	0.1619	0.2169	0.9118
S	−0.1930	0.0182	0.9408	−0.1712	0.0414	0.9412
C	−0.0814	−0.1850	0.9711	−0.1303	−0.2245	0.9706
			$\Delta_{SC} = 0.050$			

a) Hexagonal.
b) At 0 K. Reported by Kaidu [228].
c) Reported by Carazzolo and Valle [61].
d) Hydrogen atoms were also included in the optimization but are omitted here.

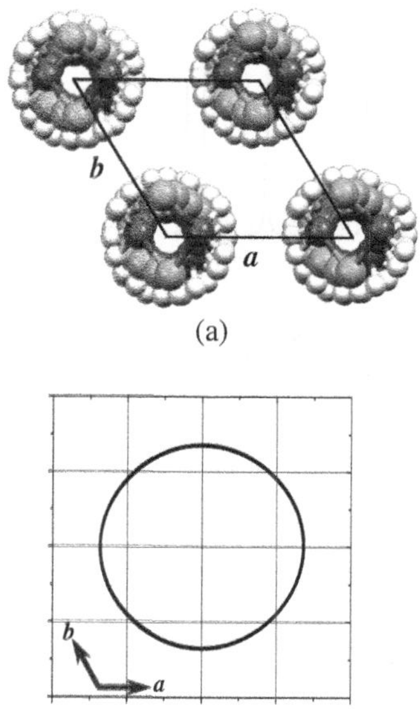

Figure 20.1 (a) Optimized PMS crystal structure (*ab*-plane): hexagonal, $a = b = 5.127$ Å and $c =$ 38.21 Å. (b) Young's modulus distribution on the *ab*-plane. The grid spacing corresponds to 10 GPa.

(Table 20.4), from which bond conformations around the CH_2—CH_2 bond were obtained with Eqs. (15.18) and (15.19), into which J_Ts and J_Gs of 2-(1,1,-dimethylethyl)-1,4-dithiane (DMEDT) were substituted (Table 9.2). The p_t values thus obtained are shown in Table 20.4. For the S—CH_2 bond, the $^3J_{CH}$ values of BMTE were analyzed as described in Section 15.3. As the J_T and J_G coefficients, those determined for 2-methyl-1,3,5-trithiane (MTT) were

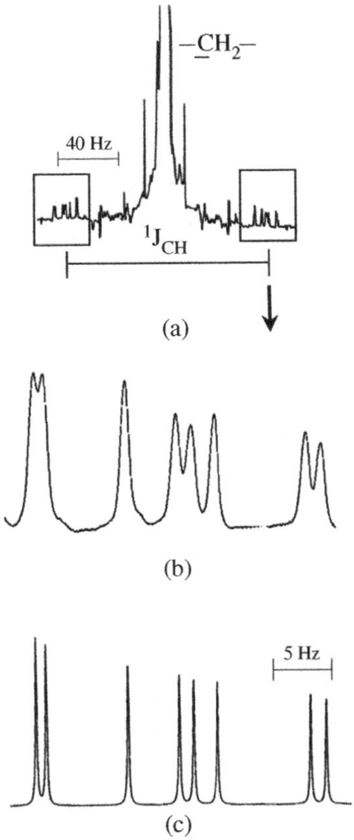

Figure 20.2 ^1H NMR spectra of BMTE in C_6D_6 at 15 °C: (a) methylene part and (b) observed and (c) simulated satellite peaks. Source: [412]/ Reproduced with permission of American Chemical Society.

Table 20.4 Observed vicinal ^1H–^1H and ^{13}C–^1H coupling constants and trans fractions of BMTE.[a)]

Medium	$^3J_{HH}$ [b)]	$^3J'_{HH}$ [b)]	$^3J_{CH}$ [b)]	p_t	
				CH_2–CH_2	S–CH_2
		NMR experiment			
Cyclohexane-d_{12}	5.25	10.96	4.64	0.71	0.10
Benzene-d_6	5.41	10.50	4.59	0.68	0.13
Chloroform-d	5.52	10.31	4.35	0.66	0.22
		MO calculation			
gas[c)]				0.83	0.21

a) At 25 °C.
b) In Hz.
c) At the MP2/6-311+G(3df,2p)//HF/6-31G(d) level.

adopted. The p_t values of the S—CH_2 bond are also shown in Table 20.4. The NMR data are consistent with the MO calculations, and both NMR and MO data show that the CH_2—CH_2 and S—CH_2 bonds have trans and gauche preferences, respectively.

The Gibbs free energies were divided into some conformational energies. The interaction energies are defined similar to those of PEO (Figure 17.3), except that E_κ was introduced to express the difference between the $g^\pm tg^\mp$ and $g^\pm tg^\pm$ conformations:

$$E_\kappa = \Delta G_{g^\pm tg^\mp} - \Delta G_{g^\pm tg^\pm} \tag{20.10}$$

Of the conformational energies, only E_ρ and E_κ were obtained to be negatives, −0.41 and −0.19 kcal mol^{-1}, respectively. The other energies are positive, thus expressing repulsions: $E_\sigma = 0.89$, $E_\omega = 0.45$, and $E_\chi = 0.50$ kcal mol^{-1}. As attempted for PMS, the bond dipole moments of PES were determined as $m_{S-C} = 1.22$ D and $m_{C-C} = 0.00$ D.

The RIS calculations with the above conformational energies yielded the characteristic ratio of PES at 25 °C of 3.1, and its temperature coefficient is -0.89×10^{-3} K^{-1}. The dipole moment ratio is 0.22, and the temperature coefficient is 6.4×10^{-3} K^{-1}. The S_{conf} value at T_m^0 is 6.0 cal K^{-1} mol^{-1}. As mentioned above, PES is so poorly soluble that neither the unperturbed chain dimension nor the dipole moment ratio is known. The ΔS_u value at T_m^0 was reported as 6.9 cal K^{-1} mol^{-1} [1]; therefore, the $S_{conf}/\Delta S_u$ ratio is 0.87.

The natural bond orbital (NBO) analysis of BMTE suggested that both S—CH_2 and CH_2—CH_2 bonds essentially prefer to be gauche rather than trans owing mainly to the $n_S \rightarrow \sigma^*_{C-C}$ and $\sigma_{C-H} \rightarrow \sigma^*_{C-S}$ interactions, respectively [412]. However, both MO and NMR results show the trans preference of the CH_2—CH_2 bond. This is probably because, to avoid strong S· · ·S repulsion, the CH_2—CH_2 bond is forced into the trans state. The E_κ stabilization of the $g^\pm tg^\mp$ states may be due to an intramolecular dipole–dipole interaction between antiparallel m_{S-C}s.

The crystal structure of PES was determined by X-ray diffraction: orthorhombic; space group *Pbcn*; lattice constants, $a = 8.50$, $b = 4.95$, and c (fiber axis) = 6.70 Å; four monomeric units per unit cell [467]. The S–CH_2–CH_2–S bonds lie in the $g^\pm tg^\mp$ form. The periodic DFT-D calculations were carried out for the PES crystal with the experimental structure set initially. The optimized lattice constants and atomic positions are listed in Table 20.5 and compared with the experimental data, and the crystal structure is depicted in Figure 20.3. The differences between theory and experiment in lattice constants and atomic coordinates are small: $\Delta_{LC} = 0.74\%$ and

Table 20.5 Optimized and experimental crystal structures of PES.[a)]

	Optimized[b)]			Experimental[c)]		
			Lattice constant (Å)			
	a	b	c	a	b	c
	8.457	4.908	6.832	8.50	4.95	6.70
			$\Delta_{LC} = 0.74\%$			
			Fractional coordinates			
	x/a	y/b	z/c	x/a	y/b	z/c
S	0.000	0.299	0.250	0.000	0.302	0.250
C	0.068	0.061	0.059	0.054	0.075	0.048
H	0.146	0.182	−0.034			
H	0.139	−0.098	0.128			
			$\Delta_{SC} = 0.011$			

a) Orthorhombic, space group *Pbcn*.
b) At 0 K. [416, 541]
c) Reported by Takahashi et al. [467].
Source: Adapted from [541]

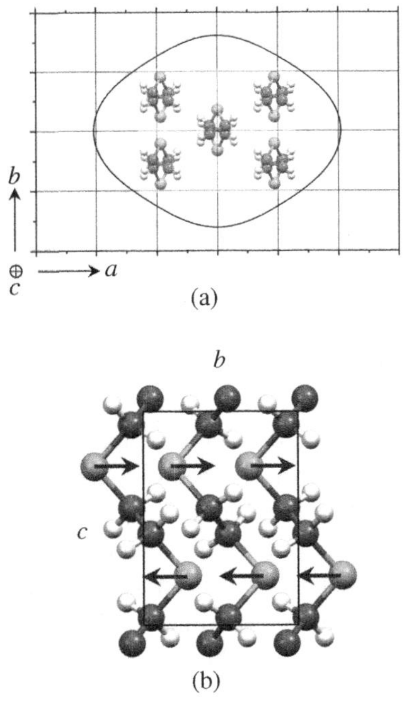

Figure 20.3 Crystal structure of PES optimized at the B3LYP-D/6-31G(d,p) level. The PES chain adopts $g^{\pm}tg^{\mp}$ conformations. (a) Young's modulus distribution of the *ab*-plane perpendicular to the *c*-axis. The grid spacing corresponds to 10 GPa. (b) The side view of the *bc*-plane. The arrow represents the dipole moment (2.066 D) per repeating unit formed in the *b*-axis direction.

$\Delta_{SC} = 0.011$. The stiffness tensor of the PES crystal was calculated to be

$$C = \begin{bmatrix} 22.453 & 6.780 & 2.833 & 0.000 & 0.000 & 0.000 \\ & 23.587 & 17.176 & 0.000 & 0.000 & 0.000 \\ & & 49.590 & 0.000 & 0.000 & 0.000 \\ & & & 9.390 & 0.000 & 0.000 \\ & & & & 2.257 & 0.000 \\ & & & & & 6.052 \end{bmatrix} (\text{GPa}) \quad (20.11)$$

From the compliance tensor, the inverse matrix of C, Young's moduli in the a-, b-, and c-axis directions were calculated to be $E_a = 20.38$, $E_b = 16.13$, and $E_c = 36.87$ GPa, respectively. The distribution of Young's modulus on the ab-plane perpendicular to the fiber axis is illustrated in Figure 20.3. The bended $g^{\pm}tg^{\mp}$ forms considerably reduce the stiffness along the fiber axis as compared with all-trans structures of, for example, polyethylene (333 GPa) and poly(glycolic acid) (451 GPa).

The Born effective charges of the individual atoms were calculated: S, −0.4126; C, +0.1778; and H, +0.0350 and −0.0064 in the unit of e [170]. These partial charges represent that a dipole moment of 2.066 D per repeating unit is formed along the bisector of ∠CSC (the arrow of Figure 20.3b). The dipole–dipole interaction energy, W, was calculated, being plotted against N_a, the number of unit cells along the a-axis in Figure 20.4 [245]. Here, $N_a = N_b = N_c$ is assumed,

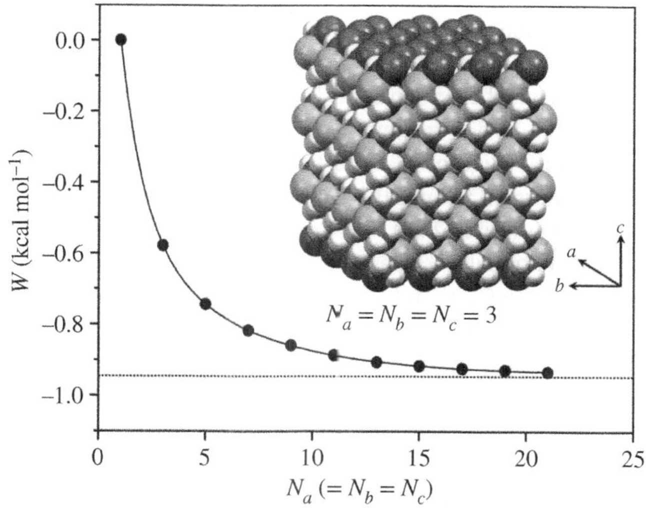

Figure 20.4 Dipole–dipole interaction energy (W) of the PES crystal, calculated under $N_a = N_b = N_c$ and plotted as a function of the number (N_a) of unit cells along the a-axis. The horizontal dotted line expresses the W value (−0.939 kcal mol^{-1}) for $N_a \to \infty$. The inserted picture represents the PES crystal of $N_a = 3$.

and W is given in the unit of kcal (mole of repeating unit)$^{-1}$. The extrapolation of W to $N_a \to \infty$ (infinitely large crystal) yields $W = -0.939$ kcal mol^{-1}.

The entropy of fusion of PES at T_m^0 (216 °C) was reported as 6.9 cal K^{-1} mol^{-1} [1, 102]. If W is independent of the temperature, the dipole–dipole interaction would rise T_m^0 by as much as $|W|/\Delta S_u = 939/6.9 = 136$ K. It is interesting to note that $T_m^0 - 136 = 80$ °C exactly corresponds to T_m^0 of PEO [293]. At least, it can be concluded that the dipole–dipole interaction is a significant factor of the high melting point of PES.

20.3 Poly(propylene sulfide) (PPS)

PPS is the analog of PPO, and its methine carbon has a methyl substituent to be a chiral center; thus, each monomeric unit is either (*R*)- or (*S*)-isomer. Herein, the (*R*)-isomer is dealt with exclusively in order to facilitate the comparison with PPO.

^{1}H and ^{13}C NMR experiments and MO calculations at the MP2/6-311+G(3df,2p) //HF/6-31G(d) level were carried out for 1,2-bis(methylthio)propane (BMTP), a model compound of PPS [404]. The Gibbs free energies of the individual conformers are listed in Table 20.6. The NMR experiments were conducted similar to those for DMP to determine the bond conformations (Table 20.7), which are compared with those evaluated from the MO calculations.

For PPO (DMP), the magnitude relation of the C–C bond conformations is generally either $p_{g^+} > p_t > p_{g^-}$ in nonpolar media or $p_{g^+} > p_{g^-} > p_t$ in a polar solvent, DMSO [397, 398]; thus, in either case, the gauche$^+$ state is the most dominant, and this tendency was termed the gauche oxygen effect [4], which was shown to be due to the C–H$\cdots$O attractions of −1 to −2 kcal mol^{-1}. In addition to the intramolecular interactions defined for PPO (Figures 15.2 and 15.5), two interactions, τ and ζ, were introduced to PPS (Figure 20.5), and the Gibbs free energies were divided into the 10 interaction energies (Table 20.8).

Furthermore, the conformational energies were adjusted so as to reproduce (set I) the NMR data on BMTP in C_6D_6 and dipole moment ratios of isotactic and atactic PPS chains in benzene [385, 386] or (set II) the same NMR data, the dipole moment ratios, and the characteristic ratio of atactic PPS in a Θ solution (*n*-hexane (31%) and toluene at 25 °C) [337]. The dipole moment and characteristic ratios were calculated by the RIS scheme with $m_{S\text{–}C}$ of 1.21 D that was determined similar to those of PMS and PES. The atactic PPS chains were prepared according to the Bernoulli trials based on the random-number generation. The conformational energies thus optimized are also listed in Table 20.8, and the calculated configurational properties are shown in Table 20.9. Both energy sets satisfactorily reproduced the targets: the set I parameters well reproduced the dipole moments of isotactic and atactic PPS, while the set II energies attained good agreement with

Table 20.6 Conformer Gibbs free energies of BMTP, evaluated by ab initio molecular orbital calculations.

k	Conformation	Statistical weight	ΔG_k [a)]
1	t t t	1	0.00
2	t t g$^+$	γ	0.42
3	t t g$^-$	δ	−0.46
4	t g$^+$ t	α	0.81
5	t g$^+$ g$^+$	$\alpha\gamma$	0.89
6	t g$^+$ g$^-$	$\alpha\delta\omega_1$	0.99
7	t g$^-$ t	β	1.20
8	t g$^-$ g$^+$	$\beta\gamma\omega_2$	2.15
9	t g$^-$ g$^-$	$\beta\delta$	0.72
10	g$^+$ t t	$\sigma\tau$	0.63
11	g$^+$ t g$^+$	$\sigma\gamma\tau$	1.05
12	g$^+$ t g$^-$	$\sigma\delta\tau$	0.14
13	g$^+$ g$^+$ t	$\sigma\alpha$	0.83
14	g$^+$ g$^+$ g$^+$	$\sigma\alpha\gamma\chi$	0.99
15	g$^+$ g$^+$ g$^-$	$\sigma\alpha\delta\omega_1$	0.78
16	g$^+$ g$^-$ t	$\sigma\beta\omega_2$	0.81
17	g$^+$ g$^-$ g$^+$		— [b)]
18	g$^+$ g$^-$ g$^-$	$\sigma\beta\delta\omega_2$	0.72
19	g$^-$ t t	σ	−0.55
20	g$^-$ t g$^+$	$\sigma\gamma$	−0.25
21	g$^-$ t g$^-$	$\sigma\delta$	−0.96
22	g$^-$ g$^+$ t	$\sigma\alpha\omega_1$	0.71
23	g$^-$ g$^+$ g$^+$	$\sigma\alpha\gamma\omega_1$	1.33
24	g$^-$ g$^+$ g$^-$		— [b)]
25	g$^-$ g$^-$ t	$\sigma\beta\tau$	1.91
26	g$^-$ g$^-$ g$^+$	$\sigma\beta\gamma\tau\omega_2$	2.75
27	g$^-$ g$^-$ g$^-$	$\sigma\beta\delta\tau$	1.39

a) Relative to the ΔG_k value of the all-trans conformation. At the MP2/6-311+G(3df, 2p)//HF/6-31G(d) level. At 25 °C.
b) These cyclic conformers were considered to be absent.

Table 20.7 Bond conformations of BMTP and isotactic PPS.

		Bond[a)]								
		a			b			c		
Medium	**Temp (°C)**	p_t	p_{g+}	p_{g-}	p_t	p_{g+}	p_{g-}	p_t	p_{g+}	p_{g-}
BMTP										
					NMR experiment					
Gas	150				0.83	0.10	0.07			
Cyclohexane-d_{12}	26	0.04	0.34	0.62	0.83	0.13	0.04		0.22	
Benzene-d_6	26	0.10	0.32	0.58	0.78	0.17	0.05		0.20	
Dimethyl-d_6 sulfoxide	25	0.12	0.31	0.57	0.72	0.20	0.08		0.14	
					MO calculation[b)]					
Gas	25	0.24	0.16	0.60	0.82	0.11	0.07	0.26	0.17	0.57
	150	0.29	0.22	0.49	0.70	0.18	0.12	0.29	0.21	0.50
					RIS simulation					
Set I	26	0.12	0.39	0.49	0.77	0.18	0.05	0.52	0.20	0.28
Set II	26	0.20	0.36	0.44	0.77	0.18	0.05	0.52	0.22	0.26
Isotactic PPS										
					NMR experiment[c)]					
CCl_4	17				0.71	0.21	0.08			
					RIS simulation					
Set I	26	0.17	0.36	0.47	0.77	0.19	0.04	0.59	0.12	0.29
Set II	26	0.29	0.31	0.40	0.76	0.20	0.04	0.59	0.15	0.26

a) The bond symbols are defined as S–(a)–CH_2–(b)–C*H(CH_3)–(c)–S in the (*R*)-form.
b) At the MP2/6-311+G(3df, 2p)//HF/6-31G(d) level.
c) Calculated from $^3J_{AC} = 4.8$ Hz and $^3J_{BC} = 9.0$ Hz. Reported by Sepulchre et al. [431].
Source: [420]/ with permission of American Chemical Society.

the experimental characteristic ratio. The $d \ln \langle r^2 \rangle_0/dT$ value was calculated to be positive, whereas there are both positive and negative experimental data. As far as $d \ln\langle \mu^2 \rangle/dT$ of isotactic PPS is concerned, the RIS calculations are satisfactory.

The first-order interaction energies, E_α and E_β, are positive and related by $E_\alpha < E_\beta$, which is consistent with the magnitude relation for PPO. As for the higher order interactions, the E_ω, $E_{\omega'}$, and E_χ energies of PPS are positive, whereas those of PPO are negative. In contrast to the C–H···O interaction, all the C–H···S contacts are repulsions, and, as a consequence, the C–C bond of PPS will be trans at high probabilities.

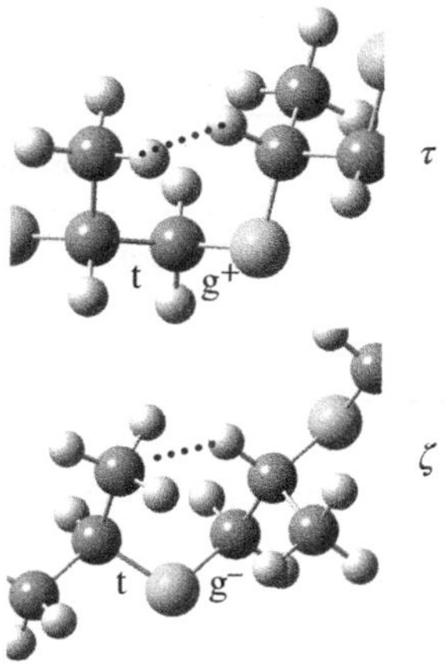

Figure 20.5 Second-order interactions (τ and ζ) introduced to PPS in addition to those defined for PPO. Source: [404]/ with permission of American Chemical Society.

Table 20.8 Conformational energies[a)] of PPS (BMTP).

	MO[b)]	Exptl	
		Set I[c)]	Set II[d)]
	First-order interaction		
E_α	0.93	0.55	0.49
E_β	1.19	1.13	1.18
E_γ	0.30	0.49	0.43
E_δ	−0.43	0.25	0.27
E_σ	−0.52	−0.99	−0.60
	Second-order interaction		
E_{ω_1}	0.55	0.64	0.89
E_{ω_2}	0.35	1.16	0.99
E_τ	1.14	0.26	0.28
E_ζ		0.35	0.35
	Third-order interaction		
E_χ	0.29	0.38	0.42

a) In kcal mol^{-1}.
b) At the MP2/6-311+G(3df, 2p)//HF/6-31G(d) level.
c) Determined from bond conformations of BMTP in benzene and dipole moment ratios of isotactic and atactic PPS in benzene at 25 °C.
d) Determined from the bond conformations, the dipole moment ratios, and the characteristic ratio of atactic PPS in the Θ solvent (*n*-hexane [31%] and toluene).

Source: [404]/ with permission of American Chemical Society.

Table 20.9 Configurational properties of PPS,[a] calculated with sets I and II energies and comparison with experimental observations.

	Isotactic			Atactic		
	Calcd		Obsd	Calcd		Obsd
	Set I	Set II		Set I	Set II	
$\langle r^2\rangle_0/nl^2$	3.3	4.0		3.2	3.9	4.0[b]
$10^3 d\ln\langle r^2\rangle_0/dT$ (K^{-1})	1.0	0.47	-2.8 ± 0.3[c]	1.6	0.87	-2.0 ± 0.3[c]
						0.51 ± 0.11[d]
$\langle\mu^2\rangle/nm^2$	0.33	0.34	0.33[e] (CCl_4)	0.38	0.38	0.37[e] 0.36[f] (CCl_4)
			0.39[e] (C_6H_6)			0.44[e] 0.44[f] (C_6H_6)
$10^3 d\ln\langle\mu^2\rangle/dT$ (K^{-1})	2.2	2.3	2.1[e] (CCl_4)	2.9	2.6	4.0[e] 2.8[f] (CCl_4)
			2.0[e] (C_6H_6)			1.5[e] 0.72[f] (C_6H_6)

a) At 25 °C.
b) Rescaled with l_{C-S} = 1.818 Å and l_{C-C} = 1.529 Å. Evaluated from the intrinsic viscosity of atactic PPS in the Θ solvent of *n*-heptane (31%) and toluene at 25 °C [337].
c) Obtained from the intrinsic viscosity of PPS in the monomer, cyclic propylene sulfide [378].
d) Estimated from the thermoelasticity measurements on networks of atactic PPS [295].
e) From isotactic PPS of weight-average molecular weight = 1.6×10^6 and atactic PPS of weight-average molecular weight = 5×10^5 [385].
f) From atactic PPS of number-average molecular weight = $5-6\times10^3$ [386].
Source: [404]/ with permission of American Chemical Society.

PPS forms an orthorhombic crystal cell of space group $P2_12_12_1$, in which the molecular chain lies in a slightly distorted all-trans form [394]. The periodic DFT-D calculations at the B3LYP-D/6-31G(d,p) level were conducted for the PPS crystal [220]. The optimized crystal structure is illustrated in Figure 20.6, and the optimized (experimental) cell constants are a = 9.74 (9.95) Å, b = 4.81 (4.89) Å, and c = 8.19 (8.20) Å. The stiffness tensor of the PPS crystal is

$$C = \begin{bmatrix} 22.018 & 5.644 & 4.602 & 0.000 & 0.000 & 0.000 \\ & 16.181 & 3.900 & 0.000 & 0.000 & 0.000 \\ & & 119.997 & 0.000 & 0.000 & 0.000 \\ & & & 3.797 & 0.000 & 0.000 \\ & & & & 4.768 & 0.000 \\ & & & & & 5.349 \end{bmatrix} \text{(GPa)} \qquad (20.12)$$

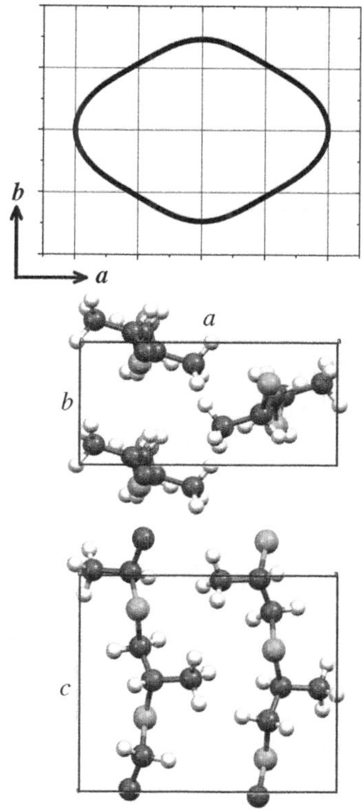

Figure 20.6 Crystal structure of PPS, optimized at the B3LYP-D/6-31G(d,p) level: (above) Young's modulus distribution of the *ab*-plane perpendicular to the *c* (fiber)-axis. The grid spacing corresponds to 10 GPa; (below) crystal cell, orthorhombic, space group $P2_12_12_1$, and lattice constants, $a = 9.74$ Å, $b = 4.81$ Å, and $c = 8.19$ Å.

Young's moduli in the *a*-, *b*-, and *c*- axis directions are $E_a = 20.0$ GPa, $E_b = 14.7$ GPa, and $E_c = 119$ GPa. The E_c value is smaller than that (208 GPa) of PPO.

20.4 Poly(trimethylene sulfide) (PTrMS)

In conformational analysis of poly(trimethylene sulfide) (PTrMS), 1,3-bis(methylthio)propane (1,3-BMTP, $CH_3SCH_2CH_2CH_2SCH_3$) was adopted as a model compound [420]. The conformer Gibbs free energies were calculated at the MP2/6-311+G(3df,2p) level. Of all the conformers, the all-gauche forms, $g^{\pm}g^{\pm}g^{\pm}g^{\pm}$, are the lowest in ΔG and also formed in the PTrMS crystal [455]. As mentioned above, the NBO analysis on BMTE, the model for PES, suggested that both S—CH_2 and CH_2—CH_2 bonds have inherent gauche preferences due to the electronic delocalization [412]. Inasmuch as PTrMS (1,3-BMTP) has three

Table 20.10 Observed vicinal $^1H-^1H$ and $^{13}C-^1H$ coupling constants and trans fractions of 1,3-BMTP[a].

Medium	$^3J_{HH}$ [b]	$^3J'_{HH}$ [b]	$^3J_{CH}$ [b]	p_t	
				CH_2-CH_2	$S-CH_2$
		NMR experiment			
Cyclohexane-d_{12}	7.40	6.68	4.60	0.34	0.22
Benzene-d_6	7.90	6.44	4.50	0.29	0.25
Chloroform-d	7.92	6.45	4.40	0.29	0.28
Methanol-d_4	7.77	6.48	4.53	0.30	0.24
Dimethyl-d_6 sulfoxide	8.00	6.34	4.52	0.28	0.24
		MO calculation[c]			
Gas				0.44	0.34

a) At 25 °C.
b) In Hz.
c) At the MP2/6-311+G(3df,2p)//HF/6-31G(d) level.

methylene units between two sulfur atoms, even if the CH_2—CH_2 bond adopts a gauche conformation, the S· · ·S close contact will be avoidable. Therefore, PTrMS (1,3-BMTP) may be most stabilized in the all-gauche form.

^1H and ^{13}C NMR experiments for 1,3-BMTP gave its bond conformations as shown in Table 20.10. For the analysis of vicinal $^1H-^1H$ coupling constants, a cyclic compound, 2-*tert*-butyl-1,3-dithiane (BDT), was prepared and underwent NMR measurements (Figure 20.7), from which J_T and J_G values were obtained (Table 20.11) and used for the NMR analysis. The vicinal coupling constants of BDT were also evaluated from the DFT calculations at the B3LYP/6-311+G(2d,p) level, being in good agreement with the experiment. In the analysis of the $^3J_{CH}$s of 1,3-BMTP, the J_T and J_G obtained from the DFT calculations on 1,3-BMTP itself were used. It can be seen that both S—CH_2 and CH_2—CH_2 bonds tend to be gauche rather than trans (Table 20.12).

The Gibbs free energies were broken down into four conformational energies (Table 20.12). The first-order interaction energies (E_ρ for S–CH_2 and E_σ for CH_2–CH_2) are negative, while the higher order interaction energies are positive (repulsive); therefore, both bonds have inherent gauche preferences. The configurational properties evaluated from the RIS calculations are also shown in Table 20.12. The characteristic ratio and dipole moment ratio agree closely with the experimental values.

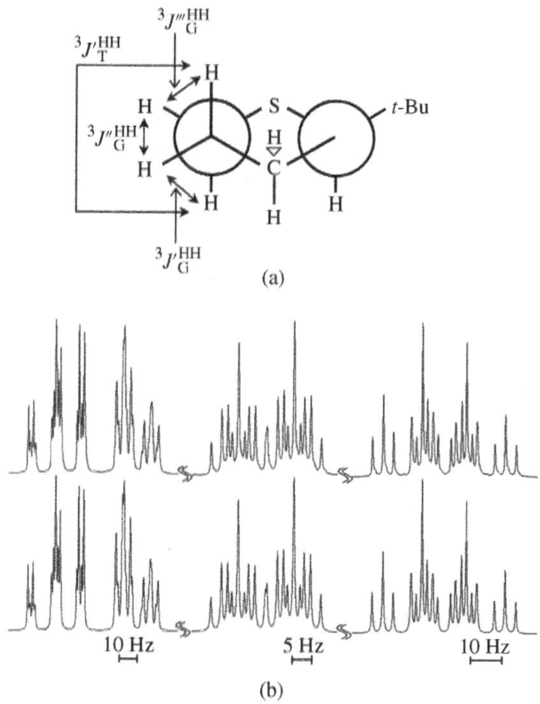

Figure 20.7 (a) 2-*tert*-Butyl-1,3-dithiane (BDT) with the definition of vicinal ^{1}H–^{1}H coupling constants.
(b) Observed (above) and calculated (below) ^{1}H NMR spectra of BDT dissolvec in methanol-d_4 at 25 °C. Source: [420]/Reproduced with permission of American Chemical Society.

Table 20.11 Vicinal ^{1}H–^{1}H coupling constants of BDT.[a)]

Solvent or method	$^3J'^{HH}_{T}$	$^3J'^{HH}_{G}$	$^3J''^{HH}_{G}$	$^3J'''^{HH}_{G}$	$^3J^{HH}_{G}$ b)
		NMR experiment[c)]			
Cyclohexane-d_{12}	12.50	2.45	4.28	3.10	3.28
Benzene-d_6	12.74	2.61	4.16	2.91	3.23
Chloroform-d	12.19	2.40	4.37	3.54	3.44
Methanol-d_4	12.57	2.51	4.19	3.10	3.27
Dimethyl-d_6 sulfoxide	12.59	2.49	4.25	3.03	3.26
		DFT calculation			
B3LYP/6-311+G(2d, p)	11.15	2.42	4.21	3.04	3.22

a) In Hz. For the definition of the coupling constants, see Figure 20.7.
b) The average $^3J^{HH}_{G}$ value.
c) At 25 °C.
Source: [420]/Reproduced with permission of American Chemical Society.

Table 20.12 Conformational energies and calculated and observed configurational properties of PTrMS.

	Calcd		Obsd
	ΔG_k [a]	Optimized [b]	
First-order interaction			
E_ρ	−0.17	−0.30	
E_σ	−0.27	−0.58	
Second-order interaction [c]			
E_ω	1.13	0.43	
$E_{\omega'}$	∞	∞	
$E_{\omega''}$	0.53	0.48	
$\langle r^2 \rangle_0/nl^2$		3.6	3.9±0.3 [d]
$10^3 d\ln\langle r^2 \rangle_0/dT$ (K^{-1})		−0.5	— [e]
$\langle \mu^2 \rangle/nm^2$		0.63	0.60 [f]
$10^3 d\ln\langle \mu^2 \rangle/dT$ (K^{-1})		−0.3	1.3 [f]
S_{conf} (cal K^{-1} mol^{-1})		7.7	
ΔS_u (cal K^{-1} mol^{-1})			6.8 [g]
T_m^0 (°C)			90.0 [g]
p_t^{S-C}		0.34	0.28 [h]
p_t^{C-C}		0.29	0.29 [h]

a) As obtained from the MO calculations.
b) Optimized so as to reproduce the experimental data.
c) See Figure 17.10.
d) Estimated from viscosity in a good solvent, $CHCl_3$ [396].
e) No experimental data available.
f) References [181, 388].
g) Quoted from reference [293].
h) Trans fractions of the S—C (p_t^{S-C}) and C—C (p_t^{C-C}) bonds, determined from NMR of 1,3-BMTP in $CDCl_3$ at 25 °C.

21

Polyselenides

Polyselenides are analogs of polyethers and polysulfides and include selenium in the backbone. For the sake of comparison with the polyethers and polysulfides discussed so far, the corresponding polyselenides, poly(methylene selenide) (PMSe), poly(ethylene selenide) (PESe), and poly(trimethylene selenide) (PTrMSe), have been treated here [421].

21.1 Poly(methylene selenide) (PMSe)

PMSe crystallizes to form either a hexagonal cell (the space group has not been determined) in which the molecular chain lies in the all-gauche (21/11) helix (h-PMSe) [60] or an orthorhombic (space group, $P2_12_12_1$) cell including the all-gauche (2/1) helical chain (o-PMSe) [59]. PMSe as well as poly(methylene oxide) (PMO) and poly(methylene sulfide) (PMS) crystallizes in the all-gauche conformation. o-PMSe changes to h-PMSe [59] immediately before melting around 185–190 °C [328].

In order to investigate the conformational characteristics of PMSe, as a model compound, 1,2-bis(methylseleno)methane (BMSeM, $CH_3SeCH_2SeCH_3$) was adopted to undergo molecular orbital (MO) calculations and NMR experiments. Table 21.1 shows Gibbs free energies evaluated by ab initio MO calculations at the MP2/6-311++G(3df,3pd) level for BMSeM and a dimeric model, $CH_3SeCH_2SeCH_2SeCH_3$. As shown in Table 21.2, the trans fractions and conformational energies for the Se—CH_2 bond were, respectively, calculated from Eqs. (20.4) and (20.5) with observed $^3J_{CH}$s (Table 21.3). Both MO calculations and NMR experiments yield the E_σ values of approximately −1 kcal mol^{-1}, which indicates a gauche preference of the Se—CH_2 bond. Between the three PMXs (X = O, S, and Se), the gauche stability can be evaluated in the order of PMO > PMS > PMSe, thus decreasing with increasing atomic number of X. The NBO analysis for BMXMs suggests that the stability of the gg form relative to tt is due to the $n_X \rightarrow \sigma^*_{C-X}$

Conformational Analysis of Polymers: Methods and Techniques for Structure-Property Relationships and Molecular Design, First Edition. Yuji Sasanuma.

Table 21.1 Gibbs free energies (ΔG_k s) of conformers of monomeric (BMSeM) and dimeric model compounds of PMSe, evaluated from MO calculations.

Conformation				Statistical weight	ΔG_k [a)] (kcal mol^{-1})
				BMSeM	
g$^+$	g$^+$			1	0.00
t	g$^+$			σ^{-1}	1.00
t	t			σ^{-2}	4.10
g$^+$	g$^-$			0	(absent)[b)]
				Dimeric model	
g$^+$	g$^+$	g$^+$	g$^+$	1	0.00
t	g$^+$	g$^+$	g$^+$	σ^{-1}	1.09
g$^+$	g$^+$	g$^+$	g$^-$	ω'	2.21
g$^+$	g$^+$	g$^-$	g$^-$	ω	2.43

a) Relative to the all-gauche conformation.
b) The local minimum of the potential was not found by the geometrical optimization.
Source: [421]/Reproduced with permission of American Chemical Society.

interaction (lone pair → antibond delocalization of electron), and the interaction energies were estimated to be 16.71 kcal mol^{-1} (PMO), 9.30 kcal mol^{-1} (PMS), and 7.26 kcal mol^{-1} (PMSe), where the positive energy means stabilization. The conformational characteristics of the three PMXs are similar to each other.

The geometrical parameters of PMSe, averaged through the RIS calculations with the conformational energies given in Table 21.4, are listed in Table 21.5, together with the configurational properties at 25 °C: characteristic ratio, 7.29; its temperature coefficient, -4.6×10^{-3} K^{-1}; and configurational entropy (3.4 cal K^{-1} mol^{-1}).

21.1.1 Crystal Structure of PMSe

The orthorhombic crystal of PMSe (o-PMSe) was optimized by DFT calculations at the B3LYP-D/6-31G(d,p) level, and the crystalline moduli were also evaluated [228].

First, the van der Waals radius ($R_{\mathrm{vdW}}^{\mathrm{Se}}$) of selenium to be used for the Grimme D2 dispersion correction was determined [175]. As far as the van der Waals radii of oxygen and sulfur are concerned, Bondi's values (1.52 and 1.80 Å, respectively)

Table 21.2 Trans fractions of BMSeM (PMSe), BMSeE (PESe), and BMSeP (PTrMSe) at 25 °C, evaluated by MO calculations, NMR experiments, and RIS calculations.

Compound	Method	Solvent	p_t Se–C	p_t C–C	E_σ of BMSeM (kcal mol^{-1})
BMSeM	MO		0.14		
	NMR	C_6D_{12}	0.11		−1.17
		C_6D_6	0.14		−1.04
		$CDCl_3$	0.15		−0.97
		CD_3OD	0.14		−1.04
		$(CD_3)_2SO$	0.16		−0.90
	RIS		0.12		
PMSe	RIS		0.18		
BMSeE	MO		0.20	0.83	
	NMR	C_6D_{12}	0.16	0.89	
		C_6D_6	0.19	0.85	
		$CDCl_3$	0.22	0.85	
		CD_3OD	0.19	0.83	
		$(CD_3)_2SO$	0.20	0.83	
	RIS		0.21	0.84	
PESe	RIS		0.29	0.83	
BMSeP	MO		0.25	0.43	
	NMR	C_6D_{12}	0.15	0.46	
		C_6D_6	0.18	0.50	
		$CDCl_3$	0.21	0.51	
		CD_3OD	0.17	0.49	
		$(CD_3)_2SO$	0.20	0.49	
	RIS		0.23	0.44	
PTrMSe	RIS		0.29	0.44	

Source: [421]/Reproduced with permission of American Chemical Society.

[45] were found to be preferable to Grimme's (1.342 and 1.683 Å) [175]. The R_{vdW}^{Se} values suggested by Bondi and Grimme are 1.90 and 1.771 Å, respectively. The crystal structure of o-PMSe [59] was optimized using either Bond's or Grimme's radius with the X-ray structure set initially. As a result, the agreement with the experiment was as follows: Bondi (1.90 Å), Δ_{LC} = 1.49%, and Δ_{CSe} = 0.005; Grimme (1.771 Å), Δ_{LC} = 0.89%, and Δ_{CSe} = 0.017. As for the lattice constants,

Table 21.3 Vicinal $^1H-^1H$ and $^{13}C-^1H$ coupling constants observed from BMSeM, BMSeE, and BMSeP at 25 °C[a)].

	Medium	Permittivity	$^3J_{HH}$	$^3J'_{HH}$	$^3J_{CH}$
BMSeM	C_6D_{12}	2.0			3.64
	C_6D_6	2.2			3.59
	$CDCl_3$	4.8			3.56
	CD_3OD	32.7			3.59
	$(CD_3)_2SO$	46.7			3.53
BMSeE	C_6D_{12}		5.14	12.05	3.86
	C_6D_6		5.25	11.65	3.77
	$CDCl_3$		5.35	11.77	3.68
	CD_3OD		5.40	11.55	3.76
	$(CD_3)_2SO$		5.32	11.35	3.73
BMSeP	C_6D_{12}		6.71	7.59	3.89
	C_6D_6		6.50	7.97	3.80
	$CDCl_3$		6.50	8.10	3.72
	CD_3OD		6.55	7.85	3.83
	$(CD_3)_2SO$		6.60	7.97	3.76

a) In Hz.
Source: [421]/Reproduced with permission of American Chemical Society.

Table 21.4 Conformational energies (kcal mol^{-1}) of BMSeM (PMSe), BMSeE (PESe), and BMSeP (PTrMSe).

	BMSeM (PMSe)	BMSeE (PESe)	BMSeP (PTrMSe)
	First-order interaction		
E_ρ		−0.29	−0.42
E_σ	−1.17	1.02	−0.11
	Second- and third-order interactions		
E_ω	2.43[a)]	0.44	0.45
$E_{\omega'}$	2.21[a)]	1.12	2.52
$E_{\omega''}$			0.77[a)]
E_κ		−0.31	

a) Evaluated from dimeric model compounds.
Source: [421]/Adapted with permission of American Chemical Society.

Table 21.5 Geometrical parameters and configurational properties of PMSe, PESe, and PTrMSe[a].

	PMSe	PESe	PTrMSe
	Bond length (Å)		
$l_{\text{Se—C}}$	1.971	1.978	1.948
$l_{\text{C—C}}$		1.516	1.528
	Bond angle (°)		
∠CSeC	97.5	98.5	97.0
∠SeCSe	117.4		
∠SeCC		112.5	112.1
∠CCC			114.0
	Dihedral angle[b] (°)		
$\phi_{g\pm}$(Se—C)	±114.8	±102.0	±96.8
$\phi_{g\pm}$(C—C)		±114.4	±114.1
	Configurational properties		
$\langle r^2 \rangle_0 / nl^2$ at $x = \infty$	7.29	5.44	5.33
$10^3 d \ln \langle r^2 \rangle_0 / dT$ (K^{-1})	−4.6	−1.1	−0.49
S_{conf} (cal K^{-1} mol^{-1})	3.4	5.1	8.0

a) At 25 °C.
b) ϕ_t(Se—C) = ϕ_t(C—C) = 0.0°.
Source: [421]/Reproduced with permission of American Chemical Society.

Grimme's is superior, whereas for the atomic position, Bondi's is preferable. The lattice constants were determined directly by X-ray diffraction, while the atomic positions were estimated from the intensity simulation. Therefore, $R_{\text{vdW}}^{\text{Se}}$ of 1.771 Å was employed throughout the DFT-D calculations on o-PMSe. The optimized data on o-PMSe are compared with the experiments in Table 21.6, and the crystal structures are illustrated in Figure 21.1a.

The stiffness tensor is

$$C = \begin{bmatrix} 28.579 & 7.315 & 8.146 & 0.000 & 0.000 & 0.000 \\ & 17.549 & 11.044 & 0.000 & 0.000 & 0.000 \\ & & 36.332 & 0.000 & 0.000 & 0.000 \\ & & & 9.172 & 0.000 & 0.000 \\ & & & & 11.241 & 0.000 \\ & & & & & 6.880 \end{bmatrix} \text{(GPa)} \tag{21.1}$$

Table 21.6 Optimized and experimental crystal structures of o-PMSe[a)].

	Optimized [b)]			Experimental[c)]		
			Lattice constant (Å)			
	a	b	c	a	b	c
	5.40	9.01	4.38	5.37	9.03	4.27
			$\Delta_{LC} = 0.89\%$			
			Fractional coordinates			
	x/a	y/b	z/c	x/a	y/b	z/c
C	0.3279	0.1018	0.8625	0.330	0.110	0.875
Se	0.0324	0.0589	0.6140	0.047	0.053	0.625
H	0.2741	0.1935	0.0104			
H	0.4771	0.1374	0.7121			
			$\Delta_{CSe} = 0.017$			

a) Orthorhombic, space group $P2_12_12_1$.
b) At 0 K.
c) Source: Adapted from [59].

Young's moduli in the a-, b-, and c-axis directions, calculated from the compliance (C^{-1}) tensor, are $E_a = 25.1$, $E_b = 13.3$, and $E_c = 28.9$, respectively. The E_c values of PMXs, representing the fiber-axis modulus, are arranged in the order of t-PMO (115 GPa) > o-PMO (82.9 GPa) > h-PMS (41.9 GPa) > o-PMSe (28.9 GPa), where t-PMO, o-PMO, and h-PMS stand for trigonal and orthorhombic PMO and hexagonal PMS crystals, respectively. The following tendencies may be found for E_cs: O > S > Se. The distribution of Young's modulus on the plane perpendicular to the fiber axis seems to be elliptical (Figure 21.1b).

The BSSE-corrected interchain interaction energies of PMXs were calculated to be (in kcal mol^{-1}) −3.17 (t-PMO), −3.38 (o-PMO), −4.51 (h-PMS), and −4.88 (o-PMSe), in the opposite order of E_c.

21.2 Poly(ethylene selenide) (PESe)

Conformational analysis of PESe was carried out via ^{1}H and ^{13}C NMR experiments and MO calculations for a model compound, 1,2-bis(methylseleno)ethane (BMSeE, $CH_3SeCH_2CH_2SeCH_3$) From NMR spectra of BMSeE, vicinal ^{1}H—^{1}H and ^{13}C—^{1}H coupling constants (Table 21.3) were obtained and analyzed to yield

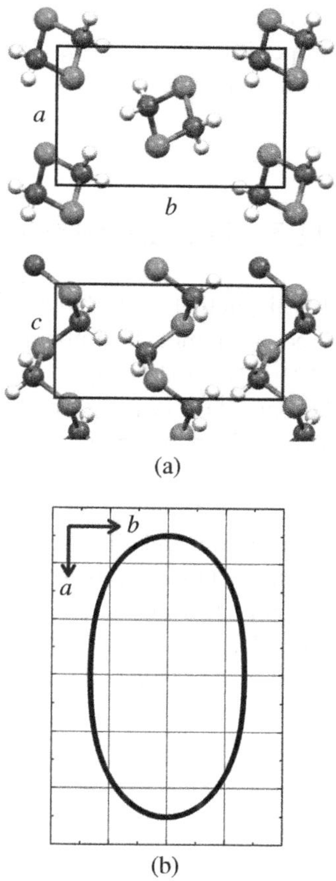

Figure 21.1 (a) Crystal structure of o-PMSe, optimized by the periodic DFT-D calculations at the B3LYP-D/6-31G(d,p) level. Crystallographic data: orthorhombic; space group, $P2_12_12_1$; lattice constants, $a = 5.40$ Å, $b = 9.01$ Å, and $c = 4.38$ Å. (b) Young' modulus distribution on the *ab*-plane perpendicular to the fiber axis. The grid spacing corresponds to 10 GPa. $E_a = 25.1$ GPa, $E_b = 13.3$ GPa, and $E_c = 28.9$ GPa.

bond conformations (Table 21.2), which show trans (p_t ~ 0.83–0.89) and gauche (p_t ~ 0.16–0.22) preferences of the C—C and Se—C bonds, respectively. The J_T and J_G values required for the NMR analysis were obtained from DFT calculations at the B3LYP/6-311++G(3df,3pd) level for BMSeE. The ab initio MO calculations at the MP2/6-311++G(3df,3pd) level indicate that the most stable conformers are $g^{\pm}tg^{\mp}$, whose Gibbs free energy is −0.90. kcal mol^{-1} relative to that of ttt. The conformer free energies were divided into conformational energies defined similar to those of BMTE (PES) (Table 21.4). The first-order interaction energies, E_ρ and E_σ, are −0.29 and 1.02 kcal mol^{-1}, respectively, while the higher order interaction energies are positive: $E_\omega = 0.44$ kcal mol^{-1} and $E_{\omega'} = 1.12$ kcal mol^{-1}, except for $E_\kappa = -0.31$ kcal mol^{-1}, which represents the $g^{\pm}tg^{\mp}$ stabilization relative to $g^{\pm}tg^{\pm}$. All the above tendencies are consistent with those of BMTE (PES).

The RIS calculations with the conformational energies yielded configurational properties of PESe at 25°C (Table 21.5) as follows: characteristic ratio, 5.44; its temperature coefficient, -1.1×10^{-3} K^{-1}; and configurational entropy, 5.1 cal K^{-1} mol^{-1}.

21.3 Poly(trimethylene selenide) (PTrMSe)

1,3-Bis(methylseleno)propane (BMSeP, $CH_3SeCH_2CH_2CH_2SeCH_3$) was used as a model compound of PTrMSe. Ab initio MO calculations at the MP2/6-311++G(3df,3pd) level showed the conformers of the lowest ΔG_k (−1.03 kcal mol^{-1}) to be $g^{\pm}ttg^{\pm}$, and $g^{\pm}g^{\pm}g^{\pm}g^{\pm}$ are the second lowest (−0.96 kcal mol^{-1}). The conformational energies defined similar to those of 1,3-BMTP (PTrMS) were obtained from ΔG_ks of all conformers: $E_\rho = -0.42$; $E_\sigma = -0.11$; $E_\omega = 0.45$; $E_{\omega'} = 2.52$; $E_{\omega''} = 0.77$ (in kcal mol^{-1}). The first-order interaction energies (E_ρ and E_σ) are negative, and all the higher order interaction energies (E_ω, $E_{\omega'}$, and $E_{\omega''}$) are positive. These tendencies were also found for 1,3-BMTP (PTrMS).

Figure 21.2 shows ^1H and ^{13}C NMR spectra of BMSeP. Small peaks caused by natural abundance ^{77}Se (7.58%, nuclear spin $I = 1/2$) are also observed. With the J_T and J_G values obtained from DFT calculations on BMSeP, the trans fractions were

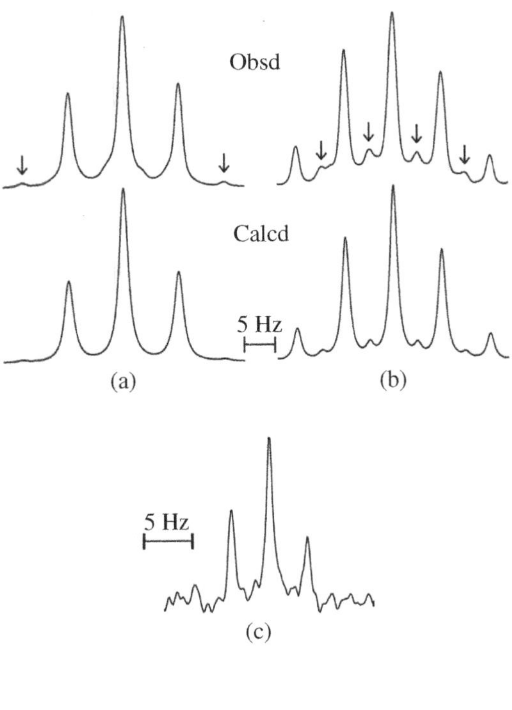

Figure 21.2 Observed (above) and calculated (below) ^1H NMR spectra of methylene protons, (a) Se−CH_2−CH_2 and (b) CH_2−CH_2−CH_2 of BMSeP dissolved in C_6D_{12} at 25°C. (c) ^{13}C NMR spectra of methyl carbons of BMSeP in C_6D_6 at 25°C. The vertical arrows indicate signals due to couplings with natural abundance ^{77}Se. Source: [421]/Reproduced with permission of American Chemical Society.

determined as in Table 21.2. The Se—C bond prefers a gauche ($p_t \sim 0.15$–0.21) form, while the C—C bond is suggested to prefer, if anything, the trans ($p_t \sim 0.46$–0.51) conformation because of $p_{g+} = p_{g-} = (1/2)p_g = (1/2)(1 - p_t)$.

The configurational properties of PTrMSe at 25°C were evaluated as follows: characteristic ratio, 5.33; its temperature coefficient, -0.49×10^{-3} K^{-1}; and configurational entropy, 8.0 cal $K^{-1}mol^{-1}$.

21.4 Summary

The polymers expressed as PMX (X = O, S, and Se) are composed of only the X—C bond, strongly prefer gauche conformations, and hence form helical structures in the crystal. The gauche stabilities are in the order of PMO > PMS > PMSe. As for conformational characteristics, PES and PESe are very close to each other but quite different from PEO. The O—C and C—C bonds of PEO prefer trans and gauche conformations, respectively; however, the E_σ energy, depending on the environment, is almost null in the gas phase but significantly negative in polar media, and the C—H···O attraction leads to negative E_ωs. The X—C and C—C bonds of PES and PESe, respectively, show gauche and trans preferences, and their most stable conformations are $g^{\pm}tg^{\mp}$ in the X—CH_2—CH_2—X bond sequence. PTrMO, PTrMS, and PTrMSe are most stabilized in $tg^{\pm}g^{\pm}t$, $g^{\pm}g^{\pm}g^{\pm}g^{\pm}$, and $g^{\pm}ttg^{\mp}$, respectively.

22

Alternating Copolymers Including Ethylene-imine, Ethylene-oxide, and Ethylene-sulfide Units

As an example of molecular design of polymers, we imagined three alternating copolymers including two units of ethylene imine, ethylene oxide, and ethylene sulfide, namely, poly(ethylene imine-*alt*-ethylene oxide) (abbreviated as P(EI-EO)) [409], poly(ethylene imine-*alt*-ethylene sulfide) (P(EI-ES)) [408], and poly(ethylene oxide-*alt*-ethylene sulfide) (P(EO-ES)) [402], and attempted to predict their conformational characteristics and configurational properties mainly by molecular orbital (MO) and rotational isomeric state (RIS) or inversional–rotational isomeric state (IRIS) calculations with the aid of nuclear magnetic resonance (NMR) experiments (Figure 22.1). The model compounds adopted here are expressed as $CH_3XCH_2CH_2YCH_3$ (X, Y = NH, O, and S, X $\neq$ Y): *N*-(2-methoxyethyl)methylamine (MEMA, X = NH and Y = O) for P(EI-EO); *N*-(2-methylthioethyl)methylamine (MTEMA, X = NH and Y = S) for P(EI-ES); and 2-methoxyethyl methyl sulfide (MEMS, X = O and Y = S) for P(EO-ES). Inasmuch as the amine group, NH, undergoes the rapid nitrogen inversion, the *l*-form has been exclusively modeled for the MO calculations and NMR experiments.

Bond conformations of the model compounds, evaluated from NMR experiments and MO calculations, are listed in Table 22.1, and the data can be interpreted as follows: (MEMA) The N—C, C—C, and C—O bonds have trans, gauche, and trans preferences, respectively, and its most stable conformer is tg^-t with a free energy of -1.76 kcal mol^{-1} relative to that of ttt. (MTEMA) The trans and gauche states are preferred in the N—C and C—S bonds, respectively. The central C—C bond may be freely rotatable because its three rotamers occur at roughly the same probability. The tg^-g^- conformer is the lowest in free energy (-1.40 kcal mol^{-1}). (MEMS) The O—C, C—C, and C—S bonds show trans, trans, and gauche preferences, respectively. Its most stable conformer is $ttg^{\pm}$ whose free energy is -0.58 kcal mol^{-1}. Both NMR and MO data show the above tendencies, which are consistent with those found for the homopolymers: (PEI) trans in N—C and gauche in C—C; (PEO) trans in O—C and gauche in C—C; and (PES) gauche in S—C and trans in C—C.

Conformational Analysis of Polymers: Methods and Techniques for Structure-Property Relationships and Molecular Design, First Edition. Yuji Sasanuma.

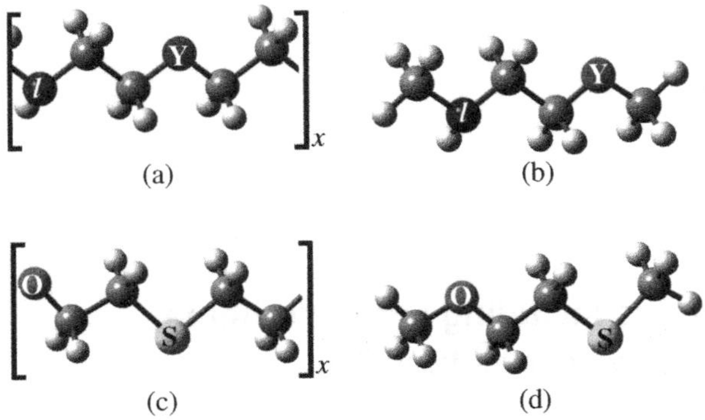

Figure 22.1 (a) Poly(ethylene imine-*alt*-ethylene oxide) (P(EI-EO), Y = O) or poly(ethylene imine-*alt*-ethylene sulfide) (P(EI-ES), Y = S). (b) Model compound: *N*-(2-methoxyethyl)methylamine (MEMA, Y = O) or *N*- (2-methylthioethyl)methylamine (MTEMA, Y = S). The NH group is assumed to be *l*-form. (c) Poly(ethylene oxide-*alt*-ethylene sulfide) (P(EO-ES)). (d) Model compound: 2-methoxyethyl methyl sulfide (MEMS).

Table 22.1 Bond conformations of MEMA, MTEMA, and MEMS.

Medium	p_t^{X-C}	p_t^{C-C}	p_t^{C-Y}
	MEMA (X = NH, Y = O)		
	MO calc		
Gas	0.78	0.08	0.90
	NMR expt		
C_6D_{12}	0.99	0.06	0.83
$CDCl_3$	0.96	0.06	0.84
CD_3OD	0.84	0.08	0.82
D_2O	0.75	0.10	0.80
	MTEMA (X = NH, Y = S)		
	MO calc		
Gas	0.70	0.26	0.25
	NMR expt		
C_6D_{12}	0.89	0.32	0.19
$CDCl_3$	0.84	0.33	0.25
CD_3OD	0.79	0.30	0.23
$(CD_3)_2SO$	0.67	0.27	0.26

(continued)

Table 22.1 (Continued)

Medium	$p_t^{X\text{-}C}$	$p_t^{C\text{-}C}$	$p_t^{C\text{-}Y}$
	MEMS (X = O, Y = S)		
	MO calc		
Gas	0.82	0.52	0.15
	NMR expt		
C_6D_{12}	0.79	0.53	0.14
C_6D_6	0.77	0.52	0.21
$CDCl_3$	0.79	0.53	0.26
CD_3OD	0.77	0.42	0.24
$(CD_3)_2SO$	0.74	0.47	0.25

The intramolecular interactions of the copolymers were defined similar to those of the homopolymers (see Figures 22.2, 22.3, and 22.4), and their energies were evaluated from conformer free energies as listed in Table 22.2. Of the first-order interaction energies, those for the N—C and O—C bonds are positive, and that for the S—C bond is negative, whereas those for the C—C bonds of MEMA and MTEMA are close to zero; therefore, the conformational preference of the C—C bonds is suggested to be determined by higher order interactions.

Of the second- and third-order interactions, the attractions, emphasized in bold, were evaluated as follows (in kcal mol^{-1}): P(EI-EO), N—H···O (η) = −1.75, C—H···O (ω) = −0.21, and C—H···N (ω') = −0.68; P(EI-ES), N—H···S (η) = −0.97 and C—H···N (ω') = −0.05; P(EO-ES), C—H···O (ω_O) = −0.38; and the

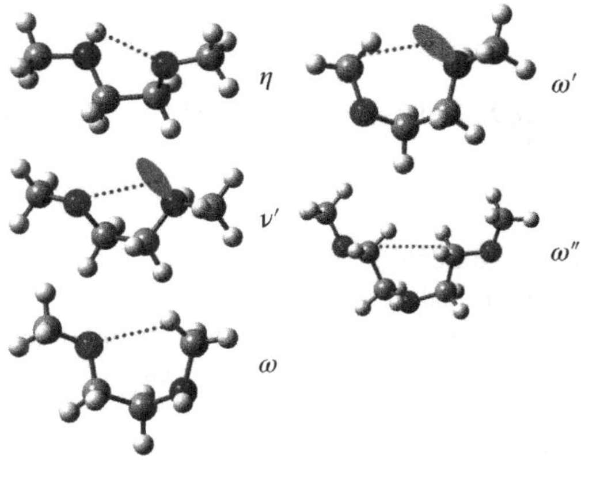

Figure 22.2 Second- and third-order intramolecular interactions defined for MEMA and P(EI-EO). Source: [409]/Reproduced with permission of American Chemical Society.

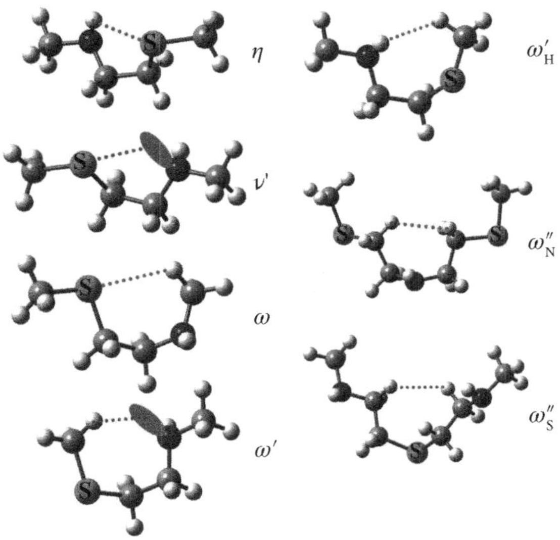

Figure 22.3 Second- and third-order intramolecular interactions defined for MTEMA and P(EI-ES). Source: [408]/Reproduced with permission of American Chemical Society.

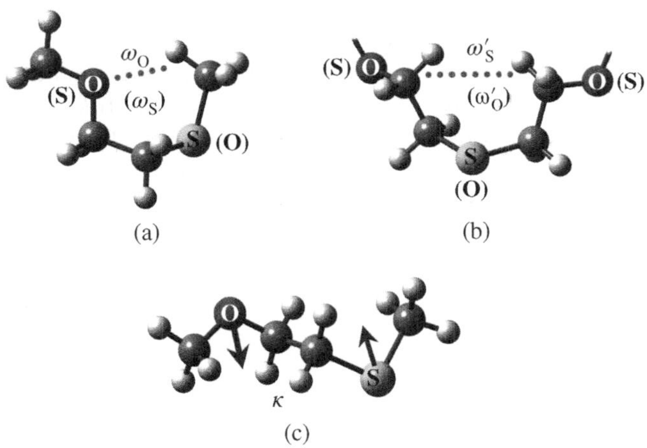

Figure 22.4 Second- and third-order intramolecular interactions defined for MEMS and P(EO-ES): (a) ω_O and ω_S; (b) ω'_S and ω'_O; (c) κ. Source: [402]/Reproduced with permission of American Chemical Society.

energy difference between $g^{\pm}tg^{\mp}$ and $g^{\pm}tg^{\pm}$ (κ) = −0.26. The other interactions are indicated to be repulsive.

The configurational properties of the three copolymers were calculated with the conformational energies listed in Table 22.2. For P(EI-EO) and P(EI-ES) undergoing the nitrogen inversion, the IRIS scheme was employed, while for P(EO-ES), the refined RIS scheme was used. The geometrical parameters were

Table 22.2 Conformational energies (kcal mol^{-1}) of MEMA and P(EI-EO), MTEMA and P(EI-ES), MEMS and P(EO-ES), derived from ab initio MO calculations.

	MEMA	MTEMA	MEMS
	P(EI-EO)	P(EI-ES)	P(EO-ES)
		First-order interaction	
E_γ [a]	1.06	0.77	
E_δ [b]	0.44	0.37	
E_σ [c]	0.05	0.01	0.50
E_ρ [d]	O—C: 1.21	S—C: −0.40	O—C: 1.04
			C—S: −0.37
		Second- and third-order interactions.[e]	
E_η	**−1.75**	**−0.97**	
$E_{\nu'}$	0.59	0.73	
E_ω	**−0.21**	1.19	
$E_{\omega'}$	**−0.68**	**−0.05**	
$E_{\omega''}$ [f]	1.24		
$E_{\omega'_H}$		1.02	
$E_{\omega''_N}$ [f]		0.59	
$E_{\omega''_S}$ [f]		0.82	
E_{ω_O}			**−0.38**
E_{ω_S} [f]			∞
$E_{\omega'_O}$ [f]			∞
$E_{\omega'_S}$			0.92
E_κ			**−0.26**

a) For the gauche$^+$ state around the N—C bond of the *l*-form.
b) For the gauche$^-$ state around the N—C bond of the *l*-form.
c) For the gauche state around the C—C bond.
d) For the gauche state around the X—C or C—Y bond.
e) See Figures 22.2, 22.3, and 22.4.
f) Determined from dimeric models

chosen from the optimized structures of the models. The characteristic ratios, their temperature coefficients, and configurational entropies thus calculated are shown in Table 22.3. In addition, the dipole moment ratio and its temperature coefficient of P(EO-ES) were calculated because the experimental values were reported only for this copolymer. The bond dipole moments were determined

Table 22.3 Configurational properties of P(EI-EO), P(EI-ES), and P(EO-ES).[a)]

		HBS[b)] (%)	x[c)]	P(EI-EO)	P(EI-ES)	P(EO-ES)
$\langle r^2\rangle_0/nl^2$	calcd	100	∞	1.33	5.15	
		0	∞	6.94	6.40	
			100			6.0
$10^3 d(\ln\langle r^2\rangle_0)/dT$ (K^{-1})	calcd	100	∞	8.6	0.96	
			100			0.22
P_m[d)]	calcd	100	∞	0.50	0.49	
		0	∞	0.50	0.50	
$\langle \mu^2\rangle/nm^2$	calcd		100			0.56
	obsd[e)]		100			0.57
$10^3 d(\ln\langle \mu^2\rangle)/dT$ (K^{-1})	calcd		100			1.1
	obsd[e)]		100			1.6

a) At 25 °C.
b) Hydrogen bond strength.
c) Degree of polymerization.
d) *Meso*-diad probability.
e) Reported by Riande and Guzmán [387]. Recalculated with m_{O-C} = 1.13 D, m_{C-C} = 0.00 D, and m_{C-S} = 1.24 D.

Source: Adapted from [402, 408]

from the MO calculations: m_{O-C} = 1.13 D, m_{C-C} = 0.00 D, and m_{C-S} = 1.24 D [402]. The refined RIS calculations (0.56) exactly reproduced the experimental dipole moment ratio (0.57) [387]. It is known that the dipole moment of polymers with a symmetrical plane, a twofold symmetry axis, or a symmetry point between the repeating units is always free from the excluded-volume effect [111, 330]. This is the case with P(EO-ES).

The characteristic ratios of P(EI-EO) and P(EI-ES) depend largely on the hydrogen-bond strength (HBS): P(EI-EO), 1.33 (HBS = 100%) and 6.94 (0%); P(EI-ES), 5.15 (100%) and 6.40 (0%). At HBS = 0%, both copolymers show the characteristic ratios close to that of polyethylene (see Table 29.4). The neighboring NH groups are separated by six bonds and hence weakly correlated with each other; therefore, the *meso*-diad probability is almost equal to 0.5 (atactic) irrespective of HBS.

Figure 13.5 of Section 13.7 shows correlations of the bond orientation in P(EI-EO). P(EI-EO) exhibits strong correlations that express the existence of helical paths of approximately 20 bonds per turn. As HBS is reduced, the correlation becomes weak and fades away; therefore, the helical path is due to the N—H· · ·O hydrogen bonds. Such helical paths may occasionally be formed in solutions and melt. In contrast, the N—H· · ·S attraction of P(EI-ES) is not strong enough for the helix formation [408].

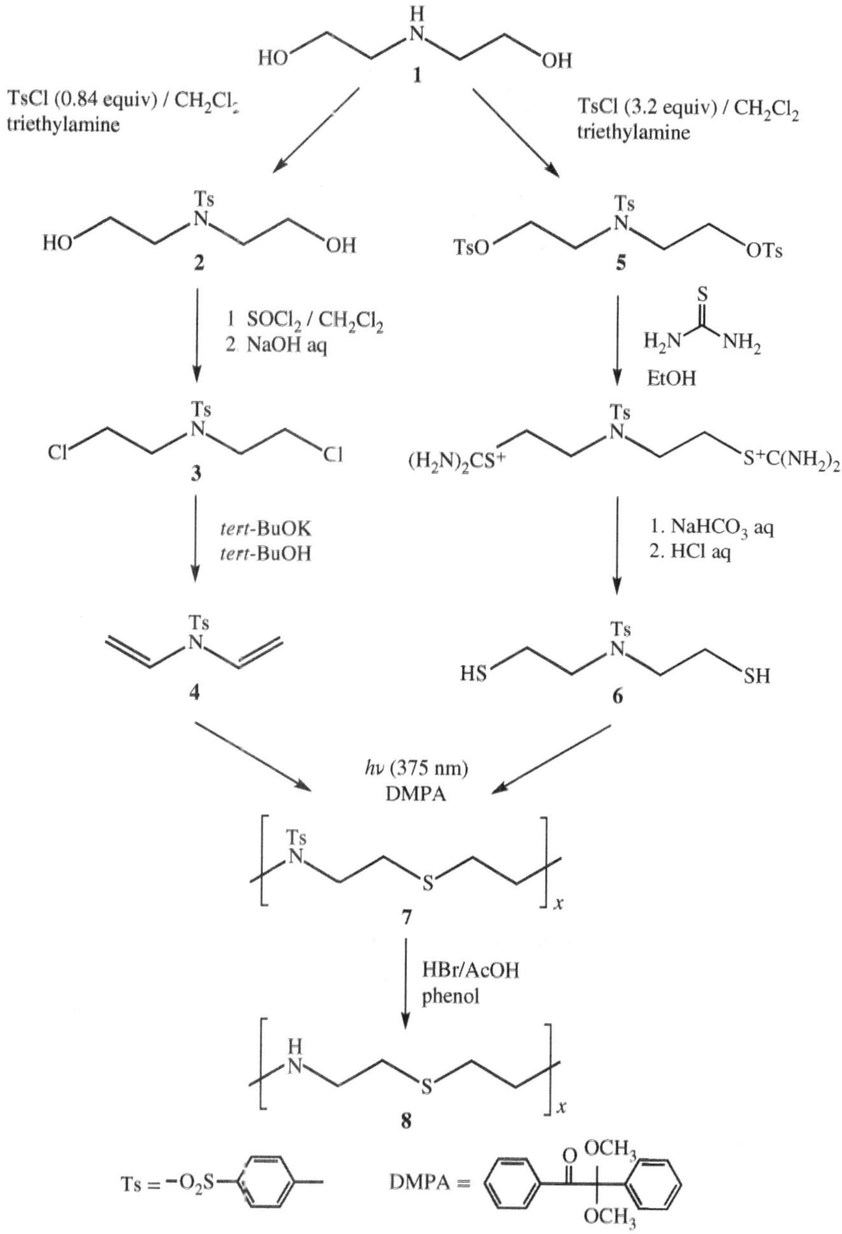

Figure 22.5 Total synthetic route of poly(*N*-tosyl-ethylene imine-*alt*-ethylene sulfide) (P(*N*-tosylEI-ES), **7**) and deprotection. Diethanolamine is the starting material of both monomers, **4** and **6**, between which thiol-ene photopolymerization was carried out to yield P(*N*-tosylEI-ES). Source: Reproduced from reference [203] with permission from the Royal Society of Chemistry.

22.1 Synthesis of P(EI-ES)

A USA patent reported that oligomeric (EI-EO) (degree of polymerization = 3–5) was synthesized by ring-opening polymerization of *N*-(2-hydroxyethyl)aziridine [482]. P(EO-ES)s (M_n = 8 and 13 kDa) were prepared by polycondensation of 2,2′-thiodiethanol and subjected to dipole moment measurements [387]. However, we had found no precedent for P(EI-ES) synthesis; accordingly, Hori et al. challenged the synthesis and successfully obtained linear P(EI-ES) [203].

The synthetic route of linear P(EI-ES) is illustrated in Figure 22.5. The starting material was diethanolamine, whose nitrogen or oxygen atom was first protected by a tosyl group. The monomers, 4-methyl-*N*,*N*-divinylbenzenesulfonamide (**4**) and *N*,*N*-bis(2-mercaptoethyl)-4-methylbenzenesulfonamide (**6**), were prepared,

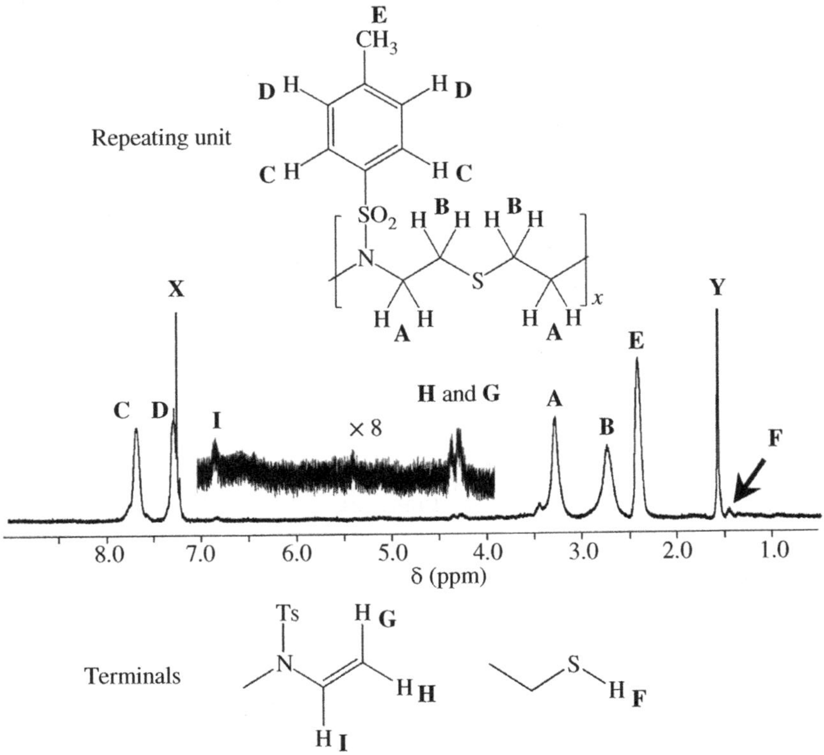

Figure 22.6 ^{1}H NMR spectrum of P(*N*-tosylEI-ES) with assignment of observed peaks: **A–E**, from the repeating unit; **F–I**, from the terminal groups; **X** and **Y**, from $CHCl_3$ and water immixed in the solvent, $CDCl_3$, respectively. Source: Reproduced from reference [203] with permission from the Royal Society of Chemistry.

and thiol-ene photopolymerization using an initiator of 2,2-dimethoxy-2-phenylacetophenone (DMPA) under irradiation of ultraviolet light of 375 nm yielded poly(*N*-tosylethylene imine-*alt*-ethylene sulfide) (P(*N*-tosylEI-ES), **7**), which was deprotected to be linear P(EI-ES).

A ^{1}H NMR spectrum of P(*N*-tosylEI-ES) is shown in Figure 22.6, and the assignment of the observed signals is also indicated there. A static light scattering experiment determined the weight-average molecular weight of P(*N*-tosylEI-ES) to be 2.77×10^4. P(*N*-tosylEI-ES) is insoluble in *n*-hexane, acetic acid, ethanol, and water; partly soluble in toluene; and readily soluble in chloroform, tetrahydrofuran, acetone, and dimethyl sulfoxide, while P(EI-ES) is insoluble in *n*-hexane, diethyl ether, and cold and hot water; partly soluble in toluene, chloroform, and acetone; and readily soluble in dimethyl sulfoxide, acetic acid, and ethanol. Thermal decompositions of P(*N*-tosylEI-ES) and P(EI-ES) begin at 545 and 517 K, respectively.

23

Aromatic Polyester (PET, PTT, and PBT)

In the early 1940s, poly(ethylene terephthalate) (PET) and poly(trimethylene terephthalate) (PTT) were patented in England [520]. Immediately, PET was moved to industrial manufacturing. In contrast, commercial production of PTT had long been left unrealized because its raw material, 1,3-propanediol (PDO), was too costly. In the early 1990s, however, continuous hydroformylation of ethylene oxide enabled economic production of PDO, and PDO can also be produced by recombinant *Escherichia coli* from corn sugar. These technical innovations have successfully made PTT commercially produced. Because PTT is superior in durability, softness, and permanent stain resistance, it has been used as activewears, carpets, and automotive fabrics [513, 547]. Inasmuch as poly(butylene terephthalate) (PBT) is rich in resilience and toughness, PBT has often been reinforced with glass fiber to be used as an engineering plastic for auto components and electric parts [513].

In the primary structure, the difference among PET, PTT, and PBT is merely the number of methylene units in the O—$(CH_2)_y$—O part (designated herein as spacer) between benzene rings: PET, $y = 2$; PTT, $y = 3$; PBT, $y = 4$. Therefore, there is no doubt that all differences in structures and properties among them stem only from the number of methylene groups.

The conformational characteristics of PET, PTT, and PBT were investigated via MO calculations and NMR experiments on their model compounds, and the refined rotational isomeric state (RIS) calculations for the polyesters and discussed in terms of (1) bond conformations, (2) molecular Kerr constants and dipole moments of the models, (3) configurational properties of the polyesters, and (4) weak attractive intramolecular interactions. In addition, (5) the crystal structures of the three aromatic polyesters were optimized by the periodic DFT calculations, and (6) the crystalline moduli were evaluated and interpreted in terms of the crystal conformations [400, 414, 419].

As models for PET, PTT, and PBT, ethylene glycol dibenzoate (EGDB, $y = 2$), trimethylene glycol dibenzoate (TriMGDB, $y = 3$), and tetramethylene glycol

Conformational Analysis of Polymers: Methods and Techniques for Structure-Property Relationships and Molecular Design, First Edition. Yuji Sasanuma.

dibenzoate (TetMGDB, $y = 4$) were, respectively, adopted and subjected to MO calculations and NMR experiments. For NMR measurements, the models were labeled partly with carbon-13: C_6H_5—$^{13}C(=O)O$—$(CH_2)_y$—$OC(=O)$—C_6H_5 (EGDB-^{13}C, TriMGDB-^{13}C, and TetMGDB-^{13}C).

Since the benzene ring and ester group are coplanar, only the O—$(CH_2)_y$—O bonds are internally rotatable. On the assumption that each of the bonds adopts three rotational isomeric states (t, g^+, and g^-), there is the possibility that EGDB, TriMGDB, and TetMGDB have $3^3 = 27$, $3^4 = 81$, and $3^5 = 243$ conformers, respectively; however, the molecular symmetry reduces the numbers of irreducible conformers of EGDB, TriMGDB, and TetMGDB to 10, 25, and 70, respectively. By way of example, Gibbs free energies of the 10 conformers of EGDB, evaluated from MO calculations at the MP2/6-311+G(2d,p)//B3LYP/6-311+G(2d,p) level, are listed in Table 8.2 (Section 8.5), from which the most stable state can be seen to be $tg^{\pm}t$.

^1H and ^{13}C NMR spectra of EGDB-^{13}C, shown in Figure 15.12, yielded vicinal ^1H—^1H and ^{13}C—^1H coupling constants, from which bond conformations of the O—CH_2 and CH_2—CH_2 bonds were derived, being compared with the MO calculations in Table 15.7. Both NMR and MO results are consistent with each other and indicate that the O—CH_2 and CH_2—CH_2 bonds prefer trans ($p_t \approx 0.4$–0.5) and gauche ($p_t \approx 0.05$–0.2) conformations, respectively.

23.1 Correction for MP2 Energy of π-π Interaction

The MO calculations on TriMGDB at the MP2 level of theory show extraordinarily low free energies for conformers whose two central CH_2—CH_2 bonds of the spacer adopt $g^{\pm}g^{\mp}$ conformational pairs: tg^+g^-t (−1.06); $tg^+g^-g^+$ (−1.92); $tg^+g^-g^-$ (−2.24); and $g^+g^+g^-g^-$ (−0.76) (see Figure 23.1), where the parenthesized values represent the free energies (in kcal mol^{-1}) in the gas phase relative to that of the all-trans state. The four conformers are bended at the center, and the two benzene rings closely face each other, which suggests that the π–π attractions may act between them. Therefore, the extraordinary stabilities of the bended conformers are probably due to the π–π attractions.

Probably owing to the π–π attractions, two bended conformers of TetMGDB also have very low free energies (in kcal mol^{-1}) relative to that of the all-trans form: $tg^+g^+g^-g^-$ (−1.53) and $tg^+g^-g^+t$ (−1.98) (Figure 23.2). Its bond conformations derived from MO calculations are inconsistent with those from NMR experiments. For example, the trans fractions at 25 °C were obtained as follows: for OCH_2—CH_2, $p_t^{MO} = 0.12$ and $p_t^{NMR} = 0.24$; for CH_2CH_2—CH_2CH_2, $p_t^{MO} = 0.30$ and $p_t^{NMR} = 0.53$. It has occasionally been pointed out that the MP2 level of theory tends to overestimate the π–π and C—H$\cdots\pi$ interaction energies [275, 327, 496].

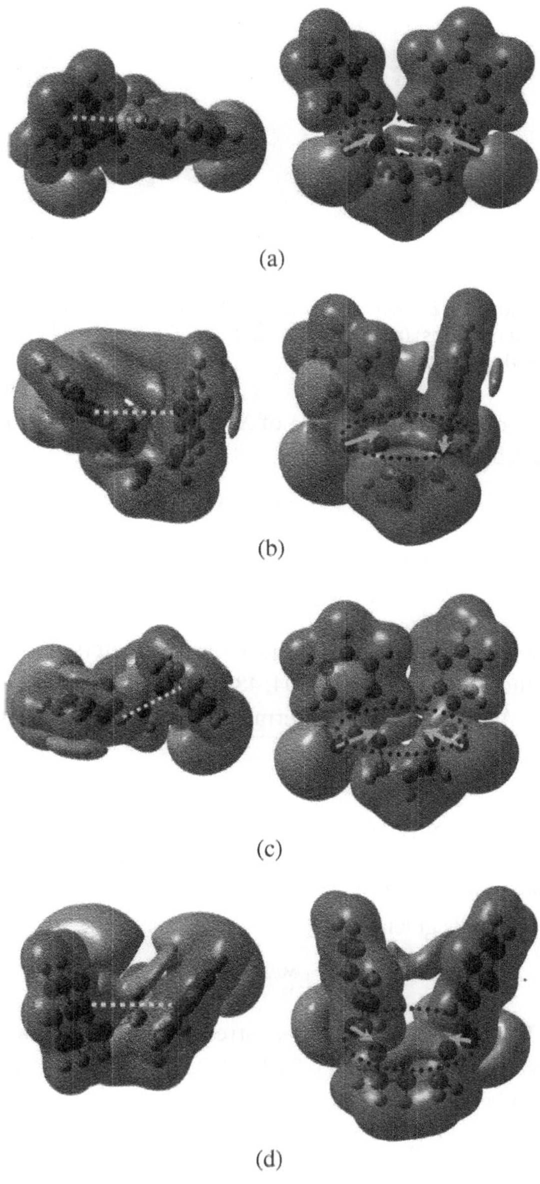

Figure 23.1 Four folding conformers of TriMGDB, (a) tg^+g^-t, (b) $tg^+g^-g^+$, (c) $tg^+g^-g^-$, and (d) $g^+g^+g^-g^-$, stabilized by intramolecular attractive interactions: top (left) and side (right) views. The dashed lines, dotted circles, and arrows represent the $\pi-\pi$ interactions of the terephthaloyl rings, electrostatic attractions between the ester groups, and dipole moments of the ester groups, respectively. Source: [414], Figure 5 (p.7208)/ Reproduced with permission of American Chemical Society.

The corrections for the MP2 overestimation were proposed, being composed of two steps: (1) extrapolation to the complete basis sets: (2) addition of higher order correlation energies. When, for example, the cc-PVXZ basis sets [123] are used, the Hartree–Fock energy (E_{∞}^{HF}) for the complete basis set ($X \rightarrow \infty$) may

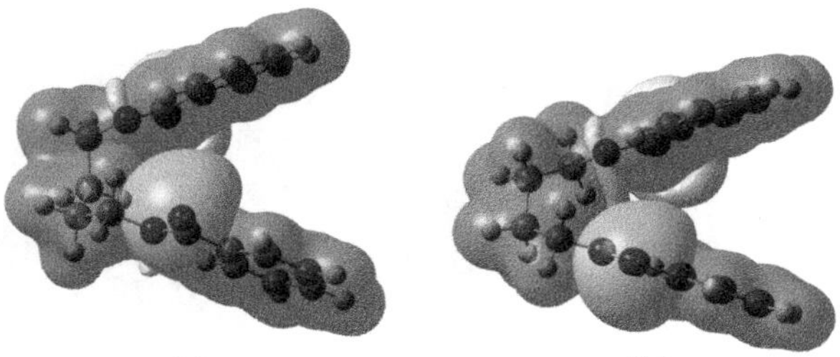

Figure 23.2 Two conformers of TetMGDB with intramolecular close contacts between phenyl rings: (a) $tg^+g^+g^-g^-$; (b) $tg^+g^-g^+t$. Source: [419], Figure 4 (p. 3909)/Reproduced with permission of Elsevier.

be obtained from those of $X = 2$ (D, double), 3 (T, triple), and 4 (Q, quadruple) according to

$$E_X^{\mathrm{HF}} = E_\infty^{\mathrm{HF}} + B\exp(-\alpha X) \tag{23.1}$$

and the corrected MP2 energy (E_∞^{MP2}) may also be derived from

$$E_X^{\mathrm{MP2}} = E_\infty^{\mathrm{MP2}} + CX^{-\beta} \tag{23.2}$$

where B and C are the adjustable parameters, and α and β were suggested to be 3.4 and 2.2, respectively [194, 489].

As the higher order term, half of the MP3 term is added to the MP2 energy:

$$E^{\mathrm{higher}} = \frac{1}{2}E^{\mathrm{MP3}} \tag{23.3}$$

It is also suggested that, even if small basis sets are employed for this correction, the modified energy would be as reliable as the coupled cluster single–double and perturbative triple (CCSD(T)) energy [327, 367]. To sum up, the totally corrected energy is given by

$$E^{\mathrm{total}} = E_\infty^{\mathrm{HF}} + E_\infty^{\mathrm{MP2}} + \frac{1}{2}E^{\mathrm{MP3}} \tag{23.4}$$

The Gibbs energies thus corrected are as follows (relative to that of the all-trans state, in kcal mol^{-1}): TriMGDB, tg^+g^-t (0.27); $tg^+g^-g^+$ (0.05); $tg^+g^-g^-$ (−0.78); and $g^+g^+g^-g^-$ (0.98); TetMGDB, $tg^+g^+g^-g^-$ (0.20) and $tg^+g^-g^+t$ (0.22). All the corrected energies were considerably enlarged as compared with the precorrected values. The bond conformations were calculated with these corrected energies, and the other energies were set as they were. The trans fractions are in good agreement with the NMR data as shown in Table 23.1.

Table 23.1 Trans fractions at 25 °C of TriMGDB and TetMGDB, model compounds of PTT and PBT, respectively.

Medium	Bond		
	$O-CH_2$	$OCH_2-CH_2CH_2$	$CH_2-CH_2CH_2$
	TriMGDB		
	MO calc[a]		
Gas	0.52	0.27	
Benzene	0.54	0.28	
	NMR expt		
Cyclohexane-d_{12}	0.50	0.28	
Benzene-d_6	0.52	0.28	
Chloroform-d	0.55	0.28	
Methanol-d_4	0.52	0.28	
Dimethyl-d_6 sulfoxide	0.53	0.27	
	TetMGDB		
	MO calc[b]		
Gas	0.52	0.20	0.51
Benzene	0.54	0.23	0.60
	NMR expt		
Cyclohexane-d_{12}	0.51	0.24	0.53
Benzene-d_6	0.53	0.27	0.57
Chloroform-d	0.54	0.24	0.57
Dimethyl-d_6 sulfoxide	0.52	0.23	0.54

a) Calculated with the corrected energies of the four bended conformers.
b) Calculated with the corrected energies of the two bended conformers.

23.2 Dipole Moment and Molar Kerr Constant

As described in Part III, the dipole moment can be calculated and compared with the experimental observations. In parallel to the dipole moment, the molar Kerr constant has also been evaluated from the RIS calculations to be used as an objective parameter for the conformational analysis of polymers [388]. Dipole moments (μs) and molecular Kerr constants (${}_mK$s) of the three model compounds here were observed with nonpolar solvents [311].

The molar Kerr constant is expressed as [388]

$$ {}_mK = \frac{2\pi N_A}{15k_B T}\left[\frac{\mu^T\hat{\alpha}\mu}{k_B T} + \mathrm{tr}\,(\hat{\alpha}\hat{\alpha})\right] \tag{23.5}$$

where N_A is Avogadro's number, k_B is Boltzmann's constant, and μ^T is the transpose of dipole moment, namely, $\mu^T = (\mu_x, \mu_y, \mu_z)$. The anisotropic part $\hat{\alpha}$ of the polarizability tensor α is given by

$$ \hat{\alpha} = \alpha - \frac{1}{3}\,\mathrm{tr}(\alpha)\,I_3 \tag{23.6}$$

where

$$ \alpha = \mathrm{diag}\left(\alpha_x, \alpha_y, \alpha_z\right) \tag{23.7}$$

and I_3 is the identity matrix of size 3. The Gaussian program [159] yields the μ^{MO} vector and α^{MO} tensor with respect to the so-called standard orientation [154]. The α^{MO} tensor is diagonalized to give its eigenvalues and the principal-axis system (PAS), and the μ^{MO} vector is simultaneously transformed to the μ vector defined for the PAS. The dipole moment and molar Kerr constant corresponding to the experimental values may be calculated from

$$ \langle\mu\rangle = \sum_k \mu_k\, M_k \exp\left(-\frac{\Delta G_k}{RT}\right) \Big/ \sum_k M_k \exp\left(-\frac{\Delta G_k}{RT}\right) \tag{23.8}$$

and

$$ \langle{}_mK\rangle = \sum_k {}_mK_k\, M_k \exp\left(-\frac{\Delta G_k}{RT}\right) \Big/ \sum_k M_k \exp\left(-\frac{\Delta G_k}{RT}\right) \tag{23.9}$$

where M_k is the multiplicity of conformer k. For the ΔG_ks of TriMGDB and TetMGDB, the free energies corrected for the MP2 overestimation were used.

The μ^{MO} vectors and α^{MO} tensors of EGDB, calculated at the B3LYP/6-311+G(2d,p) level and transformed to the PAS, are listed in Table 23.2. The $\langle\mu\rangle$ and $\langle{}_mK\rangle$ values of EGDB, TriMGDB, and TetMGDB were calculated with respect to both the PAS and the standard orientation (see Table 23.3). Even in the standard orientation whose origin is located at the center of nuclear charges, the off-diagonal elements of α^{MO} are much smaller than the diagonal ones, and hence, the α^{MO} tensor may be approximately diagonalized. For all the three model compounds, the calculated μ values are in good agreement with the experiments, and the observed molar Kerr constants are found to fall between the two $\langle{}_mK\rangle$ values.

Table 23.2 Dipole moments (μs) and polarizabilities (αs) of conformers of EGDB, evaluated by ab initio MO calculations and diagonalized with respect to the PAS (x, y, and z)[a].

				μ (D)				α (Å^3)		
k	**Conformation**			x	y	z	**Total**[b]	x	y	z
1	t	t	t	0.000	0.000	0.000	0.000	43.99	30.27	16.79
2	t	t	g$^+$	−0.198	−1.301	−2.029	2.418	44.21	23.80	23.14
3	t	g$^+$	t	0.000	2.644	0.000	2.644	36.16	30.88	22.66
4	t	g$^+$	g$^+$	−1.649	3.225	1.212	3.820	34.66	32.74	21.91
5	t	g$^+$	g$^-$	1.495	0.153	1.062	1.840	33.38	29.85	25.68
6	g$^+$	t	g$^+$	0.000	3.448	0.000	3.448	44.67	29.14	17.49
7	g$^+$	t	g$^-$	0.002	0.001	0.002	0.003	44.70	29.48	17.16
8	g$^+$	g$^+$	g$^+$	0.000	3.203	0.000	3.203	36.74	32.23	20.45
9	g$^+$	g$^+$	g$^-$	−2.775	−0.288	−1.915	3.384	35.04	31.05	22.67
10	g$^+$	g$^-$	g$^+$				(absent)[c]			

a) At the B3LYP/6-311+G(2d, p) level.
b) $\mu_{total}^2 = \mu_x^2 + \mu_y^2 + \mu_z^2$.
c) The local minimum of the potential was not found by the geometrical optimization.
Source: [400], Table 6 (p.2859)/Reproduced with permission of American Chemical Society.

Table 23.3 Calculated and observed dipole moments and molar Kerr constants of EGDB, TriMGDB, and TetMGDB.

			$\langle {}_mK \rangle$ (10^{-25} m^5 V^{-2} mol^{-1})	
	$\langle \mu \rangle$ (D)	$\langle \mu^2 \rangle^{1/2}$ (D)	PAS	Standard orientation
		EGDB		
Calcd[a]	2.93	3.05	32.3	10.6
Obsd[b]		2.75		29.1 ± 1.6
		TriMGDB		
Calcd[a]	2.36	2.58	20.4	6.31
Obsd[b]		2.59		6.9 ± 3.9
		TetMGDB		
Calcd[a]	2.40	2.63	22.7	6.29
Obsd[b]		2.68		11.6 ± 1.5

a) At the B3LYP/6-311+G(2d,p) level. In benzene at 25 °C and 1 atm.
b) Dipole moments in benzene at 25 °C and molar Kerr constants in cyclohexane at 25 °C [311].

23.3 Configurational Properties

The configurational properties of PET, PTT, and PBT were calculated by the refined RIS scheme described in Section 12.1 (see Table 12.2). The unperturbed chain dimensions of PET have been experimentally estimated from the Kurata–Stockmayer [262] and Stockmayer–Fixman [447] plots for viscosities observed with good solvents. PET is hardly soluble in common organic solvents but soluble in some acidic solvents listed in Table 23.4, where the characteristic ratios thus estimated are also shown. The $\langle r^2 \rangle_0/nl^2$ ratios calculated from the refined RIS calculations are 2.63 at 25 °C and 2.84 at 250 °C, being close to those determined from small-angle neutron scattering (SANS) experiments (2.7 from the Zimm plot and 3.1 from the Kratky plot at 250 °C and 3.1 at 285 °C, see Section 16.3.1) rather than those from the acidic solutions. The agreement with the SANS results supports Flory's assertion that molten polymers ought to stay unperturbed (in the Θ state) (Section 3.2.4). Poly(ethylene-2,6-naphthalate) (PEN) has the same spacer as that of PET; therefore, the conformational characteristics of PEN are expected to be similar to those of PET, and the configurational properties of PEN may also be evaluated from the conformational energies of EGDB. On this basis, the characteristic ratios and configurational entropies of PEN were calculated as follows: $\langle r^2 \rangle_0/nl^2$, 2.43 at 25 °C and 2.55 at 610 K (equilibrium melting point) [76]; S_{conf}, 7.08 cal K^{-1} mol^{-1} at 25 °C and 7.55 cal K^{-1} mol^{-1} at 610 K. The entropy of fusion was estimated experimentally as 9.8 cal K^{-1} mol^{-1} [76].

Table 23.4 Experimental characteristic ratios of PET[a)].

Method	$\langle r^2 \rangle_0/nl^2$ [b)]	Solvent, process	Temp (°C)	Reference
Viscosity	5.2	PH/TriCPH	30	[256, 257]
	4.0, 4.1	TFA	30	[515]
	4.3, 4.5	PH/TetCE	30	[515]
	4.2	PH/TetCE	25	[15]
	2.3, 3.5, 3.7	o-Chlorophenol	25	[315]
	4.0	DCA / CH	25	[499]
SANS	2.7, 3.1	Melt quench	250	[171]
	3.1	Melt quench	285	[215]

a) Abbreviations: TFA, trifluoroacetic acid; PH, phenol; TriCPH, trichlorophenol; TetCE, tetrachloroethane; DCA, dichloroacetic acid; CH, cyclohexane; SANS, small-angle neutron scattering.

b) The $\langle r^2 \rangle_0/nl^2$ values were calculated with nl^2/M (43.5/192 Å^2 g^{-1} mol) of the repeating unit.

Source: [400], Table 1 (p.2855)/Reproduced with permission of American Chemical Society.

As discussed in Section 12.1, the aromatic polyesters show fluorescent (dimer) emissions. The intensity and wavelength depend on the number of methylene units of the spacer and closely correlate with the frequency of close contacts between the benzene rings (see Figure 12.4).

23.4 Crystal Structure

The crystal structures of the three aromatic polyesters were optimized by periodic DFT-D calculations at the B3LYP-D/6-31G(d,p) level with the dispersion-force correction, being illustrated in Figure 23.3 [266], and the optimized lattice constants and geometrical parameters are listed in Table 23.5, together with

Table 23.5 Optimized crystal structures of PET, PTT, and PBT, compared with experiments.

	PET			PTT			PBT (α form)		
	exp I[a]	exp II[b]	opt	exp I[c]	exp II[d]	opt	exp I[e]	exp II[f]	opt
				Lattice constant (Å, °)					
a	4.56	4.51	4.45	4.64	4.59	4.38	4.83	4.89	4.59
b	5.94	5.88	5.93	6.27	6.21	6.00	5.94	5.95	5.88
c	10.75	10.79	10.76	18.64	18.31	18.35	11.59	11.67	11.81
α	98.5	100.0	101.6	98.4	98.0	101.9	99.7	98.9	100.9
β	118	118.4	119.2	93.0	90.0	90.1	115.2	116.6	114.3
γ	112	110.6	112.7	111.1	111.7	103.8	110.8	110.9	111.2
				Dihedral angle[g] (°)					
ϕ_1	159	126.0	141.9	180.0	−152.8	−166.0	−88	−92.5	−94.4
ϕ_2	180	180.0	180.0	−67.0	−61.9	−58.0	−68	−76.2	−66.7
ϕ_3	−159	−126.0	−141.9	−67.0	−73.7	−67.0	180	180	180.0
ϕ_4				180.0	167.4	−169.9	68	76.2	66.7
ϕ_5							88	92.5	94.4

a) By Daubeny et al. [101].
b) By Fu et al. (sample, PET4B) [160].
c) By Poulin-Dandurand et al. [370]
d) By Desborough et al. [107].
e) By Yokouchi et al. [540].
f) By Hall and Pass [183] and Desborough and Hall [106].
g) The dihedral angles of the spacer, O—$(CH_2)_y$—O, are designated in sequence as ϕ_1, ϕ_2, ϕ_3, etc.

the corresponding experimental values. It is known that PBT exhibits crystal polymorphs (α and β structures), and that, under tension, the α form reversibly changes to the β one [25]. Accordingly, the β structure is regarded as a nonequilibrium structure and hence was omitted from the structural optimization.

The crystalline PET chain adopts a distorted all-trans conformation; the ϕ_1 angle (141.9°) around the O—CH_2 bond somewhat deviates from the normal value (~ 180°). The spacers of PTT and PBT lie in a t–g–g–t and a $g^+ - g^+ - t - g^- - g^-$ conformations, respectively.

The compliance tensors of PET, PTT, and PBT were obtained as follows:

For PET,

$$S = \begin{bmatrix} 133.9799 & -12.9476 & -97.2459 & 8.9759 & -12.7626 & -32.4059 \\ & 44.9248 & -13.9544 & -27.9498 & -18.7392 & -25.9893 \\ & & 122.6210 & 29.0338 & 54.1337 & 46.4807 \\ & & & 146.4850 & 12.8333 & 58.9166 \\ & & & & 65.7181 & 41.6233 \\ & & & & & 171.3724 \end{bmatrix} \quad (\mathrm{TPa}^{-1}) \tag{23.10}$$

The above S tensor yields the crystalline moduli in the a-, b-, and c-axis directions as $E_a = 7.20$, $E_b = 22.26$, and $E_c = 182.4$ GPa, respectively.

For PTT,

$$S = \begin{bmatrix} 157.5523 & -13.4128 & -113.4348 & 42.2358 & 31.6339 & 19.4907 \\ & 54.4058 & -18.4426 & -10.6197 & -12.1124 & -44.7592 \\ & & 137.8137 & -16.5997 & -15.4925 & 17.4331 \\ & & & 195.8449 & 42.0008 & -21.8234 \\ & & & & 93.9245 & 40.5603 \\ & & & & & 253.4620 \end{bmatrix} \quad (\mathrm{TPa}^{-1}) \tag{23.11}$$

The crystalline moduli in the a-, b-, and c-axis directions were obtained as $E_a = 6.89$, $E_b = 18.38$, and $E_c = 7.08$ GPa, respectively.

For PBT,

$$S = \begin{bmatrix} 205.7408 & -55.7681 & -100.8172 & 105.1023 & 61.2697 & -62.1162 \\ & 86.1013 & -16.9861 & -82.5952 & -52.4611 & -24.8502 \\ & & 115.9620 & -8.5330 & 15.7754 & 74.2349 \\ & & & 199.5793 & 81.9417 & 38.8995 \\ & & & & 123.8354 & 31.3516 \\ & & & & & 209.3272 \end{bmatrix} \quad (\mathrm{TPa}^{-1}) \tag{23.12}$$

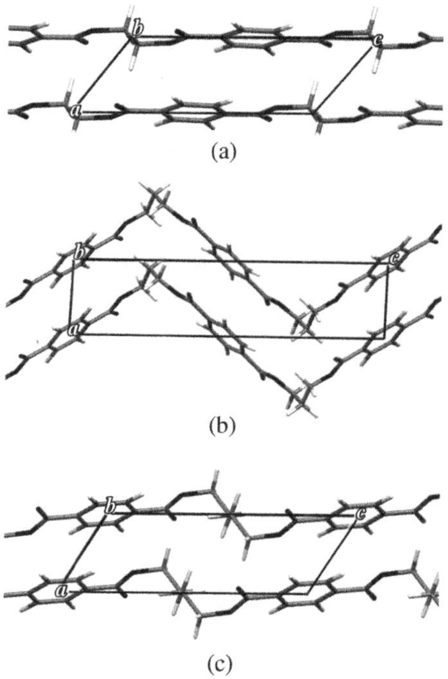

Figure 23.3 Optimized crystal structures of PET, PTT, and PBT (α form), projected on the plane perpendicular to the *b*-axis. Source: [266], Figure 3 (p.4829)/with permission of American Chemical Society/Licensed under CC BY 4.0.

The crystalline moduli in the *a*-, *b*-, and *c*-axis directions were obtained as $E_a = 4.84$, $E_b = 11.61$, and $E_c = 20.84$ GPa, respectively.

The E_c value (7.08 GPa) of PTT at 0 K is comparable to the corresponding experimental value of 5.39 GPa at −255 °C [340]. Nishino et al. have pointed out that PTT has the smallest E_c of all polymers that they have investigated so far [340]. The calculated E_c magnitude of PBT at 0 K is 20.84 GPa, and the experimental value at room temperature is 13.2 GPa [335]. The crystalline PET chain, being somewhat distorted all-trans, exhibits the E_c value (182 GPa) much smaller than that (333 GPa) of PE lying in the regular all-trans zigzag form, while the distorted PTT (tggt) and PBT ($g^+g^+tg^-g^-$) spacers show much smaller E_cs (see Figure 23.3).

24

Aliphatic Polyesters

Aliphatic polyesters were synthesized in the earliest stage of polymer chemistry [62–65] but nevertheless had long been left nonindustrialized. This is partly because, at first, aliphatic polyesters of large molecular weights enough for fiber spinning could not be produced and partly because aromatic polyesters superior to the aliphatic polyesters in physical properties appeared [520]. Nowadays, "polyester" usually means an aromatic polyester, PET. Since the late twentieth century, however, aliphatic polyesters have attracted attention because some of them are biosynthetic and/or biodegradable, and some can be produced from plant-derived substances [28, 115]. Carbon neutrality is requisite for prevention of global warming and maintenance of sustainable development. All aliphatic polyesters treated herein are biodegradable: poly(glycolic acid) (PGA); [422]; poly(lactic acid) (PLA) aka poly(lactide) [418, 422]; poly(2-hydroxybutyrate) (P2HB) [422]; poly(3-hydroxybutyrate) (P3HB) [407]; poly(ϵ-caprolactone) (PCL) [238]; poly(ethylene succinate) (PES) [410]; and poly(butylene succinate) (PBS) [410].

24.1 Poly(glycolic acid) (PGA) and Poly(2-hydroxybutyrate) (P2HB)

PGA has a symmetric repeating unit with three skeletal bonds (Figure 24.1), but nevertheless its equilibrium melting point is as high as 231.4 °C [334]. Furthermore, PGA has outstanding mechanical properties comparable to those of super engineering plastics and high gas-barrier properties and hence has been used as sutures for surgical stitching, orthopedic parts for bone fixation, interlayers between thin films of PET for bottles, and moving tools in wellbores for shale gas exploration [317, 534].

P2HB has often been used as copolymers or blends mainly with PLA to modify the original properties of PLA rather than as homopolymers [490, 491, 494].

Conformational Analysis of Polymers: Methods and Techniques for Structure-Property Relationships and Molecular Design, First Edition. Yuji Sasanuma.

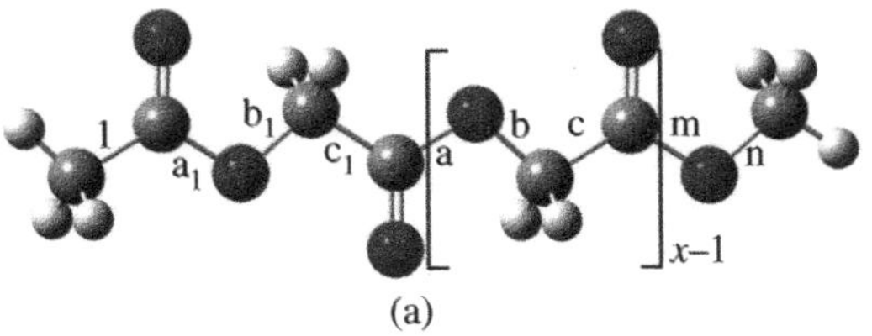

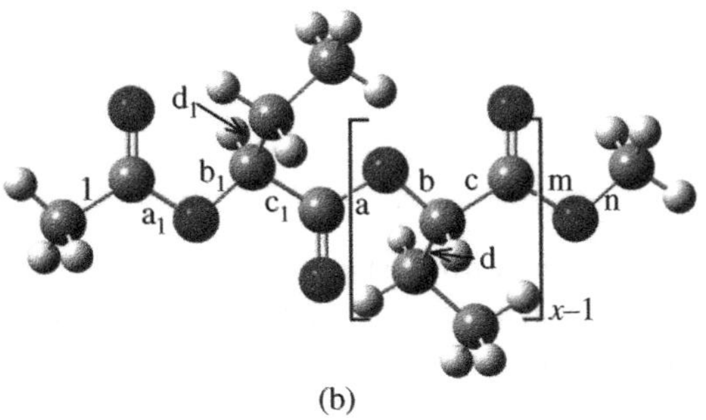

Figure 24.1 (a) Poly(glycolic acid) (PGA). (b) Poly(2-hydroxybutyrate) (P2HB). The bonds are designated as indicated, and *x* is the degree of polymerization. Source: [422], Figure 1 (p. 3731), Sasanuma et al./ Reproduced with permission of American Chemical Society

The equilibrium melting point (T_m^0) of P2HB is 130.3 °C [169], much lower than that (207 °C) of PLA [293, 510], but the stereocomplex of equimolar (*S*)- and (*R*)-forms of P2HB exhibits a much higher T_m^0 of 239 °C [493], and that of the stereo complex of PLA is still higher (279 °C) [492].

24.1.1 MO Calculation and NMR Experiment

In conformational analysis of PGA, as a monomeric model compound, methyl 2-acetoxyacetate (abbreviated as MAA: $CH_3C(=O)—O—CH_2=C(=O)—OCH_3$) was used. In preliminary molecular orbital (MO) calculations on MAA, the following facts were found: The C(=O)—O bond (bond a) may be considered to be fixed in the trans state because the other form, cis, has a very high energy (~7 kcal mol^{-1}). The O—CH_2 bond (bond b) can adopt trans, gauche$^+$, and gauche$^-$ conformations, and the CH_2—C(=O) bond (bond c) lies in either trans or synperiplanar (represented as s) state. To cover all rotational isomeric states (RISs, that is local potential minima) of MAA, the density functional theory (DFT) calculations at the B3LYP/6-311+G(2d,p) level were repeated with dihedral angles of bonds b (φ_b) and c (φ_c) being changed from −180° to 180° at intervals of 10°. At each (φ_b, φ_c) position, the structural optimization was carried out with both φ_b and φ_c being fixed. In Figure 24.2, the DFT energies thus obtained are plotted as a function of (φ_b, φ_c). In the energy map, the central high *mountain* around (0°, 0°) corresponds to strong repulsion between the C=O and -O- parts, and the still higher *mountains* around (0°, ±180°) represent close contact between two C=O groups. The energy minima can be

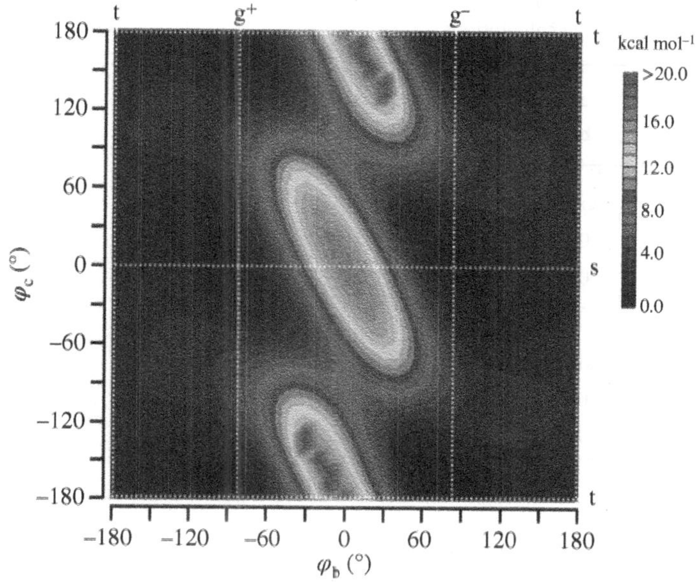

Figure 24.2 Electronic energy contour map of MAA as a function of dihedral angles of bonds b (φ_b) and c (φ_c). The electronic energies were calculated at the B3LYP/6-311+G(2d,p) level over −180° to 180° at intervals of 10° of φ_b and φ_c. Source: Reproduced with permission from Sasanuma et al. [422]/American Chemical Society.

found as bottoms of the *valleys* at intersections (g$^{\pm}$t and g$^{\pm}$s) of the horizontal and vertical dotted lines. The tt (±180°, ±180°) and ts (±180°, 0°) positions also show low energies but seem to be rather *saddles* (transition zone).

The energy map of MAA shows five energy minima: ttt, tg$^{\pm}$t, tg$^{\pm}$s. However, a dimeric model, 2-methoxy-2-oxoethyl 2-acetoxyacetate (MOAA, Figure 24.3), exhibits an additional energy minimum: tts. On these grounds, it has been assumed that PGA has six RISs in the repeating unit. The conformational energies of the six states relative to that of the most stable tg$^{\pm}$t states were derived from internal energies rather than Gibbs free energies. This is because the Gibbs free energy includes the entropy contribution due to the molecular vibrations, thus being affected by the bulky group(s) and polarization and occasionally depending on the chain length, namely whether the model is monomeric, dimeric, or trimeric. In contrast, the internal energy is almost free from such effects, and the internal energy relative to that of a given conformation is mostly close to the relative free energy [411]. The trans fractions of bonds b and c, calculated from the conformational energies for nonpolar gas phase and polar dimethyl sulfoxide (DMSO), are listed in Table 24.1

Table 24.1 Trans fractions of MAA at 25 °C, evaluated from NMR experiments and MO calculations.

	Bond	
Medium	**b**	**c**
	NMR exptl	
CD_3OD	0.01	
$(CD_3)_2SO$	0.01	
D_2O	0.08	
	MO calcd	
Gas	0.01	0.74
DMSO	0.03	0.73

MAA-^{13}C, MAA whose carbonyl carbon is labeled with carbon-13, was prepared and subjected to nuclear magnetic resonance (NMR) measurements. From vicinal coupling constants between the $^{13}C(=O)$ carbon and CH_2 protons, bond conformations around bond b were determined (Table 24.1); however, the NMR experiment yields no vicinal coupling constants regarding conformations of bond c. At least, as for bond b, the MO calculations reproduced the NMR results: bond b strongly prefers gauche conformations.

In the repeating unit, P2HB includes an asymmetrical methine carbon to which an ethyl group is bonded; therefore, there are two optical isomers: (*S*)- and (*R*)-forms. As monomeric and dimeric models of P2HB, methyl (*S*)-2-acetoxybutanoate ((*S*)-MAB) and (*S*)-1-methoxy-1-oxobutan-2-yl (*S*)-2-acetoxybutanoate ((*S*,*S*)-MOAB) have been adopted, respectively (Figure 24.3). (*S*)-MAB has four degrees of internal rotations: bonds a–c in the skeleton and bond d in the ethyl side chain. Bond a can be considered to be fixed in the trans state, while bonds b–d may have three RISs: (bond b) t, g^+, and g^-; (bond c) t, s, and g^-; and (bond d) t, g^+, and g^-.

In Table 24.2, bond conformations of (*S*)-MAB, evaluated from MO calculations, are compared with those determined from NMR experiments for (*S*)-MAB-^{13}C. In the NMR analysis for bond b, there are three unknowns and two equations; thus, the solution is indefinite. However, since the MO calculations indicate that the trans fraction (p_t) is nearly equal to zero, the g^+ and g^- fractions (p_{g^+} and p_{g^-}) were evaluated under the assumption of $p_t = 0$. Although the NMR experiment does not provide bond conformations of bond c, at least, for bonds b and d, the MO calculations reproduced the NMR data. Bonds b and c prefer gauche$^+$ and

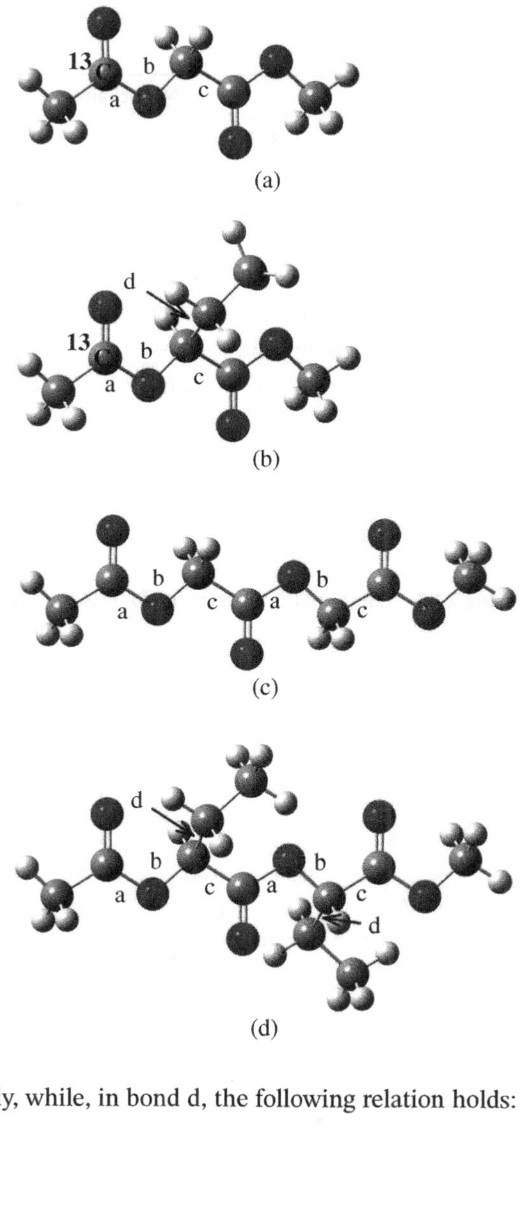

Figure 24.3 Model compounds used here: (a) methyl 2-acetoxyacetate (MAA and MAA-^{13}C), monomeric model for PGA; (b) methyl (*S*)-2-acetoxy-butanoate ((*S*)-MAB and (*S*)-MAB-^{13}C), monomeric model for P2HB; (c) 2-methoxy-2-oxoethyl 2-acetoxyacetate (MOAA), dimeric model for PGA; (d) (*S*)-1-methoxy-1-oxobutan-2-yl (*S*)-2-acetoxybutanoate ((*S*, *S*)-MOAB), dimeric model for P2HB. The bonds are designated as indicated, and the carbonyl carbon atoms of MAA-^{13}C and MAB-^{13}C were selectively labeled with carbon-13 as shown. Source: [422], Figure 2 (p. 3731) Sasanuma et al./Reproduced with permission of American Chemical Society.

trans conformations, respectively, while, in bond d, the following relation holds: $p_{g^+} \geq p_{g^-} \gg p_t$.

24.1.2 RIS Calculation

In RIS calculations on PGA and P2HB, the energy parameters listed in Tables 24.3 and 24.4 were used, respectively. For PGA, 9 × 9 statistical weight matrices including intramolecular interactions dependent on conformations of three bonds (a–c)

Table 24.2 Bond conformations of (*S*)-MAB at 25 °C, evaluated from NMR experiments and MO calculations.

	Main chain						Side chain		
	Bond b			Bond c			Bond d		
Medium	p_t	p_{g+}	p_{g-}	p_t	p_s	p_{g-}	p_t	p_{g+}	p_{g-}
	NMR exptl								
C_6D_{12}	(0.00)	0.94	0.06				0.15	0.45	0.40
$CDCl_3$	(0.00)	0.93	0.07				0.15	0.45	0.40
CD_3OD	(0.00)	0.95	0.05				0.16	0.44	0.40
$(CD_3)_2SO$	(0.00)	0.95	0.05				0.17	0.45	0.38
	MO calcd								
Gas	0.00	0.98	0.02	0.70	0.28	0.02	0.15	0.44	0.41
DMSO	0.00	0.98	0.02	0.71	0.28	0.01	0.20	0.45	0.35

Table 24.3 Conformational energies for PGA, derived from MOAA.[a]

bond[b]			gas		DMSO	
a	b	c	25 °C	m.p.[c]	25 °C	m.p.[c]
t	t	t	2.27	2.15	1.74	1.66
t	t	s	1.58	1.55	1.39	1.36
t	g±	t	0.00	0.00	0.00	0.00
t	g±	s	0.67	0.73	0.44	0.55

a) In kcal mol^{-1}.
b) Abbreviations: t, trans; g, gauche; s, synperiplanar.
c) At the equilibrium melting point, 231.4 °C.
Source: [422], Table 1 (p. 3732), Sasanuma et al./Reproduced with permission of American Chemical Society.

were formulated, while those of P2HB include statistical weights as a function of conformations around four bonds (a–d). The RIS scheme for P2HB is described in detail (Section 14.1). The configurational sequence of P2HB, namely the arrangement of (*S*)- and (*R*)-units, was built according to the Bernoulli trials for a given (*S*)-form ratio (P_S) or the Markov process for a given *meso*-diad probability (P_{meso}) as explained in Chapter 14.

Table 24.4 Conformational energies for (*S*)-form of P2HB.[a)]

Bond				ΔE_k (kcal mol^{-1})	
a	b	c	d	Gas	DMSO
t	t	s	(t)	2.25	2.38
t	t	g^-	(t)	1.99	2.68
t	g^+	t	(t)	0.00	0.00
t	g^+	t	(g^+)	−0.76	−0.55
t	g^+	t	(g^-)	−0.71	−0.47
t	g^+	s	(t)	0.77	0.58
t	g^+	s	(g^+)	0.12	0.03
t	g^+	s	(g^-)	0.23	0.22
t	g^+	g^-	(t)	2.68	3.10
t	g^-	t	(t)	2.88	3.13
t	g^-	t	(g^+)	2.03	2.24
t	g^-	g^-	(t)	1.97	2.44
t	g^-	g^-	(g^+)	1.41	1.70

a) In kcal mol^{-1}. The energies of the conformers with the g^+ state in bond b were derived from (*S*, *S*)-MOAB, and the other energies were obtained from (*S*)-MAB. This is because the structural optimization for (*S*, *S*)-MOAB always rendered the conformation of bond b gauche$^+$, even though bond b was initially set in the t or g^- conformation.
Source: [422], Table 2 (p. 3733), Sasanuma et al./Reproduced with permission of American Chemical Society.

Tables 24.5 and 24.6 present the results of the RIS calculations on PGA and P2HB, respectively. The characteristic ratios of PGA and P2HB in the gas phase (or DMSO) at 25 °C are 5.31 (5.14) and 6.44 (4.74), respectively. The P2HB chain shows larger environmental dependence. The configurational entropies (S_{conf}s) of PGA and P2HB in the gas phase at 25 °C (or T_m^0) are 2.74 (3.15) and 3.15 (3.67) cal K^{-1} mol^{-1}, respectively. The P2HB chain shows larger S_{conf} values than those of PGA. This is because P2HB has an additional internal rotation in the side chain and hence a larger degree of freedom in conformation. As a result, P2HB has a larger ΔS_u, which leads to the lower T_m^0 (130.3 °C) compared with that (231.4 °C) of PGA because of $T_m^0 = \Delta H_u / \Delta S_u$. The configurational properties of PGA, P2HB, and PLA will be compared and discussed in the following section. Inasmuch as the statistical weight matrix of bond c of P2HB also depends on conformations of

Table 24.5 Configurational properties of PGA, evaluated from RIS calculations.

	Energy parameter			
	Gas		DMSO	
	25.0 °C	231.4 °C[a)]	25.0 °C	231.4 °C[a)]
$\langle r^2 \rangle_0/nl^2$	5.31	5.34	5.14	5.27
$d \ln \langle r^2 \rangle_0/dT \times 10^3$ (K^{-1})	−0.082	−0.050	−0.042	−0.039
S_{conf} (cal K^{-1} mol^{-1})	2.74	3.15	2.96	3.28
U_{conf}[b)] (kcal mol^{-1})	−2.06	−1.76	−1.53	−1.30

a) The equilibrium melting point.
b) Relative to the crystal conformation, the all-trans form.
Source: Sasanuma et al. [422], Table 8 (p. 3736)/Reproduced with permission of American Chemical Society.

Table 24.6 Configurational properties of isotactic (*S*)-P2HB, evaluated from RIS calculations.

	Energy parameter			
	Gas		DMSO	
	25.0 °C	130.3 °C[a)]	25.0 °C	130.3 °C[a)]
$\langle r^2 \rangle_0/nl^2$	6.44	4.35	4.74	3.77
$d \ln \langle r^2 \rangle_0/dT \times 10^3$ (K^{-1})	−5.1	−2.7	−3.0	−1.6
S_{conf} (cal K^{-1} mol^{-1})	3.15	3.67	3.37	3.74
U_{conf}[b)] (kcal mol^{-1})	0.34	0.52	0.33	0.46

a) The equilibrium melting point.
b) Relative to the tg^+t (g^+) conformation.
Source: [422]/Table 9 (p. 3737), Sasanuma et al. /Reproduced with permission of American Chemical Society.

bond d, fractions of the conformational pairs could be evaluated as in Table 14.1 (Section 14.1).

In Figure 14.4a, the characteristic ratios of atactic P2HB chains generated by the Bernoulli trials are plotted against the (*S*)-form ratio (P_S). The curve is symmetrical with the central line of $P_S = 0.5$ and minimized there to be 2.17 (gas) or 2.38 (DMSO). The maxima are found at $P_S = 0$ and $P_S = 1$; the isotactic chains are most extended, and their characteristic ratios are given in Table 24.6. Figure 14.4b shows the $\langle r^2 \rangle_0/nl^2$ vs. P_{meso} plots, where the $\langle r^2 \rangle_0/nl^2$ ratio increases monotonously with P_{meso}: the isotactic ($P_{meso} = 1$) and syndiotactic ($P_{meso} = 0$, i.e. $P_{racemo} = 1$) chains exhibit the maximum and minimum, respectively. At $P_{meso} = 0$, the $\langle r^2 \rangle_0/nl^2$ value comes close to unity: 0.91 (gas) or 1.33 (DMSO). The reason for the unusually small $\langle r^2 \rangle_0/nl^2$ value can be explained as follows.

For the perfectly syndiotactic chain, the transition probability matrix can be written as

$$\mathrm{P} = \begin{array}{cc} i-1\backslash i & S \quad R \\ S & \\ R & \end{array} \begin{pmatrix} 0 & 1 \\ 1 & 0 \end{pmatrix} \tag{24.1}$$

The alternation of chiralities (...*SRSRSRSRSRSR*....) continues throughout the syndiotactic P2HB chain. In poly((*S*)-2HB -*alt*-(*R*)-2HB), dihedral angles of bonds a–c of the (*R*)-unit become opposite in sign to those of the (*S*)-unit, and consequently, the syndiotactic chain would be considerably contracted (almost crushed!).

24.1.3 Periodic DFT Calculation on PGA Crystal

As shown above, the most stable conformation of PGA is tgt, which is lower in energy by approximately 2 kcal mol^{-1} than ttt, but nevertheless the PGA chain crystallizes in the all-trans form. In the periodic DFT calculations, the crystal structure determined by X-ray diffraction [73] was initially set to be optimized at the B3LYP-D/6-31G(d,p) level. The optimized crystal structure is compared with the experiment in Table 24.7 and illustrated in Figure 24.4. The difference between theory and experiment is slight: $\Delta_{LC} = 0.0551\%$; $\Delta_{CHO} = 1.45 \times 10^{-4}$. The analysis by the crystallographic software PLATON [442] detected no classical hydrogen bond. However, an interchain C—H···O close contact (C···O distance of 3.459 Å) between the CH_2 group and —O— atom and an electrostatic attraction (C···O distance of 3.038 Å) between the carbonyl $C^{\delta+}$ and $= O^{\delta-}$ atoms are suggested to exist.

The Born effective charges [170, 304] were calculated to be as follows (in the unit of e): $\underline{C}H_2$, +0.422; $C\underline{H_2}$, +0.088; $\underline{C}$=O, +1.626; C=$\underline{O}$, −1.029; and —$\underline{O}$—, −1.195. From the effective charges, the dipole moment ($\vec{\mu}_{mono}$) per monomeric unit, —O—CH_2C(=O)—, was calculated from

$$\vec{\mu}_{mono} = \sum_{\text{monomeric unit}} q_m \vec{r}_m \tag{24.2}$$

where q_m and $\vec{r}_m$ are the electrical charge and position of atom m, respectively. The dipole moment was obtained as $\vec{\mu}_{mono} = (\mu_a, \mu_b, \mu_c) = (\pm 8.180, 0.000, \pm 10.630)$ in the unit of Debye (D). Here, μ_a, μ_b, and μ_c are the components along the orthogonal a, b, and c axes; therefore, $\mu_{mono} = |\vec{\mu}_{mono}| = 13.413$ D, and the $\vec{\mu}_{mono}$ vectors make angles of ±52.42° and ±127.58° with the a-axis (see Figure 24.4). The large dipole moments stem from the maldistributed electron density but counterbalance each other; therefore, both electronic charges and dipole moments are summed up over the unit cell to be null.

The interchain interaction energy (ΔE^{CP}) in the PGA crystal was evaluated with the BSSE corrected by the counterpoise (CP) method as described in Section 10.8

Table 24.7 Comparison between optimized and experimental crystal structures of PGA.[a)]

	Optimized			**Experimental**[b)]		
	Lattice constant (Å)					
	a	b	c	a	b	c
	5.154	5.951	7.032	5.22	6.19	7.02
	$\Delta_{LC} = 0.0551^{c)}$ (%)					
	Fractional coordinates					
	x/a	y/b	z/c	x/a	y/b	z/c
C	0.351	0.250	0.188	0.348	0.250	0.193
C	0.337	0.250	0.523	0.326	0.250	0.521
O	0.201	0.250	0.359	0.194	0.250	0.361
O	0.571	0.250	0.532	0.557	0.250	0.535
H	0.473	0.399	0.181	0.464	0.395	0.195
	$\Delta_{CHO} = 1.45 \times 10^{-4 d)}$					
	Geometrical parameters (Å, °)					
C(O)—O	1.345			1.32		
O—CH_2	1.433			1.43		
CH_2—C(O)	1.510			1.51		
C=O	1.211			1.21		
C—H	1.091					
∠C(O)—O—CH_2	115.97			114		
∠O—CH_2—C(O)	107.63			109		
∠CH_2—C(O)—O	108.84			111.5		

a) An orthorhombic cell of space group P_{cmn}.
b) Chatani et al. [73].
c) $\Delta_{LC} = \frac{1}{3}\left[\left(\frac{a_{calc}-a_{expt}}{a_{expt}}\right)^2 + \left(\frac{b_{calc}-b_{expt}}{b_{expt}}\right)^2 + \left(\frac{c_{calc}-c_{expt}}{c_{expt}}\right)^2\right]^{1/2} \times 100\ (\%)$
d) $\Delta_{CHO} = \frac{1}{N_{atom}}\sum_{atom}\left\{\left[\left(\frac{x}{a}\right)_{calc} - \left(\frac{x}{a}\right)_{expt}\right]^2 + \left[\left(\frac{y}{b}\right)_{calc} - \left(\frac{y}{b}\right)_{expt}\right]^2 + \left[\left(\frac{z}{c}\right)_{calc} - \left(\frac{z}{c}\right)_{expt}\right]^2\right\}^{1/2}$
where N_{atom} is the number of atoms in the asymmetric unit.
Source: [422]/Table 11 (p. 3739), Sasanuma et al./Reproduced with permission of American Chemical Society.

to be −15.47 kcal (mole of monomer)$^{-1}$, which can be converted to −5.16 kcal (mole of bond)$^{-1}$ or −267 cal g^{-1}. The ΔE^{CP} values of some representative polymers are listed in Table 10.3. Of them, the ΔE^{CP} value (in kcal [mole of bond]$^{-1}$) of PGA is outstandingly large, 1.44, 1.70, 3.25, and 3.05 times as much as those of the α forms of nylon 4 and nylon 6 and 9/5 and 2/1 helices of PMO, respectively. The nylon chains are jointed by N—H···O=C hydrogen bonds,

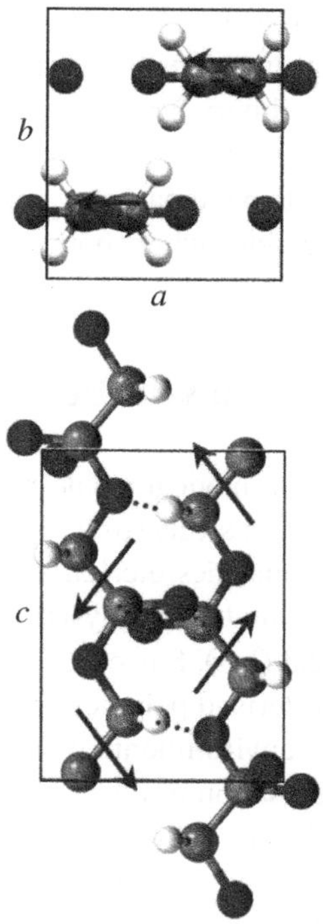

Figure 24.4 Crystal structure of PGA, optimized by the periodic DFT-D calculations at the B3LYP-D/6-31G(d,p) level with the dispersion force correction. The arrows schematically illustrate dipole moments per monomeric unit. The dotted lines express the interchain close contacts of $C^{+0.422}-H^{+0.088}\cdots O^{-1.195}$ and $C{=}O^{-1.029}\cdots C^{+1.626}({=}O)$, where the superscripts represent the Born effective charges. Source: [422], Figure 8 (p. 3740), Sasanuma et al./Reproduced with permission of American Chemical Society.

whereas the PMO helices are packed by weak dispersion forces. As elucidated above, the dipole–dipole and electrostatic interactions of PGA are so strong as to cause the high T_m^0 of 231.4 °C.

The stiffness (C) and compliance (S) tensors of the PGA crystal at 0 K were calculated at the B3LYP-D/6-31G(d,p) level to be

$$C = \begin{bmatrix} 30.367 & 11.101 & 9.268 & 0.000 & 0.000 & 0.000 \\ & 33.372 & 3.003 & 0.000 & 0.000 & 0.000 \\ & & 453.815 & 0.000 & 0.000 & 0.000 \\ & & & 5.810 & 0.000 & 0.000 \\ & & & & 11.267 & 0.000 \\ & & & & & 7.978 \end{bmatrix} (\mathrm{GPa}) \tag{24.3}$$

and

$$S = \begin{bmatrix} 37.7018 & -12.4790 & -0.6874 & 0.0000 & 0.0000 & 0.0000 \\ & 34.1134 & 0.0291 & 0.0000 & 0.0000 & 0.0000 \\ & & 2.2174 & 0.0000 & 0.0000 & 0.0000 \\ & & & 172.1173 & 0.0000 & 0.0000 \\ & & & & 88.7558 & 0.0000 \\ & & & & & 125.3416 \end{bmatrix} (\mathrm{TPa}^{-1}) \tag{24.4}$$

From the compliance tensor, Young's moduli along the crystal axes were evaluated as in Table 10.2. The E_c value (451 GPa) in the *c*-axis (chain-axis) direction is extraordinary large, and the E_a and E_b values are 26.5 and 29.3 GPa, respectively. Compared with the other all-trans polymers such as PE, nylon 4, and nylon 6, PGA seems to be remarkably stiff. This is the reason why PGA has been used mainly in applications where both biodegradability and outstandingly rigidity are required.

24.2 Poly(lactic acid) (Poly(lactide), PLA)

PLA is a biodegradable polyester that, of all biodegradable polymers, has been most extensively studied and early put to practical use as fibers, nonwoven fabric, thermoplastics, etc. [28]. Starch extracted from corn, potato, or sugar cane undergo an enzymatic hydrolysis to be glucose, which is fermented by lactobacilli to be lactic acid (LA). Since the LA thus produced is the raw material of PLA, PLA is a plant-derived polymer fulfilling carbon neutrality.

The environmentally friendly PLA has been manufactured either by direct polycondensation of LA in azeotropic solutions or by ring-opening polymerization of lactide. LA includes an asymmetric carbon atom, thus being assigned to either (*S*)- or (*R*)-isomer. The lactate fermentation yields (*S*)-LA exclusively, and the chirality will be kept in the polymerization. Hence, poly((*S*)-lactic acid) aka poly(L-lactic acid) is common, and poly((*R*)-lactic acid), i.e. poly(D-lactic acid), is produced by chemical synthesis.

Lactide includes two asymmetric carbon atoms: L-lactide has two *S* chiral centers, and D-lactide includes two *R* ones (see Figure 24.5). Racemates of L- and D-lactides are often designated as *rac*-lactide. If *x* is defined as the degree of polymerization in lactide unit, 2*x* corresponds to the number of LA repeating units obtained from the ring-opening polymerization of lactide (see Figure 24.6), Here, in principle, the lactide unit is considered to be the monomer, that is the repeating unit. The DD and LL diads of lactide are designated as *iso*, and DL and LD as *syn*. The *iso* and *syn* diads comprise *SSSS* or *RRRR* and *RRSS* or *SSRR* LA tetramers, respectively.

Polymerization of L-lactide yields isotactic poly(L-lactide), which melts around 170–180 °C (melting point, T_m). Equimolar mixtures of poly(L-lactide)

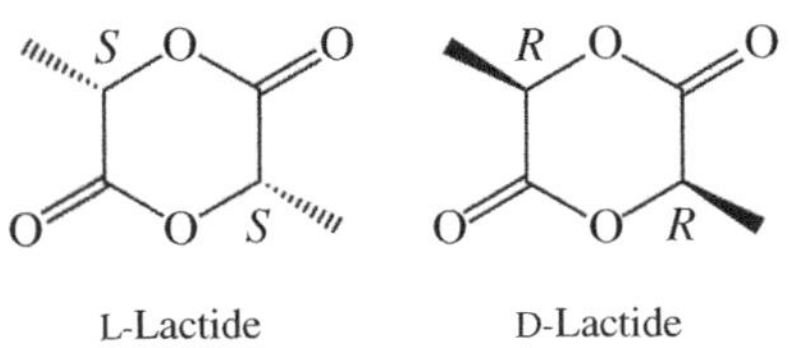

Figure 24.5 L-Lactide and D-lactide. Source: [418]/Figure 1 (p. 1902), Sasanuma and Touge/Reproduced with permission of Elsevier.

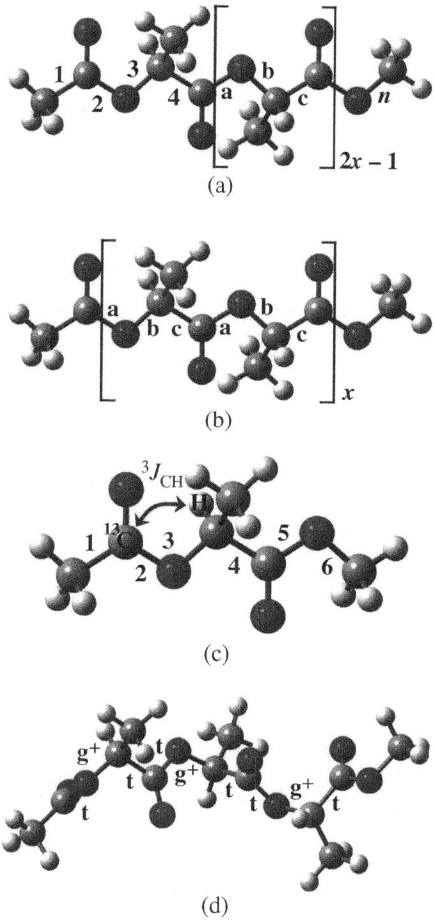

Figure 24.6 (a) Poly((*S*)-lactic acid), (b) poly(L-lactide), (c) (*S*)-methyl 2-acetoxypropanoate (designated as monomer), and (d) 1-methoxy-1-oxopropan-2-yl 2-((2-acetoxypropanoyl)-oxy)propanoate (timer) depicted in the $tg^+ttg^+ttg^+t$ conformation of the *SSS* form. The bonds are numbered or designated as indicated, and *x* is the degree of polymerization in lactide unit. $^3J_{CH}$ is the vicinal coupling constant between the ^{13}C-labeled carbonyl carbon and methine proton of monomer–^{13}C. Source: [418]/ Figure 2 (p. 1902), Sasanuma and Touge/Reproduced with permission of Elsevier.

and poly(D-lactide) crystallize to form stereocomplexes of $T_m = 220 - 230\,°C$ [495]. *rac*-Lactide is polymerized to form different stereoregularities, which depend on catalyst. For example, tin (II) octoate [473] and lithium *tert*-butoxide [234] yield *syn*-rich poly(DL-lactide)s. Single-site metal alkoxide catalysts give poly(DL-lactide)s with different stereoregularities [377, 441]. A racemate of chiral aluminum alkoxides results in *iso*-rich stereoblock poly(DL-lactide) [349, 350].

24.2.1 MO Calculation and NMR Experiment

Table 24.8 shows Gibbs free energies of (*S*)-methyl 2-acetoxypropanoate (abbreviated as monomer) calculated at the MP2/6-311+G(2d,p) and the B3LYP/6-311+G(2d,p) levels with molecular geometries optimized at the B3LYP/6-311+G(2d,p)

Table 24.8 Conformational free energies of monomer.

					ΔG_k[a)] (kcal mol^{-1})			
					MP2[b)]		B3LYP[c)]	
k	Conformation[d)]				Gas	Chloroform	Gas	Chloroform
1	t	t	g$^+$	c	12.26	10.88	11.66	10.35
2	t	t	g$^-$	c	9.32	8.35	9.28	8.19
3	t	g$^+$	t	t	0.00	0.00	0.00	0.00
4	t	g$^+$	t	c	8.18	7.14	8.45	7.14
5	t	g$^+$	g$^+$	t	0.53	0.49	0.56	0.52
6	t	g$^+$	g$^+$	c	11.95	10.20	11.50	9.67
7	t	g$^-$	t	t	2.98	2.92	3.53	3.59
8	t	g$^-$	t	c	15.46	13.80	15.26	13.41
9	t	g$^-$	g$^+$	c	18.55	17.53	18.30	17.44
10	t	g$^-$	g$^-$	t	2.15	2.12	2.86	2.99
11	t	g$^-$	g$^-$	c	12.07	10.11	13.17	11.18
12	c	t	t	c	22.78	18.57	22.12	17.07
13	c	t	g$^-$	c	18.63	16.05	18.76	15.76
14	c	g$^+$	t	t	8.53	6.71	8.58	6.56
15	c	g$^+$	t	c	18.69	15.30	19.05	15.08
16	c	g$^+$	g$^+$	t	8.54	6.89	8.68	6.89
17	c	g$^+$	g$^+$	c	17.70	15.94	17.48	15.39
18	c	g$^-$	t	t	14.11	12.22	13.57	11.87
19	c	g$^-$	t	c	23.37	20.17	22.77	19.36
20	c	g$^-$	g$^-$	t	12.04	10.18	12.53	10.90
21	c	g$^-$	g$^-$	c	19.35	17.32	19.88	17.85

a) Relative to the most stable tg$^+$tt conformer.
b) At the MP2/6-311+G(2d,p)//B3LYP/6-311+G(2d,p) level.
c) At the B3LYP/6-311+G(2d,p)//B3LYP/6-311+G(2d,p) level.
d) Conformations of bonds 2–5. Although 36 ($= 2 \times 3 \times 3 \times 2$) conformers may be enumerated, the geometrical optimization yielded the above 21 ones. The rotational isomeric states of bonds 4 and c are termed t g$^+$, and g$^-$ here; however, because the dihedral angle (163.6°) of the tg$^+$g$^+$t state falls within the synperiplanar (abbreviated s) range (150–210°), the conformation would rather be expressed as tg$^+$st.

Source: [418], Table 2 (p. 1903), Sasanuma and Touge/Reproduced with permission of Elsevier.

level. The Gibbs free energy is represented as the difference from that of the tg^+tt conformer. Under the RIS approximation, the C(=O)—O bond is assigned to either the trans or cis conformation; however, the latter has an energy (6.5–8.6 kcal mol^{-1} in the gas phase) higher than that of the former, and hence, the bond is essentially fixed in the trans state. Of the optimized conformers (Table 24.8), only tg^+tt (0.0 kcal mol^{-1}) and tg^+st (see footnote d of Table 24.8, approximately 0.5 kcal mol^{-1}) have extraordinarily low energies, tg^-tt ($\geq$3 kcal mol^{-1}) and tg^-g^-t (2–3 kcal mol^{-1}) have moderate energies, and the others have much higher energies and hence rarely occur. Therefore, the two conformers, tg^+tt and tg^+st, are dominant over the others. Table 24.9 shows the conformational fractions of bonds 3 and 4 calculated from the free energies.

^1H and ^{13}C NMR experiments for monomer-^{13}C, one of whose carbonyl carbon atoms is labeled with carbon-13, yielded vicinal ^{13}C–^1H coupling constants ($^3J_{CH}$ of Figure 24.6), from which conformational fractions of bond 3 were derived

Table 24.9 Bond conformations of monomer and poly(L-lactide) at 25 °C.

Medium, method, or configuration	Bond 3 or b			Bond 4 or c		
	p_t	p_{g+}	p_{g-}	p_t	p_{g+}(p_s)	p_{g-}
	NMR experiment on monomer[a)]					
C_6D_{12}	(0.00)	0.95	0.05			
$CDCl_3$	(0.00)	0.95	0.05			
CD_3OD	(0.00)	0.97	0.03			
$(CD_3)_2SO$	(0.00)	0.97	0.03			
	MO calculation on monomer[b)]					
MP2 (gas)	0.00	0.98	0.02	0.70	0.28	0.02
MP2 (chloroform)	0.00	0.98	0.02	0.68	0.30	0.02
B3LYP (gas)	0.00	0.99	0.01	0.71	0.28	0.01
B3LYP (chloroform)	0.00	0.99	0.01	0.71	0.29	0.00
	RIS calculation on poly(lactide)[c)]					
SSS	0.00	1.00	0.00	0.81	0.19	0.00
RRS	0.17	0.81	0.02	0.57	0.42	0.01
RSS	0.00	1.00	0.00	0.93	0.07	0.00

a) The trans fraction (p_t) was assumed to be null.
b) For bonds 2 and 5, the MO calculations yielded p_{trans} = 1.00 and p_{cis} = 0.00.
c) The bond conformations of the third (*S*)-LA unit. In bond a, p_{trans} = 1.00 and p_{cis} = 0.00. From the ΔG_k values at the B3LYP (gas) level.

Source: [418]/Table 3 (p. 1903), Sasanuma and Touge/Reproduced with permission of Elsevier.

under the assumption of $p_t = 0$. This supposition is based on the MO calculations. The origin of $p_t = 0$ is a close contact between the C(=O) oxygen and adjoining —O— atoms; the electronegative oxygen atoms repel each other and force bond 3 to turn to the gauche$^+$ conformation. In addition, the gauche$^+$ conformation allows the pendent methyl group to escape from steric repulsion with the surroundings. The gauche$^+$ and gauche$^-$ fractions of bond 3 are in good agreement with those obtained from the MO calculations. Inasmuch as there is no NMR vicinal coupling relating to conformations of bond 4, the bond conformations were not determined by NMR.

Gibbs free energies of *SSS*, *RRS*, and *RSS* LA triads were calculated from an LA trimer (Figure 24.6). Table 24.10 shows the conformer free energies of the third (*S*)-LA units of the three triads. From the free energies, those of the (*R*)-LA units of their mirror images can be derived by a symmetric operation: *SSS* → *RRR*; *RRS* → *SSR*; *RSS* → *SRR*. Here, the ring-opening polymerization of L-lactides

Table 24.10 Conformer free energies (ΔG_ks) of trimers, evaluated by ab initio MO calculations.

				ΔG_k (kcal mol^{-1})			
				MP2[a]		B3LYP[b]	
Configuration	Conformation[c]			Gas	Chloroform	Gas	Chloroform
SSS	t	g$^+$	t	0.00	0.00	0.00	0.00
	t	g$^+$	g$^+$(s)	0.67	0.53	0.84	0.66
	t	g$^-$	t	4.72	4.32	4.62	4.14
	t	g$^-$	g$^-$	2.91	2.89	3.73	3.68
RRS	t	t	t	0.77	0.73	0.48	0.45
	t	g$^+$	t	0.00	0.00	0.00	0.00
	t	g$^+$	g$^+$(s)	−0.73	−0.39	−0.05	0.33
	t	g$^-$	t	1.22	1.69	2.45	3.01
	t	g$^-$	g$^-$	0.62	1.04	2.13	2.63
RSS	t	g$^+$	t	0.00	0.00	0.00	0.00
	t	g$^+$	g$^+$(s)	1.51	1.32	1.54	1.29
	t	g$^-$	t	4.36	3.93	4.51	3.98
	t	g$^-$	g$^-$	2.42	2.55	3.63	3.74

a) At the MP2/6-311+G(2d,p) level and 25 °C.
b) At the B3LYP/6-311+G(2d,p) level and 25 °C.
c) In bonds a, b, and c of the third (*S*)-LA unit. The first and second units were set in the stable tg$^+$t ((*S*)-form) or tg$^-$t ((*R*)-form) conformation.

Source: [418]/Table 5 (p. 1904), Sasanuma and Touge/Reproduced with permission of Elsevier.

yields only poly(L-lactide), that is poly(*S*-lactic acid). From a racemic mixture of D- and L-lactide, poly(DL-lactide) including LDL and DLD lactide triads will be generated. The former corresponds to the *SSRRSS* LA hexad, which includes *SSR*, *SRR*, *RRS*, and *RSS* LA triads, and the latter comprises the *RRSSRR* LA hexad, which can be divided into *RRS*, *RSS*, *SSR*, and *SRR* LA triads. The above six LA triads cover all formed in poly(L-lactide), poly(D-lactide), and poly(DL-lactide) because *SRS* and *RSR* triads never appear in poly(DL-lactide).

24.2.2 RIS Calculation

In the refined RIS calculations on PLAs, the conformational energies prepared as above were used, together with the geometrical parameters chosen from the trimer model. The stereosequences were generated for the lactide unit (L or D) by either the Bernoulli trial or the Markov chain. The process of the stereosequence generation is described in Chapter 14. Both degree (x) of polymerization and the number (n_c) of chain were set equal to 300, because the random-number generation under these conditions was proved to fully reproduce the intended L-unit / D-unit and *iso*-diad / *syn*-diad ratios.

A number of experimental studies have reported unperturbed chain dimensions of PLA of different L-unit fractions (f_Ls). The experimental methods are light scattering (LS), viscosity measurement (η), Raman scattering, and small-angle neutron scattering (SANS). In the RIS calculations, the f_L value and temperature were set equal to the experimental values. The computational results are shown in Table 24.11. In the case where the RIS calculation reasonably or satisfactorily reproduced the experimental $\langle r^2 \rangle_0/nl^2$ value, the theoretical value of Table 24.11 is boldfaced. The experimental data are rather scattered, but the characteristic ratios of the fully isotactic PLA chains seem to stay within a range from 9 to 11. The RIS calculations at the MP2 and B3LYP levels for the gas phase (in chloroform) yielded the $\langle r^2 \rangle_0/nl^2$ values of 9.01 (7.26) and 11.84 (8.94), respectively, being fairly consistent with the experimental data.

The configurational properties of poly(DL-lactide) were calculated with the B3LYP (gas) energies that reproduced the experimental $\langle r^2 \rangle_0/nl^2$ data better than the MP2 ones. Figure 24.7 shows the characteristic ratio of atactic PLA as a function of f_L. The $\langle r^2 \rangle_0/nl^2$ vs. f_L curve is symmetrical with respect to the $f_L = 0.5$ line, minimized there to be 6.22, and maximized to be 11.84 at $f_L = 0.0$ and 1.0. The characteristic ratios of the Markov PLA chains are plotted against P_{LL} and P_{DD} in Figure 24.8. The contour map exhibits the minimum of 3.89 at $P_{LL} = P_{DD} = 0.0$ (syndiotactic) and the maximum of 11.84 at $P_{LL} = P_{DD} = 1.0$ (isotactic). The tacticity dependence of $\langle r^2 \rangle_0/nl^2$ is similar to that of P2HB.

When a common catalyst for ring-opening polymerization of lactides, lithium tert-butoxide, is used, syndiotactic-rich (76%) PLA is produced [234].

Table 24.11 Experimental characteristic ratios of unperturbed poly(lactide) chains, compared with the RIS calculations.

Experiment						$\langle r^2 \rangle_0 / nl^2$ by RIS calculation[a)]			
						MP2		B3LYP	
Author[b)]	L%[c)]	Method[d)]	Temp (°C)	Medium[d)]	$\langle r^2 \rangle_0 / nl^2$[e)]	Gas	Chloroform	Gas	Chloroform
Tonelli–Flory	100	LS, η	85	C_6H_5Br	4.4±0.4		(6.35)[f)]		(9.46)[f)]
Joziasse et al.	100	LS, η	25	CH_3CN	11.2	9.01	7.26	**11.84**	8.94
	95				11.2	8.13	6.77	**10.58**	8.49
	85				10.5	6.68	5.96	8.69	7.77
	80				10.5	6.10	5.64	8.00	7.48
	71				9.0	5.27	5.20	7.06	7.12
	54				9.2	4.53	4.77	6.28	6.77
Yang et al.	100	Raman		(bulk)	7–12	**9.01**	**7.26**	**11.84**	**8.94**
Kang et al.	100	LS, η	25	CH_3CN/ CH_2Cl_2	11.8	9.01	7.26	**11.84**	8.94
	98.7				11.0	8.77	7.13	**11.48**	8.82
	98.2				9.5	**8.67**	7.08	11.38	**8.76**
	94.3				9.3	8.04	6.72	10.42	**8.42**
	87.0				9.0	6.95	6.11	**9.02**	7.88
	80.0				8.1	6.10	5.63	**7.99**	**7.48**
Anderson and Hillmyer	100	SANS	200	(melt)	7.85 ± 0.58	6.23	5.53	**7.25**	6.17
	72				6.39 ± 0.30	5.23	5.24	**6.46**	**6.42**
	72		30	(amorphous)	8.06 ± 0.33	5.32	5.21	**7.13**	**7.09**
Dorgan et al.	100	LS, η	30	CH_3Cl	6.63 ± 0.66	8.85	**7.16**	11.57	8.79
	50	Rheology	140	(melt)	6.6 ± 0.7	4.98	5.22	**6.52**	**6.76**
Saito et al.	100	LS, η	25	CH_3Cl, HFIP	13.8 ± 0.9	9.01	7.26	**11.84**	8.94

a) Calculated under the refined RIS scheme with the Bernoulli process at the experimental L% and temperature. The boldfaced values fall within ±10% margins with respect to the experimental data.
b) References: Tonelli–Flory [486]; Joziasse et al. [225]; Yang et al. [537]; Kang et al. [230]; Anderson and Hillmyer [24]; Dorgan et al. [118, 119]; Saito et al. [393]
c) L-Lactide ((*S*)-lactic acid) content.
d) Abbreviations: LS, light scattering; η, viscosity measurement; Raman, Raman scattering; SANS, small-angle neutron scattering; rheology, melt rheology (plateau modulus); HFIP, hexafluoroisopropanol.
e) The original values reported in the literature were modified with the bond lengths shown in Table 24.12.
f) Calculated with ΔG_ks of the trimers in C_6H_5Br at 85 °C.

Source: [418]/Table 1 (p. 1903), Sasanuma and Touge /Reproduced with permission of Elsevier.

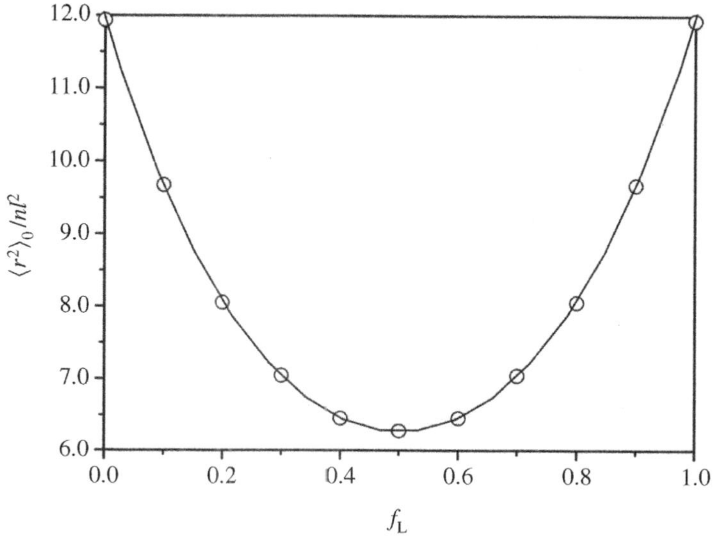

Figure 24.7 Characteristic ratio ($\langle r^2 \rangle_0/nl^2$) of the Bernoulli chain as a function of f_L. The number (n_c) of chains was set equal to 300. The open circles represent the calculated data, and the solid line is the fitted curve. Source: [418], Figure 7 (p. 1907), Sasanuma and Touge/Reproduced with permission of Elsevier.

Figure 24.8 Contour map of the characteristic ratio of the Markov chain as a function of P_{LL} and P_{DD}. The dotted line corresponds to $P_{DD} = P_{LL}$. Source: [418]/Figure 8 (p. 1908), Sasanuma and Touge/Reproduced with permission of Elsevier.

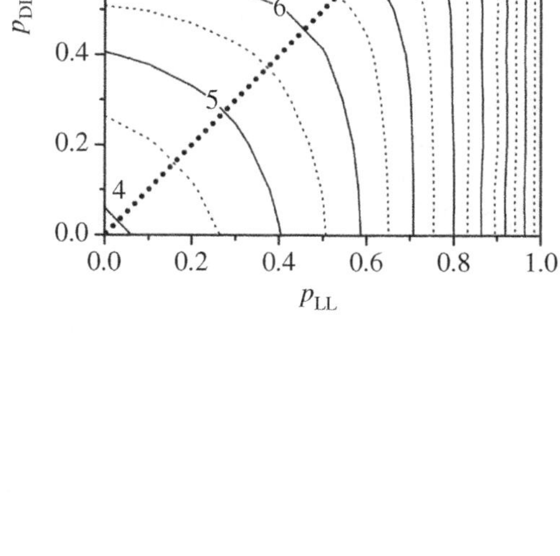

Table 24.12 Configurational properties and thermodynamic quantities of poly(lactide)s, evaluated from refined RIS calculations with the Markov process.

		Poly(L-lactide)		Poly(DL-lactide)			
		25 °C	207 °C[a)]	8.7 syn%[b)]	50 syn%	76 syn%[c)]	100 syn%
RIS calcd[d)]	$\langle r^2 \rangle_0 / nl^2$	11.84	7.16	10.35	6.22	4.83	3.89
	$d \ln \langle r^2 \rangle / dT \times 10^3$ (K^{-1})	−4.6	−1.8	−3.6	−0.45	0.57	1.1
	S_{conf}[e)] (cal K^{-1} mol^{-1})	2.02	2.68	2.09	2.41	2.60	2.75
	L-lactide content (%)	100	100	49.6	50.2	50.1	50.0
	n_{block}[f)]	∞	∞	11.1	1.99	1.32	1.00
Exptl	ΔS_u[g)] (cal K^{-1} mol^{-1})		5.83				
	ΔH_u[h)] (kcal mol^{-1})		2.80				
	$d \ln \langle r^2 \rangle_0 / dT \times 10^3$[i)] ($K^{-1}$)		−1.4				

a) The equilibrium melting point [293].
b) The stereoblock copolymer of $n_{block} \approx 11$, synthesized from *rac*-lactide with a racemic aluminum alkoxide catalyst [349, 350].
c) Synthesized from *rac*-lactide with lithium *tert*-butoxide [234].
d) With the Gibbs free energies at the B3LYP/6-311+G(2d,p) level. The averaged bond lengths (in Å) (angles in °) of bonds a–c are 1.360 (115.8), 1.436 (110.2), and 1.532 (110.5), respectively.
e) Per lactide unit.
f) The average number of LA units included in a stereoblock.
g) The entropy of fusion [293].
h) The enthalpy of fusion [293].
i) Calculated from small-angle neutron scattering data [24].
Source: Sasanuma and Touge [418], Table 6 (p. 1904)/Reproduced with permission of Elsevier.

The corresponding computational data are given in Table 24.12. It was reported that a racemic mixture of (*R*)- and (*S*)-aluminum alkoxides, used as a catalyst, yields blocks of approximately 11 successive LA units of the same chirality: –*SSSSSSSSSSS*– or –*RRRRRRRRRRR*– [349, 350]. The RIS simulation with $P_{syn} = P_{LD} = P_{DL} = 0.087$ reproduced 11 successive LA units on average. The RIS results for $P_{syn} = 0.087$, also being shown in Table 24.12, are almost the same as those of the purely isotropic PLA.

The entropy of fusion of isotactic PLA at the equilibrium melting point ($T_m^0 = 207\,°C$) was determined to be 5.83 cal K^{-1} mol^{-1}, and the RIS calculations gave $S_{conf} = 2.68$ cal K^{-1} mol^{-1} at T_m^0; therefore, the $S_{conf}/\Delta S_u$ ratio is 0.46, very small as compared with those of the other polymers [421]. The difference between ΔS_u and S_{conf} is the entropy of volume change (ΔS_v) as described in Chapter 11.

The above ΔS_u and S_{conf} values are per lactide unit. Therefore, those per LA unit are the halves. The S_{conf} values per LA unit at 25 °C are 1.01 (B3LYP) and 1.17 (MP2) cal K^{-1} mol^{-1}, being much smaller than those of PGA (2.74 cal K^{-1} mol^{-1})

and P2HB (3.15 cal K^{-1} mol^{-1}), although the three polymers have the same atomic arrangement in the backbone. The small S_{conf} of PLA is due to the restricted conformational freedom; the two conformations, tg^+t and tg^+s, are overwhelmingly dominant, whereas the symmetric PGA chain adopts tg^-t and tg^-s as well as tg^+t and tg^+s and hence shows the large S_{conf} value but nevertheless shows the high T_m^0 of 231.4 °C owing to the interchain dipole–dipole and electrostatic interactions. The P2HB chain has an additional internal rotation in the ethyl side group, which renders its S_{conf} larger and its T_m^0 lower (130.3 °C). The difference in side group, namely whether H, CH_3, or C_2H_5, results in the significant changes in the thermal properties.

24.3 Poly((*R*)-3-hydroxybutyrate) (P3HB)

Poly((*R*)-3-hydroxybutyrate) (P3HB) is a biosynthetic and biodegradable polyester that might be evolutionarily generated to be an intracellular carbon- and energy-reserve substance for microorganisms [115]. In the 1920s, when Staudinger had studied polymerization [443], a French microbiologist, Lemoigne, discovered and analyzed P3HB to detect 3-hydroxybutyrate and α-crotonic acid as its decomposed compounds [277, 278]. The glass transition temperature (T_g) of P3HB changes widely from −43 to 41 °C, depending on the environment where bacteria live, and the mode value is 4 °C [347]. It has been suggested that P3HB behaves as an amorphous elastomer in vivo above T_g, and that water would play the role of the plasticizer. Crystallization of P3HB will start by the removal of lipid components from its native granules [23, 35, 69, 236]. The repeating unit of P3HB is simple and composed of a methine, a methylene, and a methyl, and an ester groups, but nevertheless, the equilibrium melting point is moderately high, 203 °C [293]. Since P3HB was early discovered and has been extensively studied from a variety of viewpoints, a wealth of chemical information on P3HB is available. Here, conformational characteristics of P3HB are described and related to the experimental data reported thus far.

24.3.1 NMR Experiment

As a monomeric model compound, (*R*)-ethyl-3-acetoxybutanoate (termed "monomer") was adopted, and monomer-^{13}C (Figure 24.9b) was prepared and subjected to ^{1}H and ^{13}C NMR measurements. Bond conformations around bond 3 were determined from the vicinal coupling constant, $^3J_{XA}$, between the carbonyl carbon (X) and methine proton (A), and those around bond 4 were evaluated from $^3J_{AB}$ and $^3J_{AC}$. The $^3J_{XA}$ value is related to the bond conformations as

$$^3J_{XA} = J_G p_t + J'_T p_{g^+} + J'_G p_{g^-} \tag{24.5}$$

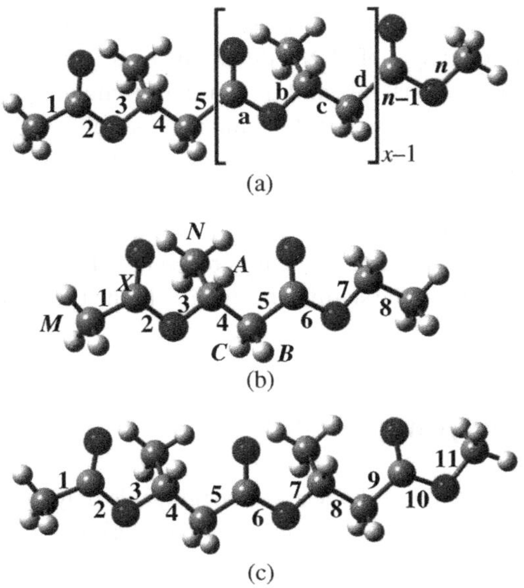

Figure 24.9 All-*trans* conformations of (a) poly((*R*)-3-hydroxybutyrate) (P3HB), (b) (*R*)-ethyl 3-acetoxybutanoate (termed "monomer" here), and (c) (*R*)-(*R*)-4-methoxy-4-oxobutan-2-yl 3-acetoxybutanoate (dimer). The skeletal bonds are labeled as indicated. The atoms of monomer are partly designated as A, B, C, M, N (^{1}H), and X (^{13}C) to represent the spin system ($ABCM_3N_3X$) according to the convention. Source: Material from reference [407], published 2013, Springer Nature.

By definition it follows

$$p_t + p_{g^+} + p_{g^-} = 1 \tag{24.6}$$

The two equations have three unknowns (p_t, p_{g^+}, and p_{g^-}), thus being indeterminate. However, under the assumption of $J_G = J'_G$, Eq. (24.5) can be changed to

$$^3J_{XA} = J_G\left(p_t + p_{g^-}\right) + J_T p_{g^+} \tag{24.7}$$

The number of the unknowns is reduced to two: (p_t + p_{g^-}) and p_{g^+}. The two equations become soluble, and the results thus obtained are shown in Table 24.13. Conformational fractions of bond 4 were obtained from $^3J_{AB}$ and $^3J_{AC}$ in the same manner as used for DMP (Section 15.1). Since there is no vicinal coupling regarding conformations of bond 5, its bond conformations are unavailable from the NMR experiment. The conformational fractions of bond 4 of monomer in a nonpolar solvent, cyclohexane, at 25 °C are $p_t = 0.36$, $p_{g^+} = 0.45$, and $p_{g^-} = 0.19$, and those in polar DMSO at 25 C are $p_t = 0.23$, $p_{g^+} = 0.58$, and $p_{g^-} = 0.19$. The bond conformations (p_t, p_{g^+}, and p_{g^-}, respectively) of P3HB and its oligomers were reported so far: 0.4, 0.6, and 0.0 (P3HB) [113, 114, 229]; 0.2, 0.7, and 0.1 (oligo((*R*)-3-hydroxybutyrate)) [280, 281]; and 0.35, 0.55, and 0.10 (16mer and 20mer of (*R*)-3-hydroxybutyrate) [517]. All the NMR experiments are consistent with each other about the magnitude relation, $p_{g^+} > p_t > p_{g^-}$.

Table 24.13 Bond conformations of monomer and polymer, derived from NMR experiments and MO and RIS calculations.

	Bond								
	3 (b)[a)]			4 (c)			5 (d)		
	($p_t + p_{g-}$)	p_{g+}							
Medium	p_t	p_{g+}	p_{g-}	p_t	p_{g+}	p_{g-}	p_t	p_{g+}	p_{g-}
				NMR expt (monomer)					
C_6D_{12}	(0.97)	0.03		0.36	0.45	0.19			
C_6D_6	(0.97)	0.03		0.28	0.53	0.19			
$CDCl_3$	(0.97)	0.03		0.29	0.52	0.19			
CD_3OD	(0.97)	0.03		0.23	0.58	0.19			
$(CD_3)_2SO$	(0.97)	0.03		0.23	0.58	0.19			
				MO calc (monomer)[b)]					
Gas	0.54	0.01	0.45	0.36	0.56	0.08	0.40	0.35	0.25
EDC	0.68	0.01	0.31	0.25	0.69	0.06	0.47	0.31	0.22
				RIS calc (polymer)[b)]					
Gas	0.75	0.01	0.24	0.16	0.69	0.15	0.19	0.69	0.12
EDC	0.86	0.01	0.13	0.09	0.83	0.08	0.19	0.73	0.08
TFE	0.86	0.01	0.13	0.08	0.83	0.09	0.16	0.76	0.08

a) For NMR (monomer) experiments, $p_t + p_{g-}$ and p_{g+}, and for MO (monomer) and RIS (polymer) calculations, p_t, p_{g+}, and p_{g-}.
b) Abbreviations: EDC, ethylene dichloride; TFE, trifluoroethanol.
Source: Adapted from Sasanuma and Katsumata [407].

24.3.2 MO Calculation

There was possibility that monomer would have 54 conformers: 2 (bond 2, trans and cis) ×3^3 (bonds 3–5, trans, gauche$^+$, and gauche$^-$). Each conformer underwent the structural optimization, and consequently, 34 conformers remained. Since the cis state of bond 2 is higher in free energy by over 7 kcal mol^{-1} than the trans form, Gibbs free energies of only conformers with the trans state in bond 2, calculated at the MP2/6-311+G(2d,p)//B3LYP/6-311+G(2d,p) level, are listed in Table 24.14. The bond conformations derived from the free energies (Table 24.13) agree well with those from the NMR experiments. The similar MO calculations were carried out for dimer (Figure 24.9) with conformations changed around bonds 6–9. The conformer of the lowest in free energy is ttg$^+$t for monomer or ttg$^+$g$^+$ for dimer.

Table 24.14 Conformational free energies of monomer and dimer.[a)]

					ΔG_k (kcal mol^{-1})				
					Monomer		Dimer		
k	Conformation[b)]				Gas	EDC	Gas	EDC	TFE
1	t	t	t	t	0.00	0.00	0.00	0.00	0.00
2	t	t	t	g$^+$					
3	t	t	t	g$^-$	−0.22	−0.03	0.05	0.09	−0.02
4	t	t	g$^+$	t	−0.74	−1.09	−0.46	−0.82	−0.89
5	t	t	g$^+$	g$^+$	−0.65	−0.84	−1.14	−1.40	−1.57
6	t	t	g$^+$	g$^-$					
7	t	t	g$^-$	t					
8	t	t	g$^-$	g$^+$	1.05	1.21	0.81	0.82	0.71
9	t	t	g$^-$	g$^-$	1.19	0.90	1.51	1.24	0.97
10	t	g$^+$	t	t	1.54	1.97	2.43	2.84	2.99
11	t	g$^+$	t	g$^+$					
12	t	g$^+$	t	g$^-$	3.32	3.43	2.91	2.92	2.92
13	t	g$^+$	g$^+$	t	2.30	2.26	2.02	1.88	1.93
14	t	g$^+$	g$^+$	g$^+$	2.36	2.48	1.58	1.73	1.59
15	t	g$^+$	g$^+$	g$^-$					
16	t	g$^+$	g$^-$	t					
17	t	g$^+$	g$^-$	g$^+$	4.07	4.29	3.55	3.63	3.56
18	t	g$^+$	g$^-$	g$^-$	4.59	4.78	3.73	4.15	4.06
19	t	g$^-$	t	t	−0.46	−0.25	−0.02	0.24	0.19
20	t	g$^-$	t	g$^+$					
21	t	g$^-$	t	g$^-$	−0.15	−0.03	0.04	−0.02	−0.18
22	t	g$^-$	g$^+$	t					
23	t	g$^-$	g$^+$	g$^+$	−0.46	−0.06	−0.76	−0.22	−0.69
24	t	g$^-$	g$^+$	g$^-$	0.12	−0.20	−0.28	−0.29	−0.64
25	t	g$^-$	g$^-$	t					
26	t	g$^-$	g$^-$	g$^+$	0.26	0.48	0.02	0.33	0.09
27	t	g$^-$	g$^-$	g$^-$	0.47	0.66	0.42	0.55	0.38

a) Relative to the all-trans conformer. The blank line indicates that the geometrical optimization did not detect the potential minimum.
b) Conformations of bonds 2–5 (monomer) or bonds 6–9 (dimer).
Source: Material from reference [407], published 2013, Springer Nature.

In these two conformers, the carbonyl C=O and methine C—H bonds are coplanar; hence, a weak C=O· · ·H—C attraction is suggested to be formed.

24.3.3 RIS Calculation and Comparison with Experiment

The refined RIS calculations were carried out with the Gibbs free energies of dimer, and the results are shown in Table 24.15. The characteristic ratios of the

Table 24.15 Configurational properties, thermodynamic quantities, and average geometrical parameters of P3HB, evaluated from refined RIS calculations.[a]

		25 °C			203 °C[b]		
	Bond	Gas	EDC	TFE	Gas	EDC	TFE
			RIS calculation				
$\langle r^2 \rangle_0/nl^2$		5.44	5.14	5.60			
S_{conf} (cal K^{-1} mol^{-1})		3.29	2.73	2.64	4.15	3.88	3.84
U_{conf} (kcal mol^{-1})		0.58	0.49	0.49	0.90	0.92	0.94
			Experiment[c]				
ΔS_u (cal K^{-1} mol^{-1})					6.67		
ΔH_u (kcal mol^{-1})					3.18		
Bond length[d] (Å)	a	1.356					
	b	1.455					
	c	1.530					
	d	1.520					
Bond angle[d] (°)	a	106.0					
	b	113.4					
	c	110.9					
	d	117.2					
Dihedral angle[d] (°)		t	g$^+$ (c)[e]	g$^-$			
	a	−0.9	176.1				
	b	−27.4	115.0	−99.2			
	c	3.3	119.0	−127.6			
	d	−18.4	124.6	−121.4			

a) Using the ΔG_k values (Table 24.14) for the individual media (gas phase, EDC, and TFE) at 25 °C.
b) The equilibrium melting point of P3HB.
c) Experimental values of entropy (ΔS_u) and enthalpy (ΔH_u) of fusion [293].
d) Using the ΔG_k values for the TFE solution at 25 °C.
e) The symbol c represents the cis conformation for bond a.
Source: Material from reference [407], published 2013, Springer Nature.

unperturbed P3HB chain at 25 °C were evaluated from the three energy sets to be 5.44 (gas), 5.14 (ethylene dichloride: EDC), and 5.60 (trifluoroethanol: TFE). The configurational entropy at the equilibrium melting point is 4.15 cal K^{-1} mol^{-1}, the entropy of fusion was derived experimentally as 6.67 cal K^{-1} mol^{-1} [293], and hence the $S_{conf}/\Delta S_u$ ratio is 0.62.

The racemic copolymer, poly(DL-β-methyl β-propiolactone) (PMPL), was found to show the Θ state at 13 °C in butyl chloride, and the $\langle S^2\rangle_0/M_w$ value was determined from its inherent viscosity to be 1.05×10^{-17} cm^2 [199]. Inasmuch as the solution properties of PMPL are essentially the same as those of P3HB, it was concluded that the Θ condition of PMPL may also be applicable to P3HB. From the $\langle S^2\rangle_0/M_w$ value, therefore, the $\langle r^2\rangle_0/nl^2$ ratio can be calculated to be 6.30 (for details of the calculations, see Section 3.3.1). In addition, the excluded volume (β), namely the binary cluster integral for segment–segment interactions, of PMPL in TFE at 25 °C was obtained as 159×10^{-24} cm^3 [199]. This is a rare case where the β value was estimated experimentally (see Section 3.2.5). By static light scattering, the M_w and $\langle S^2\rangle$ values of 14 P3HB ($M_w = 51.0 - 910 \times 10^4$) and 4 PMPL ($M_w = 8.65 - 15.0 \times 10^4$) fractions dissolved in TFE at 25 °C were determined. On the basis of the two-parameter theory and the Domb–Barrett equation, the $\langle r^2\rangle_0/nl^2$ values were calculated and averaged to be 6.19 ± 0.28 for P3HB and 6.12 ± 0.31 for PMPL, as described in Section 3.3.1. The experimental $\langle r^2\rangle_0/nl^2$ value of P3HB is comparable to that (5.60) calculated from the TFE energy set at 25 °C.

24.3.4 Crystal Structure

The crystal structure of P3HB was determined by X-ray diffraction as an orthorhombic cell of space group $P2_12_12_1$ with lattice constants, $a = 5.76$ Å, $b = 13.20$ Å, and c (fiber period) = 5.96 Å (Figure 24.10) [539]. The chain conformation in the crystal is ttg$^+$g$^+$, which is the same as the most stable state of the dimer. The dihedral angles of bonds a–d are 5° (3.0°), −28° (−28.5°), 128° (119.3°), and 133° (134.4°), where the values in the parentheses are the optimized angles of those of the ttg$^+$g$^+$ form of the dimer. Here, the dihedral angle is represented in the following manner: trans, ~0; gauche$^\pm$, ~ ±120. Therefore, these data indicate that the P3HB chain in the crystal is as unsuppressed as in the isolated state.

As mentioned at the beginning, P3HB is an intracellular carbon- and energy-reserve substance. Its biodegradability will be discussed In Section 24.6, together with those of the other biodegradable polyesters.

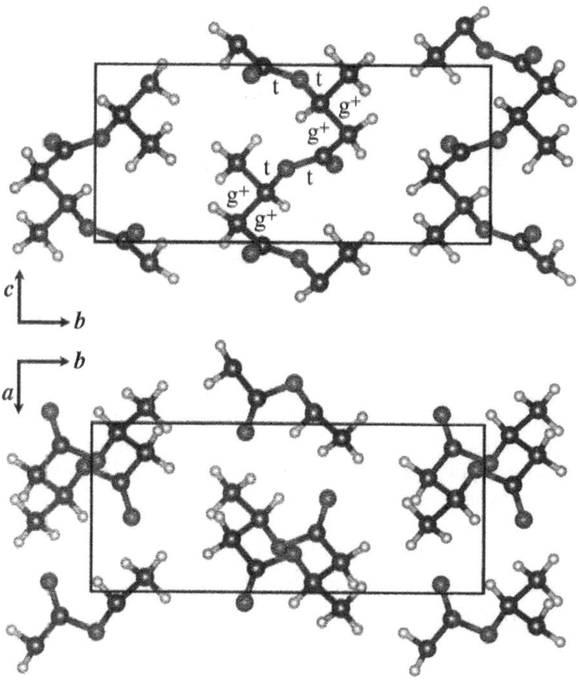

Figure 24.10 Crystal structure of P3HB [539] depicted by the VESTA software [325]. Source: Material from reference [407], published 2013, Springer Nature.

24.4 Poly(ϵ-caprolactone) (PCL)

PCL (Figure 24.11) has usually been used for applications where biodegradability is required [127, 474]. Its glass transition temperature ($T_g = -60$ °C) and melting point ($T_m = 56 - 65$ °C) of PCL are comparatively low, while PCL is soluble in a number of common solvents and compatible with different polymers, thus being blended or copolymerized with other polymers and molded into desired forms such as scaffolds in tissue engineering, long-term drug delivery microspheres, and packaging containers [127, 474, 527].

The repeating unit of PCL includes seven skeletal bonds (bonds a–g in Figure 24.11). Bond a is considered to adopt either the trans or the cis form. The latter state is higher in Gibbs free energy by over 7 kcal mol^{-1} than the former, [407]; thus, bond a can be assumed to stay trans throughout, whereas the other six (bonds b–g) may lie in trans, gauche$^+$, or gauche$^-$ conformation. Therefore, the repeating unit may possibly exhibit as many as 729(= 3^6) conformations, but the gauche$^+$ and gauche$^-$ states of each bond are equivalent, and hence, the number of irreducible conformations is, at the maximum, $(729 - 1)/2 + 1 = 365$.

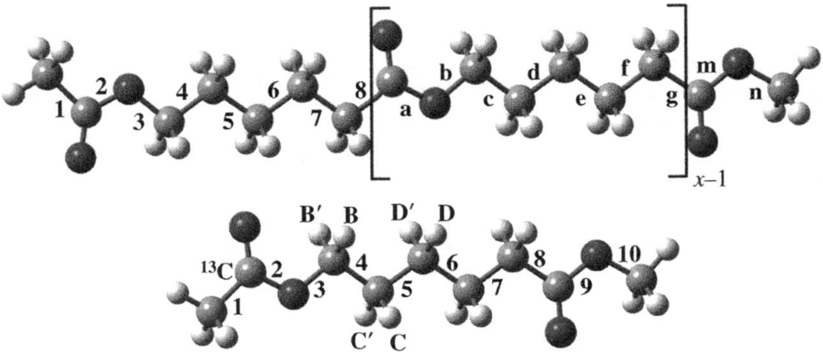

Figure 24.11 Poly(ε-caprolactone) (PCL, above) and its monomeric model compound, methyl 6-acetoxyhexanoate (MAH, below). The carbonyl carbon of MAH-^{13}C is labeled with carbon-13. The bonds and hydrogen atoms are designated as indicated, and *x* is the degree of polymerization. Source: Kawai et al. [238]/Reproduced with permission of the PCCP Owner Societies.

For a detailed conformational analysis of PCL, it would be desirable to evaluate Gibbs free energies of all the 356 conformations individually.

Methyl 6-acetoxyhexanoate (MAH) was used as a monomeric model compound (Figure 24.11). All 365 irreducible conformers of MAH were subjected to structural optimization at the B3LYP level, the electronic energies of all the optimized conformers were calculated at the MP2 level, and the bond conformations of the six bonds were evaluated from the Gibbs free energies and compared with those from the ^{1}H and ^{13}C NMR experiments for ^{13}C-labeled MAH (MAH-^{13}C, Figure 24.11). The Gibbs free energies and geometrical parameters obtained from the MO calculations were applied to the refined RIS scheme. The configurational properties thus obtained for the unperturbed PCL chain were compared with those estimated experimentally.

The crystal structure of PCL was determined by X-ray diffraction as follows [72]: orthorhombic; space group $P2_12_12_1$; two antiparallel all-trans chains packed in the unit cell; lattice constants of $a = 7.47$, $b = 4.98$, and c (fiber axis) = 17.05 Å. Between the neighboring chains, the C=O groups are longitudinally shifted by (3/14)c. The periodic DFT calculations [122, 366] with a dispersion-force correction [175, 177] were carried out for the PCL crystal. The optimized crystal structure is compared with the experiment, and its crystalline moduli were also calculated and compared with those of the representative polymers.

24.4.1 MO Calculation

All the 365 conformers underwent the geometrical optimization at the B3LYP/6-311++G(2d,p) level. Of them, 181 conformers reached the potential minima,

Table 24.16 Conformations with low Gibbs free energies of MAH.[a)]

	Bond						ΔG_k[b)] (kcal mol^{-1})		
	3	4	5	6	7	8	Gas	Benzene	DMSO
1	t	$g^{\pm}$	t	t	$g^{\pm}$	t	−0.500	−0.303	−0.044
2	$g^{\pm}$	$g^{\pm}$	t	t	$g^{\pm}$	t	−0.409	−0.144	0.215
3	t	$g^{\pm}$	t	t	$g^{\mp}$	t	−0.377	−0.234	−0.103
4	t	$g^{\pm}$	$g^{\pm}$	t	$g^{\pm}$	t	−0.341	−0.107	0.091
5	$g^{\pm}$	$g^{\mp}$	t	t	$g^{\pm}$	t	−0.290	−0.038	0.273
6	$g^{\pm}$	$g^{\pm}$	$g^{\pm}$	t	$g^{\pm}$	t	−0.153	0.050	0.163
7	t	t	t	t	$g^{\pm}$	t	−0.152	−0.005	0.175
8	t	$g^{\pm}$	t	t	t	t	−0.025	0.037	0.112
9	t	t	t	t	t	t	0.000	0.000	0.000
10	$g^{\pm}$	$g^{\pm}$	t	t	$g^{\mp}$	t	0.010	0.267	0.551
11	t	$g^{\pm}$	t	t	$g^{\mp}$	$g^{\mp}$	0.014	0.012	0.061
12	t	$g^{\pm}$	$g^{\pm}$	t	$g^{\mp}$	t	0.047	0.261	0.544
13	t	t	t	$g^{\pm}$	$g^{\pm}$	$g^{\pm}$	0.066	0.135	0.274
14	t	$g^{\pm}$	$g^{\pm}$	t	t	t	0.082	0.154	0.213
15	$g^{\pm}$	$g^{\pm}$	t	t	t	t	0.116	0.255	0.354
16	$g^{\pm}$	$g^{\pm}$	$g^{\pm}$	t	$g^{\mp}$	t	0.119	0.412	0.769
17	t	t	t	t	$g^{\pm}$	$g^{\pm}$	0.130	0.085	0.067
18	$g^{\pm}$	t	t	t	$g^{\mp}$	t	0.137	0.407	0.809
19	t	$g^{\pm}$	t	$g^{\pm}$	$g^{\pm}$	$g^{\pm}$	0.161	0.213	0.248
20	$g^{\pm}$	$g^{\pm}$	t	$g^{\mp}$	$g^{\mp}$	$g^{\mp}$	0.190	0.485	0.901

a) Arranged in ascending order of the gas-phase ΔG_k value.
b) Relative to the free energy of the all-trans conformation.
Source: Kawai et al. [238]/Reproduced with permission of the PCCP Owner Societies.

which suggests the possibly that the $(181 - 1) \times 2 + 1 = 361$ conformers would be formed. Only 20 conformers of low ΔG_ks are arranged in Table 24.16. The most stable conformer is tg$^{\pm}$ttg$^{\pm}$t in the gas phase or tg$^{\pm}$ttg$^{\mp}$t in DMSO. The ΔG_k difference between the first and the 20th conformers stays within 0.690 kcal mol^{-1} (gas) or 0.945 kcal mol^{-1} (DMSO), which indicates that a number of conformers will be formed in a small ΔG_k range. The trans fractions of bonds 3–8 at 25 °C, calculated from the ΔG_ks, are listed in Table 24.17. Bonds 4 and 7 prefer gauche conformations ($p_g \sim 0.65$–0.77), and the others show trans preferences ($p_t \sim 0.50$–0.65).

Table 24.17 Trans fractions (p_t) of MAH, derived from MO calculations and NMR experiments.

			Bond					
Medium	**dielectric Constant**[a)]	**Temp**[b)] **(°C)**	**3**	**4**	**5**	**6**	**7**	**8**
			MO calc[c)]					
Gas	1.0	25	0.50	0.30	0.57	0.63	0.23	0.68
		35	0.50	0.30	0.56	0.63	0.23	0.67
		45	0.49	0.31	0.55	0.62	0.23	0.67
		55	0.49	0.31	0.55	0.61	0.24	0.66
		64	0.49	0.31	0.54	0.61	0.24	0.66
Benzene	2.3	25	0.55	0.32	0.59	0.64	0.25	0.66
DMSO	47	25	0.60	0.34	0.60	0.65	0.29	0.64
			NMR expt					
Chloroform-*d*	4.8	25	0.59	0.27	0.57	0.59	0.23	

a) At 20 °C.
b) The equilibrium melting point is 64 °C [96, 293]
c) At the MP2/6-311++G(2d,p)//B3LYP/6-311++G(2d,p) level.
Source: Kawai et al. [238]/Reproduced with permission of the PCCP Owner Societies.

24.4.2 NMR Experiment

Figure 24.12 shows the ^{1}H and ^{13}C NMR spectra observed from MAH-^{13}C dissolved in $CDCl_3$ at 25 °C. The trans fractions of bonds 3–7, derived from vicinal coupling constants, agree well with those from the MO calculations (Table 24.17), which means that the Gibbs free energies derived from the MO calculations are so reliable as to be used for the RIS calculations.

24.4.3 RIS Calculation

In the statistical weight matrix of bond g, the elements corresponding to the 361 existent conformations were filled with the Boltzmann factors of their Gibbs free energies, and the others with zero. The temperature was set at 25 °C (standard temperature) or 64 °C (equilibrium melting point T_m^0 of PCL). The calculated results are shown in Table 24.18.

The characteristic ratio ($\langle r^2\rangle_0/nl^2$) at 25 °C is 4.09 (gas), 4.35 (benzene), or 4.53 (DMSO). The experimental $\langle r^2\rangle_0/nl^2$ values of PCL were estimated from the Stockmayer–Fixman (S–F) plots: $[\eta]/M^{1/2}$ vs. $M^{1/2}$ [447]. Here, $[\eta]$ is the intrinsic

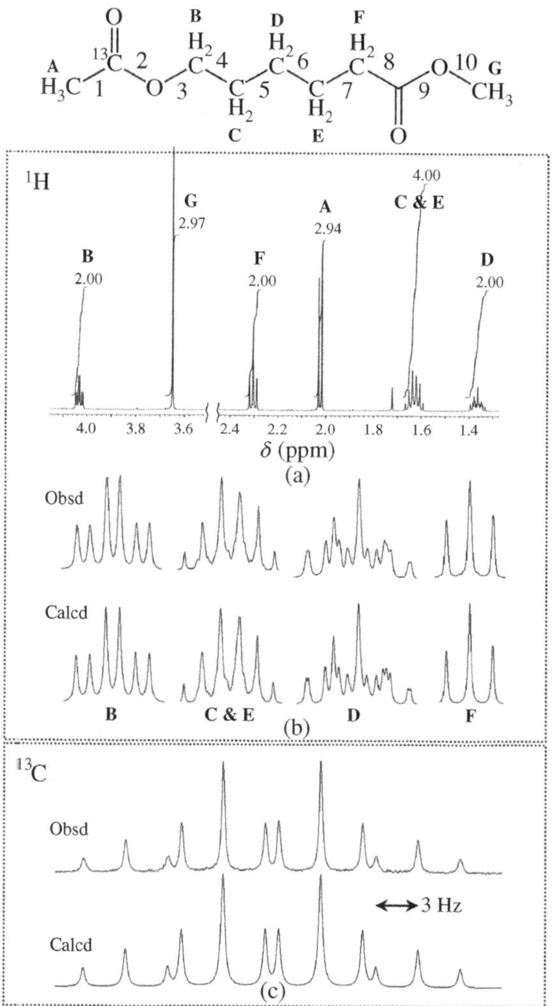

Figure 24.12 ^{1}H and ^{13}C NMR spectra observed from MAH-^{13}C dissolved in $CDCl_3$ at 25 °C (above), compared with simulations (below): (a) ^{1}H NMR and the assignment (the numerical value expresses the integrated intensity, that is the number of hydrogen atoms); (b) its enlargement; (c) ^{13}C NMR of the carbonyl carbon labeled with carbon-13. Source: [238]/Kawai et al./Reproduced with permission of the PCCP Owner Societies.

viscosity, M is the molecular weight. The intercept of the S—F plot at $M^{1/2} = 0$ yields

$$K_{\Theta} = \Phi_0 \left(\frac{\langle r^2 \rangle_0}{M} \right)^{3/2} \tag{24.8}$$

where Φ_0 is the viscosity constant. From the K_{Θ} value, the $\langle r^2 \rangle_0 / M$ value at the Θ point may be obtained and converted to the characteristic ratio according to

$$\frac{\langle r^2 \rangle_0}{nl^2} = \frac{M_0}{\left(nl^2\right)_0} \frac{\langle r^2 \rangle_0}{M} \tag{24.9}$$

Table 24.18 Configurational properties of PCL, derived from RIS calculations.

		Gas		Benzene		DMSO	
		25 °C	64 °C[a)]	25 °C	64 °C[a)]	25 °C	64 °C[a)]
$\langle r^2 \rangle_0 / nl^2$		4.09	4.03	4.35	4.27	4.53	4.44
$d \ln \langle r^2 \rangle_0 / dT \times 10^3$	(K^{-1})	−0.40	−0.37	−0.50	−0.45	−0.53	−0.48
S_{conf}[b)]	(cal K^{-1} mol^{-1})	10.6	10.8	10.5	10.7	10.4	10.6
ΔS_u[c)]	(cal K^{-1} mol^{-1})		11.5				
ΔH_u[d)]	(kcal mol^{-1})		3.90				

a) The equilibrium melting point.
b) The configurational (conformational) entropy.
c) The entropy of fusion [96, 293].
d) The enthalpy of fusion.
Source: Kawai et al. [238]/Reproduced with permission of the PCCP Owner Societies.

where M_0 and $(nl^2)_0$ are the formula mass and the sum of the square bond lengths of the repeating unit, respectively: for PCL, $M_0 = 114.14$ g mol^{-1} and $(nl^2)_0 = 15.572$ $Å^2$. The experimental $\langle r^2 \rangle_0 / nl^2$ values thus obtained are shown in Table 24.19. The $\langle r^2 \rangle_0 / nl^2$ ratios range widely from 3.9 to 6.0, depending also on Φ_0. Koleske and Lundberg [255] and Hung et al. [204] obtained the same K_Θ value of 1.2×10^{-3} but adopted different Φ_0s and determined $\langle r^2 \rangle_0 / nl^2$s of 4.1 and 4.5, respectively. These experimental values exactly agree with the RIS calculations.

The configurational entropy (S_{conf}) at T_m^0 was calculated to be 10.6–10.8 cal K^{-1} mol^{-1}, and the experimental entropy of fusion (ΔS_u) is 11.5 cal K^{-1} mol^{-1} [96, 293]; the $S_{conf}/\Delta S_u$ ratio is as much as 92–94%.

24.4.4 Crystal Structure

In the DFT-D optimization for the PCL crystal, the lattice constants and atomic coordinates determined by X-ray diffraction were set initially. The resultant crystal structure is depicted in Figure 24.13, and the optimized lattice constants and fractional coordinates are compared with the X-ray data in Table 24.20.

The DFT-D optimization yields the crystal structure at 0 K, while the X-ray diffraction expresses the ambient-temperature structure. As for the lattice constants, *a* is somewhat shorter, and *c* is slightly longer than the experiment. This is probably because the PCL chain at 0 K is free from thermal disorder, more extended along the *c* (fiber) axis and more densely packed than that at room

Table 24.19 Characteristic ratios of PCL, estimated from the Stockmayer–Fixman (S–F) plots.

Study[a)]	Solvent[b)]	Temp (°C)	$\Phi_0 \times 10^{-21}$	$K_\Theta \times 10^3$	$(\langle r^2\rangle_0/M)^{1/2} \times 10^{11}$ (cm g$^{-1/2}$ mol$^{1/2}$)	$\langle r^2\rangle_0/nl^2$
Koleske	Benzene	30	2.86	1.1		3.9
and Lundberg	DMF	30		1.2		4.1
Knecht	THF	18–35	2.87		883, 907	5.7, 6.0
and Elias[c)]	Dioxane					
	Chlorobenzene					
Jones et al.	Dioxane	25	2.66	1.9		5.9
	Benzene					
	Ethyl acetate					
Huang et al.[d)]	THF	25	2.5	1.2		4.5

a) Quoted from the following references: Koleske and Lundberg [255]; Knecht and Elias [248]; Jones et al. [224]; and Huang et al. [204]
b) Abbreviations: DMF, N,N′-dimethyl formamide; THF, tetrahydrofuran.
c) The $(\langle r^2\rangle_0/M)^{1/2}$ values are averages of those determined by five different methods including the S–F plot. A number of solvents were used, but only solvents employed for the S–F plot are written here.
d) From the radii of gyration obtained via light scattering, a somewhat different $\langle r^2\rangle_0/nl^2$ value (4.9) was also estimated.

Source: Kawai et al. [238]/Reproduced with permission of the PCCP Owner Societies.

temperature: the densities by the DFT-D calculation and X-ray diffraction are, respectively, 1.288 and 1.195 g cm^{-3}. However, the difference between the DFT-D and X-ray structures is slight: Δ_{LC} = 2.2% in lattice constants and Δ_{CO} = 0.039 in atomic coordinates.

24.4.5 Crystal Elasticity

From the compliance tensor, Young's moduli in the *a*-, *b*-, and *c*-axis directions were evaluated [342], and the Young's modulus distribution can be expressed three-dimensionally as in Figure 24.13. In general, the Young's modulus in the fiber-axis direction is the largest and depends principally on the chain conformation. The PCL chain crystallizes in the all-trans form, and hence, its E_c value is as large as 252 GPa. All-trans polymers such as PGA, PE, and nylons 4 and 6 (α form) also exhibit large fiber-axis moduli (Table 10.2 in Section 10.5).

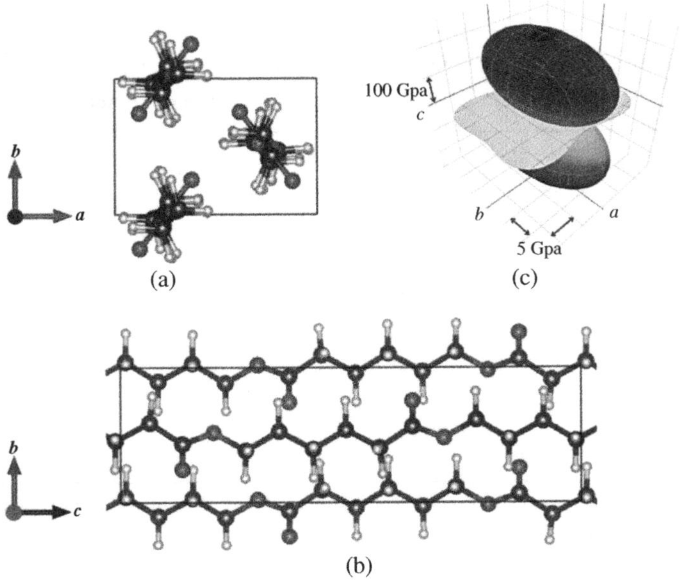

Figure 24.13 Crystal structure optimized at the B3LYP-D/6-31G(d,p) level: (a) *ab*-plane; (b) *bc*-plane; (c) three-dimensional Young's modulus distribution depicted by the ELATE software [167]. The grid spacings along the *a*, *b*, and *c* axes are 5, 5, and 100 GPa, respectively. The crystallographic data: orthorhombic; space group $P2_12_12_1$; $a = 7.00$ Å, $b = 4.88$ Å, and $c = 17.21$ Å. Source: [238]/Kawai et al./Reproduced with permission of the PCCP Owner Societies.

The three-dimensionally averaged E_{av} value, corresponding to the maximum Young's modulus of nonoriented PCL crystallites, was evaluated to be 10.7 GPa. To our knowledge, the largest experimental modulus reported so far for PCL may be that (3.7 ± 0.7 GPa) observed from its single fibers [97], which also include amorphous phases, thus being much smaller than the true crystalline modulus. The fiber-axis modulus is correlated with the chain conformation, being arranged in the order of all-trans > distorted all-trans > helical > bended. Of all three-dimensionally averaged moduli (E_{av}s) shown in Table 10.2, that of PCL is the smallest. This is probably because the PCL crystal would be deformed more readily than the others owing to the inherent flexibility of the PCL chain.

Table 24.20 Optimized and experimental crystal structures of PCL.[a]

	Optimized[b]			**Experimental[c]**		
			Lattice constant (Å)			
	a	b	c	a	b	c
	7.0[illegible]	4.88	17.21	7.47	4.98	17.05
			$\Delta_{LC} = 2.2\%$			
			Density (g cm^{-3})			
		1.288			1.195	
			Fractional coordinates			
	x/a	y/b	z/c	x/a	y/b	z/c
C	0.729	0.619	0.351	0.721	0.566	0.366
C	0.740	0.649	0.214	0.732	0.615	0.228
C	0.811	0.481	0.146	0.810	0.471	0.156
C	0.736	0.596	0.069	0.733	0.604	0.082
C	0.808	0.442	−0.003	0.813	0.459	0.011
C	0.783	0.454	0.422	0.761	0.408	0.437
O	0.645	0.838	0.353	0.621	0.755	0.366
O	0.787	0.501	0.285	0.786	0.481	0.298
H	0.585	0.675	0.212			
H	0.807	0.850	0.217			
H	0.761	0.270	0.152			
H	0.968	0.478	0.146			
H	0.773	0.814	0.064			
H	0.580	0.585	0.070			
H	0.778	0.223	0.003			
H	0.964	0.459	−0.007			
H	0.748	0.239	0.410			
H	0.939	0.453	0.426			
			$\Delta_{CO} = 0.039$			

a) An orthorhombic cell of space group $P2_12_12_1$.
b) At 0 K.
c) Reported by Chatani et al. [72]. At room temperature.
Source: Kawai et al. [238]/Reproduced with permission of the PCCP Owner Societies.

24.5 Poly(ethylene succinate) (PES) and Poly(butylene succinate) (PBS)

PES (Figure 24.14b) and PBS (Figure 24.14d) are produced from petrochemicals [63] but nevertheless biodegradable. The commercial production of PBS was begun in the 1990s and has been expanding since the 2000s [161].

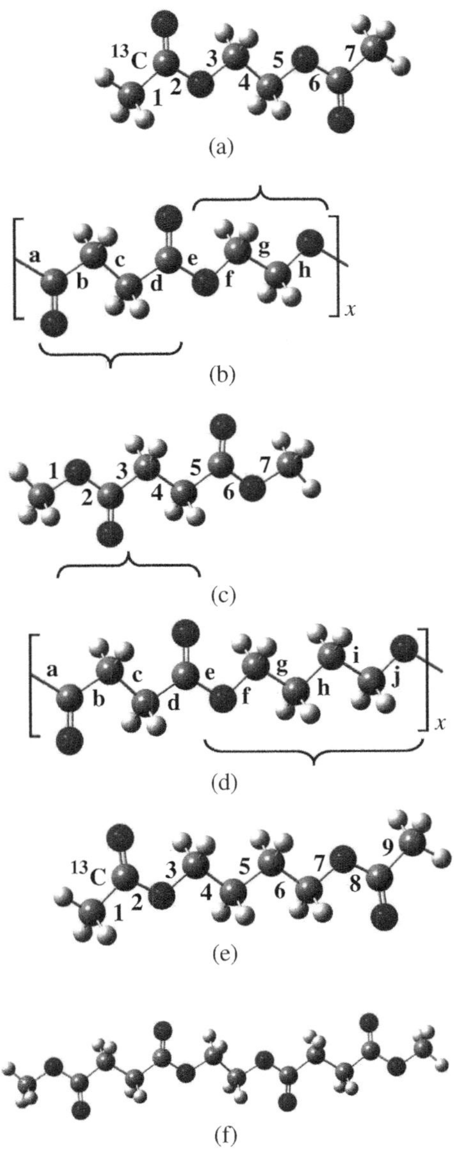

Figure 24.14 Polyesters and model compounds: (a) ethylene glycol diacetate (EGDA); (b) poly(ethylene succinate) (PES); (c) dimethyl succinate (DMS); (d) poly(butylene succinate) (PBS); (e) butylene glycol diacetate (BGDA); and (f) ethylene glycol di(methyl succinate) (EGDMS). The repeating units of PES and PBS are divided into two parts, each of which is investigated by the corresponding model compound. The skeletal bonds are numbered or termed as indicated, and *x* is the degree of polymerization. Source: [410]/Figure 1 (p. 329), Sasanuma et al./Reproduced with permission of Elsevier.

24.5.1 NMR Experiment

As a model for the repeating unit of PES, ethylene glycol di(methyl succinate) (EGDMS, Figure 24.14f) was adopted, and the repeating unit was divided at the ester linkage into two subunits, which may be represented by two smaller models: dimethyl succinate (DMS, Figure 24.14c) and ethylene glycol diacetate (EGDA, Figure 24.14a). In addition, EGDA-^{13}C, one of whose carbonyl carbons is selectively labeled by carbon-13, were prepared and underwent ^{1}H and ^{13}C NMR measurements, and all spectra were simulated to yield chemical shifts and coupling constants. Vicinal ^{1}H–^{1}H and ^{13}C–^{1}H coupling constants were analyzed, and bond conformations of the skeletal bonds were determined (Table 24.21) [410].

The trans fractions of the $COCH_2—CH_2$ bond of DMS range from 0.48 to 0.59 and agree with those of the corresponding bond of EGDMS. The small p_t values (0.05–0.14) of the $OCH_2—CH_2O$ bond of EGDA express a strong gauche preference, being in close agreement with those of the EGDMS. These results clearly indicate that the dicarboxylic and diol units, separated by the ester bond, little interfere with each other's conformation and can be treated separately. Accordingly, DMS and EGDA rather than EGDMS were used as models for PES because EGDMS is too large to undergo expensive MO calculations.

Similarly, DMS and butylene glycol diacetate (BGDA) were employed as model compounds of PBS. BGDA-^{13}C was also prepared and subjected to NMR measurements. In Table 24.22, the p_t values of three bonds of BGDA, obtained from the NMR experiments, are listed: the $OCH_2—CH_2C$ bond exhibits a moderate gauche preference, and the other two bonds prefer the trans conformation.

Table 24.21 Trans fractions (p_ts) of DMS, EGDMS, and EGDA at 25 °C, evaluated from NMR experiments and MO calculations.

	DMS		EGDMS		EGDA	
Solvent	C(=O)−CH$_2$	COCH$_2$−CH$_2$	COCH$_2$−CH$_2$	OCH$_2$−CH$_2$O	OCH$_2$− CH$_2$O	O−CH$_2$
			NMR expt			
C_6D_{12}		0.48			0.12	0.44
C_6D_6		0.59	0.58	0.06	0.07	0.53
CD_3OD		0.59	0.58	0.07	0.06	0.53
			MO calc[a)]			
Gas	0.80	0.49			0.10	0.52
C_6H_6	0.77	0.48			0.08	0.63
$CHCl_3$	0.76	0.49			0.07	0.69

a) At the MP2/6-311+G(2d,p)//B3LYP/6-311+G(2d,p) level.
Source: Sasanuma et al. [410], Table 2 (p. 331)/Reproduced with permission of Elsevier.

Table 24.22 Trans fractions (p_ts) of BGDA at 25 °C, determined from NMR experiments and MO calculations.

Solvent	O–CH$_2$	OCH$_2$–CH$_2$C	CCH$_2$–CH$_2$C
		NMR expt	
C_6D_{12}	0.53	0.27	0.54
C_6D_6	0.56	0.28	0.57
$CDCl_3$	0.58	0.28	0.58
CD_3OD	0.57	0.27	0.55
$(CD_3)_2SO$	0.56	0.28	0.55
		MO calc[a)]	
Gas	0.52	0.27	0.55
C_6H_6	0.56	0.29	0.60
$CHCl_3$	0.58	0.33	0.66

a) At the MP2/6-311+G(2d,p)//B3LYP/6-311+G(2d,p) level.

24.5.2 MO Calculation

In MO calculations, the repeating unit of PES was modeled by DMS and EGDA, each of which underwent MO calculations to derive conformer free energies. Similarly, free energies of PBS were obtained from DMS and BGDA. Since the C(=O)—O bond of esters is fixed in the trans state, three internal rotations are defined for DMS and EGDA (bonds 3–5 in Figure 24.14), and five for BGDA (bonds 3–7). For each of them, three RIS states (t, g$^+$, and g$^-$) may be assumed; therefore, under the RIS approximation, DMS, EGDA, and BGDA may form 27 (3^3), 27, and 243 (3^5) conformers. However, the molecular symmetry reduces the number of independent conformers to 10 (DMS and EGDA) and 70 (BGDA). The Gibbs free energies (ΔGs) of the irreducible conformers in the gas, benzene, and chloroform environments at 25 °C were calculated at the MP2/6-311+G(2d,p) level using the geometrical parameters optimized at the B3LYP/6-311+G(2d,p) level.

The MP2 calculations on DMS showed the all-trans and tgt conformations to be the first and second most stable states, respectively. The ΔG difference between the two conformers is as small as 0.19–0.26 kcal mol^{-1}. For EGDA, the tgt conformer is the lowest in ΔG (−1.1 to −1.5 kcal mol^{-1}), and ΔGs of ttg were also obtained as small values (0.3–0.5 kcal mol^{-1}). Of 70 possible conformers of BGDA, as many as 52 conformers were suggested to exist, and the tg$^{\pm}$tg$^{\mp}$t conformers are the most stable; its ΔG value for the gas phase was obtained as −0.56 kcal mol^{-1}. A number of conformers also show low ΔGs comparable to that of tg$^{\pm}$tg$^{\mp}$t. Trans fractions of

DMS, EGDA, and BGDA were calculated from the ΔGs, being in good agreement with the NMR experiments (Tables 24.21 and 24.22).

24.5.3 RIS Calculation

The configurational properties and thermodynamic parameters of PES and PBS were calculated according to the refined RIS scheme with the ΔG values of DMS, EGDA, and BGDA. Table 24.23 shows the results: characteristic ratio; its temperature coefficient; configurational entropy; and averaged geometrical parameters. The characteristic ratios of the two polyesters in chloroform at 25 °C were obtained as 6.2 (PES) and 7.1 (PBS), and their temperature coefficients are negative.

Table 24.23 Configurational properties and averaged geometrical parameters of PES and PBS, evaluated from RIS calculations with MO parameters.[a]

		PES			PBS		
		Gas	$CHCl_3$		Gas	$CHCl_3$	
		25 °C	25 °C	114 °C[b]	25 °C	25 °C	133.5 °C[b]
$\langle r^2\rangle_0/nl^2$		6.19	6.23	5.43	6.65	7.13	6.26
$d\ln\langle r^2\rangle_0/dT \times 10^3$ (K^{-1})		−1.9	−2.3	−1.0	−1.4	−1.7	−0.85
S_{conf} (cal K^{-1} mol^{-1})		9.58	9.16	9.96	14.0	14.0	14.6
ΔS_u[c] (cal K^{-1} mol^{-1})				16.0			21.2
Average geometry[d]	bond	$\bar{l}$	$\bar{\theta}$	$\overline{\phi_{g^+}}$	$\bar{l}$	$\bar{\theta}$	$\overline{\phi_{g^+}}$
	a	1.356	111.1		1.356	111.1	
	b	1.513	112.9	133.7	1.513	112.9	133.7
	c	1.525	112.9	111.2	1.525	112.9	111.2
	d	1.513	111.1	133.7	1.513	111.1	133.7
	e	1.356	116.2		1.356	116.4	
	f	1.441	109.3	92.6	1.447	109.5	91.3
	g	1.508	109.3	111.4	1.520	113.7	115.8
	h	1.441	116.2	92.6	1.533	113.7	112.1
	i				1.520	109.5	115.8
	j				1.447	116.4	91.3

a) At the MP2/6-311+G(2d,p)//B3LYP/6-311+G(2d,p) level.
b) The equilibrium melting point [354].
c) The entropy of fusion [354].
d) The geometrical parameters averaged at 25 °C with the MO energies including the solvent effect of $CHCl_3$. Symbols: $\bar{l}$, averaged bond length (in Å); $\bar{\theta}$, averaged bond angle (in °); $\overline{\phi_{g^+}}$, average dihedral angle (in °) of the g$^+$ conformation. $\overline{\phi_t} = 0°$ and $\overline{\phi_{g^-}} = -\overline{\phi_{g^+}}$.

Source: Sasanuma et al. [410], Table 9 (p. 336)/Reproduced with permission of Elsevier.

For the sake of comparison among different polymers, the S_{conf} value must be divided by the number of skeletal bonds in the repeating unit to be expressed in the unit of cal K^{-1} (mole of bond)$^{-1}$. The S_{conf} values at the equilibrium melting point T_m^0 (written in the parenthesis) of the major biodegradable polyesters are 0.49 (PLA, 207 °C), 1.04 (P3HB, 203 °C), 1.05 (PGA, 231.4 °C), 1.22 (P2HB, 130.3 °C), 1.25 (PES, 114 °C), 1.46 (PBS, 133.5 °C), and 1.54 (PCL, 64 °C) cal K^{-1} (mole of bond)$^{-1}$, and those at the standard temperature (25 °C) are 0.39 (PLA), 0.82 (P3HB), 0.91 (PGA), 1.05 (P2HB), 1.20 (PES), 1.40 (PBS), and 1.51 (PCL) cal K^{-1} (mole of bond)$^{-1}$.

Of them, PLA shows the smallest S_{conf}. The repeating unit of PLA has three skeletal bonds. The C(=O)—O bond is essentially fixed in the trans state, and hence, conformational changes are due to internal rotations of the other two bonds, O—CH(CH_3)—C(=O), whose rotations are severely restricted by the O· · ·O repulsion and the methyl side group bonded to the asymmetrical carbon. Consequently, only four conformations are allowed: g^+t, g^+s, g^-t, and g^-g^-. Of them, g^-t and g^-g^- are so high in free energy as to occur quite rarely, and hence, only two conformations, g^+t and g^+s, are the sources of its configurational properties. In contrast, the symmetric polyesters, PES, PBS, and PCL, are so flexible as to form a large number conformations and hence yield the large S_{conf}s.

24.5.4 Crystal Structure

The crystal structure of PES was determined by X-ray diffraction as follows: lattice constants, $a = 7.60$, $b = 10.75$, c (fiber axis) = 8.33 Å; orthorhombic cell; space group of $P_{bnb} - D_{2h}^{10}$; four molecular chains packed in the unit cell [505]. The conformations of bonds a–h are ttg^-tttg^+t. Hereafter, the part of bonds a–d of PES is referred to as DMS moiety, and that of bonds e–h as the EGDA one. The ttg^-t conformation of DMS moiety has a free energy of about 0.2 kcal mol^{-1} relative to that of the all-trans state, and the ttg^+t conformation of EGDA moiety is the most stable ($\Delta G = -1.2$ to -1.4 kcal mol^{-1}).

PBS exhibits two crystal structures: α and β forms [209–211]. The β form appears only under strain, thus considered to be a nonequilibrium state. The crystal structure of the α form was determined as follows: monoclinic cell; space group of $P2_1/n$; $a = 5.23$, $b = 9.12$, c (fiber axis) = 10.90 Å, and $\beta = 123.9°$; and two molecular chains packed in the unit cell. The PBS chain adopts the $ttttttg^+tg^-t$ conformations in bonds a–j. Similarly, the part of bonds a–d of PBS is designated as DMS moiety, and that of bonds e–j as the BGDA one. The all-trans state of DMS is the lowest in ΔG, and the ttg^+tg^-t state in the BGDA moiety corresponds to the most stable conformer of BGDA ($\Delta G = -0.6$ to -0.1 kcal mol^{-1}). The $g^{\pm}tg^{\mp}$ sequences keep the molecular chain oriented in the fiber-axis direction. In the α form, therefore, the PBS chain is allowed to lie in the most favorable state.

The crystal structures of PES and PBS were optimized by the periodic DFT at the B3LYP-D/6-31G(d,p) level with Grimme's D2 dispersion force correction, and the X-ray structures were initially set. The optimized structures of PES and the α from of PBS are compared with the X-ray experiments in Tables 24.24 and 24.25, respectively, and illustrated in Figure 24.15. The differences between theory and experiment are slight: for PES, $\Delta_{LC} = 2.3\%$ and $\Delta_{CHO} = 0.016$; for PBS, $\Delta_{LC} = 1.2\%$ and $\Delta_{CHO} = 0.068$.

The Young's moduli in the *a*-, *b*-, and *c*-axis directions were evaluated as follows (in the unit of GPa): PES, $E_a = 13.7$, $E_b = 11.1$, and $E_c = 48.6$; PBS, $E_a = 7.5$, $E_b = 14.3$, and $E_c = 67.2$. Two-dimensional Young's modulus distributions on the plane perpendicular to the fiber axis of PES and PBS are graphically displayed in Figure 24.15. Three-dimensionally averaged Young's moduli (E_{av}) of PES and PBS, expressing the Young's moduli of nonoriented crystallites at 0 K, are 23.2 and 18.5 GPa, respectively. Roughly speaking, E_{av} tends to be inversely correlated to the fiber-axis modulus.

Table 24.24 Optimized and experimental crystal structures of PES.[a)]

	Optimized[b)]			Experimental[c)]		
	Lattice constant (Å)					
	a	*b*	*c*	*a*	*b*	*c*
	7.42	10.08	8.48	7.60	10.75	8.33
	$\Delta_{LC} = 2.3\%$					
	Fractional coordinates					
	x/*a*	*y*/*b*	*z*/*c*	*x*/*a*	*y*/*b*	*z*/*c*
C	0.721	−0.027	0.664	0.719	−0.013	0.663
C	0.806	0.083	0.569	0.787	0.094	0.571
C	0.799	0.201	0.328	0.798	0.198	0.331
O	0.731	0.094	0.425	0.724	0.099	0.422
O	0.929	0.153	0.614	0.903	0.161	0.615
H	0.760	−0.122	0.609	0.760	−0.097	0.600
H	0.575	−0.021	0.651	0.575	−0.013	0.665
H	0.944	0.191	0.311	0.939	0.190	0.316
H	0.772	0.295	0.386	0.765	0.282	0.396
	$\Delta_{CHO} = 0.016$					

a) Orthorhombic, space group *Pbnb*.
b) At 0 K.
c) By Ueda et al. [505]. At room temperature.
Source: Kawai et al. [238]/Reproduced with permission of the PCCP Owner Societies.

Table 24.25 Optimized and experimental crystal structures of α form of PBS.[a)]

	Optimized[b)]				Experimental[c)]			
	Lattice constant (Å, °)							
	a	b	c	β	a	b	c	β
	5.21	8.71	10.88	125.7	5.23	9.12	10.90	123.9
	Δ_{LC} = 1.2%							
	Fractional coordinates							
	x/a	y/b	z/c		x/a	y/b	z/c	
C	0.1351	0.0571	0.5489		0.0839	0.0711	0.5389	
C	0.0890	0.1524	0.6513		−0.0117	0.1387	0.6360	
C	0.0074	0.0964	0.8420		−0.0195	0.0832	0.8352	
C	0.0634	−0.0237	0.9551		0.0690	−0.0216	0.9564	
O	0.1262	0.0508	0.7670		0.0696	0.0351	0.7495	
O	−0.1270	0.2189	0.8202		−0.1536	0.1964	0.8164	
H	0.1556	0.1359	0.4765		0.0392	0.1534	0.4566	
H	0.3600	−0.0048	0.6175		0.3308	0.0504	0.6104	
H	0.2639	0.2440	0.7082		0.1104	0.2419	0.6821	
H	−0.1473	0.2021	0.5889		−0.2591	0.1590	0.5696	
H	0.3171	−0.0467	1.0308		0.3190	−0.0237	1.0322	
H	−0.0486	−0.1293	0.8911		−0.0138	−0.1308	0.9106	
	Δ_{CHO} = 0.068							

a) Monoclinic, space group $P2_1/n$.
b) At 0 K.
c) By Ichikawa et al. [209].
Source: Kawai et al. [238]/Reproduced with permission of the PCCP Owner Societies.

24.6 Biodegradability of Polyesters

Microbiologists have investigated biodegradation behaviors of polyesters using a variety of bacteria and enzymes extracted therefrom and found the following experimental facts [115, 371, 372, 479–481]: (1) The polymer-degrading microorganisms populate in the order of P3HB > PCL > PLA degraders. (2) In the natural environment, P3HB degraders are widely distributed among the families of *Pseudonocardiaceae* and the related genera, *Micromonosporaceae*, *Thermonosporaceae*, *Streptosporangiaceae*, and *Streptomycetaceae*. (3) Lipases hydrolyze aliphatic polyesters with many methylene groups, such as PCL, PES, and PBS but cannot degrade optically active polyesters such as P3HB and PLA.

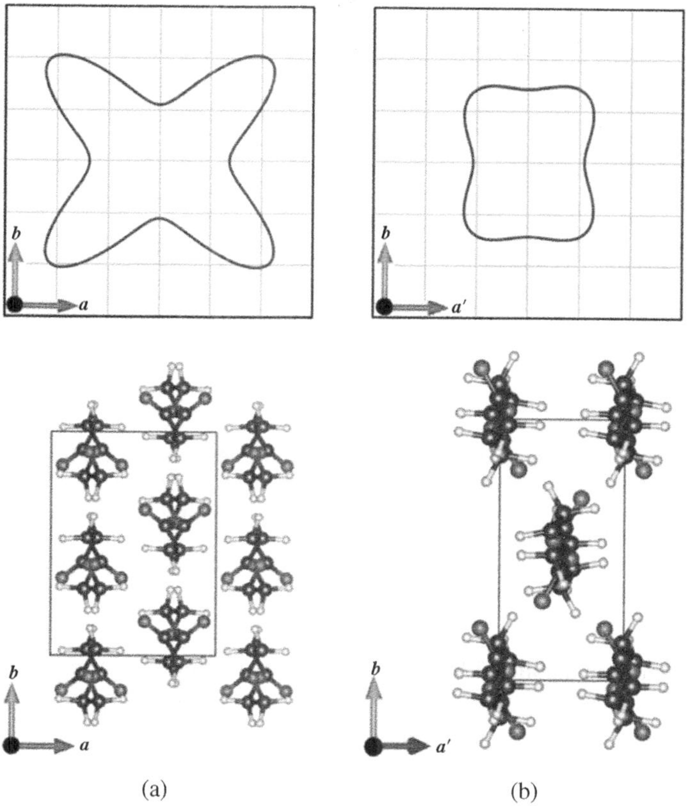

Figure 24.15 The optimized crystal structures (below) and Young's modulus distributions on the plane perpendicular to the fiber axis (above): (a) PES; (b) α-form of PBS. The grid spacing corresponds to 10 GPa. Source: [238]/Kawai et al./Reproduced with permission of the PCCP Owner Societies.

(4) A serine protease, proteinase K, which selectively cleaves the alanine–alanine link of silk fibroins, also degrades PLA. (5) The melting point of the polyesters is closely related to the enzymatic degradability.

A P3HB depolymerase was extracted from *Penicillium funiculosum*, complexed with a substrate of (*R*)-3-hydroxybutyrate (3HB), crystallized, and subjected to synchrotron X-ray diffraction, and its crystal structure was determined as shown in Figure 24.16a [200]. On the surface of the depolymerase, a hollow is shaped, and the 3HB substrate was captured therein. Figure 24.16b illustrates the conformation and hydrogen bonds of the bound hydroxybutyrate substrate, which contains two sets of bonds a–d (designated as units A and B). The conformations of bonds a–d were determined as follows: unit A, t, g^-, t, and t; unit B, t, t, g^+, and g^+; Unit B lies in the most stable conformation (ttg^+g^+), and unit A lies in a metastable

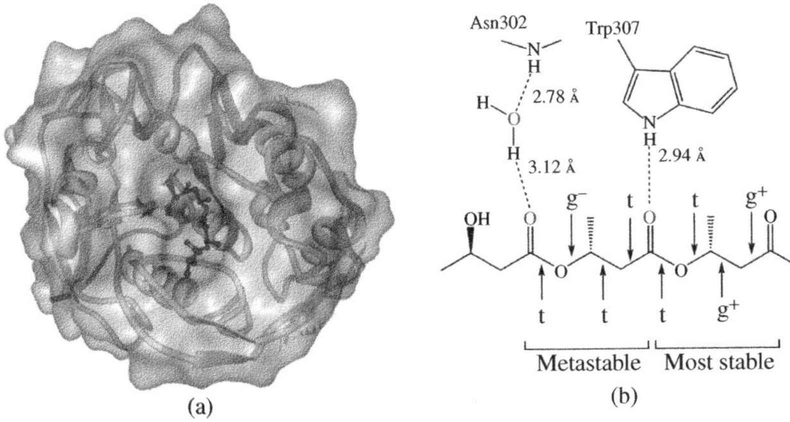

Figure 24.16 Crystal structure of a complex of a P3HB depolymerase extracted from *Penicillium funiculosum* with an (*R*)-3-hydroxybutyrate (HB) substrate (located at the center) [200]: (a) The structural data stored in the Protein Data Bank (PDB ID: 2D81) were visualized by the Protein Workshop software [326]; (b) The schematic diagram shows conformations of the skeletal bonds and N···O and O···O distances of the hydrogen bonds. Source: [407]/with permission of Springer Nature.

conformation (tg^-tt) and stabilized by an N—H···O=C hydrogen bonding with Trp307 and an O—H···O=C one with water. Only the (*R*)-isomer of 3HB can readily adopt the two stable conformations. If the substrate were the (*S*)-isomer, the mirror images (ttg^-g^- and tg^+tt) would be stabilized but unfit for the crevice and hence fail to approach the active site of the depolymerase. This may be the very reason why only (*R*)-form of P3HB exists in nature. In fact, Bachmann and Seebach [31] have investigated the degradation behaviors of 3HB oligomers composed of various stereosequences of (*R*)- and (*S*)-repeating units under the presence of a P3HB depolymerase extracted from *Alcaligenes faecalis* and found that the cleavage of the 3HB oligomers occurs only between the (*R*) units; at least two successive (*R*) units are requisite for the biodegradation.

P3HB is a truly biosynthetic and biodegradable polymer; therefore, it may be natural that many bacteria can accept P3HB. On the other hand, PCL, PES, and PBS are artificial products but nevertheless exhibit biodegradability, because these polyesters may adapt their conformations to the active site of, for instance, lipases. Figure 24.17 shows the crystal structure of a lipase, *Aspergillus oryzae* cutinase [283]. The arrow indicates its active-site crevice. For the enzymatic hydrolysis, the molecular chain is required to be completely embedded in the fissure and reach the catalytic site. For example, the PCL chain has neither asymmetric carbon nor side chain, in addition, being so flexible as to adopt a number of somewhat extended conformations. Accordingly, the PCL chain can fit itself in the narrow

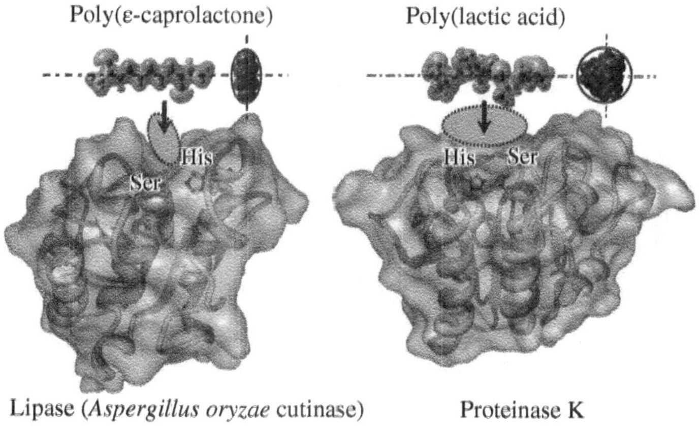

Figure 24.17 Affinities between degradable enzymes and polyesters: (left) *Aspergillus oryzae* cutinase (lipase) (PDB ID: 3GBS) [283] and PCL; (right) proteinase K extracted from *Tritirachium album limber* (PDB ID: 1IC6) [43] and PLA. The affinity depends largely on the fitness of the molecular shape of the polyester for the vacant space (indicated by the dotted circle) around the active site of the enzyme. Source: [238]/Kawai et al./Reproduced with permission of the PCCP Owner Societies.

crevice and access the active site. In contrast, the asymmetric PLA chain that can adopt only tg^+t and tg^+s conformations is too rigid, distorted, and thick to intrude into the fissure of lipases.

Proteinase K is a serine protease in which a serine residue acts as the nucleophilic center at the active site. It is known that the enzyme selectively cleaves the alanine–alanine link in silk fibroins [372]. Figure 24.17 shows proteinase K extracted from *Tritirachium album limber* [43]. Its active site is located at the center of a basin rather than a crevice; therefore, despite the distorted and thick shape, the PLA chain may access the active site.

The conformational preference of the alanine residue free from intra- and intermolecular hydrogen bonds has been investigated and expressed as the Ramachandran diagram [182, 355, 430, 433]. Its most stable conformation was found at the ϵ_L (polyproline II) position (see Figure 24.18); the existing probability was estimated to be 79% [182]. The dihedral angles ($\phi = -70°$, $\psi = 130°$) are comparable to those (−72.4°, 164.1°) of the most stable g^+t conformation of (*S*)-methyl 2-acetoxypropanoate, the monomeric model of LA, whose second most stable g^+s state is formed at (−73.1°, −14.9°), which is quite close to the position (−75°, −20°) of the α_L conformation of alanine. Consequently, the two highly populated conformations of LA monomer are so similar to those of alanine in both molecular shape and charge distribution that proteinase K would misidentify PLA as a polypeptide chain and hydrolyze PLA.

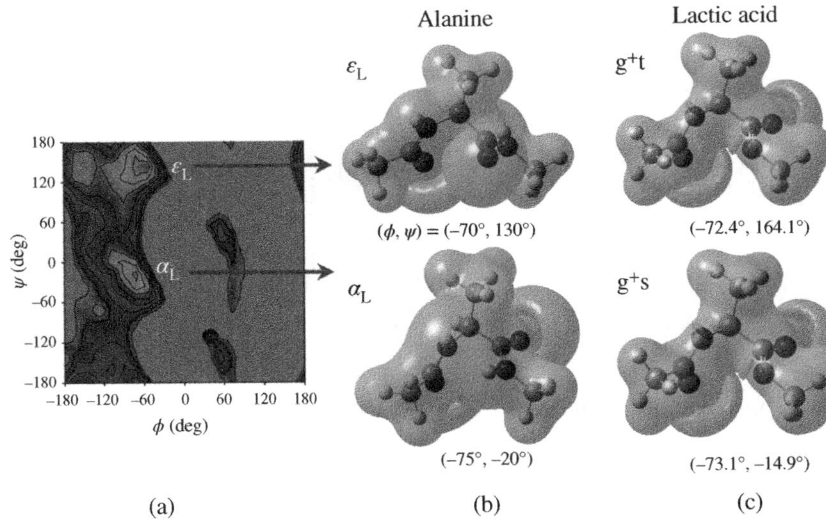

Figure 24.18 (a) Ramachandran diagram of unfolded alanine. Source: Adapted with permission from Parchaňský et al. [355]/American Chemical Society. (b) The first most stable conformations ($\varepsilon_L \approx g^+t$) and the second stable ones ($\alpha_L \approx g^+s$) of unfolded alanine and lactic acid model. The dihedral angle pairs (ϕ, ψ) are written in the parentheses. Source: [410]/Figure 7 (p. 337), Sasanuma et al./Reproduced with permission of Elsevier.

As mentioned above, the microbiologists have pointed out that the melting points of the biodegradable polyesters are closely related to the enzymatic degradabilities [372, 479–481]. The relationship may be interpreted in terms of chain flexibility (or rigidity) that can be quantified by S_{conf}. The configurational entropy, S_{conf}, amounts for a large percentage of the entropy of fusion, ΔS_u, and the equilibrium melting point (T_m^0) is given by $T_m^0 = \Delta H_u / \Delta S_u$, where ΔH_u is the enthalpy of fusion.

As shown in the previous section, the S_{conf} values at the equilibrium melting point T_m^0 (written in the parenthesis) of the biodegradable polyesters of interest here were evaluated to be 0.49 (PLA, 207 °C), 1.04 (P3HB, 203 °C), 1.05 (PGA, 231.4 °C), 1.22 (P2HB, 130.3 °C), 1.25 (PES, 114 °C), 1.46 (PBS, 133.5 °C), and 1.54 (PCL, 64 °C) cal K^{-1} (mole of bond)$^{-1}$. It can be seen that T_m^0 is inversely correlated to S_{conf} except for that of PGA enhanced by interchain interactions (ΔH_u). As has been revealed above, the affinity between polyester and depolymerase is significantly influenced by the chain flexibility. Therefore, the melting point is closely related with the enzymatic degradability via S_{conf}. All the five statements propounded by the microbiologists have been fully elucidated in terms of polymer physicochemistry, especially the conformational characteristics of the polyesters.

25

Polycarbonates

"Polycarbonate" is a generic term representing polymers that contain carbonate groups (–O–(C=O)–O–) between repeating units. However, "polycarbonate" mostly indicates a specific thermoplastic polycarbonate produced from bisphenol A by polycondensation with phosgene or by transesterification with diphenyl carbonate. The polycarbonate is amorphous and superior in transparency, thermal stability, and impact strength, thus being used for electronic compounds, construction materials, 3D printing, data storage, organic glasses, etc. [292].

However, this section exclusively treats "green" polycarbonates produced from epoxides and carbon dioxide: poly(ethylene carbonate) (PEC); poly(propylene carbonate) (PPC); and poly(cyclohexene carbonate) (PCHC). Owing to the growing worldwide interest in global warming, these polycarbonates, into which carbon dioxide may be fixed effectively, have been paid particular attention to. Inasmuch as monomeric units of PPC and PCHC include chiral centers, their stereo- and regioregularities are determined by the employed polymerization method. Therefore, stereospecific catalysts have been developed to control the primary and higher order structures and physical properties.

Inoue et al. [217, 218] synthesized alternating copolymers from epoxides and carbon dioxide (Figure 25.1) using diethylzinc and water as catalysts. It is of particular interest that carbon dioxide, the most thermodynamically stable (inactive) carbon compound, is reactivated by the ring-opening free energy of the distorted cyclic ethers. However, PEC and PPC are difficult to utilize as either hard plastics or flexible rubbers because these polycarbonates give rise to glass transitions around room temperature: PEC, ca. 20 °C and PPC, 35–40 °C (Figure 25.2) [504].

Inoue et al. [219] also synthesized PCHC from cyclohexene oxide (CHO) and CO_2, while, interestingly, Nozaki et al. [341] employed diethylzinc and chiral amino alcohol as catalysts and obtained PCHCs of up to enantiomeric excess = 73% from the same starting compounds. Since then, a variety of stereospecific catalysts have been developed to produce stereo- and regioregular polycarbonates [77, 100, 286, 289, 460]. The glass transition temperature and melting point of

Conformational Analysis of Polymers: Methods and Techniques for Structure-Property Relationships and Molecular Design, First Edition. Yuji Sasanuma.

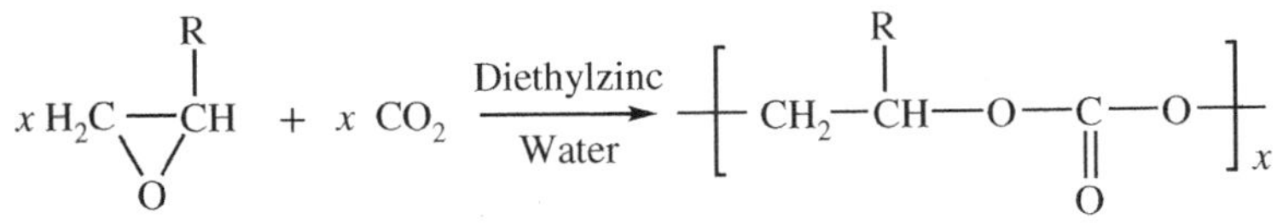

Figure 25.1 Polymerization of aliphatic polycarbonates from epoxides and carbon dioxide: R = H, poly(ethylene carbonate) (abbreviated as PEC); R = CH_3, poly(propylene carbonate) (PPC).

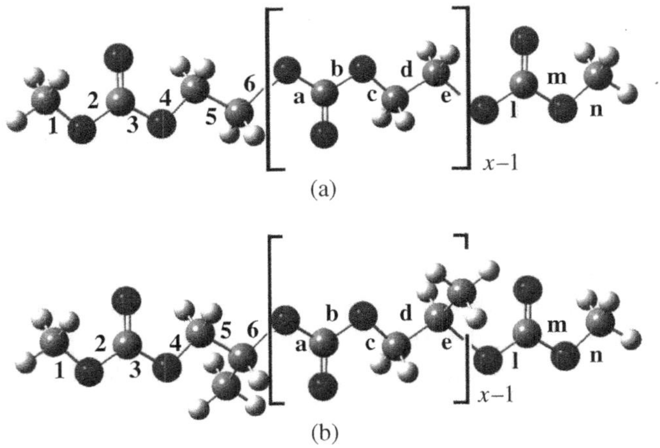

Figure 25.2 (a) Poly(ethylene carbonate) (PEC) and (b) poly(propylene carbonate) (PPC). The bonds are designated as shown, and *x* is the degree of polymerization. To facilitate the RIS calculations, the polymeric chains are terminated by methyl groups.
Source: [415]/Figure 1 (p. 4809), Sasanuma and Takahashi/Reproduced with permission of American Chemical Society.

perfectly isotactic PCHC are 130 and 248 °C, respectively [178]. Probably, the thermal properties allow us to use the PCHC as a hard resin. Furthermore, PCHC can be depolymerized into cyclohexene carbonates (CHCs), which will be recycled and transformed into PCHC again [178, 219, 432]. This means that a closed recycling process without CO_2 release may be accomplished by the production, decomposition, and reproduction of PCHC.

25.1 Poly(ethylene carbonate) (PEC) and Poly(propylene carbonate) (PPC)

Similar to poly(propylene oxide) (PPO), PPC shows two kinds of configurations: stereoisomers, (*R*)- and (*S*)-forms (Figure 25.3) and regioisomers, H—H, H—T,

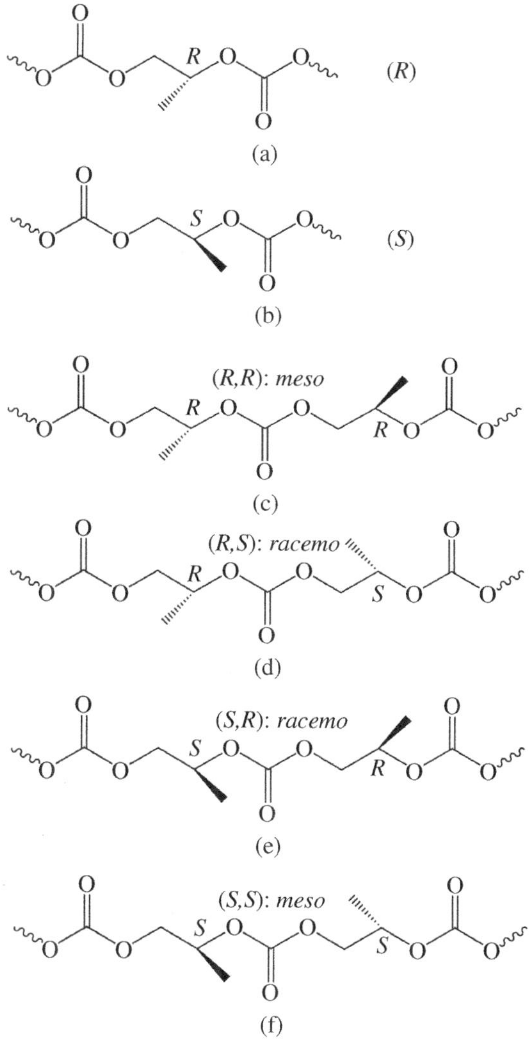

Figure 25.3 Stereosequences of PPC: (a) (*R*)- and (b) (*S*)-isomers; (c) (*R*, *R*)-, (d) (*R*, *S*)-, (e) (*S*, *R*)-, and (f) (*S*, *S*)-diads. (*R*, *R*) and (*S*, *S*) combinations are designated as *meso*, and (*R*, *S*) and (*S*, *R*) ones as *racemo*. When the polymeric chain is composed of only *meso* (*racemo*) couplings, the stereoregularity is termed isotactic (syndiotactic). Source: [415]/Figure 2 (p. 4809), Sasanuma and Takahashi/American Chemical Society.

and T—T linkages (Figure 25.4). Conformational analysis of PEC and PPC was carried out via MO calculations and nuclear magnetic resonance (NMR) experiments on the individual model compounds: PEC, ethylene glycol bis(methyl carbonate) (abbreviated as E_model, Figure 25.5) and PPC, propylene glycol bis(methyl carbonate) (P_model). The configurational properties of the two polycarbonates were evaluated from the refined rotational isomeric state (RIS) calculations using the MO energies. For PPC, the regio- and stereosequences were generated according to the Bernoulli trial and the Markov chain.

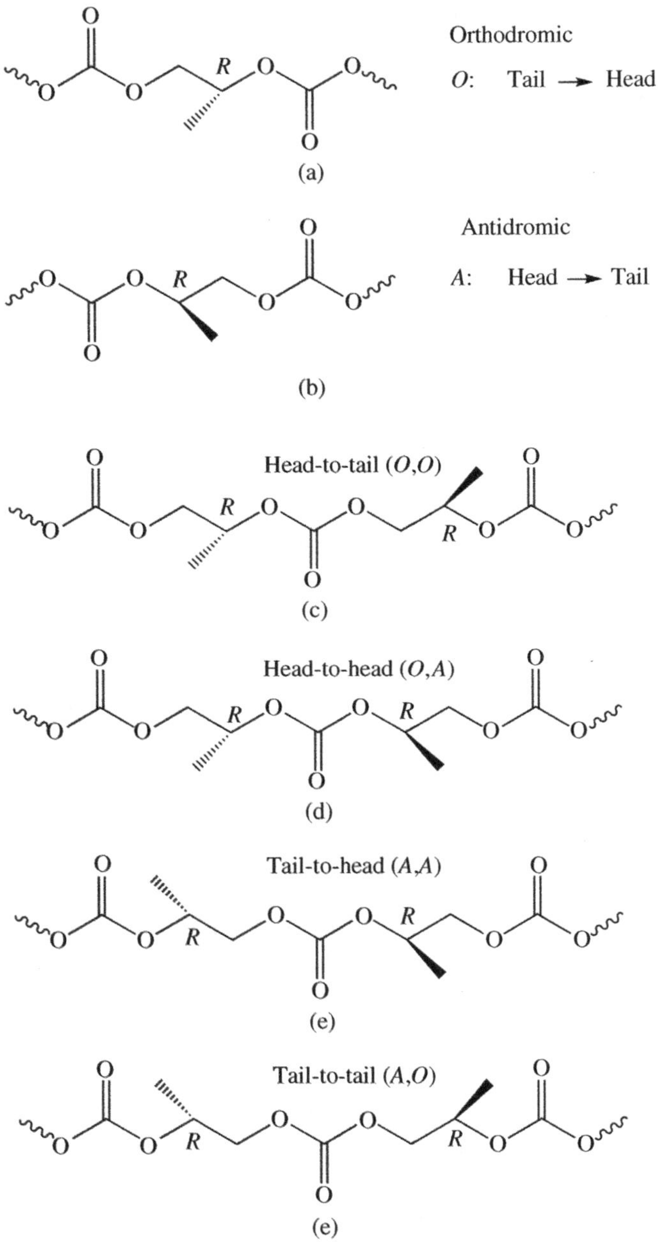

Figure 25.4 Regiosequences of PPC. Definition of (a) orthodromic (abbreviated as *O*) and (b) antidromic (*A*) directions. There are four possible combinations of (*O*) and (*A*) directions between neighboring units: (c) (*O*, *O*), head-to-tail (represented as H–T); (d) (*O*, *A*), head-to-head (H–H); (e) (*A*, *A*), tail-to-head (T–H); and (f) (*A*, *O*), tail-to-tail (T–T). The tail-to-head linkage is included in H–T; therefore, the three expressions, H–T, H–H, and T–T, are used. Source: [415]/Figure 3 (p. 4809), Sasanuma and Takahashi/American Chemical Society.

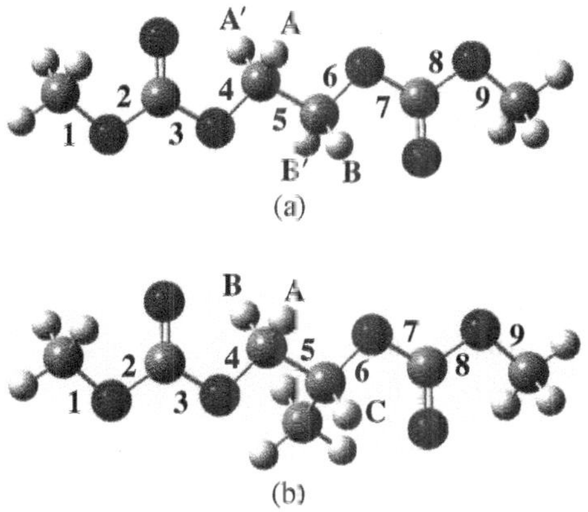

Figure 25.5 Model compounds of PEC and PPC: (a) for PEC, ethane-1,2-diyl dimethyl bis(carbonate) (abbreviated as E_model) and (b) for PPC, dimethyl propane-1,2-diyl bis(carbonate) (P_model). As indicated, the bonds are numbered, and the methylene and methine protons are designated for NMR analysis. (*R*)- and (*S*)-P-models yield the identical NMR spectra; therefore, (*R*)-P-model is exclusively employed herein.
Source: [415]/Figure 4 (p. 4810), Sasanuma and Takahashi/Reproduced with permission of American Chemical Society.

25.1.1 NMR Experiment

Vicinal ^1H–^1H coupling constants observed from E_model and P_model were analyzed similar to those of 1,2-dimethoxyethane (DME) and 1,2-dimethylpropane (DMP) to yield conformational fractions of bond 5. The small p_t values of E_model (Table 25.1) indicate a strong gauche preference and tend to decrease slightly with solvent polarity. The bond conformations of bond 5 of P_model are listed in Table 25.2, and the magnitude relation of $p_t \ll p_{g-} < p_{g+}$ is always kept regardless of the solvent. In the conformational preference, E_model is close to DME, and P_model is similar to DMP dissolved in polar dimethyl sulfoxide (DMSO, Sections 17.2 and 15.1).

25.1.2 MO Calculation

Table 25.3 shows the conformer free energies of the E_model. The most stable conformation is tg$^{\pm}$t. Both aromatic and aliphatic esters with the O—CH_2—CH_2—O bond sequence exhibit strong gauche preferences in the central CH_2—CH_2 bond. The free energies of the tg$^{\pm}$t conformation were obtained as, for example, −1.1 kcal mol^{-1} for poly(ethylene terephthalate) (PET) [400] and −1.2 kcal mol^{-1} for poly(ethylene succinate) (PES) [410]. These ΔG_k values are comparable to that of the E_model (Table 25.3). Its ttg$^{\pm}$ conformer also has a relatively small free energy of 0.1–0.3 kcal mol^{-1}. The ttg$^{\pm}$ conformation forms a short

Table 25.1 Trans fractions (p_ts) of E_model at 25 °C: comparison between MO calculations and NMR experiments.

	Bond 4 (6)	Bond 5	
Medium	**MO**	**MO**	**NMR**
Gas	0.39	0.12	
Chloroform	0.50	0.07	0.06
Acetone	0.54	0.05	0.05
Methanol	0.55	0.04	0.04
DMSO	0.55	0.04	0.03

Source: Adapted from Sasanuma and Takahashi [415].

Table 25.2 Bond conformation of (*R*)-P_model at 25 °C: comparison between NMR experiments and MO calculations.

	Bond 4			Bond 5			Bond 6		
Medium	p_t	p_{g+}	p_{g-}	p_t	p_{g+}	p_{g-}	p_t	p_{g+}	p_{g-}
				NMR expt					
Chloroform				0.11	0.50	0.39			
Acetone				0.09	0.49	0.42			
Methanol				0.09	0.49	0.42			
DMSO				0.07	0.49	0.44			
				MO calc					
Gas	0.45	0.41	0.14	0.20	0.38	0.42	0.59	0.01	0.40
Chloroform	0.49	0.40	0.11	0.11	0.48	0.41	0.67	0.00	0.33
Acetone	0.51	0.39	0.10	0.08	0.50	0.42	0.68	0.00	0.32
Methanol	0.52	0.38	0.10	0.07	0.50	0.43	0.68	0.00	0.32
DMSO	0.52	0.38	0.10	0.07	0.50	0.43	0.68	0.00	0.32

C=O···C—H contact (2.40 Å) between the carbonyl and methylene hydrogen, and the *O=C—O—C—H* part forms a plane (* represents the contact point). In Table 25.1, the trans fractions calculated from the ΔG_k values are compared with those from nuclear magnetic resonance (NMR). At least, as for bond 5, the MO and NMR data are fully consistent with each other.

The charge distribution of the most stable tg$^+$t conformation of the E_model, based on the MO calculations, shows the negative charges due to the carbonate

Table 25.3 Conformer free energies (ΔG_ks) of E_model for PEC, evaluated from MO calculations[a)].

				PEC[b)]				
k	Conformation[c)]			Gas	Chloroform	Acetone	Methanol	DMSO
1	t	t	t	0.00	0.00	0.00	0.00	0.00
2	t	t	g^+	0.11	0.23	0.28	0.29	0.29
3	t	g^+	t	−0.81	−1.35	−1.59	−1.62	−1.64
4	t	g^+	g^+	−0.59	−1.00	−1.20	−1.23	−1.25
5	t	g^+	g^-			(absent)[d)]		
6	g^+	t	g^+	1.90	1.89	1.85	1.84	1.84
7	g^+	t	g^-	0.59	0.82	0.87	0.88	0.88
8	g^+	g^+	g^+	−0.42	−0.45	−0.48	−0.49	−0.49
9	g^+	g^+	g^-	−0.72	−0.81	−0.89	−0.90	−0.91
10	g^+	g^-	g^+			(absent)[d)]		

a) In kcal mol^{-1}. Relative to the all-trans conformation.
b) From E_model. At the MP2/6-311+G(2d,p)//B3LYP/6-311+G(2d,p) level.
c) In the O—CH_2—CH_2—O bond sequence.
d) The local minimum of the potential was not found by the geometrical optimization.
Source: [415]/Table 5 (p. 4812), Sasanuma and Takahashi/Reproduced with permission of American Chemical Society.

oxygen atoms to be distributed helically on the molecular surface; accordingly, the PEC chains electrostatically repel each other and fail to form a regular crystal. In addition, PEC and PPC prefer distorted conformations including gauche bonds. These may be the origin of their amorphous nature.

Owing to the asymmetric carbon, P_model may have 27 (= 3^3) conformers. A number of conformers have negative ΔG_k values. In particular, those of g^+g^+t and tg^+t, and $g^+g^-g^-$ are large negative; therefore, the P_model (PPC) also adopts distorted shapes due to the gauche conformations. The bond conformations of the P_model, calculated from the ΔG_k values (not shown), are listed in Table 25.2, where the magnitude relation of $p_t \ll p_{g-} < p_{g+}$ can be found for bond 5 (CH_2-$CH(CH_3)$), thus being consistent with the NMR experiment.

25.1.3 RIS Calculation

Characteristic ratios, their temperature coefficients, configurational entropies, and averaged geometrical parameters of unperturbed PEC and PPC chains at 25 °C are presented in Table 25.4. The PPC chain is assumed to be isotactic and composed of only the H—T linkage.

Table 25.4 Configurational properties and averaged geometrical parameters of PEC and isotactic (*R*)-PPC at 25 °C, evaluated from the refined RIS calculations with MO parameters including solvent effects.

		Gas	Chloroform	Acetone	Methanol	DMSO
		PEC				
$\langle r^2\rangle_0/nl^2$		2.42	2.52	2.54	2.54	2.54
$d\ \ln\langle r^2\rangle_0/dT \times 10^3$ (K^{-1})		0.32	−0.12	−0.25	−0.26	−0.27
S_{conf} (cal K^{-1} mol^{-1})		5.51	5.16	4.97	4.94	4.93
$f_e/f \times 10^3$		95	−35	−73	−78	−80
Geometry[a)](chloroform)	Bond	$\bar{l}$	$\bar{\theta}$	$\bar{\phi}_t$	$\bar{\phi}_{g+}$	$\bar{\phi}_{g-}$
	a	1.339	107.9	0.0		
	b	1.339	115.5	0.0		
	c	1.440	110.0	0.0	95.5	−95.5
	d	1.510	110.0	0.0	111.9	−111.9
	e	1.440	115.5	0.0	95.5	−95.5
		Isotactic (*R*)-PPC of 100% H—T				
$\langle r^2\rangle_0/nl^2$		2.36	2.26	2.26	2.26	2.26
$d\ \ln\langle r^2\rangle_0/dT \times 10^3$ (K^{-1})		1.07	0.44	0.24	0.20	0.18
S_{conf} (cal K^{-1} mol^{-1})		4.34	4.12	4.05	4.04	4.03
$f_e/f \times 10^3$		320	130	72	60	52
Geometry[a)](chloroform)	Bond	$\bar{l}$	$\bar{\theta}$	$\bar{\phi}_t$	$\bar{\phi}_{g+}$	$\bar{\phi}_{g-}$
	a	1.337	107.9	0.0		
	b	1.340	115.5	−0.1		
	c	1.440	110.1	0.1	87.7	−72.3
	d	1.517	107.3	3.4	111.9	−113.0
	e	1.453	116.5	−31.0	116.5	−96.8

a) The geometrical parameters averaged at 25 °C with the MO energies including the solvent effect of chloroform. Symbols: $\bar{l}$, averaged bond length (in Å); $\bar{\theta}$, averaged bond angle (in deg); and $\bar{\phi}_\xi$, average dihedral angle (in deg) of the ξ conformation.
Source: [415]/Table 8 (p. 4814), Sasanuma and Takahashi/American Chemical Society.

The $\langle r^2\rangle_0/nl^2$ values are 2.42–2.54 (PEC) and 2.26–2.36 (PPC), much smaller than, for example, those of PEO (5.2 at 34.5 °C) [7, 237, 412], PPO (6.0 at 50 °C) [22, 398], and polyethylene (6.4–8.3 around 140 °C) [6, 132, 141, 282, 294, 424]. This is because both PEC and PPC show strong gauche preferences in all bonds. The $d \ln\langle r^2\rangle_0/dT$ values of PEC and PPC decrease with increasing solvent polarity. The temperature coefficient of PEC changes its sign between gas and chloroform, while that of PPC is always positive.

As described in Section 4.4, the temperature coefficient is related to the rubber elasticity according to $Td\ln\langle r^2\rangle_0/dT = f_e/f$, where the tension ($f$) due to the rubber elasticity $f = f_e + f_S$, with f_e and f_S being the energy (f_e) and entropy (f_S) terms, respectively. The PPC chain shows positive f_e/f values because PPC by nature prefers distorted conformations such as g^+g^+t, $g^+g^-g^-$, tg^+t, tg^-g^-, and tg^-t. As temperature increases, the chain distortion will be released with the gauche bonds being shifted to trans: $d\ln\langle r^2\rangle_0/dT > 0$. The stretching of the PPC chain increases the internal energy (stable → unstable change): $(\partial U/\partial L) > 0$; therefore, $f_e/f > 0$. The positive $d\ln\langle r^2\rangle_C/dT$ value of PPC suggests the possibility of rubber elasticity. On the other hand, PEC changes the sign of f_e/f, depending on the environment.

It was reported that PEC with a low T_g behaves as an elastomer at room temperature with an elongation at break greater than 600% and completely recovers to the initial length after removal of the load [478]. Although PPC is brittle below 20 °C, an effective plasticizer (for example, 10 wt% of 1,6-bis(methyl urethane)hexane) reduces T_g of PPC and enhances the mechanical properties: elongation at break, approximately 700%; tensile strength, 30 MPa [75].

The characteristic ratios of the PPC chains with different regio- and stereosequences were also calculated with the ΔG_k values on the chloroform environment because the dielectric constant (approximately 3) of PPC [291] is comparatively close to that (4.8) of chloroform. According to the Bernoulli trial, the $\langle r^2\rangle_0/nl^2$ value of PPC was calculated as a function of p_R and p_{ortho} (Figure 25.6a), where p_R and p_{ortho} are event probabilities of (R)-isomer and the orthodromic direction, respectively. For definitions of (R)- and (S)-isomers and orthodromic (O) and antidromic (A) directions, see Figures 25.3 and 25.4. The stereoinversion (R) → (S) or (S) → (R) and regioinversion ((O) → (A) or (A) → (O)) were assumed to occur independently (independent-event model). For atactic chains generated by the Bernoulli trial, the calculated quantities of $f(p)$s always satisfy $f(p_{\mathrm{ortho}}) = f(1 - p_{\mathrm{ortho}})$ and $f(p_R) = f(1 - p_R)$. This means that the $f(p)$ is symmetric with respect to $p_{\mathrm{ortho}} = 0.5$ and $p_R = 0.5$.

In Figure 25.6a, therefore, the $\langle r^2\rangle_0/nl^2$ value is plotted within ranges of $0.0 \le p_{\mathrm{ortho}} \le 0.5$ and $0.0 \le p_R \le 0.5$. On the line of $p_R = 0.0$ (isotactic), the characteristic ratio increases from 2.26 to 2.66 with an increase in p_{ortho}. In the range of $p_R = 0.4$ –0.5 (atactic), the $\langle r^2\rangle_0/nl^2$ data converge to ~ 2.54. This means that the stereochemically irregular PPC chains have almost the same average dimension irrespective of the regioregularity.

After polymerization, PPC usually keeps the original chirality of propylene oxide and includes the H—T linkage predominantly. However, propylene oxide rarely undergoes abnormal ring opening, and consequently, the H—H and T—T linkages

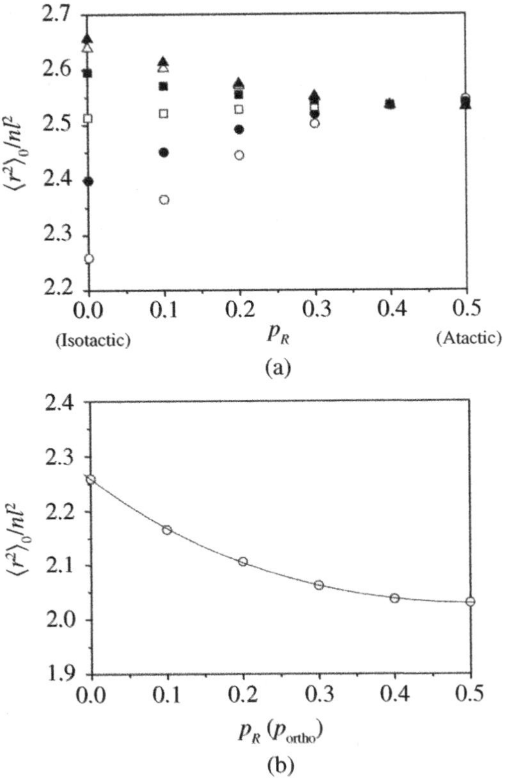

Figure 25.6 Characteristic ratios ($\langle r^2 \rangle_0/nl^2$s) of PPC, derived from the refined RIS calculations with the Bernoulli trial. (a) (Independent-event model) the regio- and stereosequences were generated independently of each other, and the $\langle r^2 \rangle_0/nl^2$ values are plotted against p_R for different p_{ortho} and (p_{H-T}, p_{H-H}, p_{T-T}) values: 0.0 and (1.00, 0.00, 0.00) (open circle); 0.10 and (0.82, 0.09, 0.09) (filled circle); 0.20 and (0.68, 0.16, 0.16) (open square); 0.30 and (0.58, 0.21, 0.21) (filled square); 0.40 and (0.52, 0.24, 0.24) (open triangle); 0.50 and (0.50, 0.25, 0.25) (filled triangle). (b) (Synchronous model) the regio- and stereoforms are changed synchronously: (*O*, *R*)→(*A*, *S*), (*A*, *R*)→(*O*, *S*), (*O*, *S*)→(*A*, *R*), or (*A*, *S*)→(*O*, *R*). The solid line represents a cubic function fitted to the calculated date (open circle). Source: [415]/Figure 7 (p. 4814), Sasanuma and Takahashi/American Chemical Society.

occur before and behind the defective unit, at which the chiral form is forced to be different from those of the neighbors [81, 395]: both stereoinversion (*R*) → (*S*) or (*S*) → (*R*) and regioinversion (*O*) → (*A*) or (*A*) → (*O*) occur simultaneously, but the defect probability stays as small as several percent. Thus, the defect incidence may be represented by the Bernoulli trial, both regio- and stereoinversions being synchronized (synchronous model). In Figure 25.6b, the characteristic ratios thus calculated are plotted as a function of p_R (= p_{ortho}). The curve is also symmetric with respect to $p_R = 0.5$ ($p_{ortho} = 0.5$) and goes down with increasing p_R. The synchronous inversions render the PPC chain more contracted: the $\langle r^2 \rangle_0/nl^2$ value decreases by 10% between $p_R = 0.0$ and 0.5.

In Chapter 14, the chain dimensions of PPC, calculated by the refined RIS scheme with the Markov stochastic model, are drawn as a contour map as a function of $p_{H\text{-}T}$ and p_{meso} (Figure 14.2), where $p_{H\text{-}T}$ and p_{meso} are the event probabilities of the H—T linkage and *meso* diad, respectively. On the line of $p_{meso} = 0.0$ (syndiotactic), the $\langle r^2 \rangle_0/nl^2$ value increases from 1.98 to 2.80 with increasing $p_{H\text{-}T}$, while on the line of $p_{meso} = 1.0$ (isotactic), it decreases from 3.08 to 2.26. Therefore, the maximum can be found at $p_{meso} = 1.0$ and $p_{H\text{-}T} = 0.0$: either (O, R)- and (A, R)-units or (O, S)- and (A, S)-units are arranged alternately. The minimum is essentially equal to that (≈ 2.0) of the above synchronous model of $p_R = p_{ortho} = 0.5$ and located at the origin, $p_{meso} = p_{H\text{-}T} = 0.0$, where either (O, R)- and (A, S)-units or (A, R)- and (O, S)-units are arranged alternately (synchronous inversions).

As shown above, the characteristic ratio of PPC depends only a little on the regio- and stereoregularities. This is probably because the carbonate group separates the neighboring O–CH_2–CH(CH_3)–O parts with two rigid O—C(=O)—O bonds; therefore, the conformational correlations between adjacent repeating units are weak, and the individual units are allowed to change the conformation almost freely from the neighbors. Accordingly, it is preferable that PEC and PPC should be prepared without paying particular attention to the regio- and stereoregularities, processed so as to lower the glass transition temperatures, and used as amorphous flexible materials.

25.2 Poly(cyclohexene carbonate) (PCHC)

PCHC includes two chiral centers in the repeating unit (see Figure 25.7). The chirality stems from the precursor, cyclohexene oxide (CHO). In the polymerization, (R, S)-*cis*-CHO is exclusively used because *cis*-CHO undergoes a stereo inversion of either $R \rightarrow S$ or $S \rightarrow R$ [341], passes through (R, R)- or (S, S)-*trans*-intermediate, and becomes *trans*-PCHC. Then, there is the possibility that two kinds of diads may be generated: *meso*, $(R, R)(R, R)$ and $(S, S)(S, S)$, and *racemo*, $(R, R)(S, S)$ and $(S, S)(R, R)$. If the PCHC chain includes only the *meso* (*racemo*) diad, the stereoregularity is termed isotactic (syndiotactic) (Figure 25.7). Almost all PCHCs synthesized so far are rich in *meso* diads [125, 284, 530], and PCHCs rich in *racemo* diads are quite rare: exceptionally, syndio-rich PCHCs of up to *racemo* probability (P_{racemo}) = 0.81 were reported [89]. Here, the PCHCs including (R, R)- and/or (S, S)-repeating units are designated as *trans*-PCHC, in which two O—C bonds connected to the same cyclohexane ring appear on the different sides of the ring plane.

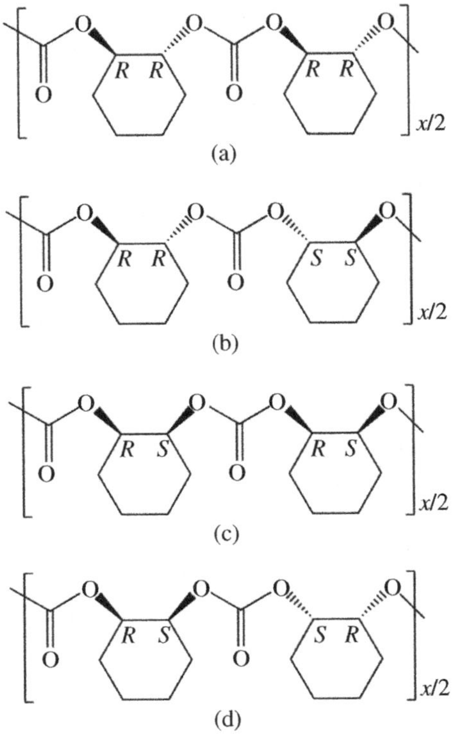

Figure 25.7 (a) Isotactic and (b) syndiotactic *trans*-poly(cyclohexene carbonate) (abbreviated herein as PCHC) and (c) isotactic and (d) syndiotactic *cis*-PCHC. *x* is the degree of polymerization. Source: [542]/Figure 1 (p. 9363), Yoshida et al./Reproduced with permission of American Chemical Society.

Probably, stereoregular PCHCs composed of (*R*, *S*)- and/or (*S*, *R*)-*cis*-repeating units have not been synthesized yet. Nevertheless, we can define *meso* and *racemo* linkages for such PCHCs: *meso*, (*R*, *S*)(*R*, *S*) and (*S*, *R*)(*S*, *R*); *racemo*, (*R*, *S*)(*S*, *R*) and (*S*, *R*)(*R*, *S*). On this basis, isotactic, syndiotactic, and atactic PCHCs are also defined (Figure 25.7). Here, the PCHCs including (*R*, *S*)- and/or (*S*, *R*)-repeating units are designated as *cis*-PCHC, in which the two O–C bonds appear on the same side of the cyclohexane plane.

(1*R*,2*R*)-*trans*-Di(methoxycarbonyloxy)cyclohexane (*trans*-DMCC) and (1*R*,2*S*)-*cis*-di(methoxycarbonyloxy)cyclohexane (*cis*-DMCC) were adopted as model compounds of *trans*- and *cis*-PCHCs, respectively (Figure 25.8). Conformational analysis of *trans*- and *cis*-PCHCs were conducted via ^{1}H and ^{13}C NMR experiments and molecular orbital (MO) calculations on the two models, and the configurational properties of the *trans*- and *cis*-PCHCs were evaluated from the RIS calculations with the Bernoulli and Markov stochastic processes.

25.2.1 MO Calculation

(1*R*,2*R*)-*trans*-DMCC and (1*R*,2*S*)-*cis*-DMCC were used exclusively as the models because (1*S*,2*S*)-*trans*-DMCC and (1*S*,2*R*)-*cis*-DMCC are, respectively, their

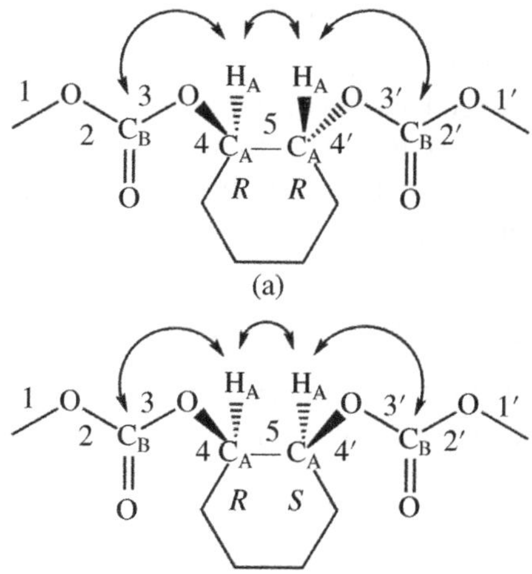

(a)

(b)

Figure 25.8 Model compounds: (a) (1*R*,2*R*)-*trans*-di(methoxycarbonyloxy)-cyclohexane (*trans*-DMCC); (b) (1*R*,2*S*)-*cis*-di(methoxycarbonyloxy)cyclohexane (*cis*-DMCC). The bonds are numbered as shown. The curved arrows indicate couples of $^3J_{HH}$ (between H_A and H_A) and $^3J_{CH}$ (between C_B and H_A).
Source: [542]/Figure 2 (p. 9363), Yoshida et al./Reproduced with permission of American Chemical Society.

mirror images, which exhibit NMR spectra identical with those of the prototypes, and, in the RIS calculations, the original conformational energies and geometrical parameters can be transformed to those of the mirror image by proper symmetry operations. Under the RIS approximation, 288 conformers may be defined for *trans*-DMCC (*cis*-DMCC); however, the number of asymmetric conformers is 156 (144). Here, the numbers and symbol in the parentheses are related to *cis*-DMCC. Bonds 2 and 3 or 2′ and 3′ are not allowed to adopt the cis conformation simultaneously owing to the steric hindrance between the cyclohexane ring and terminal methyl group; therefore, the remaining 90 (81) conformers underwent the MO calculations, and, finally, 73 (62) conformers reached the potential minima.

Of the 73 (62) conformers, the most stable state is tttg$^+$ttt (ttg$^-$g$^-$g$^+$tt). The conformational fractions of bonds 2–5, calculated form the ΔG_ks (not presented herein), are shown in Table 25.5. Bonds 2–4 of *trans*-DMCC strongly prefer trans

Table 25.5 Bond conformations of (1*R*,2*R*)-*trans*-DMCC at 25 °C, evaluated from MO calculations and NMR experiments.

	Bond number						
	2	3	4			5	
	MO						^{1}H NMR
Medium	p_t	p_t	p_t	p_{g+}	p_{g-}	p_t	p_t
Gas	0.98	0.99	0.87	0.01	0.12	0.32	
Chloroform	0.94	0.97	0.90	0.01	0.09	0.13	0.13
DMSO	0.91	0.96	0.89	0.01	0.10	0.09	0.04

conformations. For bond 5, namely the C_A—C_A bond of the cyclohexane ring (Figure 25.8), only two rotamers are defined: trans, viz. axial-axial and gauche$^+$, viz. equatorial-equatorial. The latter is more stable than the former. Bonds 2–4 of *cis*-DMCC show trans, trans, and gauche$^-$ preferences, respectively. Its bond 5 can adopt gauche$^+$ and gauche$^-$ states, and both are geometrically equivalent; thus, $p_{g+} = p_{g-} = 1/2$.

25.2.2 NMR Experiment

^{1}H NMR of the methine proton H_A of *trans*-DMCC gave the vicinal coupling constant between two H_As, $^3J_{HH}$, which may be expressed as

$$^3J_{HH} = J_G\, p_t + J_T\, p_{g+} \tag{25.1}$$

where p_t and p_{g+} are trans and gauche$^+$ fractions of bond 5 and satisfy

$$p_t + p_{g+} = 1 \tag{25.2}$$

The coefficients, J_G and J_T, were taken from a Karplus equation $^3J_{HCCH}(\phi^{CC})$ of the O—CH—CH—O bond sequence of α-D-arabinofuranoside [459]:

$$^3J_{HCCH}(\phi^{CC}) = 5.23 + 0.02\cos(\phi^{CC}) + 4.67\cos(2\phi^{CC}) \tag{25.3}$$

This equation yields $J_T = 9.50$ Hz (optimized $\phi^{CC} = 168.49^\circ$) and $J_G = 2.53$ Hz ($\phi^{CC} = 62.74^\circ$), where ϕ^{CC}s correspond to the optimized dihedral angles of bond 5 of *trans*-DMCC. The p_t values thus obtained are seen to be comparable to the MO calculations (Table 25.5); both MO and NMR data show a strong gauche$^+$ (equatorial–equatorial) preference.

^{13}C NMR of carbon B yields the vicinal $^{13}C_B$–^1H_A coupling constant, $^3J_{CH}$, which may provide conformational information on bonds 4 and 4′. The dihedral angles of bonds 4 and 4′ vary to a large extent, depending on the conformations of their neighboring bonds. In addition, the chirality of carbon C_A makes g$^+$ and g$^-$ conformations of bond 4 nonequivalent; thus, three unknowns, p_t, p_{g+}, and p_{g-}, cannot be uniquely determined from only two equations: e.g. for bonds 4 and 4′ of *trans*-DMCC, $^3J_{CH} = J_G\, p_t + J_T\, p_{g+} + J_G\, p_{g-}$ and $p_t + p_{g+} + p_{g-} = 1$. Instead, to confirm the MO data, the vicinal $^{13}C_B$–^1H_A coupling constants were calculated from the free energies and dihedral angles that the MO calculations gave and compared with the experimental $^3J_{CH}$ values. The procedures and results are described in detail (Section 9.2.2). In short, a Karplus-type equation expressing $^3J_{CH}(\phi^{OC})$ as a function of ϕ^{OC} was formulated from DFT calculations, and the observed $^3J_{CH}$ values were satisfactorily reproduced from the Karplus equation and the ΔG_k values.

25.2.3 RIS Calculation

The difference in Gibbs free energy between trans and cis states of bonds a and b was evaluated from MO calculations for a dimeric model, dicyclohexyl carbonate (C_6H_{11}–O–C(=O)–O–C_6H_{11}) at the MP2/6-311+G(2d,p)//B3LYP/6-311+G(2d,p) level (Table 25.6). For the bond symbols, see Figure 25.9.

Table 25.6 Conformational energies for the RIS calculations on (*R*, *R*)-*trans*- and (*R*, *S*)-*cis*-PCHCs.

Conformation		ΔG_k (kcal mol^{-1})		
		Gas	Chloroform	DMSO
Bonds a and b				
cis[a)]		2.93	2.15	1.84
Bonds c—e				
		(*R*, *R*)-*trans*-PCHC[b)]		
ttt		0.00	0.00	0.00
tg^+	g^+tt	8.29	8.54	8.63
ttg^-	g^-tt	0.03	−0.02	−0.04
tg^+t		−1.03	−1.71	−1.99
tg^+g^+	g^+g^+t	2.30	1.67	1.34
g^+tg^+		12.15	12.68	12.85
g^+tg^-	g^-tg^+	8.05	8.14	8.10
$g^+g^+g^+$		4.15	4.14	4.11
$g^+g^+g^-$	$g^-g^+g^+$	1.89	1.49	1.25
		(*R*, *S*)-*cis*-PCHC[c)]		
tg^+t	tg^-t	0.00	0.00	0.00
tg^+g^-	g^+g^-t	3.70	3.52	3.32
$g^+g^+g^+$	$g^-g^-g^-$	6.62	7.24	7.44
$g^+g^+g^-$	$g^+g^-g^-$	10.36	10.94	11.05
$g^-g^+g^+$	$g^-g^-g^+$	−2.64	−2.39	−2.29

a) From dicyclohexyl carbonate. Relative to the trans state.
b) From *trans*-DMCC. Relative to the ttt conformation.
c) From *cis*-DMCC. Relative to the tg^+t (tg^-t) conformation.
Source: [542]/Table 4 (p. 9367), Yoshida et al./Reproduced with permission of American Chemical Society.

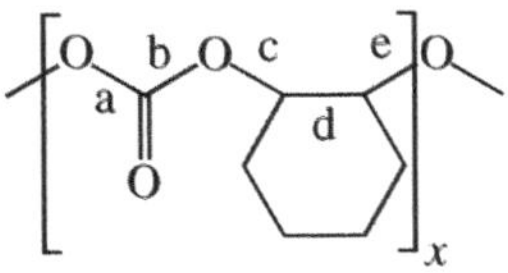

Figure 25.9 Poly(cyclohexene carbonate) (PCHC). As indicated, the skeletal bonds are designated.

The conformational energies for bonds c—e of *trans*- and *cis*-PCHCs were, respectively, determined from ΔG_ks of (*R*,*R*)-*trans*- and (*R*,*S*)-*cis*-DMCCs. Table 25.6 shows those of the conformations optimized normally, and the other states are regarded as extremely unstable or nonexistent. The statistical weight matrices and geometrical parameters of (*S*,*S*)-*trans*-PCHC and (*S*,*R*)-*cis*-PCHC were derived by symmetry operations from those of the (*R*,*R*)-*trans*- and (*R*,*S*)-*cis*-PCHCs, respectively.

Atactic *trans*-PCHC chains were generated by repeating Bernoulli trials as a function of the event probability (P_{RR}) of the (*R*, *R*)-unit, and isotactic, syndiotactic, and in-between tacticities were generated by the Markov process as a function of the probability (P_{meso}) of *meso* diad: (*R*, *R*)(*R*, *R*) and (*S*, *S*)(*S*, *S*). The probabilities were normalized to fulfill $P_{RR} + P_{SS} = 1$ and $P_{meso} + P_{racemo} = 1$, where P_{SS} and P_{racemo} are event probabilities of (*S*, *S*)-unit and *racemo* diad, respectively. On the basis of previous studies [415, 418, 422], both the degree of polymerization (x) and the number (n_c) of generated chains were set equal to 300.

Table 25.7 shows the results of the RIS calculations on *trans*-PCHC in the gaseous, chloroform, and DMSO environments at 25 °C. In addition, only for the medium environment (chloroform), the data at the glass transition temperature (T_g, 130 °C) and the melting point (T_m, 248 °C) of the isotactic chain were calculated. The $\langle r^2 \rangle_0/nl^2$ values of isotactic *trans*-PCHC are very large and depend on both medium and temperature: 36.76 (gas); 28.95 (chloroform, 25 °C); 12.82 (chloroform, 130 °C); 8.19 (chloroform, 248 °C); and 21.09 (DMSO, 25 °C). In the (*R*, *R*)-unit, the tg$^+$t conformation with the (bent) equatorial–equatorial (eq–eq) form is the lowest in energy, and metastable ttt and ttg$^-$ and g$^-$tt states lie in the (extended) axial–axial form (Table 25.6). As the environment becomes more polar, the eq–eq form becomes more stable. With an increase in temperature, unstable conformations containing cis in bonds a and b and gauche in bonds c and e will be somewhat more populated. These variations reduce the chain dimension of isotactic *trans*-PCHC.

In contrast, the $\langle r^2 \rangle_0/nl^2$ values of syndiotactic *trans*-PCHCs are extremely small: 1.65 (gas); 0.83 (chloroform); 0.74 (DMSO). Such phenomena have been found for syndiotactic alternating copolymers of L- and D-alanines [319] and poly(2-hydoxybutyrate) [422]. Figure 25.10 illustrates *meso* (isotactic) and *racemo* (syndiotactic) dimers lying in the most stable conformation. In the figure, the arrow represents the end-to-end vector (**r**). The *meso* dimer has a long **r**, whereas

Table 25.7 Configurational properties of isotactic and syndiotactic *trans*- and *cis*-PCHCs, evaluated from RIS calculations.

	Energy parameter				
	gas	Chloroform			DMSO
	25 °C	25 °C	130 °C[a)]	248 °C[b)]	25 °C
	Isotactic *trans*				
$\langle r^2\rangle_0/nl^2$	36.76	28.95	12.82	8.19	21.09
$d\ \ln\langle r^2\rangle_0/dT \times 10^3$ (K^{-1})	−6.8	−10.7	−5.2	−2.7	−10.1
S_{conf}[c)] (cal K^{-1} mol^{-1})	2.35	1.79	3.11	4.23	1.74
	Syndiotactic *trans*				
$\langle r^2\rangle_0/nl^2$	1.65	0.83			0.74
$d\ \ln\langle r^2\rangle_0/dT \times 10^3$ (K^{-1})	4.7	9.6			10.5
	Isotactic *cis*				
$\langle r^2\rangle_0/nl^2$	0.52	0.59			0.65
$d\ \ln\langle r^2\rangle_0/dT \times 10^3$ (K^{-1})	1.08	2.12			2.63
S_{conf}[c)] (cal K^{-1} mol^{-1})	1.68	2.05			2.30
	Syndiotactic *cis*				
$\langle r^2\rangle_0/nl^2$	59.85	24.86			16.11
$d\ \ln\langle r^2\rangle_0/dT \times 10^3$ (K^{-1})	−8.7	−9.8			−9.0

a) The glass transition temperature observed from *trans*-PCHC of P_{meso} = 0.99 [178].
b) The melting point observed from *trans*-PCHC of P_{meso} = 0.99 [178].
c) The statistical weight matrices of the (*S*, *S*)-unit ((*S*, *R*)-unit) are derived from those of (*R*, *R*)-unit ((*R*, *S*)-unit) by symmetry operations, and hence, the partition function (*Z*) and the S_{conf} value obtained therefrom are independent of P_{RR} and P_{meso}. The isotactic and syndiotactic PCHC chains have the same S_{conf} value.

Source: [542]/Table 5 (p. 9368), Yoshida et al./Reproduced with permission of American Chemical Society.

the *racemo* one has a very short **r**. The pictures clearly elucidate the geometrical differences between isotactic and syndiotactic *trans*-PCHCs.

The relationship between configurational properties and tacticity of *cis*-PCHC is completely opposite to those of *trans*-PCHC. For example, isotactic *cis*-PCHCs show very small characteristic ratios, whereas the syndiotactic chains yield very large ones: isotactic, 0.52 (gas), 0.59 (chloroform), and 0.65 (DMSO); syndiotactic, 59.85 (gas), 24.86 (chloroform), and 16.11 (DMSO). It seems that fully stereoregular *cis*-PCHCs have not been reported yet; however, because the configurational properties of isotactic and syndiotactic *cis*-PCHCs are similar to syndiotactic and isotactic *trans*-PCHCs, respectively, these facts suggest that it would not be necessary to aim for stereoregular *cis*-PCHCs.

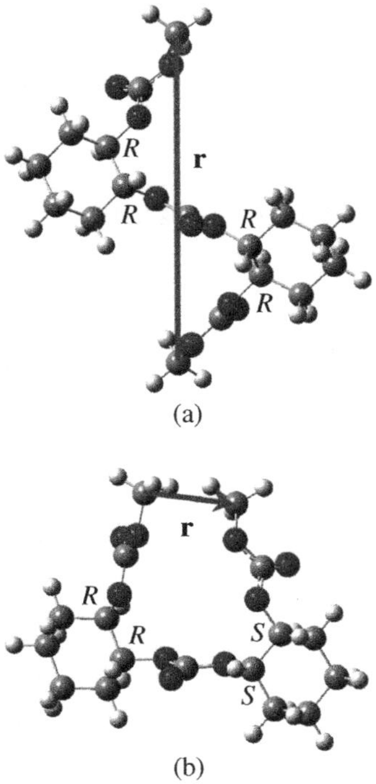

Figure 25.10 Dimer models with (a) *meso* and (b) *racemo* linkages in the most stable conformation: bonds a and b adopt the trans state, and bonds c–e lie in the tgt conformation ($tg^{+}t$ in (R, R) and $tg^{-}t$ in (S, S)). The arrow expresses the end-to-end vector, **r**. Source: [542]/Figure 10 (p. 9369), Yoshida et al./ Reproduced with permission of American Chemical Society.

25.2.4 Coherence Number

The coherence number (n_{coh}), the average number of successive repeating units with the same stereoisomer, is defined to represent the stereo-continuity. For example, if an oligo(cyclohexene carbonate) composed of

$$(R,R)(R,R)(R,R)(R,R)(S,S)(S,S)(R,R)(R,R)(R,R)$$

is supposed, that is it includes 4 (R, R)-units + 2 (S, S)-units + 3 (R, R)-units, then n_{coh} is evaluated to be $(4+2+3)/3 = 3.00$. It is obvious that the minimum n_{coh} is unity of syndiotactic chains, and that the maximum n_{coh} is the degree of polymerization (x) for a single isotactic chain or $x \times n_c$ for n_c isotactic chains. The coherence number gives a clue as to whether the stereopolymer may form stable crystallites. When stereo-polymeric chains crystallize, the surrounding units of the neighboring chains must be the same stereoisomer to form a homogeneous crystal or the other enantiomer to form a stereocomplex. Probably, with increasing coherence number, the crystallite or stereocomplex becomes well ordered and highly crystalline.

Figure 25.11 shows the coherence numbers calculated on the basis of the Markov process as a function of P_{meso}. The n_{coh} vs. P_{meso} curve goes up very slowly with increasing P_{meso}: e.g., 1.99 at $P_{meso} = 0.5$; 4.93 at $P_{meso} = 0.80$. It was found that X-ray diffraction peaks can be observed only from *trans*-PCHCs of $P_{meso} > 0.90$

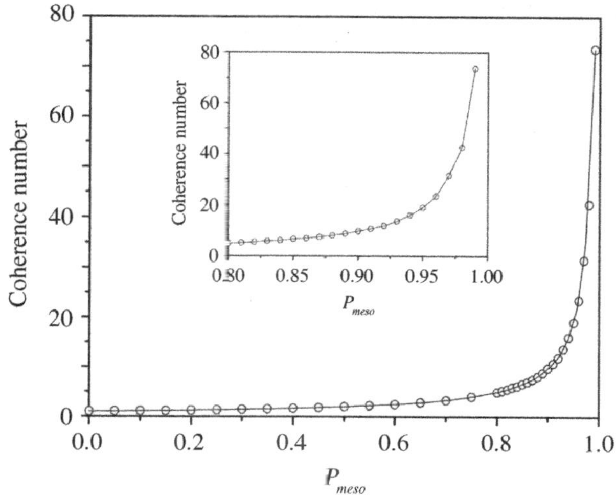

Figure 25.11 Coherence numbers (n_{coh}s) evaluated at 0.05 intervals of P_{meso} between 0.0 and 0.80 and at 0.01 intervals between 0.80 and 1.00 (inset). Source: [542]/Figure 13 (p. 9370), Yoshida et al./Reproduced with permission of American Chemical Society.

[460, 528]. At P_{meso} = 0.90, 0.98, and 0.99, the n_{coh} values are 9.74, 42.65, and 73.65, respectively. A *trans*-PCHC sample of P_{meso} > 0.99 was found to exhibit intense X-ray diffraction peaks even without annealing [125]. A blend of (*R*, *R*)- and (*S*, *S*)-*trans* PCHCs of P_{meso} = 0.98 seems to form a stereocomplex [285, 528].

Almost all PCHCs synthesized so far are iso-rich *trans*-polymers. In order to produce semicrystalline plastics of PCHC, it is appropriate to aim for *trans*-PCHCs of as high isotacticity as possible. Exceptionally, syndio-rich *trans*-PCHCs of up to P_{racemo} = 0.81 were prepared [89], and the configurational properties of the *trans*-PCHC at 25 °C can be calculated as follows: $\langle r^2\rangle_0/nl^2 = 1.48$, $d \ln \langle r^2\rangle_0/dT \times 10^3 = 5.56\ \text{K}^{-1}$, $S_{conf} = 1.79\ \text{cal K}^{-1}\ \text{mol}^{-1}$, and $n_{coh} = 1.23$. These parameters suggest that the syndio-rich *trans*-PCHC chains form amorphous aggregates and hence are far from semicrystalline. In fact, no fusion was observed therefrom. However, the temperature coefficient ($d \ln \langle r^2\rangle_0/dT$) at 25 °C is large positive; thus, as discussed in Section 4.4, the syndio-rich *trans*-PCHC chain would possibly behave like an elastomer if incorporated into a network structure.

It has been pointed out that PCHCs show low elongations at break and brittleness [448]. Isotactic *trans*-PCHCs with high T_gs cannot be expected to be flexible around room temperature. However, block copolymerizations with flexible polymer(s) such as poly(ethylene carbonate) and poly(propylene carbonate) of low T_gs [415] or if a small amount of ethylene-carbonate or propylene-carbonate unit is randomly inserted into the PCHC chain, the drawbacks would be somewhat improved.

26

Nylon 4

Nylon 4 is produced from 2-pyrrolidone, which can be prepared by fermentation of biomass using *Escherichia coli* from biobased γ-glutamic acid via γ-aminobutyric acid [535]; therefore, nylon 4 is a biobased and carbon-neutral polymer. In addition, nylon 4 is hydrolyzed to be γ-aminobutyric acid in the presence of *Pseudomonas* sp. ND-11 inhabiting widely in activated sludge and lastly broken down to CO_2, H_2O, and NO_3^- [185–187, 240, 454, 536]. Therefore, nylon 4 is also a biodegradable polymer.

Nylon 6 is superior in mechanical strength, rigidity, thermal stability, and chemical residence [168, 451]. These advantages stem mainly from interchain N—H···O=C hydrogen bonds Nylon 4 also forms strong hydrogen bonds so as to melt at temperatures (260–265 °C) higher than T_m (225 °C) of nylon 6 (Figure 26.1) [99, 168, 293]. Therefore, nylon 4 is expected to be practically used as a biobased and biodegradable polyamide. Section 10.8 describes the evaluation of the interchain interaction energies of the two polyamides from the periodic DFT-D calculations at the B3LYP/6-31G(d,p) level: nylon 4 (α form), −210 cal g^{-1}; nylon 6 (α form), −188 cal g^{-1}; and nylon 6 (γ form), −181 cal g^{-1}. In addition, nylon 4 is superior in mechanical properties to nylon 6 (Table 10.2). The chain-axis Young's moduli (E_bs) of the polyamides are 334 GPa (nylon 4, α); 316 GPa (nylon 6, α); and 120 GPa (nylon 6, γ), and the three-dimensionally averaged Young's moduli (E_{av}s) are 15.1 GPa (nylon 4, α); 13.7 GPa (nylon 6, α); and 24.8 GPa (nylon 6, γ).

The crystal structure and solid properties of nylon 4, revealed by the periodic DFT-D computations, are described in Sections 10.5 and 10.8. Here, the chain characteristics of nylon 4 are elucidated via molecular orbital (MO) calculations and nuclear magnetic resonance (NMR) experiments on its small model compound, *N*-acetyl-γ-aminobutyric acid *N*'-methylamide (designated herein as ABAMA, Figure 26.1) [165].

Conformational Analysis of Polymers: Methods and Techniques for Structure-Property Relationships and Molecular Design, First Edition. Yuji Sasanuma.

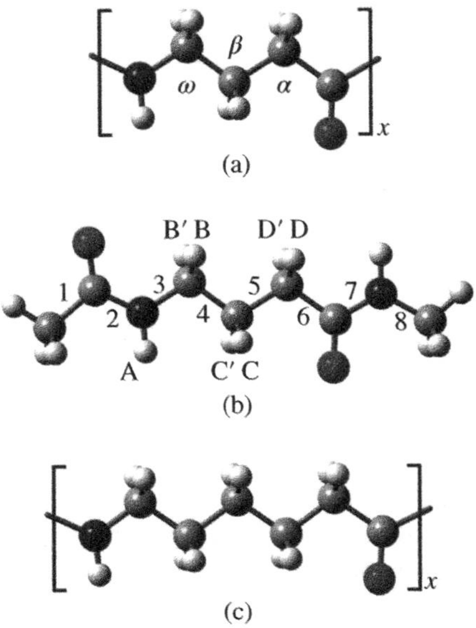

Figure 26.1 (a) Nylon 4 and (b) its model compound, *N*-acetyl-γ-aminobutyric acid *N*′-methylamide (designated herein as ABAMA) with designations of carbon atoms (α, β, and ω) of nylon 4 and hydrogen atoms (A, B, B′, C, C′, D, and D′) and bond numbers (1–8) of ABAMA. (c) Nylon 6. Source: [165], Figure 1 (p.9545)/with permission of American Chemical Society/Licensed under CC BY 4.0.

26.1 MO Calculation

The C(=O)—NH atoms of ABAMA lie on a plane, and the C—N bond is irrotatable; therefore, only internal rotations around bonds 3—6 are considered (Figure 26.1). Under the RIS approximation (trans, gauche$^+$, and gauche$^-$), 81 (= 3^4) staggered conformers may be possible; however, since the g^+ and g^- conformations of each bond are equivalent, the number of irreducible conformers is 41. All the 41 conformers underwent the geometrical optimization at the B3LYP/6-311+G(2d,p) level, and consequently, 15 conformers were successfully optimized (Table 26.1). The dihedral angles around the CH_2—C(=O) bond (bond 6) are distributed to a large extent, and those of the HN—CH_2 bond (bond 3) are also scattered, whereas those of bonds 4 and 5 keep normal (see Table 26.1).

Figure 26.2 illustrates three conformers of low ΔG_ks, in which intramolecular N—H· · ·O=C hydrogen bonds seem to be formed. Their ΔG_k values affected by the solvation of 2,2,2-trifluoroethanol (TFE) and H· · ·O distances are, respectively, −3.92 kcal mol^{-1} and 2.00 Å (Figure 26.2a, k = 15), −3.59 kcal mol^{-1} and 1.94 Å (b, k = 3), and −2.95 kcal mol^{-1} and 2.65 Å (c, k = 14).

The trans fractions (p_ts) of bonds 3—5 were calculated with the ΔG_ks according to the conformational classification shown in Table 26.1. All the p_t values are so small as to indicate that ABAMA strongly prefers distorted shapes rich in gauche states as in Figure 26.2. The three bonds show a common tendency: the p_t values increase with medium polarity. This is probably because the intramolecular hydrogen bonds would be disturbed or cleaved by the polar solvents.

Table 26.1 Results of MO calculations on ABAMA.

	Dihedral angle[a] (°)				Conformation[b]			ΔG_k[c] (kcal mol^{-1})			
	Bond[c]				Bond[d]			Medium			
k	3	4	5	6	3	4	5	Gas	$CHCl_3$	TFE[e]	DMSO[f]
1[g]	180.0	180.0	180.0	180.0	t	t	t	0.00	0.00	0.00	0.00
2	141.0	−61.7	−64.8	−29.8	t	g^+	g^+	−0.31	−0.99	−1.46	−1.52
3	−175.4	−66.6	91.1	149.2	t	g^+	g^-	−4.65	−3.95	−3.59	−3.54
4	−91.4	177.7	172.2	125.9	g^+	t	t	−2.51	−2.41	−2.34	−2.33
5	−74.7	160.8	−71.6	−6.6	g^+	t	g^+	−3.80	−2.95	−2.57	−2.53
6	−88.7	179.3	67.9	−153.1	g^+	t	g^-	−3.30	−2.67	−2.37	−2.34
7	−84.1	−61.7	−172.3	−117.9	g^+	g^+	t	−1.78	−1.80	−1.98	−2.00
8	−96.8	−66.7	175.9	120.8	g^+	g^+	t	−0.98	−1.30	−1.68	−1.73
9	−79.8	−59.1	−65.3	160.6	g^+	g^+	g^+	−3.01	−2.34	−2.22	−2.21
10	−97.3	−48.5	−49.3	−103.1	g^+	g^+	g^+	−4.31	−2.93	−2.36	−2.30
11	−99.7	−70.8	73.6	−121.0	g^+	g^+	g^-	−1.93	−1.71	−1.86	−1.89
12	−106.8	64.6	−172.1	−138.6	g^+	g^-	t	−2.97	−2.51	−2.31	−2.29
13	103.8	−67.0	−177.2	−158.9	g^+	g^-	t	−3.11	−2.57	−2.42	−2.41
14	−104.2	64.0	−81.9	132.6	g^+	g^-	g^+	−5.25	−3.64	−2.95	−2.87
15	−99.6	69.6	74.2	−100.2	g^+	g^-	g^-	−5.65	−4.42	−3.92	−3.87

a) The dihedral angle is defined here according to the IUPAC recommendation [308]: trans (t) ~ 180 ± Δ°; gauche$^+$ (g^+) ~ −60 ± Δ°; gauche$^-$ (g^-) ~ +60 ± Δ°, where Δ stands for the allowance.
b) The dihedral angles of bond 6 are distributed too widely to be classified into a few rotational isomeric states.
c) At the MP2/6-311+G(2d,p)//B3LYP/6-311+G(2d,p) level. Relative to ΔG_k of the all-trans conformation.
d) For the bond numbers, see Figure 26.1.
e) 2,2,2-Trifluoroethanol.
f) Dimethyl sulfoxide.
g) The molecular geometry was optimized with the dihedral angles fixed at 180° so that the all-trans conformation would be the control for ΔG_k and ^{13}C NMR chemical shifts.

Source: [165], Table 1 (p. 9545)//with permission of American Chemical Society/Licensed under CC BY 4.0.

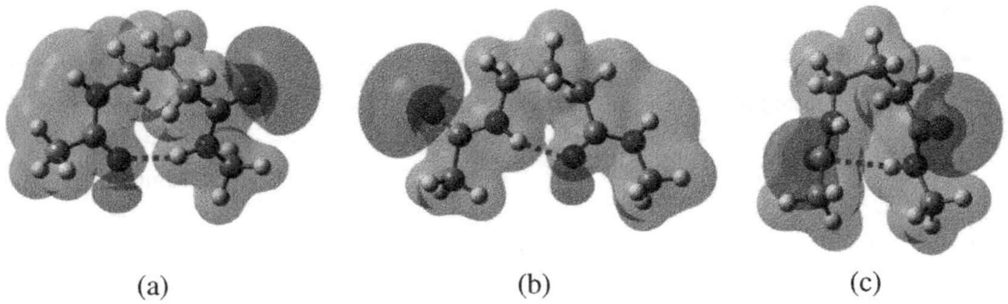

Figure 26.2 Stable conformers of ABAMA with the electrostatic potential surface: (a) $k = 15$, $g^+g^-g^-(\sim g^+)$; (b) $k = 3$, $tg^+g^-(\sim t)$; (c) $k = 14$, $g^+g^-g^+(\sim t)$. k is the conformer number (see Table 26.1), and approximate conformations of bond 6 are written the parentheses. The dotted lines represent intramolecular N—H···O=C hydrogen bonds. Source: [165], Figure 2 (p.9546)/with permission of American Chemical Society/Licensed under CC BY 4.0.

26.2 NMR Experiment

From the ^{1}H NMR spectra observed from three kinds of methylene protons, BB′, CC′, and DD′ (see Figure 26.1) of ABAMA, the chemical shifts and spin–spin coupling constants were derived. Of them, vicinal ^{1}H—^{1}H coupling constants were used in the conformational analysis. The trans fractions (p_ts) of the individual bonds, determined from the NMR experiments, are in good agreement with the MO calculations (Table 26.2). As described above, the MO calculations suggest that the small p_t values reflect the intramolecular N—H···O=C hydrogen bonds, whose existence is also supported by the NMR experiments.

Table 26.2 Trans fractions of the HN–CH_2 (bond 3), NCH_2–CH_2 (bond 4), and CH_2–$CH_2C(=O)$ (bond 5) bonds of ABAMA at 25 °C, determined from MO calculations and NMR experiments.

Medium	3: HN–CH_2	4: NCH_2–CH_2	5: CH_2–$CH_2C(=O)$
		MO calc	
Gas	0.10	0.04	0.02
Chloroform	0.22	0.08	0.06
Trifluoroethanol	0.25	0.10	0.12
Dimethyl sulfoxide	0.25	0.11	0.12
		NMR expt	
Chloroform-*d*	0.21	0.10	0.11
Trifluoroethanol[a)]	0.21	0.12	0.18
Dimethyl-d_6 sulfoxide	0.30	0.11	0.17

a) Using nondeuterated trifluoroethanol and chloroform-*d* as the solvent and external standard, respectively.

To investigate whether the isolated nylon 4 chain as well as ABAMA forms the hydrogen bonds, ^{1}H NMR spectra of nylon 4 itself were measured, and it was investigated whether the observed spectra could be reproduced with the same values of spin–spin coupling constants as obtained from ABAMA. In Figure 26.3c, as an example, the spectra of nylon 4 in TFE at 25 °C are compared with the simulation. Both spectra are seen to be essentially identical, which indicates that nylon 4 has the vicinal coupling constants close to those of ABAMA; hence, nylon 4 has similar p_ts reflecting the hydrogen bonding.

Solid-state ^{13}C NMR of nylon 4 was measured via the cross-polarization/magic-angle spinning (CP/MAS), ^{13}C pulse saturation transfer (PST) /MAS, and ^{13}C low-power decoupling (LD) /MAS techniques; determined ^{13}C chemical shifts of the crystalline and noncrystalline chains separately [259]; and plotted the chemical shift differences ($\Delta\delta^{i}$s, i, carbon species) between the crystalline and amorphous phases against carbon species (Figure 26.4). The ^{13}C chemical shift differences of nylon 4 between in the crystalline phase and in a 1,1,1,3,3,3-hexafluoro-2-propanol (HFIP) solution are also plotted in Figure 26.4. Both $\Delta\delta^{i}$ values, depending largely on carbon species, oscillate similarly.

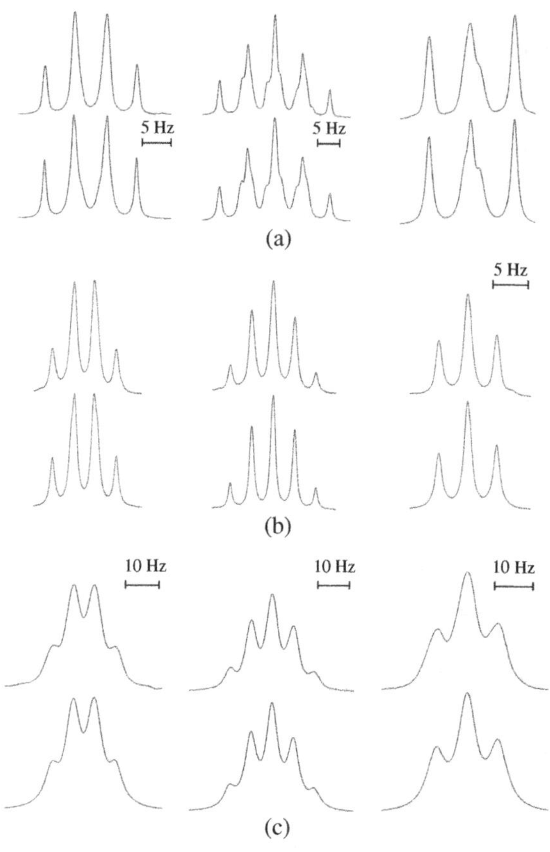

Figure 26.3 Observed (above) and calculated (below) ^{1}H NMR spectra of three kinds of methylene protons (from left to right, BB′, CC′, and DD′; for the proton designations, see Figure 26.1): (a) ABAMA dissolved in chloroform-*d* at 45 °C; (b) ABAMA in TFE at 25 °C; (c) nylon 4 in TFE at 25 °C. The number- and weight-average molecular weights of the nylon 4 sample were determined to be 10.3 and 67.7 kDa, respectively. Source: [165], Figure 3 (p. 9546)/with permission of American Chemical Society/Licensed under CC BY 4.0.

The chemical shift differences ($\Delta\delta^i$s) of ABAMA between in the all-trans conformation and in the free state were calculated. The chemical shift $\langle\delta^i\rangle$ of carbon i of ABAMA in the free state was calculated as the weight average over all the possible conformations:

$$\langle\delta^i\rangle = \frac{\sum_k \delta_k^i \exp\left(-\frac{\Delta G_k}{RT}\right)}{\sum_k \exp\left(-\frac{\Delta G_k}{RT}\right)} \tag{26.1}$$

where δ_k^i is the ^{13}C chemical shift of carbon i of conformer k. The $\Delta\delta^i$ value of ABAMA is defined as the difference from that ($\delta^{\text{all-trans}}$) of the all-trans state:

$$\Delta\delta^i = \delta^{\text{all-trans}} - \langle\delta^i\rangle \tag{26.2}$$

This is because nylon 4 crystallizes in the all-trans conformation. The $\Delta\delta^i$ values, calculated from the δ_k^i and ΔG_k data on the gas phase (filled circle) and TFE solution (filled triangle), are plotted in Figure 26.4. The $\Delta\delta^i$ plots of ABAMA oscillate similar to those for the nylon 4. On the theoretical basis of the γ and δ effects [47, 365, 485] described in Section 15.2, nylon 4 and ABAMA are indicated to have analogous conformational distributions. As shown above, ABAMA prefers the bent shapes due to the intramolecular N—H···O=C hydrogen bonds; therefore, the nylon 4 chain also lies in such distorted conformations and forms the hydrogen bonds even in the free states such as solute, melt, and amorphous solid.

In general, nylon 4 is suggested to be superior in thermal and mechanical properties to nylon 6, probably owing to the denser intra- and intermolecular hydrogen bonds. Nylon 4, satisfying both carbon neutrality and biodegradability, is expected to be used for tough environmentally friendly materials.

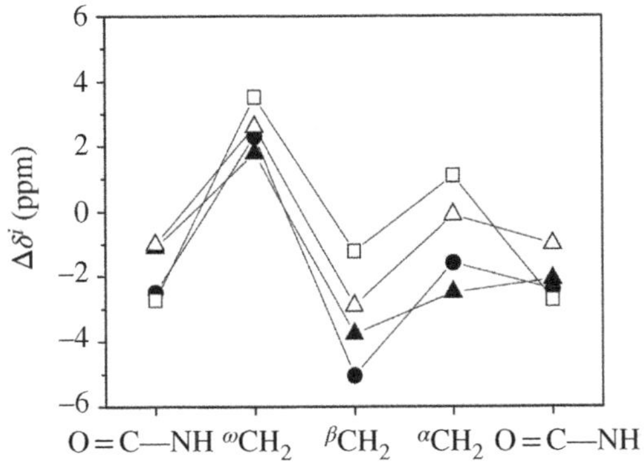

Figure 26.4 ^{13}C NMR chemical shift differences of ABAMA and nylon 4: (●) ABAMA in the gas phase, from Eq. (26.2); (▲) ABAMA in TFE, from Eq. (26.2); (△) nylon 4 in the amorphous phase, from $\Delta\delta^i = \delta^i_{\text{crystalline}} - \delta^i_{\text{amorphous}}$ [259]; (□) nylon 4 in the HFIP solution, from $\Delta\delta^i = \delta_{\text{crystalline}} - \delta^i_{\text{solution}}$ [259]. i indicates the carbon species (C=O, $^{\alpha}$CH$_2$, $^{\beta}$CH$_2$, and $^{\omega}$CH$_2$), see Figure 26.1a. Source: [165], Figure 5 (p. 9549)/with permission of American Chemical Society/Licensed under CC BY 4.0.

27

Aromatic Polyester, Polythionoester, Polythioester, Polydithioester, Polyamide, and Polythioamide

The polymers expressed by the chemical formula shown in Figure 27.1 are structural homologues but classified into different polymer families, depending on the atoms (atomic group) positioned at the X and Y sites: X = Y = O, polyester; X = S and Y = O, polythionoester; X = O and Y = S, polythioester; X = Y = S,

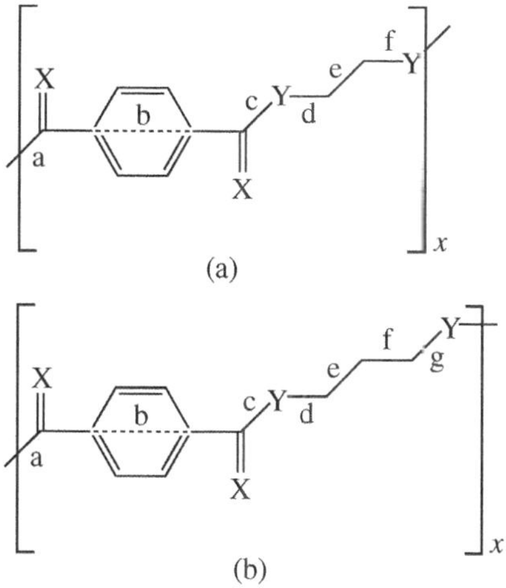

Figure 27.1 Aromatic polymers treated here and expressed as P*y*XY: polyester (X = Y = O); polythionoester (X = S and Y = O); polythioester (X = O and Y = S); polydithioester (X = Y = S); polyamide (X = O and Y = NH); and polythioamide (X = S and Y = NH). (a) $y = 2$, poly(ethylene terephthalate) (P2OO); poly(ethylene thionoterephthalate) (P2SO); poly(ethylene dithioterephthalate) (P2OS); poly(ethylene tetrathioterephthalate) (P2SS); poly(ethylene terephthalamide) (P2ONH); and poly(ethylene terephthalthioamide) (P2SNH). (b) $y = 3$, poly(trimethylene terephthalate) (P3OO); poly(trimethylene thionoterephthalate) (P3SO); poly(trimethylene dithioterephthalate) (P3OS); poly(trimethylene tetrathioterephthalate) (P3SS); poly(trimethylene terephthalamide) (P3ONH); and poly(trimethylene terephthalthioamide) (P3SNH).

Conformational Analysis of Polymers: Methods and Techniques for Structure-Property Relationships and Molecular Design, First Edition. Yuji Sasanuma.

polydithioester; X = O and Y = NH, polyamide; and X = S and Y = NH, polythioamide. The structures and properties of these polymers may be compared with each other and interpreted in terms of the conformational characteristics due to the heteroatoms at X and Y. Here, the polymers are represented as P*y*XY: Polymer (P) + the number of methylene groups (*y*) between two Ys + atom at X + atom(s) at Y. For the details, see Figure 27.1. The Y–$(CH_2)_y$–Y part is termed "spacer", and only the polymers of $y = 2$ and 3 are treated here.

As has been proven so far, the conformational characteristics of a given polymer can be investigated via its small model compound that has the same skeletal bond sequence as that of the polymer (see Figure 27.2). The model for the polymer P*y*XY is expressed as M*y*XY. In order to determine the bond conformations of the spacers of P*y*XYs by NMR, ^{13}C-labeled M*y*XYs, which provide conformational fractions around the Y—CH_2 as well as CH_2—CH_2 bonds, were synthesized [10–12, 331, 416].

P2OO (poly(ethylene terephthalate), PET) and P3OO (poly(trimethylene terephthalate), PTT) have been extensively studied and manufactured industrially,

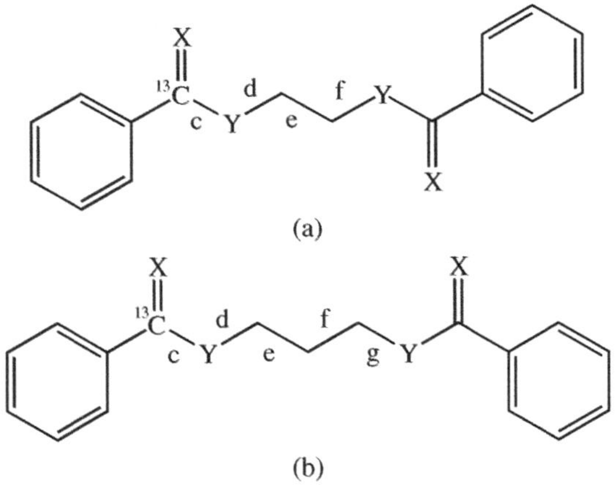

Figure 27.2 Model compounds adopted here and designated as M*y*XY. (a) $y = 2$, ethane-1,2-diyl dibenzoate (M2OO); *O,O′*-(ethane-1,2-diyl) dibenzothioate (M2SO); *S,S′*-(ethane-1,2-diyl) dibenzothioate (M2OS); ethane-1,2-diyl dibenzodithioate (M2SS); *N,N′*-(ethane-1,2-diyl) dibenzamide (M2ONH); and *N,N′*-(ethane-1,2-diyl) dibenzothioamide (M2SNH). (b) $y = 3$, propane-1,3-diyl dibenzoate (M3OO); *O,O′*-(propane-1,3-diyl) dibenzothioate (M3SO); *S,S′*-(propane-1,3-diyl) dibenzothioate (M3OS); propane-1,3-diyl dibenzodithioate (M3SS); *N,N′*-(propane-1,3-diyl)dibenzamide (M3ONH); and *N,N′*-(propane-1,3-diyl) dibenzothioamide (M3SNH). The bonds are designated as indicated.

whereas the other PyXYs have little or never been paid attention to. When we began this series of studies, we could not find any reports on PySO and PySS. This is because the thionoester –C(=S) –O– and dithioester –C(=S)–S– groups in particular are unstable. Tanaka attempted a solid-phase reaction between M2OO and Lawesson's reagent under microwave irradiation and obtained M2SO (the yield was as small as 0.9%), and the main product was the half-sulfurated compound, M2(0.5O:0.5S)O, i.e. $C_6H_4C(=O)O(CH_2)_2OC(=S)C_6H_4$ (yield 23%) [416, 468]. Nevertheless, NMR experiments could be conducted for ^{13}C-labeled M2SO and yielded its bond conformations, from which conformational characteristics of P2SO were predicted. In addition, single-crystal X-ray diffraction experiments revealed the crystal structures of M2SO and M2(0.5O:0.5S)O. As a result of laborious work, Abe established a synthetic scheme for PySSs as illustrated in Figure 9.1 (Section 9.1.1) [10, 12], where PySS is represented as $PyTS_4$. By the synthetic method, he prepared PySSs of $y = 2–5$ and investigated their conformational characteristic and crystal structures. Although PyOS, PyONH, and PySNH were synthesized already, their structures and physical properties had been left unrevealed.

27.1 MO Calculation

The first, second, and third most stable conformers of MyXYs, together with their Gibbs free energies obtained from MO calculations, are listed in Table 27.1. The results may be classified according to the Y species. In the spacer of Y = O and $y = 2$, i.e. O–CH_2–CH_2–O, the tg$^{\pm}$t conformation is the most stable, whereas it adopts the all-trans form in the crystal. The S–CH_2–CH_2–S spacer (Y = S and $y = 2$) strongly prefers the g$^{\pm}$tg$^{\mp}$ conformation as found for poly(ethylene sulfide), whose g$^{\pm}$tg$^{\mp}$ states are stabilized in both solutions and crystal by dipole–dipole interactions (Section 20.2). The NH—CH_2—CH_2—NH spacer forms intramolecular attractions such as C–H···O (2.54 Å, tg$^{\pm}$g$^{\mp}$ of X = O), N–H···O (2.07 Å, tg$^{\pm}$g$^{\mp}$ of X = O), and C–H···S (2.82 Å, g$^{\pm}$g$^{\pm}$g$^{\pm}$ of X = S), where the numerical value represents the H···X distance. These weak hydrogen bonds stabilize the respective conformations. The trimethylene spacer ($y = 3$) of Y = O is most stabilized in the tg$^{\pm}$g$^{\pm}$g$^{\pm}$ conformations and adopts the second tg$^{\pm}$g$^{\pm}$t states in the crystal. In the trimethylene spacer of Y = S, the most stable states are g$^{\pm}$ttg$^{\pm}$, and the spacers of Y = NH and $y = 3$ prefer the tttg$^{\pm}$ (X = O) and tttt (X = S) conformations.

Table 27.1 First-third most stable conformers and their Gibbs free energies of models and crystal conformations of models and polymers.

				Stable conformer[b]						Crystal[a]		
										Model		Polymer
Model	Polymer	X	Y	First	ΔG_k	Second	ΔG_k	Third	ΔG_k	Conformation	ΔG_k	Conformation
y = 2												
M2OO	P2OO	O	O	$tg^{\pm}t$	−1.12	$tg^{\pm}g^{\pm}$	−0.96	$g^{\pm}g^{\pm}g^{\pm}$	−0.84	ttt	0.00	ttt
M2SO	P2SO	S	O	$tg^{\pm}t$	−2.07	$tg^{\pm}g^{\pm}$	−1.13	$g^{\pm}g^{\pm}g^{\pm}$	−1.05	ttt	0.00	
M2OS	P2OS	O	S	$g^{\pm}tg^{\mp}$	−3.13	$g^{\pm}g^{\pm}g^{\pm}$	−2.35	$g^{\pm}g^{\pm}g^{\mp}$	−1.64	$g^{\pm}tg^{\mp}$	−3.13	
M2SS	P2SS	S	S	$g^{\pm}tg^{\mp}$	−2.08	$g^{\pm}g^{\pm}g^{\mp}$	−1.81	$g^{\pm}g^{\pm}g^{\pm}$	−1.81	$g^{\pm}tg^{\mp}$	−2.08	
M2ONH	P2ONH	O	NH	$tg^{\pm}g^{\mp}$	0.00	$g^{\pm}tg^{\pm}$	0.49	$g^{\pm}g^{\pm}g^{\pm}$	0.59	$g^{\pm}tg^{\mp}$	0.89	
M2SNH	P2SNH	S	NH	$g^{\pm}g^{\pm}g^{\pm}$	−0.76	$g^{\pm}tg^{\mp}$	−0.73	$tg^{\pm}g^{\mp}$	−0.7	$g^{\pm}g^{\pm}g^{\pm}$	−0.76	
y = 3												
M3OO	P3OO	O	O	$tg^{\pm}g^{\pm}g^{\pm}$	−1.32	$tg^{\pm}g^{\pm}t$	−1.18	$tg^{\pm}tg^{\pm}$	−0.94	$tg^{\pm}g^{\pm}t$	−1.18	$tg^{\pm}g^{\pm}t$
M3OS	P3OS	O	S	$g^{\pm}ttg^{\pm}$	−2.71	$g^{\pm}g^{\pm}g^{\pm}g^{\pm}$	−2.52	$g^{\pm}tg^{\pm}g^{\pm}$	−2.25	$tttg^{\pm}$	−1.23	
M3SS	P3SS	S	S	$g^{\pm}ttg^{\pm}$	−2.10	$g^{\pm}tg^{\pm}g^{\pm}$	−1.78	$g^{\pm}g^{\pm}g^{\pm}g^{\pm}$	−1.74	(liquid)		
M3ONH	P3ONH	O	NH	$tttg^{\pm}$	−0.54	$g^{\pm}ttg^{\mp}$	−0.34	$g^{\pm}ttg^{\pm}$	−0.32	$ttg^{\pm}g^{\pm}$	0.32	
M3SNH	P3SNH	S	NH	tttt	0.00	$g^{\pm}ttg^{\mp}$	0.97	$tg^{\pm}g^{\pm}t$	1.03	$tttg^{\pm}$	1.19	

a) The crystallographic data were quoted from the following references: M2OO [351]; M2SO [468]; M2OS [104]; M2SS [13]; M2ONH [353]; M2SNH [332]; M3OO [352]; M3OS [274]; M3ONH [52]; M3SNH [333]; P2OO [101]; and P3OO [107, 370].

b) For $y = 2$, conformations of bonds d–f, and for $y = 3$, conformations of bonds d–g. The ΔG_k value of each conformation is the difference from the Gibbs free energy of the all-trans form. Because M2ONH does not form the all-trans conformation, its ΔG_k of $tg^{\pm}g^{\mp}$ is set null.

27.2 Bond Conformation

Bond conformations of MyXYs are listed in Table 27.2. The MO and NMR data reasonably agree with each other. M2OO and M2SO prefer trans and gauche states in the O—CH_2 and CH_2—CH_2 bonds, respectively. The high stability of the $g^{\pm}tg^{\mp}$ conformations of M2OS and M2SS results in the small and large trans fractions in the S—CH_2 and CH_2—CH_2 bonds, respectively. Two bonds of M2ONH and M2SNH would rather adopt gauche states. In general, M3XYs except M3XNHs are similar in bond conformation to M2XYs; MyXNHs are subject to the N–H· · ·X hydrogen bonds.

The rotational isomeric states (RISs) around bond b (the so-called virtual bond) that connects two carbonyl or thiocarbonyl carbon atoms are divided into two types: two states (trans and cis) of PyOO, PySO, and PyOS; six states (trans–trans, trans–cis$^{\pm}$, cis–trans$^{\pm}$, and cis–cis) of PySS, PyONH, and PySNH. For the definition of the six rotamers, see Figure 27.3; for numerical values of their dihedral angles, Gibbs free energies, and fractions, see Table 27.3. The PyXYs of the six states with similar Gibbs energies may perform nearly free rotations around bond b.

Table 27.2 Trans fractions (p_ts) of Y–CH_2 and CH_2–CH_2 bonds of models at 25 °C, evaluated from MO calculations and NMR experiments.

			p_t				
			MO (gas)		NMR		
Model	X	Y	Y–CH_2	CH_2–CH_2	Solvent	Y–CH_2	CH_2–CH_2
$y = 2$							
M2OO	O	O	0.45	0.06	Benzene	0.46	0.04
M2SO	S	O	0.74	0.05	Benzene	0.69	0.05
M2OS	O	S	0.01	0.70	Benzene	0.11	0.56
M2SS	S	S	0.05	0.35	Benzene	0.13	0.27
M2ONH	O	NH	0.29	0.19	DMSO	0.30	0.26
M2SNH	S	NH	0.32	0.31	DMSO	0.44	0.28
$y = 3$							
M3OO	O	O	0.52	0.27	Benzene	0.52	0.28
M3OS	O	S	0.12	0.44	Benzene	0.16	0.47
M3SS	S	S	0.19	0.38	Benzene	0.22	0.48
M3ONH	O	NH	0.22	0.61	DMSO	0.31	0.49
M3SNH	S	NH	0.55	0.62	DMSO	0.40	0.53

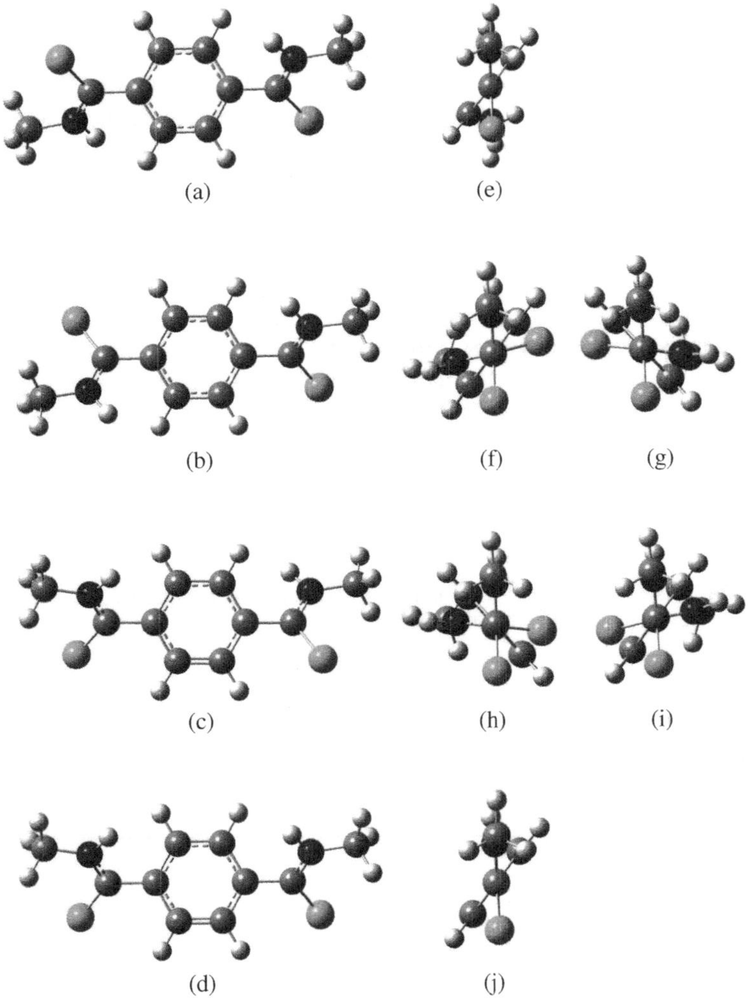

Figure 27.3 Six rotational isomeric states around the benzene ring: top views (a) trans–trans (abbreviated as t–t), (b) trans–cis (t–c), (c) cis–trans (c–t), and (d) cis–cis (c–c); side views (e) t–t, (f) $(t–c)^+$, (g) $(t–c)^-$, (h) $(c–t)^+$, (i) $(c–t)^-$, and (j) c–c. The model here is *N,N′*-dimethylterephthalthioamide (X = S and Y = NH). The first (left) symbol of the six states represents the relative orientation between the C=X bonds, and the second (right) expresses whether the two C=X bonds appear on the same (cis) side or opposite (trans) sides with respect to the benzene plane. The signs, + and −, stand for the rotational directions of the dihedral angle, being similar to those of g^+ and g^-. Source: [331]/Reproduced with permission of Royal Society of Chemistry.

Table 27.3 Dihedral angles (ϕ), free energies (ΔGs), and fractions (p_{conf}s) of the ξ orientations around the benzene ring of the aromatic polyester, polythionoester, polythioester, polydithioester, polyamides, and polythioamides at 25 °C[a].

			Orientation between −Y−C=X groups: ξ			
		(PyXY)	**trans–trans**	**trans–cis±**	**cis–trans±**	**cis–cis**
Polyester[b]	ϕ_ξ	P2OC	0.00			180.0
		P3OC	0.00			180.0
	ΔG_ξ		0.00			0.15
	p_ξ		0.56			0.44
Polythionoester[b]	ϕ_ξ	P2SO	0.00			180.0
	ΔG_ξ		0.00			0.03
	p_ξ		0.51			0.49
Polythioester[b]	ϕ_ξ	P2OS	0.00			180.0
		P3OS	0.00			180.0
	ΔG_ξ		0.00			0.15
	p_ξ		0.56			0.44
Polydithioester	ϕ_ξ	P2SS	0.00	±71.5	±106.3	180.0
		P3SS	0.00	±70.8	±106.3	180.0
	ΔG_ξ		0.00	−0.10	0.15	0.05
	p_ξ		0.17	0.20	0.13	0.16
Polyamide	ϕ_ξ	P2ONH	0.00	±53.6	±123.0	180.0
		P3ONH	0.00	±46.9	±129.2	180.0
	ΔG_ξ		0.00	−0.10	−0.38	0.08
	p_ξ		0.12	0.15	0.24	0.11
Polythioamide	ϕ_ξ	P2SNH	0.00	±79.6	±98.0	180.0
		P3SNH	0.00	±75.8	±102.5	180.0
	ΔG_ξ		0.00	−0.19	−0.32	−0.14
	p_ξ		0.12	0.16	0.20	0.15

a) Parameters: ϕ_ξ, dihedral angle according to the convention in polymer science (°); ΔG_ξ, free energy difference from that of the trans–trans orientation (kcal mol^{-1}); and p_ξ, the existing probability. Because of the rounding error, the sum of p_ξs does not exactly agree with unity.
b) The –Y–C=X group lies on the benzene plane; thus, only two states, trans and cis orientations, are defined for the polyester, polythionoester, and polythioester.

Source: [331]/Reproduced with permission of Royal Society of Chemistry.

27.3 RIS Calculation, Thermal Properties, and Solubility

The RIS calculations were carried out with the Gibbs free energies, and the geometrical parameters derived from the MO computations. In the $\langle r^2 \rangle_0/nl^2$ calculations, the virtual bond (bond b) was used; however, the structural similarity between PyXYs allows us to compare the RIS results as they are (Table 27.4).

Table 27.4 Configurational and thermal properties of polymers at 25 °C[a].

			RIS calculation			Thermal analysis		
Polymer (PyXY)	X	Y	$\langle r^2 \rangle_0/nl^2$	S_{conf} (cal K^{-1} mol^{-1})	Solid state[b]	T_g (°C)	T_m (°C)	T_d (°C)
$y = 2$								
P2OO[c]	O	O	2.63	7.12	SC	77	253	399
P2SO[d]	S	O	2.05	5.02				
P2OS	O	S	16.70	4.87	SC	36	NOB	346
P2SS	S	S	5.20	8.87	AM	71	NOB	220
P2ONH	O	NH	3.15	8.37	LC	52	NOB	421
P2SNH[e]	S	NH	3.55	9.38	AM	23	NOB	261
$y = 3$								
P3OO[c]	O	O	4.14	8.93	SC	52	226	376
P3OS	O	S	10.50	8.40	SC	40	202	350
P3SS	S	S	6.06	11.50	SC	44	143,[f] 153[g]	260
P3ONH	O	NH	3.65	10.50	LC	30	NOB	391
P3SNH[e]	S	NH	5.66	10.60	AM	10	NOB	250

a) Symbols and abbreviations: T_g, glass transition temperature; T_m, melting point as observed; T_d, thermal decomposition temperature; SC, semicrystalline; AM, amorphous; LC, low crystallinity; NOB, not observed.
b) Confirmed by X-ray diffraction.
c) For P2OO and P3OO, the averages of thermal data provided by the PolyInfo database [347]. For the other polymers, the experimental values detected by differential scanning calorimetry or thermogravimetry are shown.
d) P2SO has not been synthesized yet, and hence, no experimental data are available.
e) Number- (M_n) and weight-average (M_w) molecular weights, respectively: P2SNH, 10.7 kDa and 17.3 kDa; P3SNH, 13.9 kDa and 29.3 kDa.
f) Observed from the sample polymerized in *N,N*-dimethylformamide.
g) Observed from the sample polymerized in dimethyl sulfoxide.

At a glance, one may find that P2OS shows a very large $\langle r^2\rangle_0/nl^2$ value of 16.70 and, probably, also its small S_{conf} value. This is because its $g^{\pm}tg^{\mp}$ states are so stable that the P2OS chain is extended. The X-ray diffraction of P2OS exhibits a number of peaks, which shows the P2OS sample to be semicrystalline. Nevertheless, no fusion was observed from P2OS below T_d (346 °C). The equilibrium melting point is given as $T_m^0 = \Delta H_u/\Delta S_u$, where ΔH_u and ΔS_u are, respectively, the enthalpy and entropy of fusion, and S_{conf} accounts for a large fraction of ΔS_u; therefore, T_m^0 of P2OS would be higher than T_d. In addition, S_{conf} can be interpreted as the entropy difference between amorphous (solution and molten) and crystalline states. The small S_{conf} value suggests that P2OS would be hardly soluble in solvents. Actually, solubility tests showed that P2OS is insoluble in common organic solvents and even in trifluoroacetic acid (TFA), 1,1,1,3,3,3-hexafluoro-2-propanol (HFIP), and hot concentrated sulfuric acid (100 °C). Therefore, P2OS is an insoluble and infusible polymer and hence difficult to mold but extremely resistant to chemicals. P3OS also shows a large characteristic ratio of 10.50.

Of P2XYs, P2SS, P2ONH, and P2SNH show large S_{conf}s. This is because their six-state rotation around bond b leads to a large degree of conformational freedom. As a consequence, P2SS is insoluble in common solvents but soluble in TFA/dichloromethane (DCM) (1 : 3 in volume), HFIP/DCM (1 : 3), and o-dichlorobenzene (DCB)/phenol (3 : 2), while P2ONH and P3ONH are soluble in HFIP, TFA, and sulfuric acid, and P2SNH is soluble in pyridine, *N,N*-dimethylformamide, *N,N*-dimethylacetamide, *N*-methylpyrrolidone, and the abovementioned solvents of P2ONH. In general, the solubility of PyOS seems to be enhanced with an increase in *y*: P3OS is soluble in phenol, DCB/phenol (3 : 2), and sulfuric acid, and P5OS is also soluble in TFA, HFIP, TFA/DCM (1 : 3), and HFIP/DCM (1 : 3). P3SNH is soluble in solvents similar to those of P2SNH.

In general, P2XY has a glass transition temperature higher than that of P3XY. Except for PyOOs, PyXYs are poor in crystallinity. The six-state PyXYs are amorphous or of low crystallinity and hence exhibit no melting. This may be because the nearly free rotation around bond b would disturb well-ordered molecular packing. The T_ds are divided into two ranges according to the X atom: X = O, 350–420 °C (high) and X = S, 220–260 °C (low). PySYs are reddish in color. Whether the coloration will be advantageous or disadvantage depends on the usage.

In this series of studies, conformational characteristics and configurational properties of PyXYs were predicted or revealed via theoretical treatments, which may be considered to be a molecular design, and the predictions were actually confirmed by the experiments, which may be an experimental verification.

28

Polysilanes

Polysilanes (PSs) have been expected to be applied for future devices because of their unique optical properties such as thermochromism, solvatochromism, ionochromism, and piezochromism [321]. The wavelengths of UV absorption and emission shift with changes in the main-chain conformation.

Figure 16.7 of Section 16.1.2 shows the UV spectra observed from the Θ solutions and solid films of the three PSs: poly(di-*n*-butylsilane) (PDBS), poly(di-*n*-hexylsilane) (PDHS), and poly(methyl-*n*-propylsilane) (PMPrS) [235]. The PDBS, PDHS, and PMPrS chains of the Θ state show the absorption maxima (λ_{max} s) at 314, 314, and 306 nm, respectively. Fresh solid films of PDBS and PDHS exhibit the same λ_{max} s as those of the Θ solutions; however, the mature PDHS film exhibits another absorption (at 363 nm), which comes from the all-anti conformation (AAAA...) due to side-chain crystallization [268, 288, 376]. In the crystalline state, the PDBS chain lies in the all-deviant $D_+D_+D_+D_+\ldots$ (or $D_-D_-D_-D_-\ldots$) structure with the same sign and forms a 7/3 helix [427]. The λ_{max} of PMPrS moves from 306 to 329 nm by solidification. Here, the conformational terminology is based on that proposed by PS researchers [316]: anti (A), dihedral angle $\phi \approx 180°$; transoid (T), $\phi \approx 165°$; deviant (D), $\phi \approx 150°$; ortho (O), $\phi \approx 90°$; and gauche (G), $\phi \approx 60°$, where ϕ is defined according to the International Union of Pure and Applied Chemistry (IUPAC) recommendation [308].

To reveal the intrinsic nature of the PS conformation, the Θ conditions of PDBS, PDHS, and PMPrS were explored by static light scattering with the aid of size-exclusion chromatography (SEC) and SEC-multiangle laser light scattering (MALS) [235]. The experimental details are described in Section 16.1.2. The Θ conditions (solvent composition and temperature) and the unperturbed $\langle r^2 \rangle_0/nl^2$ ratios were, respectively, determined as follows: PDBS, *n*-hexane at 19.1 °C and 42.3; PDHS, *n*-hexane (58.2%) and 2-propanol (41.8%) at 25 °C and 42.5; and PMPrS, *n*-hexane (62.6%) and 2-propanol (37.4%) at 25°C and 19.9.

By reference to the experimental $\langle r^2 \rangle_0/nl^2$ values, conformational analysis of the PSs was carried out via the rotational isomeric state (RIS) scheme combined with

Conformational Analysis of Polymers: Methods and Techniques for Structure-Property Relationships and Molecular Design, First Edition. Yuji Sasanuma.

molecular dynamics (MD) simulations to evaluate their conformational energies, predict configurational properties, and, furthermore, find relationships of conformational characteristics with higher order structures and physical properties [406]. More specially, dynamic motions of the long side chains were simulated by MD with the main chain frozen at the given dihedral angles (ϕs) to evaluate the energy there, and the energy distribution was drawn as a two-dimensional contour map, from which conformational energies were picked up and applied to the RIS scheme to derive the statistical averages of the physical properties and interpret the relevant experimental observations in terms of conformational characteristics of the PS chains.

28.1 Molecular Dynamics

If an MD simulation for a PS with long alkyl side chains is performed with the main chain being frozen at the given ϕs, then the total energy averaged over a fully long time may be considered as the conformational energy weighted by probabilities of side-chain conformations. From the MD simulations for oligomeric model compounds, the positions (ϕs) and energies of the potential minima were determined.

28.1.1 General Procedures

The MD simulations were carried out with the Cerius^2 package (Molecular Simulation, Inc.), and the polymer-consistent force fields optimized for PSs [449] and alkanes [208]. The charge equilibration method [380] and the Ewald method [157] with a 9-Å cutoff were used to calculate charge distributions and long-range electrostatic interactions, respectively. The NVT canonical ensemble (the number of atoms, temperature, and volume are fixed) was assumed, and the time interval was set to 10 fs. The temperature was initially set at 600 K for 1 ps and decreased to 298.15 K. After 30 ps, the total energies were recorded at 10-fs intervals for 50 ps to outline the energy maps, and for 10 ns to search for the potential minima, at which the data collected during the last 5 ns were averaged to be the $E_{\eta\eta'}$ value (η and η' are conformations).

28.1.2 PDBS and PDHS

In the MD simulations for PDBS and PDHS, their decameric and undecameric model compounds, $Me[SiR_2]_{10}Me$ and $Me[SiR_2]_{11}Me$ (Me, terminal methyl group; R, *n*-butyl or *n*-hexyl group), were adopted. The decamers were used to search for the potential minima around the central Si—Si bond with the outer bonds being rotatable. For every 30 degrees of dihedral angles, ϕ_{i-1} and ϕ_i, of the central bonds of the undecamers, the MD simulation was performed, and the time-average energy was plotted as a function of ϕ_{i-1} and ϕ_i. The local minima were explored on the energy map, and additional MD simulations were carried out in the vicinity at ϕ intervals of 10°. The time-average energy was considered the first- plus second-order interaction energy, $E_{\eta\eta'}$, of the $\eta\eta'$ conformational pair.

Energy contour maps obtained from the undecamers of PDBS and PDHS are shown in Figure 28.1. The local minima are located around ϕ_i (or ϕ_{i-1}) = ±150°

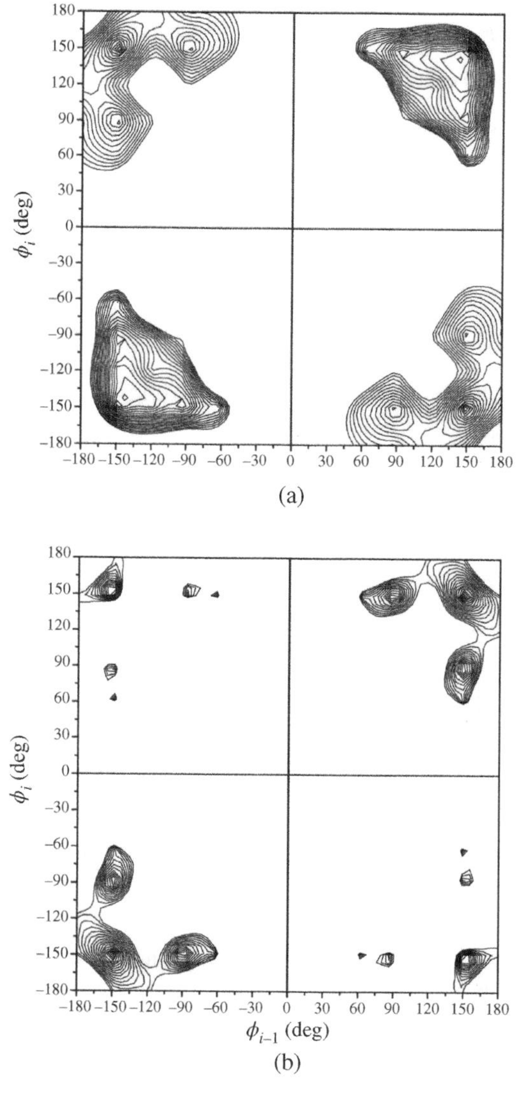

Figure 28.1 Energy contour maps of undecamers of (a) PDBS and (b) PDHS. The contour lines are drawn at intervals of 0.5 kcal mol^{-1}. Source: [406], Figure 5 (p.11855)/Reproduced with permission of American Chemical Society.

Table 28.1 Dihedral angles and energy parameters of PDBS and PDHS, as evaluated from MD simulations, and the characteristic ratios calculated therefrom.

	PDBS	PDHS
Dihedral angle, deg		
$\phi_{D_\pm}$	±151.5	±151.7
$\phi_{O_\pm}$	±92.3	±92.3
$\phi_{G_\pm}$	±71.5	±72.3
MD energy,[a)]kcal mol^{-1}		
First-order		
$E_{D_\pm}$	0.00	0.00
$E_{O_\pm}$	1.85	2.00
$E_{G_\pm}$	2.18	2.52
First- plus second-order		
$E_{D_\pm D_\pm}$	0.00	0.00
$E_{D_\pm D_\mp}$	0.11	0.23
$E_{D_\pm O_\pm}$	2.12	2.39
$E_{D_\pm O_\mp}$	2.59	2.95
$E_{D_\pm G_\pm}$	2.91	2.78
$E_{D_\pm G_\mp}$	2.59	3.58
$E_{O_\pm O_\pm}$	>10	>10
$E_{O_\pm O_\mp}$	>10	>10
$E_{O_\pm G_\pm}$	>10	>10
$E_{O_\pm G_\mp}$	>10	>10
$E_{G_\pm G_\pm}$	>10	>10
$E_{G_\pm G_\mp}$	>10	>10
$\langle r^2\rangle_0/nl^2$ [b)]	42.0 [c)]	54.0 [d)]

a) Relative to $E_{D_\pm}$ or $E_{D_\pm D_\pm}$.
b) Calculated from the dihedral angles, MD energies, l_{Si-Si} of 2.35 Å, and ∠SiSiSi of 115.4°.
c) At 19.1 °C.
d) At 25.0 °C.
Source: [406], Table 2 (p.11856)/Reproduced with permission of American Chemical Society.

($D_{\pm}$), ±90° ($O_{\pm}$), and ±70° ($G_{\pm}$) on the line of ϕ_{i-1} (or ϕ_i) = ±150° ($D_{\pm}$). Potential wells of PDHS are much steeper than those of PDBS; an increase in side-chain length by two methylene units significantly reduces allowable angular ranges of ϕ_{i-1} and ϕ_i. Dihedral angles and conformational energies, evaluated from more detailed computations, are listed in Table 28.1. Only conformations related to $D_{\pm}$ states have the energies lesser than 10 kcal mol^{-1}; hence, the other states may be negligible. Most of the two consecutive Si—Si bonds are suggested to adopt one of four states, D_+D_+, D_-D_-, D_+D_-, and D_-D_+.

28.1.3 PMPrS

Inasmuch as PMPrS has methyl and *n*-propyl side chains on the same silicon atom, the first- and second-order interaction energies are assumed to depend on up to tetrad and pentad configurations, respectively. The ϕ_η and $E_{\eta\eta'}$ values were determined from the 6 tetramers and 10 pentamers of different configurational sequences, respectively; according to the concept of pseudoasymmetry, 16 (2^4) tetrads and 32 (2^5) pentads were generated. Of the 32 pentads, only 10 (*dldld*, *dldll*, *dlddl*, *dlddd*, *dlldd*, *dllld*, *dllll*, *ddldd*, *ddlll*, and *ddddd*) are irreducible. For each tetrad or pentad, the MD simulation was performed as described above, and the obtained energies are plotted as a contour map, from which the conformational energies were extracted for the RIS calculations.

28.2 RIS Calculation

Characteristic ratios of PDBS and PDHS, calculated from the ϕ_η and $E_{\eta\eta'}$ values, are shown in Table 28.1. The bond length $l_{Si—Si}$ and bond angle ∠SiSiSi were set to 2.35 Å and 115.4°, respectively [321, 519]. The calculated $\langle r^2\rangle_0/nl^2$ value of PDBS at 19.1 °C (the Θ point) is 42.0, and that of PDHS at 25.0 °C (the Θ point) is 54.0. The experimental $\langle r^2\rangle_0/nl^2$ values of PDBS and PDHS are 42.3 ± 4.2 and 42.5 ± 3.9, respectively. Therefore, the RIS calculations reproduced the experiment exactly for PDBS and reasonably for PDHS.

Most Si—Si bond pairs adopt either $D_{\pm}D_{\pm}$ or $D_{\pm}D_{\mp}$ conformations. Accordingly, the energy parameter, $E_{D_{\pm}D_{\mp}}$ could be determined so as to reproduce the experimental $\langle r^2\rangle_0/nl^2$ values of PDBS and PDHS: $E_{D_{\pm}D_{\mp}}$ = 0.12 kcal mol^{-1}. On the basis of the significant figures, it is preferable that $E_{D_{\pm}D_{\mp}}$ should be represented as 0.1 kcal mol^{-1}.

For PMPrS, virtual chains including a variety of stereosequences were arranged according to the Markov chain. This procedure was repeated n_c times (n_c chains were generated), and their $\langle r^2\rangle_0/nl^2$ values were calculated for the individual chains. The $\langle r^2\rangle_0/nl^2$ values of the n_c chains are averaged to be compared with

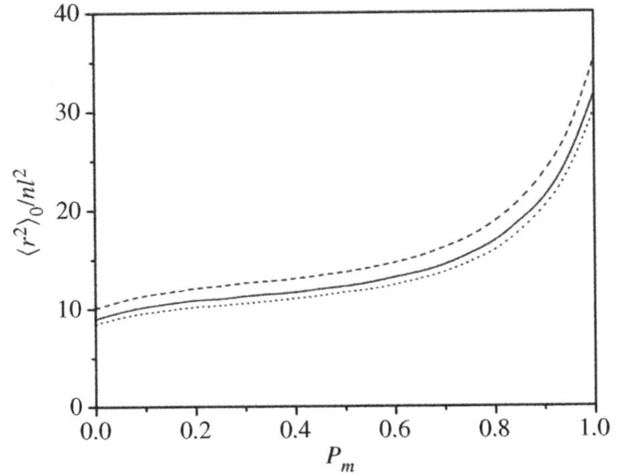

Figure 28.2 Characteristic ratio of PMPrS of x (degree of polymerization = number of bonds) = ∞ and n_c = 256 as a function of P_m: dotted line, ∠SiSiSi = 114.0°; solid line, ∠SiSiSi = 115.4°; dashed line, ∠SiSiSi = 118.0°. Source: [406], Figure 10 (p.11857)/Reproduced with permission of American Chemical Society.

the experimental value. The bond angle ∠SiSiSi was set to 114.0°, 115.4°, or 118.0° because an X-ray diffraction study determined ∠SiSiSi as 116±2° [85]. In Figure 28.2, the characteristic ratio is expressed as a function of *meso*-diad probability (P_m). The $\langle r^2\rangle_0/nl^2$ value gradually increases with P_m: $P_m = 0.0$ (syndiotactic), 8.5–10.1 and $P_m = 1.0$ (isotactic), 30.1–35.3. The allowances of $\langle r^2\rangle_0/nl^2$ are due to that of the bond angle. The *meso* linkage renders the asymmetric PS chain extended. The $\langle r^2\rangle_0/nl^2$ vs. P_m curves suggest that the experimental values of 19.9 ± 2.2 correspond to P_m = 0.76–0.92; the *meso* linkage, which avoids the steric repulsion between the longer *n*-propyl chains, may be preferred.

28.3 Physical Properties

It is known that PSs show an order–disorder transition before degradation [321, 509]. For example, poly(dimethylsilane) (PDMS), PDBS, poly(di-*n*-pentyl)silane (PDPS), and PDHS change the crystal structure (Phase I) to a conformational-disordering (condis) state (Phase II) at 157, 87, 65, and 50°C, respectively [287, 425, 427, 509]. At the order–disorder transition, an endothermic change is detected by differential scanning calorimeter (DSC). Wunderlich's group [509] estimated the disordering entropies (ΔS_ds) of PDPS and PDHS of crystallinity 100% to be 3.04 and 18.0 cal K^{-1} mol^{-1}, respectively. The net contribution of the backbone to ΔS_d was suggested to be about 2 cal K^{-1} mol^{-1}. The large ΔS_d value of PDHS undoubtedly stems from the melting of crystalline side chains.

The S_{conf} values of PDBS and PDHS, evaluated from the RIS calculations with the MD energies, are 1.5 cal K^{-1} mol^{-1}. Then, the $E_{D_\pm D_\mp}$ values were set to 0.1 kcal mol^{-1}. The net ΔS_d value (about 2 cal K^{-1} mol^{-1}) includes the entropy of volume change as well as S_{conf} (1.5 cal K^{-1} mol^{-1}). The small S_{conf} value represents the inherent rigidity of PDBS and PDHS chains, whose Si—Si bonds are allowed to adopt either D_+ or D_- conformation even in the Θ state. This is the origin of the liquid crystallinity of Phase II; steric repulsions between the long side chains force the backbone to be extended and rotate along the molecular axis with ϕ_i kept around ±150°. The most stable state is either $D_+D_+ \ldots D_+$ or $D_-D_- \ldots D_-$ conformation, which forms the 7/3 helix. In Phase II, the PS chains may be somewhat deviated from the all-deviant structure with the same sign. Because of the small energy difference of 0.1 kcal mol^{-1} between $D_\pm D_\pm$ and $D_\pm D_\mp$, the $D_\pm D_\mp$ states must be left in the crystal, act as a defect, and reduce the crystallinity.

The average dihedral angle $\langle|\phi|\rangle$ may be calculated from

$$\langle|\phi|\rangle = \frac{1}{n_c(x-2)} \sum_{k=1}^{n_c} \sum_{i=2}^{x-1} \sum_{\eta} |\phi_{i;\eta}^k| \, p_{i;\eta}^k \tag{28.1}$$

where $\phi_{i;\eta}^k$ and $p_{i;\eta}^k$ are the dihedral angle and probability ($\sum_\eta p_{i;\eta}^k = 1$) of the η state of the ith Si—Si bond of the kth chain, respectively. The conformational probability can be calculated from the statistical weight matrices. Inasmuch as Si—Si bonds of PMPrS have different stereosequences, the summation of Eq. (28.1) is performed over $(x-2)$ bonds of n_c chains, where x is the number of bonds. For PDBS and PDHS, Eq. (28.1) can be simplified to

$$\langle|\phi|\rangle = \sum_{\eta} |\phi_\eta| \, p_\eta \tag{28.2}$$

where p_η is the average fraction (bond conformation) of the η state.

The $\langle|\phi|\rangle$ values for the Θ state were obtained as follows: PDBS, 150.6°; PDHS, 151.2°; and PMPrS, e.g. 139.6° for ∠SiSiSi = 115.4° and $P_m = 0.5$ and 146.0° for ∠SiSiSi = 118.0° and $P_m = 0.83$. The former and latter parameter sets of PMPrS yielded $\langle r^2\rangle_0/nl^2$ values of 12.3 and 19.9, respectively. Because crystallized PDBS and PDHS chains adopt the 7/3 helical and all-anti structures, respectively, the corresponding $\langle|\phi|\rangle$ values are 151.5° and 180.0°. X-ray diffraction studies on PMPrS [85, 222, 231] indicated that all the Si—Si bonds adopt T conformations. If the dihedral angles in the Θ state are kept even in the solid, the $\langle|\phi|\rangle$ value of solid PMPrS chains is 165.5°. In Figure 28.3, the λ_{max} values obtained from Figure 16.7 of Section 16.1.2 are plotted against $\langle|\phi|\rangle$. On the vertical line of $\langle|\phi|\rangle = 180$, the λ_{max} value of 374 nm is plotted instead of $\lambda_{max} = 363$ nm [84, 376]. The data, collected from three PSs, appear to form a master curve.

The S_{conf} value of PDMS at the disordering transition (157 °C) was calculated to be 2.7–2.9 cal K^{-1} mol^{-1} using energy parameters offered by Michl's group [348], staying within the ΔS_d range (1.7–2.9 cal K^{-1} mol^{-1}) estimated by Wunderlich's group [509]. This is reasonable, because the short methyl substituent does not yield

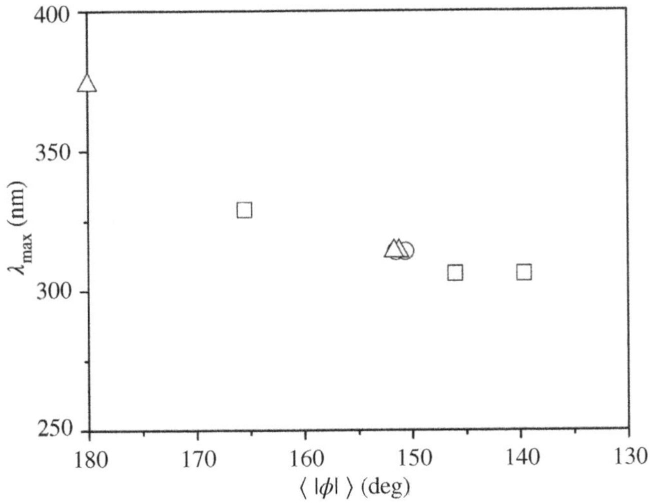

Figure 28.3 Correlation between UV absorption maximum (λ_{max}) and average dihedral angle ($\langle|\phi|\rangle$): circle, PDBS; triangle, PDHS; square, PMPrS. Source: [406], Figure 11 (p.11858)/Reproduced with permission of American Chemical Society.

configurational entropy. It is suggested that the isolated PDMS chain can adopt $G_{\pm}$ conformations [19, 339, 348]; therefore, PDMS is more flexible than PDHS (S_{conf} = 1.5 cal K^{-1} mol^{-1}). Nevertheless, PDMS is insoluble in almost all solvents, whereas PDHS is soluble in a number of solvents. The solubility of PDHS is due to the configurational entropy of the long side chain. *n*-Hexane is a good solvent for PDHS and would feel as if the PDHS chain were a cluster of *n*-hexane molecules of low mobility, and the mixing is athermal as if hexane is dissolved in hexane. Thus, the driving force of solubility is not enthalpy but entropy generated chiefly from the side chains.

Conformational characteristics of PDBS, PDHS, and PMPrS have been investigated by the RIS scheme with MD simulation. The characteristic ratios thus calculated for PDBS, PDHS, and atactic PMPrS (P_m = 0.5) are, respectively, 42.0, 54.0, and 12.3, being comparable to static light scattering experiments: PDBS, 42.3; PDHS, 42.5; and PMPrS, 19.9. The physical properties exhibited by PSs, such as order–disorder transitions, UV absorptions, and solubilities, are elucidated in terms of the characteristic parameters derived from the RIS–MD calculations.

29

Polyethylene (PE)

Conformational analysis of polyethylene (PE) has a long history. As far as the rotational isomeric state (RIS) scheme on PE is concerned, as many as 76 examples were collected in a book edited by Rehn, Mattice, and Suter and published in 1997 [383]. Of the studies introduced there, the work published in 1966 by Abe, Jernigan, and Flory (referred to as AJF1966) [6] is representative and also described in Flory's book [141]. In their treatment, three rotamers (trans, gauche$^+$, and gauche$^-$) were assumed for each CH_2—CH_2 bond, two conformational energies (E_σ and E_ω) were defined, and the 3 × 3 statistical weight matrix was formulated as shown in Eq. (11.1) of Chapter 11. The geometrical parameters were set to $l_{C-C} = 1.53$ Å, $\angle$CCC = 112°, $\phi_t = 0°$, and $\phi_{g\pm} = \pm 112.5°$. As a result of the RIS simulation, the energies were optimized to be $E_\sigma = 0.4 \pm 0.1$ kcal mol^{-1} and $E_\omega = 2.0$–2.4 kcal mol^{-1}, which well reproduced the experimental values of the characteristic ratio and its temperature coefficient.

In 1982, one of the authors, Abe, revisited the RIS calculation on PE (Abe1982) [2]; he adopted the preexponential factors (σ_0 and ω_0) that had been originally introduced by Suter and Flory [452] to express the broadness of the potential minimum. The statistical weights of PE were modified to

$$\sigma = \sigma_0 \exp(-E_\sigma/RT) \tag{29.1}$$

and

$$\omega = \omega_0 \exp(-E_\omega/RT) \tag{29.2}$$

The preexponential factors, σ_0 and ω_0, were set to 0.9 and 1.1, respectively. The conformational energies were determined as E_σ = 0.56 kcal mol^{-1} and E_ω = 1.94 kcal mol^{-1}. The dihedral angles for the gauche$^\pm$ states were ±116.5°. The RIS calculations using the modified parameters yielded the following results: $\langle r^2 \rangle_0/nl^2 = 7.84$; $d \ln \langle r^2 \rangle_0/dT = -1.23 \times 10^{-3}$ K^{-1}; and trans fraction, 0.626.

Gibbs free energies of *n*-butane and *n*-pentane were calculated by the CCSD(T) method with large cc-pVQZ basis sets and geometrical parameters optimized at

Conformational Analysis of Polymers: Methods and Techniques for Structure-Property Relationships and Molecular Design, First Edition. Yuji Sasanuma.

the B3LYP/6-311+G(2d,p) level. It is well known that the CCSD(T) method yields very accurate energies but is so expensive as to be used only for small molecules. The free energies thus obtained were compared with those calculated at the MP2/cc-pVQZ level, and both energies exactly agree with each other (see Table 8.1); therefore, the MP2/cc-pVQZ computation could be considered to be fully reliable for *n*-alkanes and was also applied to *n*-hexane, *n*-heptane, and *n*-octane in the gas phase and *n*-hexadecane solution at 25 and 140 °C. The *n*-hexadecane solution may nearly represent the nonpolar environment, and the Θ conditions of PE are mostly nonpolar solutions at temperatures close to 140 °C. According to the two-energy RIS scheme [6], the conformer free energies may be approximated by sums of E_σ and E_ω. By the least-squares method, the two conformational energies were optimized for each *n*-alkane as shown in Table 29.1. As seen from the table, E_σ ranges from 0.58 to 0.73 kcal mol^{-1}, and E_ω from 1.58 to 2.68 kcal mol^{-1}, and both energies tend to decrease with an increase in chain length, being closer to those of Abe1982 than those of AJF1966.

Here, a different RIS approach has been introduced: the PE chain is divided into segments of three methylene groups, and each segment is composed of three CH_2—CH_2 bonds and treated as if being *n*-hexane lacking in both methyl terminals (see Figure 29.1). The energy parameters were taken from Gibbs free energies of *n*-hexane at the MP2/cc-pVQZ level (see Table 29.2), and the geometrical parameters were also chosen from those of *n*-hexane (Table 29.3). From Table 29.2, the free energies are seen to be almost independent of the environment and temperature. Inasmuch as each segment is composed of three CH_2—CH_2 bonds, there are 27 (3^3) conformations, which are reduced to 10 by the molecular symmetry

Table 29.1 First- and second-order interaction energies of *n*-alkanes[a)].

	25 °C						140 °C					
	Gas			*n*-Hexadecane			Gas			*n*-Hexadecane		
	E_σ	E_ω	SD	E_σ	E_ω	SD	E_σ	E_ω	SD	E_σ	E_ω	SD
n-Butane	0.73			0.73			0.74			0.75		
n-Pentane	0.69	2.68		0.69	2.68		0.70	3.17		0.70	3.17	
n-Hexane	0.64	1.83	0.04	0.64	1.83	0.06	0.66	1.85	0.06	0.66	1.85	0.06
n-Heptane	0.58	1.58	0.10	0.59	1.58	0.10	0.60	1.55	0.11	0.60	1.55	0.11
n-Octane	0.63	1.65	0.12	0.63	1.65	0.12	0.66	1.66	0.15	0.66	1.66	0.15

a) In kcal mol^{-1}. SD stands for standard deviation.

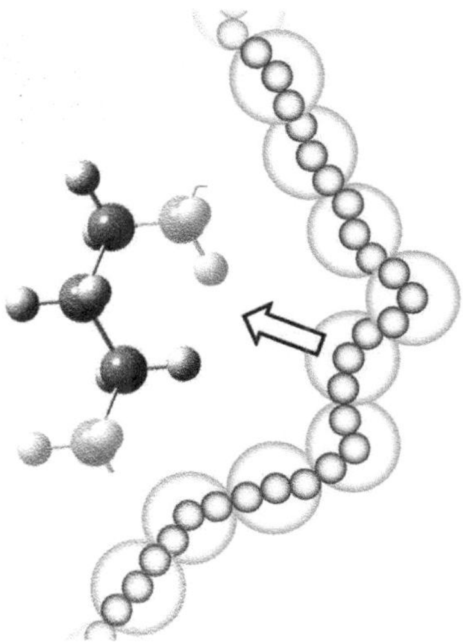

Figure 29.1 The segment model of PE adopted here. The PE chain is divided into small segments. One segment is composed of three methylene group and three CH_2-CH_2 bonds.

Table 29.2 Conformer free energies of *n*-hexane.

	Conformation			ΔG_k [a] (kcal mol^{-1})				Interaction [b]	
				Gas		*n*-Hexadecane			
k	a	b	c	25 °C	140 °C	25 °C	140 °C	σ	ω
1	t	t	t	0.00	0.00	0.00	0.00	0	0
2	t	t	g^+	0.56	0.54	0.56	0.54	1	0
3	t	g^+	t	0.73	0.77	0.74	0.77	1	0
4	t	g^+	g^+	1.34	1.41	1.34	1.41	2	0
5	t	g^+	g^-	3.05	3.07	3.06	3.08	2	1
6	g^+	t	g^+	1.19	1.15	1.20	1.16	2	0
7	g^+	t	g^-	1.41	1.40	1.41	1.40	2	0
8	g^+	g^+	g^+	1.87	2.00	1.87	2.00	3	0
9	g^+	g^+	g^-	3.74	3.86	3.75	3.87	3	1
10	g^+	g^-	g^+	5.64	5.77	5.65	5.78	3	2

a) At the MP2/cc-pVQZ//B3LYP/6-311+G(2d,p) level.
b) The numbers of gauche (σ) and $g^{\pm}g^{\mp}$ (ω) conformations included in the conformer.

Table 29.3 Geometrical parameters of the CH_2-CH_2 bonds[a].

Conformation					
a	b	c	Bond length (Å)	Bond angle (°)	Dihedral angle (°)
t	t	t	1.531	113.7	0.0
t	g^+	t	1.535	114.9	113.3
t	g^-	t	1.535	114.9	−113.3
g^+	t	t	1.532	113.3	3.7
g^+	g^+	t	1.536	114.8	115.9
g^+	g^-	t	1.536	116.3	−115.6
g^-	t	t	1.532	113.3	−3.7
g^-	g^+	t	1.536	116.3	115.6
g^-	g^-	t	1.536	114.8	−115.9
t	t	g^+	1.532	114.9	3.7
t	g^+	g^+	1.536	116.1	115.9
t	g^-	g^+	1.536	116.1	−115.6
g^+	t	g^+	1.534	114.5	5.5
g^+	g^+	g^+	1.537	116.0	119.7
g^+	g^-	g^+	1.539	117.7	−114.4
g^-	t	g^+	1.534	114.6	0.0
g^-	g^+	g^+	1.539	117.5	114.6
g^-	g^-	g^+	1.539	116.0	−114.6
t	t	g^-	1.532	114.9	−3.7
t	g^+	g^-	1.536	116.1	115.6
t	g^-	g^-	1.536	116.1	−115.9
g^+	t	g^-	1.534	114.6	0.0
g^+	g^+	g^-	1.539	116.0	114.6
g^+	g^-	g^-	1.539	117.5	−114.6
g^-	t	g^-	1.534	114.5	−5.5
g^-	g^+	g^-	1.539	117.7	114.4
g^-	g^-	g^-	1.537	116.0	−119.7

a) For the methyl terminal, the following parameters were used: $l_{C\text{-}C} = 1.529$ Å; ∠CCC = 113.4°; $\phi_t = 0°$; $\phi_{g\pm} = \pm120.0°$.

Figure 29.2 The PE chain here is treated as poly(trimethylene), whose repeating unit includes three CH_2-CH_2 bonds designated as a, b, and c. The conformational energies of poly(trimethylene) are taken from Gibbs free energies of conformers of *n*-hexane.

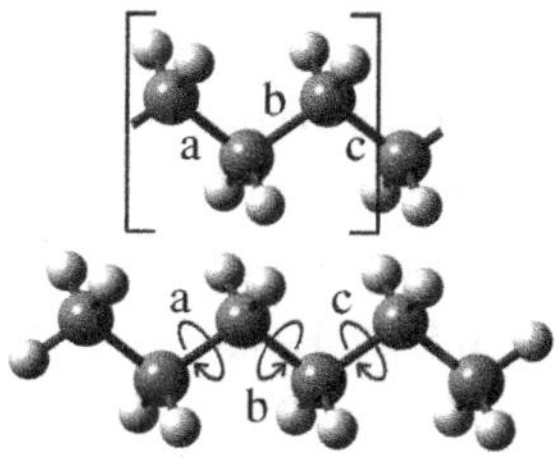

(Figure 29.2). The statistical weight matrices can be formulated similar to those of poly(ethylene terephthalate) (PET): Eq. (12.30) in Section 12.1.

For the refined RIS calculation, the computer program was written in FORTRAN, and the source code is presented in Appendix A. The input data are computational conditions (temperatures and calculated items) and geometrical parameters of the CH_2—CH_2 bond in the format of Table 29.3 along with those of the terminal CH_3—CH_2 bonds. Figure 29.3 shows the output on 1000 mer (molecular weight $M = 42$ kDa) of poly(trimethylene) at 140 °C. The characteristic ratio is 7.96, and the root-mean-square end-to-end distance ($\langle r^2 \rangle_0^{1/2}$) is 237.1 Å. Figure 29.4 shows the $\log_{10} \langle r^2 \rangle_0^{1/2}$ vs. $\log_{10} M$ plot of poly(trimethylene). The dotted line there has a slope of 1/2. In the range of $\log_{10} M \geq 3$ ($M \geq 1000$), the calculated data are located on the dotted line. Therefore, the relation of the plot can be expressed as

$$\log_{10} \langle r^2 \rangle_0^{1/2} = \frac{1}{2} \log_{10} M + c \tag{29.3}$$

where c is a constant. Equation (29.3) is changed to

$$\langle r^2 \rangle_0^{1/2} \propto M^{1/2} \tag{29.4}$$

Equation (29.4) shows that the poly(trimethylene) chain behaves as an unperturbed chain.

The trans fractions of bonds a (c) and b are 0.563 and 0.685, respectively, and the average is 0.603. In the three-methylene segment, bonds a and c are equivalent. The geometrical parameters depend on the conformations of the previous, current, and next bonds, while the conformational energy is expressed as a function of those of the current and previous two bonds. Therefore, bond angles of bonds a and c are averaged to be slightly different from each other. The average geometrical parameters were obtained as follows: bond length, 1.534 Å (1.533) Å; bond angle, 114.7° (114.5°); dihedral angles of gauche$^{\pm}$ states, ±114.8° (±115.1°). Here, the values in and out of the parentheses correspond to bonds b and a (c), respectively.

In Table 29.4, the calculated configurational properties at 140 °C, obtained by extrapolation of $M \to \infty$, are compared with the experimental values at temperatures of 127.5–180 °C. The experimental $\langle r^2 \rangle_0 / nl^2$s are scattered, and the

Polyethylene: poly (trimethylene) model

degree of polymerization: 1000

calculated data:

charateristic ratio at 140. 00 C: 7.96372

<r^2>0 : 0.56223E+05 <r^2>0^(1/2) : 0.23711E+03

temp. coefficient (×10^3) of Cx (c.r.) (K^–1) at 140. 00 C: –1.27317

bond conformations at 140. 00 C

a:	0.56287	0.43713
b:	0.68461	0.31539
c.	0.56287	0.43713

averaged geometrical parameters at 140. 00 C

bond length & angle		dihdral angles (t, g⁺, g⁻)		
1.534	114.634	–0.000	114.847	–114.847
1.533	114.484	0.000	115.096	–115.096
1.534	114.678	–0.000	114.847	–114.847

Figure 29.3 Output of the refined RIS calculations for PE with the poly(trimethylene) model.

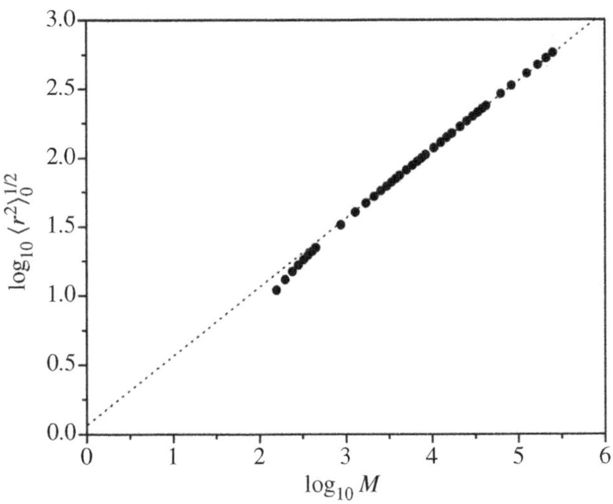

Figure 29.4 Double logarithmic plot of the root-mean-square end-to-end distance ($\langle r^2 \rangle_0^{1/2}$) vs. molecular weight (*M*), obtained from the refined RIS calculation based on the poly(trimethylene) model. The dotted line has a slope of 1/2.

Table 29.4 Configurational properties of unperturbed PE chains: comparison between theory and experiment.

State	Method[a)]	Medium	Temp (°C)	$\langle r^2\rangle_0/nl^2$	$d\ln\langle r^2\rangle_0/dT$ (10^3 K^{-1})	S_{conf} (cal K^{-1} bond^{-1})
			calcd			
	RIS	Gas	140	7.97	−1.27	1.80
		n-Hexadecane	140	7.98	−1.28	1.80
			exptl			
Solution[b)]	VA	Decalin	140	6.8		
	VT	Bis-2-ethylhexyl adipate	145	5.2		
	LT	Bis-2-ethylhexyl adipate	145	10.3		
	VT	Biphenyl	127.5	7.0		
		Dodecanol	137.5	6.8		
		Dodecanol	138	6.9		
		Diphenylmethane	142.2	6.9		
		Decanol	153.3	7.6		
		Diphenyl ether	161.4	6.4		
		Octanol	180.1	6.4		
		Biphenyl	127.5	7.1		
		Diphenylmethane	142.2	7.0		
		Diphenyl ether	163.9	6.8		
	ST	Dotriacontane	140–190		−1.15	
	VA	Hexadecane	140		−1.2	
Melt	SANS		150	7.6[c)]		
			133–163	8.3[d)]		
				6.9[d)]		
				6.9[d)]		
			140	7.5[e)]		
			140		−1.1[e)]	
	PVT		(145.5)[f)]			1.77[f)]
			(145.5)			1.90[f)]
			(145.5)			1.62[f)]

a) Abbreviations: VA, viscosity in a good solvent; LT, Zimm's plot for a Θ solution; VT, viscosity in a Θ solvent; ST, stress (from a temperature coefficient of undiluted sample); VG, viscosity (estimated from an extrapolation method); SANS, small-angle neutron scattering; PVT, pressure–volume–temperature measurement.
b) Reference [263].
c) Reference [424].
d) Reference [282].
e) Reference [258].
f) At the equilibrium melting point. Reference [293].

calculated value of 7.98 stays within the experimental range. The temperature coefficient of -1.28×10^{-3} K^{-1} is close to the experimental value of -1.1 to -1.2×10^{-3} K^{-1}, and the S_{conf} value of 1.80 cal K^{-1} bond^{-1} is in agreement with the entropy change at constant volume of 1.62–1.90 cal K^{-1} bond^{-1} [293]. The refined RIS calculations satisfactorily reproduced all the experimental data, being fully consistent with Abe1982 rather than AJF1966.

The refined RIS scheme based on the segment model using the high-precision MO calculations has provided accurate information on the physical properties of PE.

A

FORTRAN Computer Program for Refined RIS Calculations on Polyethylene

The FORTRAN source code for the refined rotational isomeric state (RIS) calculation on polyethylene (to be exact, poly(trimethylene)) is presented here. The user needs to prepare two text files: ifile and gfile. The former includes computational conditions and energy parameters, and the latter contains geometrical parameters in the form of Table 29.3. If one wishes to revise the source code for a different polymer, one must first build the statistical weight matrices (U_js) suitable for the polymeric chain, read the source code line by line by reference to the exposition and mathematical expressions of the RIS scheme, and extensively change the subroutine (u_generator) to set the U_j matrices and related subroutines by paying particular attention to array sizes. It is preferable that all real variables are defined as real*8 or real*16 to avoid overflow. However, if all variables are set as real*16, the computation would be extremely time consuming.

FORTRAN is an advanced computer language that was developed early but nevertheless has still been improved and used widely in computational sciences because superb FORTRAN compilers provide us with high-speed computing and precise results. If needed or desired, the FORTRAN source code may be translated into different languages such as C and Python.

After one revises the program, even if the compiler gives no error message, the program may be further checked by the following procedures.

- If all bond lengths (l_{CC}), bond angles (∠CCC), and dihedral angles of trans (ϕ_t) and gauche$^\pm$ ($\phi_{g\pm}$) conformations are set as follows: l_{CC} = 2 Å, ∠CCC = 120°, $\phi_t = 0°$, and $\phi_{g\pm} = \pm 120°$, then the $\langle r^2 \rangle_0/nl^2$ value of 1000 mer of the all-trans poly(trimethylene) chain terminated with methyl groups treated here can be readily calculated by hand:

$$\frac{\langle r^2 \rangle_0}{nl^2} = \frac{[2\sin(120°/2) \times (3 \times 1000 + 2)]^2}{3002 \times 2^2} = \frac{3}{4} \times 3002 = 2251.5 \qquad \text{(A.1)}$$

If all Gibbs free energies but that of the all-trans state are set to a large value of, for example, 20 kcal mol^{-1}, then, in the RIS calculation, the polymeric chain

Conformational Analysis of Polymers: Methods and Techniques for Structure-Property Relationships and Molecular Design, First Edition. Yuji Sasanuma.

would be fixed in the all-trans conformation. Under these conditions, the RIS calculations ought to yield $\langle r^2 \rangle_0 / nl^2 = 2251.5000$, which exactly agrees with the above hand calculation. Simultaneously, it follows that $p_t = 1.00000$ and $p_g = 0.00000$ in all skeletal bonds.

- If all the energy parameters are set equal to zero, the bond conformations would be $p_t = 0.33333$ and $p_g = 0.66667$.

```
c    * Refined Rotational Isomeric State (RIS) Calculation on Polyethylene
c
      implicit double precision (a-h,o-z)
      character*72 title,etitle,ifile,ofile,gfile
c
      common /obsdat/temp1,temp2,temp4,temp5
      common /obbonf/nbc,nnbc,temp3(10)
      common /fractn/fa1,fa2,fa3,fb1,fb2,fb3,fc1,fc2,fc3
      common /degree_of_poly/npoly
      common /caldat/z,a2,a3,a4r,a4d,a5
      common /tempcoe/dr_dT
c
      common /fract_geo/fr(3,3,3,3)
      common /geomtry/blen(3,3,3,3),bang(3,3,3,3),dang(3,3,3,3)
      common /geomtr1/bang1(3,3,3,3),dang1(3,3,3,3)
      common /average/avblen(3),avbang(3),avdang(3,3)
      common /conf/iconf(4,250),nconf(250),itemp_conf(250)
      common /edata/gibbs_e(100)
c
      real*16:: z,z0,z1,z2,z3,sav,uconf,a2
      dimension iope(10)
c
   10 write(*,'(a)')  ' Poly(trimethylene): P3M'
      write(*,'(a¥)') '        input    file name      : '
      read(*,'(a)',err=10) ifile
      write(*,'(a¥)') '        geometry file name      : '
      read(*,'(a)',err=10) gfile
      write(*,'(a¥)') '        output   file name      : '
      read(*,'(a)',err=10) ofile
      write(*,'(a¥)') '        degree of polymerization: '
      read(*,*,err=10) npoly
c
c     input of inital data from ifile
      open(unit=5,file=ifile)
c
      read(5,5) title
    5 format(a70)
c
      iope=0
c
      read(5,*) temp1, iope(1)    ! for characteristic ratio
c
      read(5,*) temp2, iope(2)    ! for temp. coeff. of C.R.
c
      read(5,*) nbc,   iope(3)
      do nnbc=1,nbc
            read(5,*) temp3(nnbc)
      end do
      read(5,*) temp4, iope(4)    ! for configurational entropy
      read(5,*) temp5, iope(5)    ! for averaged geometry
c     all energy and geometrical parameters to be printed
      read(5,*) ifull
c
c     energy parameter (in kcal/mol)
      read(5,5) etitle
      read(5,*) imax1
      do i=1,imax1
            read(5,*) nump, itemp_conf(i),gibbs_e(i)
      end do
c
      close(unit=5)
c
c     input of geometrical parameters from gfile
      call geopar(gfile)
c
      open(unit=6,file=ofile,access='sequential',
     +status='unknown')
c
      write(6,11) title
   11 format(1x,a70)
c
      write(6,150) npoly
  150 format(/'     degree of polymerization: ',i5)
c
      write(6,160)
```

```
  160 format(/'    calculated data:')
c
c     * characteristic ratio
      write(*,41)
41    format(/' * characteristic ratio')
      if(iope(1) /= 1) go to 1001
      call calmom(0,1,a4r,temp1)  ! characteristic ratio at temp1
c
      write(6,165) temp1,a4r
  165 format(/'    characteristic ratio at ',f6.2,' C: ',f10.5)
      write(6,166) a2, sqrt(a2)
  166 format(/'    <r^2>0 : ',e14.5,'  <r^2>0^(1/2): ',e14.5)
c
c     * temperature coefficients of c.r. at temp2
      write(*,42)
42    format(/' * temperature coefficient')
c
 1001 if(iope(2) /= 1) go to 1002
c
      call calmom(0,1,cr1,temp2-1.0)
      call calmom(0,1,cr2,temp2+1.0)
      dr_dT=(dlog(cr2/cr1))/2.0
c
      write(6,180) temp2,dr_dT*1000.0
  180 format(/'    temp. coefficient (x10^3) of Cx (c.r.) (K^-1) at '
     +,f6.2,' C: ',f10.5)
c
c     * bond conformations at temp3's
 1002 if(iope(3) /= 1) go to 1003
      write(*,43)
43    format(/' * bond conformation')

      do nnbc=1,nbc
            call calmom(1,3,dummy,temp3(nnbc))
      end do
c
c     * configurational entropy at temp4
 1003 if(iope(4) /= 1) go to 1004
      write(*,44)
44    format(/' * configurational entropy')
c
      call calmom(0,1,dummy,temp4)
      z0=z
      z1=qexp((1.0/dfloat(npoly))*qlog(z))
      call calmom(0,1,dummy,temp4-1.0)
      z2=qexp((1.0/dfloat(npoly))*qlog(z))
      call calmom(0,1,dummy,temp4+1.0)
      z3=qexp((1.0/dfloat(npoly))*qlog(z))
c
      sav  = 1.9872*( qlog(z1) + (temp4+273.15)/z1 * (z3-z2)/2.0 )
      uconf=(1.9872/1000.0)*(temp4+273.15)**2  /z1 * (z3-z2)/2.0
c
      write(6,280) sav,uconf,temp4
  280 format(/'    configurational entropy (cal/K mol): ',f10.5,
     +/'    configurational energy  (kcal/mol) : ',f10.5,
     +'  at ',f6.2,' C')
c
      write(6,227) z0,z1
  227 format(/'    partition function of whole chain  : ',d15.7
     +/'    partition function per monomer     : ',d15.7)
c
c     * geometrical parameters averaged at temp5
c
 1004 if(iope(5) /= 1) go to 1005
      write(*,45)
45    format(/' * averaged geometry')
c
      call calmom(0,1,dummy,temp5)
c
      write(6,510) temp5
  510 format(/'    averaged geometrical parameters (a-e) at ',f7.2,' C')
      write(6,515)
  515 format('    bond length & angle        dihdral angles (t,g+,g-)')
      do n=1,3
            write(6,520) avblen(n),avbang(n),(avdang(i,n),i=1,3)
```

```
      end do
  520 format(7x,f7.3,2x,f8.3,3x,3(f8.3,2x))
c
c
c   * all energy and geometrical parameters
 1005 if(ifull /= 1) go to 1006
      write(6,20)
   20 format(/' * conformational free energies in kcal/mol')
      write(6,11) etitle
      do k=1,imax1
            write(6,30) k,itemp_conf(k),gibbs_e(k)
      end do
   30 format(' #',i2,2x,i5,2x,f10.6)
c
c     output of geometrical parameters
      call geo_out
 1006 close(unit=6)
c
      stop
      end
c
c
      subroutine calmom(iend,iop,a4,temp)
      implicit double precision(a-h,o-z)
c
c     iend=1 data output as the final computation
c     iend   otherwise  no detailed output
c     iop =1 characteristic ratio
c     iop =2 dipole moment ratio (unavailable)
c     iop =3 bond conformations
c
      common /obsdat/temp1,temp2,temp4,temp5
      common /obbonf/nbc,nnbc,temp3(10)
      common /degree_of_poly/npoly
      common /caldat/z,a2,a3,a4r,a4d,a5
      common /tempcoe/dr_dT
c
      common/umatrix/u1(3,3),u2(3,3),u3(3,9),u4(9,9),
     +ua(9,9),ub(9,9),uc(9,9),un(9,9)
c
      dimension ua1(9,9),ua2(9,9),ua3(9,9),
     +ub1(9,9),ub2(9,9),ub3(9,9),
     +uc1(9,9),uc2(9,9),uc3(9,9)
c
      dimension a(3,9),b(3,9),uinit(3,9)
      dimension aa(9,9),bb(9,9),umono(9,9)
c
      common /mono/f21,f22,f23,f31,f32,f33,f41,f42,f43
c
      common /geomtry/blen(3,3,3,3),bang(3,3,3,3),dang(3,3,3,3)
      common /average/avblen(3),avbang(3),avdang(3,3)
c
      real*16:: z,part a2,a5
c
      write(*,48)
48    format(' calmom')
c
      call u_generator(temp)
c
      if(iop.eq.3) go to 998
c
      call g_maker
      call multiply(z,part,45,135)
c
      write(*,47) z, part
47    format(' z: ',d12.5,'  and  (m^2)/2: ',d12.5)
c
      call fract
c
      call geometry
c
      a2=2.0*part/z
      a5= avblen(3)**2
     a+(avblen(1)**2 + avblen(2)**2 + avblen(3)**2)*dfloat(npoly)
     b+ avblen(1)**2
```

```
      a4=a2/a5
c
      write(*,43) (avblen(kk),kk=1,3)
43    format(' bond length (a-e): ',5(f7.3))
      go to 999
c
c     ua1(9,9),ua2(9,9),ua3(9,9),ub1(9,9),ub2(9,9),ub3(9,9),
c     uc1(9,9),uc2(9,9),uc3(9,9)
c
  998 continue
      ua1=0.0
      ua2=0.0
      ua3=0.0
      ub1=0.0
      ub2=0.0
      ub3=0.0
      uc1=0.0
      uc2=0.0
      uc3=0.0
c
      do i=1,9
            do j=1,7,3
                  ua1(i,j)=ua(i,j)
                  ub1(i,j)=ub(i,j)
                  uc1(i,j)=uc(i,j)
            end do
      end do
c
      do i=1,9
            do j=2,8,3
                  ua2(i,j)=ua(i,j)
                  ub2(i,j)=ub(i,j)
                  uc2(i,j)=uc(i,j)
            end do
      end do
c
      do i=1,9
            do j=3,9,3
                  ua3(i,j)=ua(i,j)
                  ub3(i,j)=ub(i,j)
                  uc3(i,j)=uc(i,j)
            end do
      end do
c
c     a(3,9),b(3,9)
c
      call proda(u2,  u3,   a,  3,   3,   9)
      call proda(a,   u4,   b,  3,   9,   9)
      call transu(b,  uinit,    3,   9)
c
      if(npoly <= 2) go to 897
c
c     aa(9,9),bb(9,9)
c
      call proda (ua,  ub,  aa,  9,9,9)
      call proda (aa,  uc,  bb,  9,9,9)
      call transu(bb,  umono,    9,9)
c
      call fract_poly(10,uinit,umono,ua,ub,uc,zall)
c
      call fract_poly(10,uinit,umono,ua1,ub,uc,za1)
      call fract_poly(10,uinit,umono,ua2,ub,uc,za2)
      call fract_poly(10,uinit,umono,ua3,ub,uc,za3)
c
      call fract_poly(10,uinit,umono,ua,ub1,uc,zb1)
      call fract_poly(10,uinit,umono,ua,ub2,uc,zb2)
      call fract_poly(10,uinit,umono,ua,ub3,uc,zb3)
c
      call fract_poly(10,uinit,umono,ua,ub,uc1,zc1)
      call fract_poly(10,uinit,umono,ua,ub,uc2,zc2)
      call fract_poly(10,uinit,umono,ua,ub,uc3,zc3)
c
      fa1 = za1/zall
      fa2 = za2/zall
      fa3 = za3/zall
```

```
c
      fb1 = zb1/zall
      fb2 = zb2/zall
      fb3 = zb3/zall
c
      fc1 = zc1/zall
      fc2 = zc2/zall
      fc3 = zc3/zall
c
  897 if(iend.ne.1) go to 999
c
      write(6,996) temp
  996 format(/'    bond conformations at ',f6.2' C')
      write(6,994) fa1,fa2+fa3
  994 format('       a:  ',2f10.5)
      write(6,993) fb1,fb2+fb3
  993 format('       b:  ',2f10.5)
      write(6,992) fc1,fc2+fc3
  992 format('       c:  ',2f10.5)
c
  999 return
      end
c
c
      subroutine fract_poly(npoly,uinit,umono,ua,ub,uc,zall)
      implicit double precision(a-h, o-z)
c
c     uall = (unit)(umono)^(npoly/2-1)(unow)(umono)^k
c     npoly: odd number,     k = npoly/2
c          : even number,    k = npoly/2-1
c
      dimension uinit(3,9),umono(9,9)
      dimension ua(9,9),ub(9,9),uc(9,9)
      dimension uua(3,9),uub(3,9)
c
      dimension aa(9,9),bb(9,9)
      dimension unow(9,9),upre(9,9),upost(9,9)
c
      call proda (ua, ub, aa, 9,9,9)
      call proda (aa, uc, bb, 9,9,9)
      call transu(bb,unow,9,9)
c
      if(npoly.eq.3) go to 1000
c
      call transu(umono,aa,9,9)
      do j=1,npoly/2-1
            call proda(aa,umono,bb,9,9,9)
            call transu(bb,aa,9,9)
      end do
      call transu(aa,upre,9,9)
c
      call transu(umono,aa,9,9)
      if(mod(npoly,2).eq.0) then
            do j=1,npoly/2-1
                  call proda(aa,umono,bb,9,9,9)
                  call transu(bb,aa,9,9)
            end do
      else
            do j=1,npoly/2
                  call proda(aa,umono,bb,9,9,9)
                  call transu(bb,aa,9,9)
            end do
      end if
      call transu(aa,upost,9,9)

      call proda (uinit, upre,  uua, 3,9,9)
      call proda (uua,   unow,  uub, 3,9,9)
      call proda (uub,   upost, uua, 3,9,9)
      go to 2000
c
 1000 call proda(uinit,unow, uua,3,9,9)
      call proda(uua,  umono,uub,3,9,9)
      call transu(uub,uua,3,9)
c
 2000 continue
```

```
      zall=0.0
      do i=1,3
            do j=1,9
                  zall=zall+uua(i,j)
            end do
      end do
c
      return
      end
c
c
      subroutine transu(a,b,m1,m2)
      implicit double precision (a-h,o-z)
      dimension a(m1,m2),b(m1,m2)
c
      do i=1,m1
            do j=1,m2
                  b(i,j)=a(i,j)
            end do
      end do
c
      return
      end
c
c
      subroutine u_generator(temp)
      implicit double precision(a-h,o-z)
c
      common/umatrix/u1(3,3),u2(3,3),u3(3,9),u4(9,9),
     +ua(9,9),ub(9,9),uc(9,9),un(9,9)
      common /ujmatrix/uj(9,9)
c
      common /conf/iconf(4,250),nconf(250),itemp_conf(250)
      common /conf_no/ii(8),jj(8),mm(8),nn(8)
      common /matpos/mpos(9,9),ix(4),iy(4),matrix(2,4,250)
      common/edata/gibbs_e(100)
c
      dimension c1(3,1),c3(3,1)
      dimension e3(3,3)
      dimension r1(1,3),r3(1,3)

c
      dimension temp9_3(9,3)
c
      write(*,1)
    1 format('   sub: u_generator')
c
      u1=0.0
      u2=0.0
      u3=0.0
      u4=0.0
c
      ua=0.0
      ub=0.0
      uc=0.0
      un=0.0
c
      c1 =0.0
      c3 =0.0
      e3 =0.0
      r1 =0.0
      r3 =0.0
c
      c1(1,   1)=1.0
      c3(1:3,1)=(/1.0,1.0,1.0/)
      r1(1,1)=1.0
      r3(1,1:3)=(/1.0,1.0,1.0/)
c
      e3(1,1)=1.0
      e3(2,2)=1.0
      e3(3,3)=1.0
c
c  u1(3x3) and u2(3x3)
      call DirPro(c1,r1,u1,3,1,1,3)
      call DirPro(c1,r3,u2,3,1,1,3)
```

```
c
c  u3(3x9)
      call DirPro(e3,r3,u3,3,3,1,3)
c
c  u4(9x9)
      call open_9_9(temp)
      u4=uj
c
c  ua(9x9),ub(9x9),uc(9x9)
      call DirPro(c3,e3,temp9_3,3,1,3,3)
      call DirPro(temp9_3,r3,ua,9,3,1,3)
      ub=ua
c
c  uc(9x9)
      uc=u4
c
c  un(9x9)
      call DirPro(c3,e3,temp9_3,3,1,3,3)
      call DirPro(temp9_3,r1,un,9,3,1,3)
c
c
      call Norm(u4,9,9)
      call Norm(uc,9,9)
  999 return
      end
c
c
      subroutine Norm(a,m1,m2)
      implicit double precision(a-h,o-z)
c
      dimension a(m1,m2)
c
      fac=0.0
      eps=1.0d-10
c
      do i=1,m1
            do j=1,m2
                  if(a(i,j).gt.fac) fac=a(i,j)
            end do
      end do
c
      do i=1,m1
            do j=1,m2
                  a(i,j)=a(i,j)/fac
                  if(a(i,j).lt.eps) a(i,j)=0.0
            end do
      end do
c
      return
      end
c
c
      subroutine open_9_9(temp)
c     Formation of 9 X 9 U-matrix including 4th-order intramolecular
c
      implicit double precision(a-h,o-z)
      common /conf/iconf(4,250),nconf(250),itemp_conf(250)
      common /conf_no/ii(8),jj(8),mm(8),nn(8)
      common /matpos/mpos(9,9),ix(4),iy(4),matrix(2,4,250)
      common/edata/gibbs_e(100)
      common /ujmatrix/uj(9,9)
c
      write(*,1)
    1 format('   sub: open_uj')
c
      ii=0
      jj=0
      mm=0
      nn=0
c
      imax=0
      iconf=0
      mpos=0
      matrix=0
c
```

```
      do 100 ii2=1,3
            ii(2)=ii2
            do 100 ii3=1,3
                  ii(3)=ii3
                  do 100 ii4=1,3
                        ii(4)=ii4
c
                        call check(imax)
  100 continue
c
      nsum=0
      do i=1,imax
            nsum=nsum+nconf(i)
      end do
      open(unit=7,file='confcheck.txt')
      write(7,50) nsum
      do i=1,imax
            write(7,200) (iconf(k,i),matrix(1,k,i),matrix(2,k,i),k=1,4),
     +      nconf(i)
      end do
   50 format( '# conf: ',i5)
  200 format(4(i5,2x,i3,2x,i3,2x),2x,i3)
      write(7,*)
      write(7,250)
  250 format('  Matrix form 9x9')
      do i=1,9
            write(7,300) (mpos(i,j),j=1,9)
      end do
  300 format(9(i5,2x))
      write(7,*)
      do i=1,9
            write(7,350) ((mpos(i,j)-(mpos(i,j)/10)*10),j=1,9)
      end do
  350 format(9i2)
      close(unit=7)
c
      rt=0.0019872*(273.15+temp)
c
      do i=1,imax
            do k=1,4
                  uj(matrix(1,k,i),matrix(2,k,i))=dexp(-gibbs_e(i)/rt)
            end do
      end do
c
      return
      end
c
c
      subroutine check(imax)
      implicit double precision(a-h,o-z)
c
      common /conf/iconf(4,250),nconf(250),itemp_conf(250)
      common /conf_no/ii(8),jj(8),mm(8),nn(8)
      common /matpos/mpos(9,9),ix(4),iy(4),matrix(2,4,250)
      dimension no(4)
c
      do k=2,4
            if(ii(k).eq.1) jj(k)=1
            if(ii(k).eq.2) jj(k)=3
            if(ii(k).eq.3) jj(k)=2
      end do
c
      do k=2,4
            mm(k)=ii(6-k)
            nn(k)=jj(6-k)
      end do
c
c     when imax=0 (the first time), conformation is all-trans.
      if(imax.eq.0) then
            nconf(1)=1
            imax=1
            mpos(1,1)=111
            do k=1,4
                  iconf(k,imax)=111
                  matrix(1,k,imax)=1
```

```
                  matrix(2,k,imax)=1
             end do
             go to 2000
        end if
c
        no(1)=100*ii(2)+10*ii(3)+ii(4)
        no(2)=100*jj(2)+10*jj(3)+jj(4)
        no(3)=100*mm(2)+10*mm(3)+mm(4)
        no(4)=100*nn(2)+10*nn(3)+nn(4)
c
        ix(1)=(ii(2)-1)*3+ii(3)
        iy(1)=(ii(3)-1)*3+ii(4)
        ix(2)=(jj(2)-1)*3+jj(3)
        iy(2)=(jj(3)-1)*3+jj(4)
        ix(3)=(mm(2)-1)*3+mm(3)
        iy(3)=(mm(3)-1)*3+mm(4)
        ix(4)=(nn(2)-1)*3+nn(3)
        iy(4)=(nn(3)-1)*3+nn(4)
c
        do i=1,imax
             do k=1,4
                  if(iconf(1,i).eq.no(k)) go to 2000
             end do
        end do
c
        ident=0
        do iii=2,4
             do jjj=1,iii-1
                  if(no(iii).eq.no(jjj)) ident=ident+1
             end do
        end do
c
        imax=imax+1
        nconf(imax)=4-ident
        do k=1,4
             iconf(k,imax)=no(k)
             mpos(ix(k),iy(k))=no(k)
             matrix(1,k,imax)=ix(k)
             matrix(2,k,imax)=iy(k)
        end do
        if(iconf(1,imax).ne.itemp_conf(imax)) write(*,2500)
 2500   format(/' Error: Conformation number mismatch!!')
c
 2000   return
        end
c
c
        subroutine geopar(gfile)
        implicit double precision (a-h,o-z)
c
        common /geomtry/blen(3,3,3,3),bang(3,3,3,3),dang(3,3,3,3)
        character*72 boninf,gfile
c
        common /geomtr1/bang1(3,3,3,3),dang1(3,3,3,3)   ! in degree
c
        write(*,48)
48      format(' geopar')
c
        open(unit=7,file=gfile)
c
        do n=1,3    ! bond number
             read(7,*) boninf
             do i=1,3    ! (j+1)-th bond
                  do j=1,3    ! (j-1)-th
                       do k=1,3    ! j-th
             read(7,*) nconf,blen(j,k,i,n),bang1(j,k,i,n),dang1(j,k,i,n)
                       end do
                  end do
             end do
c
        end do
c
        close(unit=7)
c
        rad=3.141592654/180.0
```

```
c
      do n=1,3
            do i=1,3
                  do j=1,3
                        do k=1,3
                  bang(j,k,i,n)=(180.0-bang1(j,k,i,n))*rad
                  dang(j,k,i,n)=dang1(j,k,i,n)*rad
                        end do
                  end do
            end do
      end do
c
      return
      end
c
c
      subroutine geo_out
      implicit double precision (a-h,o-z)
c
      common /geomtry/blen(3,3,3,3),bang(3,3,3,3),dang(3,3,3,3)
      common /geomtr1/bang1(3,3,3,3),dang1(3,3,3,3)
c
      write(*,48)
48    format(' geo_out')
c
      write(6,5)
    5 format(/'   geometrical parameters')
c
      do n=1,3
            write(6,10) n
            nconf=0
            do i=1,3
                  do j=1,3
                        do k=1,3
            write(6,20) nconf+1,blen(j,k,i,n),bang1(j,k,i,n),dang1(j,k,i,n)
                              nconf=nconf+1
                        end do
                  end do
            end do
c
      end do
c
   10 format(' bond # ',i2)
   20 format(1x,i2,2x,3(f8.3,2x))
c
      return
      end
c
c
      subroutine g_maker
      implicit double precision (a-h,o-z)
c
      common/umatrix/u1(3,3),u2(3,3),u3(3,9),u4(9,9),
     +ua(9,9),ub(9,9),uc(9,9),un(9,9)
      dimension g1(15,15,3,3),g2(15,15,3,3),g3(15,45,3,3),
     +g4(45,45,3,3),
     +ga(45,45,3,3),gb(45,45,3,3),gc(45,45,3,3),
     +gn(45,45,3,3)
c
      dimension h1(45,45),h2(45,45),h3(45,135),h4(135,135),
     +ha(135,135),hb(135,135),hc(135,135),
     +hn(135,135)
c
      common /h_unit/hinit(45,135),hmono(135,135),hend(135,135)
      common /geomtry/blen(3,3,3,3),bang(3,3,3,3),dang(3,3,3,3)
c
      dimension amom(3,3),ba(3,3),da(3,3)
c
      dimension  g1t(15,15),g2t(15,15),g3t(15,45),g4t(45,45),
     +gat(45,45),gbt(45,45),gct(45,45),
     +gnt(45,45)
c
      dimension a0(45,45),a(45,135),b(45,135)
      dimension aa(135,135),bb(135,135)
```

```
c                  term. |first unit |other units(x>2) |term.
c  sequential # (n)  1    2    3    4    5    6    7    8
c  bond #            1    2    3    4    a    b    c    n
c  nbon              1    2    2    2    2    2    2    3
c        note)   nbon = # in geometrical parameter file
c
      write(*,48)
48    format(' g_maker')
c
      do 100 n=1,10
            if(n == 1) then   ! for bond 1
                  nbon=1
                  go to 10
            else if(n == 8) then
                  nbon=3
                  go to 10
            else
                  nbon=2
                  go to 10
            end if
c
   10 continue
      do 100 inow=1,3  ! j-th bond (geometry)
            do 100 i=1,3     ! (j+1)-th
                  do j=1,3
                        do k=1,3
                              amom(j,k)=blen(j,k,i,nbon)
                        end do
                  end do
c
                  do j=1,3
                        do k=1,3
                              ba(j,k)=bang(j,k,i,nbon)
                              da(j,k)=dang(j,k,i,nbon)
                        end do
                  end do
c
                  if(n==1) then
                        call g_generator(3,3,amom,ba,da,inow,u1,g1t)
                        do L=1,15
                              do M=1,15
                                    g1(L,M,inow,i)=g1t(L,M)
                              end do
                        end do
                        go to 100
                  else if(n==2) then
                        call g_generator(3,3,amom,ba,da,inow,u2,g2t)
                        do L=1,15
                              do M=1,15
                                    g2(L,M,inow,i)=g2t(L,M)
                              end do
                        end do
                        go to 100
                  else if(n==3) then
                        call g_generator(3,9,amom,ba,da,inow,u3,g3t)
                        do L=1,15
                              do M=1,45
                                    g3(L,M,inow,i)=g3t(L,M)
                              end do
                        end do
                        go to 100
                  else if(n==4) then
                        call g_generator(9,9,amom,ba,da,inow,u4,g4t)
                        do L=1,45
                              do M=1,45
                                    g4(L,M,inow,i)=g4t(L,M)
                              end do
                        end do
                        go to 100
                  else if(n==5) then
                        call g_generator(9,9,amom,ba,da,inow,ua,gat)
                        do L=1,45
                              do M=1,45
                                    ga(L,M,inow,i)=gat(L,M)
                              end do
```

```
                        end do
                        go to 100
                  else if(n==6) then
                        call g_generator(9,9,amom,ba,da,inow,ub,gbt)
                        do L=1,45
                              do M=1,45
                                    gb(L,M,inow,i)=gbt(L,M)
                              end do
                        end do
                        go to 100
                  else if(n==7) then
                        call g_generator(9,9,amom,ba,da,inow,uc,gct)
                        do L=1,45
                              do M=1,45
                                    gc(L,M,inow,i)=gct(L,M)
                              end do
                        end do
                        go to 100
                  else if(n==8) then
                        call g_generator(9,9,amom,ba,da,inow,un,gnt)
                        do L=1,45
                              do M=1,45
                                    gn(L,M,inow,i)=gnt(L,M)
                              end do
                        end do
                        go to 100
                  end if
c
 100        continue
c
            call h_fill( 3,  3,g1,h1)
            call h_fill( 3,  3,g2,h2)
            call h_fill( 3,  9,g3,h3)
            call h_fill( 9,  9,g4,h4)
            call h_fill( 9,  9,ga,ha)
            call h_fill( 9,  9,gb,hb)
            call h_fill( 9,  9,gc,hc)
            call h_fill( 9,  9,gn,hn)
c
c     a0(45,45),a(45,135),b(45,135)
c
            call proda(h1,h2,   a0,  45,   45,   45)
            call proda(a0,h3,    a,  45,   45,  135)
            call proda(a, h4,    b,  45,  135,  135)
c
            hinit=b
c
c     aa(135,135),bb(135,135)
c
            call proda(ha,hb,aa,135,135,135)
            call proda(aa,hc,bb,135,135,135)
c
            hmono=bb
c
            hend=hn
c
            return
            end
c
c
            subroutine g_generator(iy,ix,amom,ba,da,inow,u0,g)
            implicit double precision (a-h,o-z)
c
            dimension amom(3,3),ba(3,3),da(3,3)
            dimension u0(iy,ix),uz(iy,ix),g(5*iy,5*ix)
            dimension tm(3,3),tm1(3,3,3,3)
            dimension ub(iy,3*ix),uc(iy,ix),ud(3*iy,3*ix),ue(3*iy,ix)
            dimension amat(ix,ix),T(3*ix,3*ix)
c
            do i=1,5*iy
                  do j=1,5*ix
                        g(i,j)=0.0
                  end do
            end do
c
```

```
          do 100 j=1,3
                do 100 k=1,3
                      call henkan(ba(j,k),da(j,k),tm)
                      do 100 mm=1,3
                            do 100 nn=1,3
                                  tm1(mm,nn,j,k)=tm(mm,nn)
  100     continue
c     inow: conformation, 1:t, 2:g+, 3:g-
          call u_conf(iy,ix,u0,uz,inow)
c
          call T_matrix(tm1,T,ix)
          call M_matrix(amom,amat,ix)
          call U_M2_2  (uz,amat,uc,iy,ix)
          call UMxR3_T (uz,amat,T,ub,iy,ix)
          call UxE3_T  (uz,ud,T,iy,ix)
          call UMxC3   (uz,amat,ue,iy,ix)
          call super   (g,ub,uc,ud,ue,uz,iy,ix)
c
          return
          end
c
c
          subroutine u_conf(iy,ix,u0,uz,inow)
          implicit double precision (a-h,o-z)
c
          dimension u0(iy,ix),uz(iy,ix)
c
          do i=1,iy
                do j=1,ix
                      uz(i,j)=0.0
                end do
          end do
c
          if(ix==3) then
                do i=1,iy
                      uz(i,inow)=u0(i,inow)
                end do
                go to 999
          else
                do i=1,iy
                      do j=1,iy/3
                            jx=3*(j-1)+inow
                            uz(i,jx)=u0(i,jx)
                      end do
                end do
          end if
c
  999     return
          end
c
c
          subroutine henkan(th,ph,tr)
          implicit double precision (a-h,o-z)
          dimension tr(3,3)
c
          ts=dsin(th)
          tc=dcos(th)
          ps=dsin(ph)
          pc=dcos(ph)
c
          tr(1,1)=tc
          tr(1,2)=ts
          tr(1,3)=0.0
          tr(2,1)=ts*pc
          tr(2,2)=-tc*pc
          tr(2,3)=ps
          tr(3,1)=ts*ps
          tr(3,2)=-tc*ps
          tr(3,3)=-pc
c
          return
          end
c
c
          subroutine T_matrix(tm1,T,ix)
```

```
          implicit double precision (a-h,o-z)
c
          dimension tm1(3,3,3,3),T(3*ix,3*ix)
          dimension e3(3,3),t27(27,27)
c
          T=0.0
c

          do i=1,3
                do j=1,3
                      if(i==j) then
                            e3(i,j)=1.0
                      else
                            e3(i,j)=0.0
                      end if
                end do
          end do
c
          t27=0.0
c
          if(ix==3)  go to 10
          if(ix==9)  go to 20
          if(ix==27) go to 50
c
   10     continue
          do 30 k =1,3
                do 30 mm=1,3
                      do 30 nn=1,3
                T(3*(k-1)+mm,3*(k-1)+nn)=tm1(mm,nn,1,k)
   30     continue
          go to 99
c
   20     continue
          do 40 j =1,3
                do 40 k =1,3
                      do 40 mm=1,3
                            do 40 nn=1,3
                T(9*(j-1)+3*(k-1)+mm,9*(j-1)+3*(k-1)+nn)=tm1(mm,nn,j,k)
   40     continue
          go to 99
c
   50     continue
          do 60 j =1,3
                do 60 k =1,3
                      do 60 mm=1,3
                            do 60 nn=1,3
                t27(9*(j-1)+3*(k-1)+mm,9*(j-1)+3*(k-1)+nn)=tm1(mm,nn,j,k)
   60     continue
          call DirPro(e3,t27,T,3,3,27,27)
c
   99     continue
          return
          end
c
c
          subroutine M_matrix(amom,amat,ix)
          implicit double precision (a-h,o-z)
c
          dimension amom(3,3),amat(ix,ix)
          dimension e3(3,3),am(9,9)
c
          amat=0.0
          am=0.0
c
          do i=1,3
                do j=1,3
                      if(i==j) then
                            e3(i,j)=1.0
                      else
                            e3(i,j)=0.0
                      end if
                end do
          end do
c
          if(ix==3)  go to 10
```

```
      if( x==9)  go to 20
      if( x==27) go to 30
c
 10   continue
      do k=1,3
            amat(k,k)=amom(1,k)
      end do
      go to 99
c
 20   continue
      do j=1,3
            do k=1,3
                  amat(3*(j-1)+k,3*(j-1)+k)=amom(j,k)
            end do
      end do
      go to 99
c
 30   continue
      do j=1,3
            do k=1,3
                  am(3*(j-1)+k,3*(j-1)+k)=amom(j,k)
            end do
      end do
      call DirPro(e3,am,amat,3,3,9,9)
c
 99   return
      end
c
c
      subroutine U_M2_2(u,amat,uc,iy,ix)
      implicit double precision (a-h,o-z)
c
      dimension u(iy,ix),uc(iy,ix),amat(ix,ix),am2(ix,ix)
c
      do i=1,iy
            do j=1,ix
                  uc( ,j)=0.0
            end do
      end do
c
      call proda(amat,amat,am2,ix,ix,ix)
      call proda(u,am2,uc,iy,ix,ix)
c
      do i=1,iy
            do j=1,ix
                  uc(i,j)=uc(i,j)/2.0
            end do
      end do
c
      return
      end
c
c
c
      subroutine UMxR3_T(u,amat,T,ub,iy,ix)
      implicit double precision (a-h,o-z)
c
      dimension u(iy,ix),a(iy,ix),amat(ix,ix),T(3*ix,3*ix),ub(iy,3*ix),
     +   r3(1,3),b(iy,3*ix)
c
      do i=1,iy
      do j=1,3*ix
            ub(i,j)=0.0
      end do
      end do
c
      r3(1,1)=1.0
      r3(1,2)=0.0
      r3(1,3)=0.0
c
      call proda (u,amat,a,iy,ix,ix)
      call DirPro(a,r3,b,iy,ix,1,3)
      call proda (b,T,ub,iy,3*ix,3*ix)
c
      return
```

```
          end
c
c
          subroutine UxE3_T(u,ud,T,iy,ix)
          implicit double precision (a-h,o-z)
c
          dimension u(iy,ix),b(3*iy,3*ix),e3(3,3),T(3*ix,3*ix),ud(3*iy,3*ix)
c
          do i=1,3*iy
                do j=1,3*ix
                      ud(i,j)=0.0
                end do
          end do
c
          do i=1,3
                do j=1,3
                      if(i==j) then
                            e3(i,j)=1.0
                      else
                            e3(i,j)=0.0
                      end if
                end do
          end do
c
          call DirPro(u,e3,b,iy,ix,3,3)
          call proda(b,T,ud,3*iy,3*ix,3*ix)
c
          return
          end
c
c
          subroutine UMxC3(u,amat,ue,iy,ix)
          implicit double precision (a-h,o-z)
c
          dimension u(iy,ix),amat(ix,ix),ue(3*iy,ix),c3(3,1),a(iy,ix)
c
          do i=1,3*iy
                do j=1,ix
                      ue(i,j)=0.0
                end do
          end do
c
          c3(1,1)=1.0
          c3(2,1)=0.0
          c3(3,1)=0.0
c
          call proda (u,amat,a,iy,ix,ix)
          call DirPro(a,c3,ue,iy,ix,3,1)
c
          return
          end
c
c
          subroutine multiply(z,part,iy,ix)
          implicit double precision (a-h,o-z)
c
          common /h_unit/hinit(45,135),hmono(135,135),hend(135,135)
          common /degree_of_poly/npoly
c
          real*16:: h1(iy,ix),hpre(ix,ix),hnow(ix,ix),
     +    hpost(ix,ix),hn(ix,ix),hall(iy,ix),htemp(iy,ix)
          real*16:: z,part
c
          write(*,1)
    1     format('  sub: multiply')
c
          h1=hinit
c
          hpre=hmono
          hnow=hmono
c
          hn=hend
c
          if(npoly.eq.1) then
                call proda_d(h1,hn,hall,iy,ix,ix)
```

```
                 go to 400
          else if(npoly.eq.2) then
                 call proda_d(h1,hpre,htemp,iy,ix,ix)
                 call proda_d(htemp,hn,hall,iy,ix,ix)
                 go to 400
          end if
c
          do 200 mm=3,npoly
                 write(*,190) mm
  190            format('  * ',i5,' unit')
                 call proda_d(hpre,hnow,hpost,ix,ix,ix)
                 call transs_d(hpost,hpre,ix,ix)
  200     continue
c
          call proda_d(h1,hpre,htemp,iy,ix,ix)
          call proda_d(htemp,hn,hall,iy,ix,ix)
c
  400     z    =0.0q0
          part=0.0q0
          do j=1,9
                 z    = z    + hall(1,  j)    +hall(1,  j+45)+hall(1,  j+90)
     +           + hall(16,j)    +hall(16,j+45)+hall(16,j+90)
     +           + hall(31,j)    +hall(31,j+45)+hall(31,j+90)
                 part= part + hall(1,  j+36)+hall(1,  j+81)+hall(1,  j+126)
     +           + hall(16,j+36)+hall(16,j+81)+hall(16,j+126)
     +           + hall(31,j+36)+hall(31,j+81)+hall(31,j+126)
          end do
c
c
          return
          end
c
c
          subroutine h_fill(iy,ix,g,h)
          implicit double precision (a-h,o-z)
c
          dimension g(5*iy,5*ix,3,3),h(15*iy,15*ix)
c
          do i=1,15*iy
                 do j=1,15*ix
                        h(i,j)=0.0
                 end do
          end do
c
          ny=5*iy
          nx=5*ix
c
          do 30 i=1,3
                 do 30 j=1,3
                        do 30 mm=1,ny
                               do 30 nn=1,nx
                 h(ny*(i-1)+mm,nx*(j-1)+nn)=g(mm,nn,i,j)
   30     continue
c
          return
          end
c
c
          subroutine super(a,b,c,d,e,f,iy,ix)
          implicit double precision (a-h,o-z)
          dimension a(iy*5,ix*5),b(iy,ix*3),c(iy,ix),d(iy*3,ix*3),
     +    e(iy*3,ix),f(iy,ix)
c
          do i=1,iy*5
          do j=1,ix*5
                 a(i,j)=0.0
          end do
          end do
c
          do i=1,iy
                 do j=1,ix
                        a(i,j)=f(i,j)
c
                        k=i+iy*4
                        l=j+ix*4
```

```
                    a(k,l)=f(i,j)
                    a(i,l)=c(i,j)
              end do
        end do
c
        do i=1,iy
              do j=1,ix*3
                    k=j+ix
                    a(i,k)=b(i,j)
              end do
        end do
c
        do i=1,iy*3
              do j=1,ix*3
                    k=i+iy
                    l=j+ix
                    a(k,l)=d(i,j)
              end do
        end do
c
        do i=1,iy*3
              do j=1,ix
                    k=i+iy
                    l=j+ix*4
                    a(k,l)=e(i,j)
              end do
        end do
c
        return
        end
c
c
        subroutine DirPro(a,b,c,m,n,k,l)
        implicit double precision (a-h,o-z)
        dimension a(m,n),b(k,l),c(m*k,n*l)
c
        do i=1,m*k
              do j=1,n*l
                    c(i,j)=0.0
              end do
        end do
c
        do mm=1,m
              do nn=1,n
                    do i=1,k
                          do j=1,l
              c((mm-1)*k+i,(nn-1)*l+j)=a(mm,nn)*b(i,j)
                          end do
                    end do
              end do
        end do
c
        return
        end
c
c
        subroutine transs(a,b,m,n)
        implicit double precision (a-h,o-z)
        dimension a(m,n),b(m,n)
c
        do i=1,m
              do j=1,n
                    b(i,j)=a(i,j)
              end do
        end do
c
        return
        end
c
c
        subroutine transs_d(a,b,m,n)
        implicit double precision (a-h,o-z)
        real*16:: a(m,n),b(m,n)
c
        do i=1,m
```

```
                do j=1,n
                        b(i,j)=a(i,j)
                end do
        end do
c
        return
        end
c
c
        subroutine proda(a,b,c,m1,m2,m3)
        implicit double precision (a-h,o-z)
        dimension a(m1,m2),b(m2,m3),c(m1,m3)
c
        do i=1,m1
                do j=1,m3
                        c(i,j)=0.0
c
                        do k=1,m2
                                c(i,j)=c(i,j)+a(i,k)*b(k,j)
                        end do
c
        end do
        end do
c
        return
        end
c
c
        subroutine proda_d(a,b,c,m1,m2,m3)
        real*16:: a(m1,m2),b(m2,m3),c(m1,m3)
c
        do i=1,m1
                do j=1,m3
                        c(i,j)=0.0
c
                        do k=1,m2
                                c(i,j)=c(i,j)+a(i,k)*b(k,j)
                        end do
c
        end do
        end do
c
        return
        end
c
c
        subroutine fract
        implicit double precision(a-h,o-z)
c
        common /fract_geo/fr(3,3,3,3)
        common /degree_of_poly/npoly
c
        common/umatrix/u1(3,3),u2(3,3),u3(3,9),u4(9,9),
     +  ua(9,9),ub(9,9),uc(9,9),un(9,9)
c
        dimension uit(9,9),ujt(9,9),ukt(9,9)
        dimension ui(9,9),uj(9,9),uk(9,9)
c
        dimension a(3,9),b(3,9),uinit(3,9)
        dimension aa(9,9),bb(9,9),umono(9,9)
c
c     a(3,9),b(3,9)
c
        write(*,48)
48      format(' fract')
c
        call proda (u2,  u3,   a,  3,  3,  9)
        call proda (a,   u4,   b,  3,  9,  9)
        call transu(b,  uinit,     3,  9)
c
c     aa(9,9),bb(9,9)
c
        call proda (ua,  ub,  aa,  9,9,9)
        call proda (aa,  uc,  bb,  9,9,9)
        call transu(bb,  umono,    9,9)
```

```
c
          call fract_poly2(uinit,umono,ua,ub,uc,ua,ub,uc,zall)
c
          fr=0.0
c
          do 10 ip=1,3
c
                    if(ip==1) then
                         ui=uc
                         uj=ua
                         uk=ub
                         go to 100
c
                    else if(ip==2) then
                         ui=ua
                         uj=ub
                         uk=uc
                         go to 100
c
                    else if(ip==3) then
                         ui=ub
                         uj=uc
                         uk=ua
                         go to 100
c
                    end if
c
  100     continue
c
          do 10 ii=1,3
               do 10 jj=1,3
                    do 10 kk=1,3
c
                              do L=1,9
                                   do M=1,9
                                        uit(L,M)=0.0
                                        ujt(L,M)=0.0
                                        ukt(L,M)=0.0
                                   end do
                              end do
c
               ukt(3*(ii-1)+jj,3*(jj-1)+kk)=uk(3*(ii-1)+jj,3*(jj-1)+kk)
c
               do L=1,3
                    ujt(3*(L-1)+ii,3*(ii-1)+jj)=uj(3*(L-1)+ii,3*(ii-1)+jj)
               end do
c
               do L=1,3
                    do M=1,9
                         uit(M,3*(L-1)+ii)=ui(M,3*(L-1)+ii)
                    end do
               end do
c
c
c     ip: 1=a, 2=b, 3=c
c
               if(ip==1) then ! c, a, b
                    call fract_poly2(uinit,umono,ua,ub,uit,ujt,ukt,uc,z)
                    go to 200
               else if(ip==2) then ! a, b, c
                    call fract_poly2(uinit,umono,ua,ub,uc,uit,ujt,ukt,z)
                    go to 200
               else if(ip==3) then ! b, c, a
                    call fract_poly2(uinit,umono,ua,uit,ujt,ukt,ub,uc,z)
                    go to 200
c
               end if
c
  200          fr(ii,jj,kk,ip)=z/zall
               write(*,*) ii,jj,kk,ip,fr(ii,jj,kk,ip)
c
   10     continue
c
          return
          end
```

```
c
c
        subroutine fract_poly2(uinit, umono, ua1, ub1, uc1,
     +  ua2, ub2, uc2, z)
        implicit double precision(a-h, o-z)
c
        dimension uinit(3, 9), umono(9, 9)
        dimension ua1(9, 9), ub1(9, 9), uc1(9, 9)
        dimension ua2(9, 9), ub2(9, 9), uc2(9, 9)
        dimension upre(9, 9), unow(9, 9), upost(9, 9)
        dimension aa(9, 9), bb(9, 9)
        dimension uua(3, 9), uub(3, 9)
        dimension e9(9, 9)
c
c     npoly<--naver=10: bond conformations averaged over the 10mer.
        npoly=10
c
        do i=1, 9
              do j=1, 9
                    if(i==j) then
                          e9(i, j)=1.0
                    else
                          e9(i, j)=0.0
                    end if
              end do
        end do
c
        call proda (ua1, ub1,  aa,  9, 9, 9)
        call proda (aa,  uc1,  bb,  9, 9, 9)
        call proda (bb,  ua2,  aa,  9, 9, 9)
        call proda (aa,  ub2,  bb,  9, 9, 9)
        call proda (bb,  uc2,  aa,  9, 9, 9)
        call transu(aa, unow, 9, 9)
c
        if(npoly==3) go to 1000
        if(npoly==4) go to 1500
        if(npoly==5) go to 1700
c
        call transu(e9, aa, 9, 9)
        do j=1, ((npoly-2)/2)-1
              call proda (aa, umono, bb, 9, 9, 9)
              call transu(bb, aa, 9, 9)
        end do
        call transu(aa, upre, 9, 9)
c
        call transu(e9, aa, 9, 9)
        do j=1, npoly-((npoly-2)/2)-2
              call proda (aa, umono, bb, 9, 9, 9)
              call transu(bb, aa, 9, 9)
        end do
        call transu(aa, upost, 9, 9)
c
c     uua(3, 9), uub(3, 9)
        call proda (uinit, upre,  uua, 3, 9, 9)
        call proda (uua,   unow,  uub, 3, 9, 9)
        call proda (uub,   upost, uua, 3, 9, 9)
        go to 2000
c
 1000   call proda (uinit, unow,  uua, 3, 9, 9)
        go to 2000
c
 1500   call proda (uinit, unow,  uua, 3, 9, 9)
        call proda (uua,   umono, uub, 3, 9, 9)
        call transu(uub, uua, 3, 9)
        go to 2000
c
 1700   call proda (uinit, unow,  uua, 3, 9, 9)
        call proda (uua,   umono, uub, 3, 9, 9)
        call proda (uub,   umono, uua, 3, 9, 9)
c
 2000   z=0.0
        do i=1, 3
              do j=1 9
                    z=z+uua(i, j)
              end do
```

```
            end do
c
            return
            end
c
c
            subroutine geometry
            implicit double precision(a-h,o-z)
c
            common /fract_geo/fr(3,3,3,3)
c
            common /geomtry/blen(3,3,3,3),bang(3,3,3,3),dang(3,3,3,3)
            common /geomtr1/bang1(3,3,3,3),dang1(3,3,3,3)
            common /average/avblen(3),avbang(3),avdang(3,3)
c
            dimension frc(3,3)
c
            write(*,48)
48          format(' geometry')
c
            avblen=0.0
            avbang=0.0
            avdang=0.0
            frc   =0.0
c
            do 10 n=1,3
               do 10 i=1,3
                  do 10 j=1,3
                     do 10 k=1,3
c
c nonexistent conformers (with null bond length) are removed from the average
c    geometry parameters of nbon=2 are used here.
               nbon=2
               if(blen(j,k,i,nbon).lt.1.0D-10) go to 10
c
               avblen(n)=avblen(n)+blen (j,k,i,nbon)*fr(j,k,i,n)
               avbang(n)=avbang(n)+bang1(j,k,i,nbon)*fr(j,k,i,n)
c
               do iconf=1,3
                  if(k.eq.iconf) then
                     avdang(iconf,n)=avdang(iconf,n)
     +               +dang1(j,iconf,i,nbon)*fr(j,iconf,i,n)
                  frc(iconf,n)=frc(iconf,n)+fr(j,iconf,i,n)
                  end if
                  end do
c
   10       continue
c
            do 12 n=1,3
               avblen(n)=avblen(n)/(frc(1,n)+frc(2,n)+frc(3,n))
               avbang(n)=avbang(n)/(frc(1,n)+frc(2,n)+frc(3,n))
               do 12 k=1,3
                  avdang(k,n)=avdang(k,n)/frc(k,n)
   12       continue
            return
            end
c
c
```

B

Answers of Problems

Problem 1 Here, two formulae of series are used:

$$\sum_{k=1}^{n} x^k = \frac{x - x^{n-1}}{1 - x}$$

and

$$\sum_{k=1}^{n} kx^k = \frac{x - (n + 1)x^{n+1} + nx^{n+2}}{(1 - x)^2}$$

Let $x = -\cos\theta$. Then,

$$
\begin{aligned}
&2l^2 \sum_{k=1}^{n-1} (n - k)(-\cos\theta)^k \\
&= 2nl^2 \frac{x - x^n}{1 - x} - 2l^2 \left[\frac{x - nx^n}{1 - x} + \frac{x^2 - x^{n+1}}{(1 - x)^2} \right] \\
&= 2nl^2 \left\{ \frac{x - x^n}{1 - x} - \frac{1}{n} \left[\frac{x - nx^n}{1 - x} + \frac{x^2 - x^{n+1}}{(1 - x)^2} \right] \right\} \\
&= 2nl^2 \left\{ \frac{x}{1 - x} - \frac{1}{n} \left[\frac{x}{1 - x} + \frac{x^2 - x^{n+1}}{(1 - x)^2} \right] \right\} \\
&= 2nl^2 \left\{ \frac{x}{1 - x} - \frac{1}{n} \left[\frac{x - x^2 + x^2 - x^{n+1}}{(1 - x)^2} \right] \right\} \\
&= 2nl^2 \left\{ \frac{x}{1 - x} - \frac{x}{n} \left[\frac{1 - x^n}{(1 - x)^2} \right] \right\} \\
&= 2nl^2 \left\{ -\frac{\cos\theta}{1 + \cos\theta} + \frac{\cos\theta\,[1 - (-\cos\theta)^n]}{n(1 + \cos\theta)^2} \right\}
\end{aligned}
$$

Conformational Analysis of Polymers: Methods and Techniques for Structure-Property Relationships and Molecular Design, First Edition. Yuji Sasanuma.

Therefore,

$$\begin{aligned}\langle r^2\rangle/nl^2 &= 1+2\left\{-\frac{\cos\theta}{1+\cos\theta}+\frac{\cos\theta\,[1-(-\cos\theta)^n]}{n(1+\cos\theta)^2}\right\}\\ &= \frac{1-\cos\theta}{1+\cos\theta}+\frac{2\cos\theta\,[1-(-\cos\theta)^n]}{n(1+\cos\theta)^2}\end{aligned}$$

Problem 2

$$\begin{aligned}&\frac{d\left[4\pi r^2 p(r,n)\right]}{dr}\\ &= \frac{d}{dr}\left[\left(\frac{3}{2\pi nl^2}\right)^{3/2}\exp\left(-\frac{3r^2}{2nl^2}\right)4\pi r^2\right]\\ &= 4\pi\left(\frac{3}{2\pi nl^2}\right)^{3/2}\frac{d}{dr}\left[\exp\left(-\frac{3r^2}{2nl^2}\right)r^2\right]\\ &= 4\pi\left(\frac{3}{2\pi nl^2}\right)^{3/2}\left[-\frac{6r}{2nl^2}\exp\left(-\frac{3r^2}{2nl^2}\right)r^2+2r\left(-\frac{3r^2}{2nl^2}\right)\right]\\ &= 4\pi\left(\frac{3}{2\pi nl^2}\right)^{3/2}\exp\left(-\frac{3r^2}{2nl^2}\right)r\left(-\frac{3r^2}{nl^2}+2\right)=0\\ &-\frac{3r^2}{nl^2}+2=0\ (\because r\neq 0)\\ &\therefore r=\left(\frac{2}{3}nl^2\right)^{1/2}\end{aligned}$$

Problem 3 The following formula is used here:

$$\int_0^\infty \exp(-ax^2)x^{2n}dx=\frac{(2n-1)\,!!}{2^{n+1}}\left(\frac{\pi}{a^{2n+1}}\right)^{\frac{1}{2}}\quad (a>0)$$

where

$$(2n-1)\,!!=(2n-1)(2n-3)\cdots 3\cdot 1$$

$$\begin{aligned}\langle r^2\rangle &= \int_0^\infty r^2 p(r,n)4\pi r^2 dr\\ &= 4\pi\int_0^\infty r^4\left(\frac{3}{2\pi nl^2}\right)^{3/2}\exp\left(-\frac{3r^2}{2nl^2}\right)dr\\ &= 4\pi\left(\frac{3}{2\pi nl^2}\right)^{3/2}\int_0^\infty r^4\exp\left(-\frac{3r^2}{2nl^2}\right)dr\\ &= 4\pi\left(\frac{3}{2\pi nl^2}\right)^{3/2}\frac{3}{2^3}\left[\pi\left(\frac{3}{2nl^2}\right)^{-5}\right]^{1/2}\\ &= nl^2\end{aligned}$$

Problem 4 Stirling's approximation allows us to remove the factorials from the equation:

$$\begin{aligned}
\ln\frac{n!}{n_1!n_2!} &= n\ln n - n - (n_1\ln n_1 - n_1 + n_2\ln n_2 - n_2) \\
&= (n\ln n - n_1\ln n_1 - n_2\ln n_2) - (n - n_1 - n_2) \\
&= -\left[-(n_1+n_2)\ln n + n_1\ln n_1 + n_2\ln n_2\right] \quad (\because n = n_1 + n_2) \\
&= -\left[n_1(-\ln n + \ln n_1) + n_2(-\ln n + \ln n_2)\right] \\
&= -\left(n_1\ln\frac{n_1}{n} + n_2\ln\frac{n_2}{n}\right) \\
&= -(n_1\ln X_1 + n_2\ln X_2)
\end{aligned}$$

Therefore, we have

$$S_{\text{mix}} = -k(n_1\ln X_1 + n_2\ln X_2)$$

Problem 5 Stirling's approximation is used again, and it should be noted that $n = n_1 + xn_2$, i.e. $n_1 = n - xn_2$.

$$\begin{aligned}
\ln W &= \ln\frac{n!}{(n-xn_2)!\,n_2!}\left(\frac{z-1}{n}\right)^{n_2(x-1)} \\
&= \ln n! - \ln n_1! - \ln n_2! - n_2(x-1)\ln(z-1) - n_2(x-1)\ln n \;(\because n_1 = n - xn_2) \\
&= (n_1+xn_2)\ln(n_1+xn_2) - (n_1+xn_2) - n_1\ln n_1 + n_1 - n_2\ln n_2 + n_2 \\
&\quad + n_2(x-1)\ln(z-1) - n_2(x-1)\ln(n_1+xn_2) \\
&= n_1\left[\ln(n_1+xn_2) - \ln n_1\right] + n_2\left[x\ln(n_1+xn_2) - \ln n_2 - (x-1)\ln(n_1+xn_2)\right] \\
&\quad + n_2(x-1)\left[\ln(z-1) - 1\right] \\
&= -n_1\ln\frac{n_1}{n_1+xn_2} - n_2\ln\frac{n_2}{n_1+xn_2} - n_2(x-1)\ln\frac{z-1}{e}
\end{aligned}$$

Therefore, we have

$$S = -k\left[n_1\ln\frac{n_1}{n_1+xn_2} + n_2\ln\frac{n_2}{n_1+xn_2} - n_2(x-1)\ln\frac{z-1}{e}\right]$$

$$\begin{aligned}
\Delta S_{\text{mix}} &= S - \Delta S_{\text{disorientation}} \\
&= -k\left[n_1\ln\frac{n_1}{n_1+xn_2} + n_2\ln\frac{n_2}{n_1+xn_2} - n_2(x-1)\ln\frac{z-1}{e}\right] \\
&\quad + kn_2\left[\ln x + (x-1)\ln\frac{z-1}{e}\right] \\
&= -k\left[n_1\ln\frac{n_1}{n_1+xn_2} + n_2\ln\frac{n_2}{n_1+xn_2} + n_2\ln x\right] \\
&= -k\left[n_1\ln\frac{n_1}{n_1+xn_2} + n_2\ln\frac{xn_2}{n_1+xn_2}\right]
\end{aligned}$$

Problem 6

$$\Delta G_{\text{mix}} = kT(n_1 \ln v_1 + n_2 \ln v_2 + \chi_1 n_1 v_2)$$
$$= kT\left(n_1 \ln \frac{n_1}{n_1 + xn_2} + n_2 \ln \frac{xn_2}{n_1 + xn_2} + \chi_1 n_1 \frac{xn_2}{n_1 + xn_2}\right)$$

$$\frac{1}{kT}\left(\frac{\partial \Delta G_{\text{mix}}}{\partial n_1}\right)_{T,P,n_2}$$
$$= \frac{\partial}{\partial n_1}\left(n_1 \ln \frac{n_1}{n_1 + xn_2} + n_2 \ln \frac{xn_2}{n_1 + xn_2} + \chi_1 n_1 \frac{xn_2}{n_1 + xn_2}\right)$$
$$= \ln \frac{n_1}{n_1 + xn_2} + n_1 \frac{n_1 + xn_2}{n_1}\left(\frac{\partial}{\partial n_1}\frac{n_1}{n_1 + xn_2}\right) + n_2 \frac{n_1 + xn_2}{xn_2}\left(\frac{\partial}{\partial n_1}\frac{xn_2}{n_1 + xn_2}\right)$$
$$+ \chi_1\left[\frac{xn_2}{n_1 + xn_2} + n_1\left(\frac{\partial}{\partial n_1}\frac{xn_2}{n_1 + xn_2}\right)\right]$$
$$= \ln \frac{n_1}{n_1 + xn_2} + n_1 \frac{n_1 + xn_2}{n_1}\frac{xn_2}{(n_1 + xn_2)^2} + n_2 \frac{n_1 + xn_2}{xn_2}\frac{(-xn_2)}{(n_1 + xn_2)^2}$$
$$+ \chi_1\left[\frac{xn_2}{n_1 + xn_2} + n_1 \frac{(-xn_2)}{(n_1 + xn_2)^2}\right]$$
$$= \ln \frac{n_1}{n_1 + xn_2} + \frac{xn_2}{n_1 + xn_2} - \frac{1}{x}\frac{xn_2}{n_1 + xn_2} + \chi_1 \frac{(xn_2)^2}{(n_1 + xn_2)^2}$$
$$= \ln v_1 + v_2 - \frac{1}{x}v_2 - \chi_1 v_2^2$$

$$\Delta\mu_1 = RT\left[\ln(1 - v_2) + \left(1 - \frac{1}{x}\right)v_2 + \chi_1 v_2^2\right] \quad (\because v_1 + v_2 = 1)$$

Problem 7 According to the arithmetic shown in Exercise of Chapter 3, the z value of sample 2 (sample 3) of Table 3.1 is obtained as 38.9 (29.4), from which the α_r^2 and α_S^2 values can be calculated to be 7.09 (6.34) and 6.61 (5.92), respectively. The characteristic ratio is determined as 6.41 (6.83).

Problem 8 Let f be defined as

$$f = \ln(1 - v_2) + \left(1 - \frac{1}{x}\right)v_2 + \chi_1 v_2^2$$

Then, we have

$$\left(\frac{\partial f}{\partial v_2}\right)_{T,P} = -\frac{1}{1 - v_2} + 1 - \frac{1}{x} + 2\chi_1 v_2 = 0$$
$$\left(\frac{\partial^2 f}{\partial v_2^2}\right)_{T,P} = -\frac{1}{(1 - v_2)^2} + 2\chi_1 = 0$$

The latter expression gives

$$v_2 = 1 - (2\chi_1)^{-1/2}$$

Substitution of this relation into the first derivative leads to

$$-(2\chi_1)^{1/2} + 1 - \frac{1}{x} + 2\chi_1 - (2\chi_1)^{1/2} = 0$$

That is,

$$\left[(2\chi_1)^{1/2}\right]^2 - 2(2\chi_1)^{1/2} - \left(1 - \frac{1}{x}\right) = 0$$

This is a quadratic equation of $(2\chi_1)^{1/2}$; therefore, we can obtain

$$(2\chi_1)^{1/2} = 1 + \left[1 - \left(1 - \frac{1}{x}\right)\right]^{1/2} = 1 + x^{-1/2}$$

Therefore, we have

$$\chi_{1c} = \frac{1}{2}\left(1 + x^{-1/2}\right)^2 \approx \frac{1}{2} + \frac{1}{x^{1/2}}$$

and

$$v_{2c} = 1 - \left(1 + x^{-1/2}\right)^{-1/2} = \frac{1}{1 + x^{1/2}}$$

Problem 9

$$\begin{aligned}
\left[\frac{\partial \ln(f/T)}{\partial T}\right]_{V,L} &= \left[\frac{T}{f}\frac{\partial(f/T)}{\partial T}\right]_{V,L} \\
&= \frac{T}{f}\frac{1}{T^2}\left[\left(\frac{\partial f}{\partial T}\right)_{V,L} T - f\right] \\
&= \frac{1}{fT}\left[-\left(\frac{\partial U}{\partial L}\right)_{V,T}\right] \\
&= \frac{1}{fT}\left(-f_e\right)
\end{aligned}$$

Therefore,

$$-T\left[\frac{\partial \ln(f/T)}{\partial T}\right]_{V,L} = \frac{f_e}{f}$$

The tension is expressed by Eq. (4.43):

$$f = \left(\frac{\nu kT}{L_0}\right)\frac{\langle r^2\rangle}{\langle r^2\rangle_0}\left(\alpha - \alpha^{-2}\right)$$

With the relation, because only $\langle r^2\rangle_0$ depends on temperature, we have

$$\begin{aligned}\frac{\partial}{\partial T}\left(\frac{f}{T}\right) &= \frac{\partial}{\partial T}\left[\left(\frac{\nu k}{L_0}\right)\frac{\langle r^2\rangle}{\langle r^2\rangle_0}\left(\alpha-\alpha^{-2}\right)\right] \\ &= \left[\left(\frac{\nu k}{L_0}\right)\langle r^2\rangle\left(\alpha-\alpha^{-2}\right)\right]\frac{\partial}{\partial T}\langle r^2\rangle_0^{-1} \\ &= -\left[\left(\frac{\nu k}{L_0}\right)\langle r^2\rangle\left(\alpha-\alpha^{-2}\right)\right]\langle r^2\rangle_0^{-2}\frac{\partial\langle r^2\rangle_0}{\partial T} \\ &= -\left[\left(\frac{\nu k}{L_0}\right)\frac{\langle r^2\rangle}{\langle r^2\rangle_0}\left(\alpha-\alpha^{-2}\right)\right]\langle r^2\rangle_0^{-1}\frac{\partial\langle r^2\rangle_0}{\partial T} \\ &= -\frac{f}{T}\langle r^2\rangle_0^{-1}\frac{\partial\langle r^2\rangle_0}{\partial T} \\ &= -\frac{f}{T}\frac{\partial\ln\langle r^2\rangle_0}{\partial T}\end{aligned}$$

Therefore,

$$-\frac{T}{f}\frac{\partial}{\partial T}\left(\frac{f}{T}\right) = \frac{\partial\ln\langle r^2\rangle_0}{\partial T}$$

that is

$$-\frac{\partial\ln(f/T)}{\partial T} = \frac{\partial\ln\langle r^2\rangle_0}{\partial T}$$

Accordingly, we obtain

$$\frac{f_e}{f} = -T\left[\frac{\partial\ln(f/T)}{\partial T}\right]_{V,l} = T\frac{d\ln\langle r^2\rangle_0}{dT}$$

Problem 10 The partial partition function $Z'_{g^+;3}$ is calculated from

$$Z'_{g^+;3} = \mathbf{J}^*\begin{bmatrix}1 & \sigma & \sigma\\ 0 & 0 & 0\\ 0 & 0 & 0\end{bmatrix}\begin{bmatrix}0 & \sigma & 0\\ 0 & \sigma & 0\\ 0 & \sigma\omega & 0\end{bmatrix}\mathbf{J} = \sigma+\sigma^2+\sigma^2\omega$$

Therefore, the bond conformation $p_{g^+;3}$ is expressed as

$$p_{g^+;3} = \frac{Z'_{g^+;3}}{Z} = \frac{\sigma+\sigma^2+\sigma^2\omega}{1+4\sigma+2\sigma^2+2\sigma^2\omega}$$

From $E_\sigma = 0.5$ kcal mol^{-1} and $E_\omega = 2.0$ kcal mol^{-1}, we obtain $\sigma = 0.43$ and $\omega = 0.034$ at 25 °C, and consequently, $p_{g^+;3} = 0.20$.

Problem 11

$$\frac{d(\ln Z)}{dT} = \frac{1}{Z}\frac{dZ}{dT}$$

$$\begin{aligned}
S_{\text{conf}} &= \frac{R}{x}\left[\ln Z + \frac{T}{Z}\frac{dZ}{dT}\right] \\
&= \frac{R}{x}\left[\ln Z + \frac{T}{Z}\frac{d}{dT}\sum_{k}^{K}\exp\left(-E_k/RT\right)\right] \\
&= \frac{R}{x}\left[\ln Z + \frac{T}{Z}\sum_{k}^{K}\frac{E_k}{RT^2}\exp\left(-E_k/RT\right)\right] \\
&= \frac{1}{x}\left[R\ln Z + \frac{1}{T}\frac{\sum_k^K E_k \exp\left(-E_k/RT\right)}{Z}\right] \\
&= \frac{1}{x}\left[R\ln Z + \frac{1}{T}\langle E_k\rangle\right] \\
&= \frac{1}{x}\left[R\ln Z + \frac{U(T) - U(0)}{T}\right]
\end{aligned}$$

This is because

$$\frac{\sum_k^K E_k \exp\left(-E_k/RT\right)}{Z}$$

corresponds to the weight-average of E_k ($\langle E_k\rangle$), that is $U(T) - U(0)$.

Bibliography

1 Abe, A. Configuration-dependent properties of the poly(thioethylene) chain. *Macromolecules* 13 (1980), 546–549.

2 Abe, A. Comment on a paper by Oyama and Shiokawa. *Polym. J. (Tokyo, Jpn.)* 14 (1982), 427–430.

3 Abe, A., Furuya, H., Mitra, M. K., and Hiejima, T. Molecular mechanics assessment of the configurational statistics of polyoxyethylene. *Comput. Theor. Polym. Sci.* 8 (1998), 253–258.

4 Abe, A., Hirano, T., and Tsuruta, T. Conformational energies and the random-coil configuration of poly(oxypropylene). *Macromolecules* 12 (1979), 1092–1100.

5 Abe, A., Inomata, K., Tanisawa, E., and Ando, I. Conformation and conformational energies of dimethoxymethane and 1,1-dimethoxyethane. *J. Mol. Struct.* 238 (1990), 315–323.

6 Abe, A., Jernigan, R. L., and Flory, P. J. Conformational energies of *n*-alkanes and the random configuration of higher homologs including polymethylene. *J. Am. Chem. Soc.* 88 (1966), 631–639.

7 Abe, A. and Mark, J. E. Conformational energies and the random-coil dimensions and dipole moments of the polyoxides $CH_3O[(CH_2)_yO]_xCH_3$. *J. Am. Chem. Soc.* 98 (1976), 6468–6476.

8 Abe, A., Takeda, T., and Hiejima, T. Role of conformation entropy in determining the phase transitions of polymeric systems. *Macromol. Symp.* 152 (2000), 255–265.

9 Abe, A., Tasaki, K., and Mark, J. E. Rotational isomeric state analysis of poly(oxyethylene). Conformational energies and the random-coil configuration. *Polym. J. (Tokyo, Jpn.)* 17 (1985), 883–893.

10 Abe, D., Fukuda, Y., and Sasanuma, Y. Chemistry of aromatic polythioesters and polydithioesters. *Polym. Chem.* 6 (2015), 3131–3142.

11 Abe, D., Fukuda, Y., and Sasanuma, Y. Correction: Chemistry of aromatic polythioesters and polydithioesters. *Polym. Chem.* 7 (2016), 1682–1682.

12 Abe, D. and Sasanuma, Y. Molecular design, synthesis and characterization of aromatic polythioester and polydithioester. *Polym. Chem.* 3 (2012), 1576–1587.

Conformational Analysis of Polymers: Methods and Techniques for Structure-Property Relationships and Molecular Design, First Edition. Yuji Sasanuma.

13 Abe, D., Sasanuma, Y., and Sato, H. Ethane-1,2-diyl bis(benzenedithioate). *Acta Crystallogr., Sect. E* 67 (2011), o961.

14 Abragam, A. and Pryce, M. H. L. Theory of the nuclear hyperfine structure of paramagnetic resonance spectra in crystals. *Proc. R. Soc. (London)* A205 (1951), 135–153.

15 Aharoni, S. M. Unperturbed dimensions and critical molecular weight for entanglement of poly(ethylene terephthalate) and poly(ethylene isophthalate). *Makromol. Chem.* 179 (1978), 1867–1871.

16 Akcasu, A. Z., Summerfield, G. C., Jahshan, S. N., Han, C. C., Kim, C. Y., and Yu, H. Measurement of single chain neutron scattering in concentrated polymer solutions. *J. Polym. Sci., Part B: Polym. Phys.* 18 (1980), 863–869.

17 Akinc, A., Thomas, M., Klibanov, A. M., and Langer, R. Exploring polyethylenimine-mediated DNA transfection and the proton sponge hypothesis. *J. Gene Medicine* 7 (2005), 657–663.

18 Akita, S., Einaga, Y., Miyaki, Y., and Fujita, H. Solution properties of poly(D-β-hydroxybutyrate). 1. Biosynthesis and characterization. *Macromolecules* 9 (1976), 774–780.

19 Albinsson, B., Antic, D., Neumann, F., and Michl, J. Conformers of n-Si_5Me_{12}: A comparison of ab initio and molecular mechanics methods. *J. Phys. Chem. A* 103 (1999), 2184–2196.

20 Alexandrowicz, Z., and Accad, Y. Monte Carlo of long lattice chains. Variation of the excluded volume. *Macromolecules* 6 (1973), 251–255.

21 Allen, G. The molecular basis of rubber elasticity. *Proc. R. Soc. London, Ser A* 351 (1976), 381–396.

22 Allen, G., Booth, C., and Price, C. III–The unperturbed dimensions of poly(propylene oxide). *Polymer* 8 (1967), 397–401.

23 Amor, S. R., Rayment, T., and Sanders, J. K. M. Poly(hydroxybutyrate) in vivo: NMR and X-ray characterization of the elastomeric state. *Macromolecules* 24 (1991), 4583–4588.

24 Anderson, K. S. and Hillmyer, M. A. Melt chain dimensions of polylactide. *Macromolecules* 37 (2004), 1857–1862.

25 Apostolov, A. A., Fakirov, S., Stamm, M., Patil, R. D., and Mark, J. E. Alpha-beta transition in poly(butylene terephthalate) as revealed by small-angle X-ray scattering. *Macromolecules* 33 (2000), 6856–6860.

26 Astruc, D., Boisselier, E., and Ornelas, C. Dendrimers designed for functions: from physical, photophysical, and supramolecular properties to applications in sensing, catalysis, molecular electronics, photonics, and nanomedicine. *Chem. Rev.* 110 (2010), 1857–1959.

27 Astrup, E. E. An investigation of the molecular structure and conformation of 1,2-dimethoxyethane, $CH_3-O-CH_2-CH_2-O-CH_3$, in the gas phase. *Acta Chem. Scand.* 33A (1979), 665–669.

28 Aurus, R., Lim, L.-T., Selke, S. E. M., and Tsuji, H., Eds. *Poly(lactic acid): Synthesis, Structures, Properties, Processing, and Applications*. John Wiley & Sons, Hoboken, NJ, USA, 2010.

29 Austin, A., Petersson, G. A., Frisch, M. J., Dobek, F. J., Scalmani, G., and Throssell, K. A density functional with spherical atom dispersion terms. *J. Chem. Theory Comput.* 8 (2012), 4989–5007.

30 Avitabile, G., Napolitano, R., Pirozzi, B., Rouse, K. D., Thomas, M. W., and Willis, B. T. M. Low temperature crystal structure of polyethylene: results from a neutron diffraction study and from potential energy calculations. *J. Polym. Sci., Polym. Lett. Ed.* 13 (1975), 351–355.

31 Bachmann, B. M. and Seebach, D. Investigation of the enzymatic cleavage of diastereomeric oligo(3-hydroxybutanoates) containing two to eight HB units. A model for the stereoselectivity of PHB depolymerase from *Alcaligenesfaecalis* T_1. *Macromolecules* 32 (1999), 1777–1784.

32 Bak, K., Elefante, G., and Mark, J. E. Configurational properties of poly(ethylene oxide) and poly(tetramethylene oxide). *J. Phys. Chem.* 71 (1967), 4007–4011.

33 Baldwin, D. T., Mattice, W. L., and Gandour, R. D. Molecular mechanics assessment of the configurational statistics of polyoxyethylene. *J. Comput. Chem.* 5 (1984), 241–247.

34 Barfield, M. Ab initio IGLO studies of the conformational and substituent dependencies of .alpha.-, .beta.-, .gamma.-, and .delta.-effects in the 13C NMR spectra of 1-substituted butanes. *J. Am. Chem. Soc.* 117 (1995), 2862–2876.

35 Barnard, G. N. and Sanders, J. K. M. The poly-β-hydroxybutyrate granule in vivo. *J. Biol. Chem.* 264 (1989), 3286–3291.

36 Barret, A. J. Intrinsic viscosity and friction coefficients for an excluded volume polymer in the Kirkwood approximations. *Macromolecules* 17 (1984), 1566–1572.

37 Barret, A. J. Investigation of moments of intrachain distances in linear polymers. *Macromolecules* 17 (1984), 1561–1565.

38 Bartlett, R. J. and Purvis, G. D. Many-body perturbation theory, coupled-pair many-electron theory, and importance of quadruple excitations for the correlation problem. *Int. J. Quantum Chem.* 14 (1978), 561–581.

39 Becke, A. D. Density-functional thermochemistry. I. The effect of the exchange-only gradient correction. *J. Chem. Phys.* 96 (1992), 2155–2160.

40 Becke, A. D. Density-functional thermochemistry. III. The role of exact exchange. *J. Chem. Phys.* 98 (1993), 5648–5652.

41 Beech, D. R. and Booth, C. Unperturbed dimensions of poly(ethylene oxide). *J. Polym. Sci., Part A-2: Polym. Phys.* 7 (1969), 575–586.

42 Berry, G. C. Thermodynamic and conformational properties of polystyrene. I. Light-scattering studies on dilute solutions of linear polystyrenes. *J. Chem. Phys.* 44 (1966), 4550–4564.

43 Betzel, C., Gourinath, S., Kumar, P., Kaur, P., Perbandt, M., Eschenburg, S., and Singh, T. P. Structure of a serine protease proteinase K from *Tritirachiumalbumlimber* at 0.98 Å resolution. *Biochemistry* 40 (2001), 3080–3088.

44 Bluestone, S., Mark, J. E., and Flory, P. J. The interpretation of viscosity-temperature coefficients for poly(oxyethylene) chains in a thermodynamically good solvent. *Macromolecules* 7 (1974), 325–328.

45 Bondi, A. van der Waals volumes and radii. *J. Phys. Chem.* 68 (1964), 441–451.

46 Boucher, E. A. and Hines, P. M. Properties of aqueous salt solutions of poly(ethylene oxide): thermodynamic quantities based on viscosity and other measurements. *J. Polym. Sci., Polym. Phys. Ed.* 16 (1978), 501–511.

47 Bovey, F. A. and Mirau, P. A. *NMR of Polymers*, 1st ed. Academic Press, London, UK, 1996.

48 Boys, S. F. and Bernardi, F. The calculation of small molecular interactions by the differences of separate total energies. Some procedures with reduced errors. *Mol. Phys.* 19 (1970), 553–566.

49 Brandenburg, J. G., Alessio, M., Civalleri, B., Peintinger, M. F., Bredow, T., and Grimme, S. Geometrical correction for the inter- and intramolecular basis set superposition error in periodic density functional theory calculations. *J. Phys. Chem. A* 117 (2013), 9282–9292.

50 Brant, D. A., Miller, W.G., and Flory, P.J. Conformational energy estimates for statistically coiling polypeptide chains. *J. Mol. Biol.* 23 (1967), 47–65.

51 Braun, S., Kalinowski, H.-O., and Berger, S. *150 and More Basic NMR Experiments*, 2nd expanded ed. Wiley-VCH, Weinheim, Germany, 1998.

52 Brisson, J. and Brisse, F. Model compounds of aromatic nylons: the structure of N,N'-trimethylenedibenzamide and N,N'-pentamethylenedibenzamide and their relation to 3T and 5T nylons. *Macromolecules* 19 (1986), 2632–2639.

53 Budzelaar, P. H. *gNMR*, version 5.0. IvorySoft & Adept Scientific plc, Letchworth, UK, 2004.

54 Bunn, C. W. The crystal structure of long-chain normal paraffin hydrocarbons. The shape of the CH_2 group. *Trans. Faraday Soc.* 35 (1939), 482–491.

55 Bushweller, C. H. Conformational preferences and the rotation-inversion dichotomy in aliphatic amines. In *Acyclic Organonitrogen Stereodynamics*, J. B. Lambert and Y. Takeuchi, Eds. VCH Publishers, New York, USA, 1992, chapter 1: 1–55.

56 Cancès, E., Mennucci, B., and Tomasi, J. A new integral equation formalism for the polarizable continuum model: theoretical background and applications to isotropic and anisotropic dielectrics. *J. Chem. Phys.* 107 (1997), 3032–3041.

57 Carazzolo, G. A. Structure of the normal crystal form of polyoxymethylene. *J. Polym. Sci., Part A* 1 (1963), 1573–1583.

58 Carazzolo, G. and Mammi, M. Crystal structure of a new form of polyoxymethylene. *J. Polym. Sci., Part A* 1 (1963), 965–983.

59 Carazzolo, G. and Mammi, M. Crystal structure of orthorhombic polyselenomethylene. *Makromol. Chem.* 100 (1967), 28–40.

60 Carazzolo, G. and Valle, G. Crystal structure of hexagonal polyselenomethylene. *J. Polym. Sci., Part A* 3 (1965), 4013–4027.

61 Carazzolo, G. and Valle, G. Crystal structure of hexagonal polythiomethylene. *Makromol. Chem.* 90 (1966), 66–77.

62 Carothers, W. H. Linear condensation polymers, 1937. US Patent 2,071,250.

63 Carothers, W. H. and Arvin, J. A. Studies on polymerization and ring formation. II. Poly-esters. *J. Am. Chem. Soc.* 51 (1929), 2560–2570.

64 Carothers, W. H. and Hill, J. W. Studies of polymerization and ring formation. XII. Linear superpolyesters. *J. Am. Chem. Soc.* 54 (1932), 1559–1566.

65 Carothers, W. H. and Hill, J. W. Studies of polymerization and ring formation. XV. Artificial fibers from synthetic linear condensation superpolymers. *J. Am. Chem. Soc.* 54 (1932), 1579–1587.

66 Case, D., Belfon, K., Ben-Shalom, I., Brozell, S., Cerutti, D., Cheatham, T., III, Cruzeiro, V., Darden, T., Duke, R., Giambasu, G., Gilson, M., Gohlke, H., Goetz, A., Harris, R., Izadi, S., Izmailov, S., Kasavajhala, K., Kovalenko, A., Krasny, R., Kurtzman, T., Lee, T., LeGrand, S., Li, P., Lin, C., Liu, J., Luchko, T., Luo, R., Man, V., Merz, K., Miao, Y., Mikhailovskii, O., Monard, G., Nguyen, H., Onufriev, A., Pan, F., Pantano, S., Qi, R., Roe, D., Roitberg, A., Sagui, C., Schott-Verdugo, S., Shen, J., Simmerling, C., Skrynnikov, N.R., Smith, J., Swails, J., Walker, R., Wang, J., Wilson, L., Wolf, R., Wu, X., Xiong, Y., Xue, Y., York, D., and Kollman, P. *Amber2020*. University of California, San Francisco, USA, 2020.

67 Castellano, S. and Bothner-By, A. A. Analysis of NMR spectra by least squares. *J. Chem. Phys.* 41 (1964), 3863–3869.

68 Cava, M. P. and Levinson, M. I. Thionation reactions of lawesson's reagents. *Tetrahedron* 41 (1985), 5061–5087.

69 Ceccorulli, G., Pizzoli, M., and Scandola, M. Plasticization of bacterial poly(3-hydroxybutyrate). *Macromolecules* 25 (1992), 3304–3306.

70 Cesari, M., Perego, G., and Marconi, W. The crystal structure of isotactic poly(propylene oxide). *Makromol. Chem.* 94 (1966), 194–204.

71 Chatani, Y., Kobatake, T., Tadokoro, H., and Tanaka, R. Structural studies of poly(ethylenimine). 2. Double-stranded helical chains in the anhydrate. *Macromolecules* 15 (1982), 170–176.

72 Chatani, Y., Okita, Y., Tadokoro, H., and Yamashita, Y. Structural studies of polyesters. III. Crystal structure of poly-ϵ-caprolactone. *Polym. J. (Tokyo, Jpn.)* 1 (1970), 555–562.

73 Chatani, Y., Suehiro, K., Ôkita, Y., Tadokoro, H., and Chujo, K. Structural studies of polyesters. I Crystal structure of polyglycolide. *Makromol. Chem.* 113 (1968), 215–229.

74 Cheeseman, J. R., Trucks, G. W., Keith, T. A., and Frisch, M. J. A comparison of models for calculating nuclear magnetic resonance shielding tensors. *J. Chem. Phys.* 104 (1996), 5497–5509.

75 Chen, L., Qin, Y., Wang, X., Zhao, X., and Wang, F. Plasticizing while toughening and reinforcing poly(propylene carbonate) using low molecular weight urethane: role of hydrogen-bonding interaction. *Polymer* 52 (2011), 4873–4880.

76 Cheng, S. Z. D. and Wunderlich, B. Glass transition and melting behavior of poly(ethylene 2,6-naphthalenedicarboxylate). *Macromolecules* 21 (1988), 789–797.

77 Childers, M. I., Longo, J. M., Zee, N. J. V., LaPointe, A. M., and Coates, G. W. Stereoselective epoxide polymerization and copolymerization. *Chem. Rev.* 114 (2014), 8129–8152.

78 Chipman, D. M. Charge penetration in dielectric models of solvation. *J. Chem. Phys.* 106 (1997), 10194–10206.

79 Chipman, D. M. Reaction field treatment of charge penetration. *J. Chem. Phys.* 112 (2000), 5558–5565.

80 Chipman, D. M. Comparison of solvent reaction field representations. *Theor. Chem. Acc.* 107 (2002), 80–89.

81 Chisholm, M. H. and Zhou, Z. Concerning the mechanism of the ring opening of propylene oxide in the copolymerization of propylene oxide and carbon dioxide to give poly(propylene oxide). *J. Am. Chem. Soc.* 126 (2004), 11030–11039.

82 Chiu, D. S., Takahashi, Y., and Mark, J. E. Dimensions of poly(trimethylene oxide) chains in a theta-solvent. *Polymer* 17 (1976), 670–672.

83 Chu, B. *Laser Light Scattering: Basic Principles and Practice.* Dover, Mineola, NY, USA, 2007.

84 Chunwachirasiri, W., Kanaglekar, I., Lee, G., West, R., and Winokur, M. X-ray diffraction and UV-Vis absorption studies of poly(di-*n*-hexylsilane) and poly(di-*n*-octylsilane) after quenching. *J. Phys. Chem. A* 119 (2001), 31–34.

85 Chunwachirasiri, W., Kanaglekar, I., Winokur, M. J., Koe, J. C., and West, R. Structure and chain conformation in poly(methyl-*n*-alkyl)silanes. *Macromolecules* 34 (2001), 6719–6726.

86 Ciferri, A., Hoeve, C. A. J., and Flory, P. J. Stress-temperature coefficients of polymer networks and the conformational energy of polymer chains. *J. Am. Chem. Soc.* 83 (1961), 1015–1022.

87 Civalleri, B., Zicovich-Wilson, C. M., Valenzano, L., and Ugliengo, P. B3LYP augmented with an empirical dispersion term (B3LYP-D*) as applied to molecular crystals. *CrystEngComm* 10 (2008), 405–410.

88 Claridge, T. D. W. *High-Resolution NMR Techniques in Organic Chemistry*, 3rd ed. Elsevier, Amsterdam, Netherlands, 2016.

89 Cohen, C. T., Thomas, C. M., Peretti, K. L., Lobkovsky, E. B., and Coates, G. W. Copolymerization of cyclohexene oxide and carbon dioxide using (salen)Co(III) complexes: synthesis and characterization of syndiotactic poly(cyclohexene carbonate). *Dalton Trans.* (2006), 237–249.

90 Cossi, M. and Barone, V. Separation between fast and slow polarizations in continuum solvation models. *J. Phys. Chem. A* 104 (2000), 10614–10622.

91 Cossi, M., Barone, V., Cammi, R., and Tomasi, J. Ab initio study of solvated molecules: a new implementation of the polarizable continuum model. *Chem. Phys. Lett.* 255 (1996), 327–335.

92 Cotts, P. M. Thermodynamic and hydrodynamic size of a dialkyl-substituted polysilane in a mixed θ-solvent. *Macromolecules* 27 (1994), 2899–2903.

93 Cramer, C. J. *Essentials of Computational Chemistry: Theory and Models*, 2nd ed. John Wiley & Sons, Chichester, England, 2004.

94 Cremer, D. Møller-Plesset perturbation theory: from small molecule methods to methods for thousands of atoms. *WIREs Comput. Mol. Sci.* 1 (2011), 509–530.

95 Cremer, D. and Gräfenstein, J. Calculation and analysis of NMR spin-spin coupling constants. *Phys. Chem. Chem. Phys.* 9 (2007), 2791–2816.

96 Crescenzi, V., Manzini, G., Calzolari, G., and Borri, C. Thermodynamics of fusion of poly-β-propiolactone and poly-ϵ-caprolactone. Comparative analysis of the melting of aliphatic polylactone and polyester chains. *Eur. Polym. J.* 8 (1972), 449–463.

97 Croisier, F., Duwez, A.-S., Jérôme, C., Léonard, A. F., van der Werf, K. O., Dijkstra, P. J., and Bennink, M. L. Mechanical testing of electrospun PCL fibers. *Acta Biomaterialia* 8 (2012), 218–224.

98 Curtis, K. A., Miller, D., Millard, P., Basu, S., Horkay, F., and Chandran, P. L. Unusual salt and pH induced changes in polyethylenimine solutions. *PLoS One* 11 (2016), e0158147.

99 Dachs, K. and Schwartz, E. Pyrrolidone, capryllactam and laurolactam as new monomers for polyamide fibers. *Angew. Chem., Int. Ed.* 1 (1962), 430–435.

100 Darensbourg, D. J. Making plastics from carbon dioxide: salen metal complexes as catalysts for the production of polycarbonates from epoxides and CO_2. *Chem. Rev.* 107 (2007), 2388–2410.

101 Daubeny, R. P., Bunn, C. W., and Brown, C. J. The crystal structure of polyethylene terephthalate. *Proc. R. Soc. London A* 226 (1954), 531–542.

102 de Chirico, A. and Zotteri, L. Crystallization and glass transitions of polyethylene sulphide and polyisobutylene sulphide. *Macromolecules* 11 (1975), 487–490.

103 Debye, P. The intrinsic viscosity of polymer solutions. *J. Chem. Phys.* 14 (1946), 636–639.

104 Deguire, S. and Brisse, F. The methylenic sequence conformation in a series of isoelectronic molecules. *Can. J. Chem.* 66 (1988), 341–347.

105 Deng, W., Cheeseman, J. R., and Frisch, M. J. Calculation of nuclear spin-spin coupling constants of molecules with first and second row atoms in study of basis set dependence. *J. Chem. Theory Comput.* 2 (2006), 1028–1037.

106 Desborough, I. J. and Hall, I. H. A comparison of published crystalline structures of poly(tetramethylene terephthalate). *Polymer* 18 (1977), 825–830.

107 Desborough, I. J., Hall, I. H., and Neisser, J. Z. The structure of poly(trimethylene terephthalate). *Polymer* 20 (1979), 545–552.

108 Desiraju, G. R. and Steiner, T. *The Weak Hydrogen Bond.* Oxford University Press, New York, USA, 1999.

109 Dirac, P. A. M. Note on exchange phenomena in the Thomas atom. *Math. Proc. Camb. Philos. Soc.* 26 (1930), 376–385.

110 Ditchfield, R., Hehre, W. J., and Pople, J. A. Self-consistent molecular-orbital methods. IX. An extended Gaussian-type basis for molecular-orbital studies of organic molecules. *J. Chem. Phys.* 54 (1971), 724–728.

111 Doi, M. Excluded-volume effect on dipole moment of polar macromolecules. *Polym. J. (Tokyo, Jpn.)* 3 (1972), 252–253.

112 Doi, M. *Introduction to Polymer Physics.* Oxford University Press, New York, USA, 1995.

113 Doi, Y., Kunioka, M., Nakamura, Y., and Soga, K. Nuclear magnetic resonance studies on poly(β-hydroxybutyrate) and a copolyester of β-hydroxybutyrate and β-hydroxyvalerate isolated from *Alcaligeneseutrophus* H16. *Macromolecules* 19 (1986), 2860–2864.

114 Doi, Y., Kunioka, M., Nakamura, Y., and Soga, K. Proton and carbon-13 NMR analysis of poly(β-hydroxybutyrate) isolated from *Bacillusmegaterium*. *Macromolecules* 19 (1986), 1274–1276.

115 Doi, Y. and Steinbuechel, A., Eds. *Biopolymers, Biology, Chemistry, Biotechnology, Applications, Volume 3b, Polyesters II - Properties and Chemical Synthesis.* Willey-VCH, Weinheim, Germany, 2002.

116 Domb, C. and Barrett, A. J. Universality approach to the expansion factor of a polymer chain. *Polymer* 17 (1976), 179–184.

117 Domb, C. and Joyce, G. S. Cluster expansion for a polymer chain. *J. Phys. C: Solid State Phys.* 5 (1972), 956–976.

118 Dorgan, J. R., Janzen, J., Clayton, M. P., Hait, S. B., and Knauss, D. M. Melt rheology of variable L-content poly(lactic acid). *J. Rheol.* 49 (2005), 607–619.

119 Dorgan, J. R., Janzen, J., Knauss, D. M., Hait, S. B., Limoges, B. R., and Hutchinson, M. H. Fundamental solution and single-chain properties of polylactides. *J. Polym. Sci., Part B: Polym. Phys.* 43 (2005), 3100–3111.

120 Dovesi, R., Civalleri, B., Orlando, R., Roetti, C., and Saunders, V. R. Ab initio quantum simulation in solid state chemistry. *Rev. Comput. Chem.* 21 (2005), 1–125.

121 Dovesi, R., Erba, A., Orlando, R., Zicovich-Wilson, C. M., Civalleri, B., Maschio, L., Rérat, M., Casassa, S., Baima, J., Salustro, S., and Kirtman, B. Quantum-mechanical condensed matter simulations with CRYSTAL. *WIREs Comput. Mol. Sci.* 8 (2018), e1360.

122 Dovesi, R., Saunders, V. R., Roetti, C., Orlando, R., Zicovich-Wilson, C. M., Pascale, F., Civalleri, B., Doll, K., Harrison, N. M., Bush, I. J., D'Arco, P., Llunell, M., Causà, M., Noël, Y., Maschio, L., Erba, A., Rerat, M., and Casassa, S. Crystal17 user's manual. *University of Torino (Torino)* (2017).

123 Dunning, Jr., T. H. Gaussian basis sets for use in correlated molecular calculations. I. The atoms boron through neon and hydrogen. *J. Chem. Phys.* 90 (1989), 1007–1023.

124 Elias, H.-G. and Adank, G. Konstitution und Losungseigenschaften von Makromolekulen. VIII. Ungestorte Dimensionen von Polytetrahydrofuran. *Makromol. Chem.* 103 (1967), 230–243.

125 Ellis, W. C., Jung, Y., Mulzer, M., Girolamo, R. D., Lobkovsky, E. B., and Coates, G. W. Copolymerization of CO_2 and meso epoxides using enantioselective β-diiminate catalysts: a route to highly isotactic polycarbonates. *Chem. Sci.* 5 (2014), 4004–4011.

126 Erba, A., Mahmoud, A., Orlando, R., and Dovesi, R. Elastic properties of six silicate garnet end members from accurate ab initio simulations. *Phys. Chem. Miner.* 41 (2014), 151–160.

127 Espinoza, S. M., Patil, H. I., Martinez, E. S. M., Pimentel, R. C., and Ige, P. P. Poly-ϵ-caprolactone (PCL), a promising polymer for pharmaceutical and biomedical applications: Focus on nanomedicine in cancer. *Int. J. Polym. Mater. Polym. Biomater.* 69 (2020), 85–126.

128 Evans, J. M. and Huglin, M. B. Unperturbed dimensions of poly(tetrahydrofuran). *Makromol. Chem.* 127 (1969), 141–152.

129 Feldkamp, L. A., Venkataraman, G., and King, J. S. Dispersion relation for skeletal vibrations in deuterated polyethylene. In *Neutron Inelastic Scattering* (1968), vol. 2, IAEA, Vienna, pp. 159–166.

130 Fermi, E. Un metodo statistice per la determinazione di alcune proprieta dell'atomo. *Rend. Accad., Lincei* 6 (1927), 602–607.

131 Fermi, E. Über die magnetischen Momente der Atomkerne. *Z. Phys.* 60 (1930), 320–333.

132 Fetters, L. J., Lohse, D. J., and Colby, R. H. Chain dimensions and entanglement spacings. In *Physical Properties of Polymer Handbook*, J. E. Mark, Ed., 2nd ed. Springer, New York, USA, 1999, pp. 447–454.

133 Fixman, M. Excluded volume in polymer chains. *J. Chem. Phys.* 23 (1955), 1656–1659.

134 Flory, P. J. Thermodynamics of high polymer solutions. *J. Chem. Phys.* 9 (1941), 660–661.

135 Flory, P. J. Thermodynamics of high polymer solutions. *J. Chem. Phys.* 10 (1942), 51–61.

136 Flory, P. J. The configuration of real polymer chains. *J. Chem. Phys.* 17 (1949), 303–310.

137 Flory, P. J. Thermodynamics of crystallization in high polymers. IV. A theory of crystalline states and fusion in polymers, copolymers, and their mixtures with diluents. *J. Chem. Phys.* 17 (1949), 223–240.

138 Flory, P. J. *Principles of Polymer Chemistry*. Cornell University Press, Ithaca, NY, USA, 1953.

139 Flory, P. J. Mean-square moments of chain molecules. *Proc. Natl. Acad. Sci. U.S.A.* 51 (1964), 1060–1067.

140 Flory, P. J. Stereochemical equilibrium in chain molecules. *J. Am. Chem. Soc.* 89 (1967), 1798–1804.

141 Flory, P. J. *Statistical Mechanics of Chain Molecules*. John Wiley & Sons, New York, USA, 1969.

142 Flory, P. J. Fifteenth spiers memorial lecture. Thermodynamics of polymer solutions. *Discuss. Faraday Soc.* 49 (1970), 7–29.

143 Flory, P. J. Moments of the end-to-end vector of a chain molecule, its persistence and distribution. *Proc. Natl. Acad. Sci. U.S.A.* 70 (1973), 1819–1823.

144 Flory, P. J. Foundations of rotational isomeric state theory and general methods for generating configurational averages. *Macromolecules* 7 (1974), 381–392.

145 Flory, P. J. Theoretical predictions on the configurations of polymer chains in the amorphous state. *J. Macromol. Sci. Part B Phys.* 12 (1976), 1–11.

146 Flory, P. J. Conformations of macromolecules in condensed phases. *Pure Appl. Chem.* 56 (1984), 305–312.

147 Flory, P. J. and Abe, Y. Configurational averages for chain molecules: higher moments and related quantities. *J. Chem. Phys.* 54 (1971), 1351–1363.

148 Flory, P. J., Ciferri, A., and Hoeve, C. A. J. The thermodynamic analysis of thermoelastic measurements on high elastic materials. *J. Polym. Sc.* 45 (1960), 235–236.

149 Flory, P. J. and Fisk, S. Effect of volume exclusion on the dimensions of polymer chains. *J. Chem. Phys.* 44 (1966), 2243–2248.

150 Flory, P. J. and Fox, Jr., T. G. Treatment of intrinsic viscosities. *J. Am. Chem. Soc.* 73 (1951), 1904–1908.

151 Flory, P. J. and Jernigan, R. L. Second and forth moments of chain molecules. *J. Chem. Phys.* 42 (1965), 3509–3519.

152 Flory, P. J. and Krigbaum, W. R. Statistical mechanics of dilute polymer solutions. II. *J. Chem. Phys.* 18 (1950), 1086–1094.

153 Flory, P. J. and Mark, J. E. The configuration of the polyoxymethylene chain. *Makromol. Chem.* 75 (1964), 11–21.

154 Foresman, J. B. and Frisch, Æ. *Exploring Chemistry with Electronic Structure Methods*, 3rd ed. Gaussian, Inc., Wallingford, CT, USA, 2015.

155 Foresman, J. B., Keith, T. A., Wiberg, K. B., Snoonian, J., and Frisch, M. J. Solvent effects. 5. Influence of cavity shape, truncation of electrostatics, and electron correlation on ab initio reaction field calculations. *J. Phys. Chem.* 100 (1996), 16098–16104.

156 Fredericks, R. J., Doyne, T. H., and Sprague, R. S. Crystallographic studies of nylon 4. I. Determination of the crystal structure of the α polymorph of nylon 4. *J. Polym. Sci.: Part A-2* 4 (1966), 899–911.

157 Frenkel, D. and Smit, B. *Understanding Molecular Simulation: From Algorithms to Applications*, 2nd ed. Academic Press, New York, USA, 2002.

158 Friebolin, H. *Basic One- and Two-Dimensional NMR Spectroscopy*, 2nd enlarged ed. VCH, Weinheim, Germany, 1993.

159 Frisch, M. J., Trucks, G. W., Schlegel, H. B., Scuseria, G. E., Robb, M. A., Cheeseman, J. R., Scalmani, G., Barone, V., Petersson, G. A., Nakatsuji, H., Li, X., Caricato, M., Marenich, A. V., Bloino, J., Janesko, B. G., Gomperts, R., Mennucci, B., Hratchian, H. P., Ortiz, J. V., Izmaylov, A. F., Sonnenberg, J. L., Williams-Young, D., Ding, F., Lipparini, F., Egidi, F., Goings, J., Peng, B., Petrone, A., Henderson, T., Ranasinghe, D., Zakrzewski, V. G., Gao, J., Rega, N., Zheng, G., Liang, W., Hada, M., Ehara, M., Toyota, K., Fukuda, R., Hasegawa, J., Ishida, M., Nakajima, T., Honda, Y., Kitao, O., Nakai, H., Vreven, T., Throssell, K., Montgomery, Jr., J. A., Peralta, J. E., Ogliaro, F., Bearpark, M. J., Heyd, J. J., Brothers, E. N., Kudin, K. N., Staroverov, V. N., Keith, T. A., Kobayashi, R., Normand, J., Raghavachari, K., Rendell, A. P., Burant, J. C., Iyengar, S. S., Tomasi, J., Cossi, M., Millam, J. M., Klene, M., Adamo, C., Cammi, R., Ochterski, J. W., Martin, R. L., Morokuma, K., Farkas, O., Foresman, J. B., and Fox, D. J. *Gaussian16 Revision C.01*. Gaussian Inc., Wallingford, CT, USA, 2016.

160 Fu, Y., Busing, W. R., Jin, Y., Affholter, K. A., and Wunderlich, B. Poly(ethylene terephthalate) fibers. 1. Crystal structure and morphology studies with full-pattern X-ray diffraction refinement. *Macromolecules* 26 (1993), 2187–2193.

161 Fujimaki, T. Processability and properties of aliphatic polyesters, 'BIONOLLE', synthesized by polycondensation reaction. *Polym. Degrad. Stab.* 59 (1998), 209–214.

162 Fujita, H. A new method of treating light-scattering data on dilute polymer solutions. *Polym. J. (Tokyo, Jpn.)* 1 (1970), 537–541.

163 Fujita, H. *Polymer Solutions*. Elsevier, Amsterdam, The Netherlands, 1990.

164 Fukuda, Y., Abe, D., Tanaka, Y., Uchida, J., Suzuki, N., Miyai, T., and Sasanuma, Y. Solution properties of poly(*N*-methylethylene imine), a highly hydrophilic polycation. *Polym. J. (Tokyo, Jpn.)* 48 (2016), 1065–1072.

165 Fukuda, Y. and Sasanuma, Y. Computational characterization of nylon 4, a biobased and biodegradable polyamide superior to nylon 6. *ACS Omega* 3 (2018), 9544–9555.

166 Fukuda, Y. and Sasanuma, Y. Thermal and mechanical properties of poly(methylene oxide) polymorphs unraveled by periodic density functional theory. *Macromolecules* 51 (2018), 8672–8680.

167 Gaillac, R., Pullumbi, P., and Coudert, F.-X. ELATE: an open-source online application for analysis and visualization of elastic tensors. *J. Phys. Condens. Matter* 28 (2016), 275201.

168 Galanty, P. G. Nylon 6. In *Polymer Data Handbook*, J. E. Mark, Ed. Oxford University Press, New York, USA, 1999, pp. 180–185.

169 Ghanbari, A., and Prud'homme, R. E. Lamellar and spherulitic crystallization of poly(s-2-hydroxybutanoic acid) and its stereocomplexes. *Polymer* 112 (2017), 377–384.

170 Ghosez, P., Michenaud, J.-P., and Gonze, X. Dynamical atomic charges: the case of abo_3 compounds. *Phys. Rev. B: Condens. Matter Mater. Phys.* 58 (1998), 6224–6240.

171 Gilmer, J. W., Wiswe, D., Zachmann, H.-G., Kugler, J., and Fischer, E. W. Changes in molecular conformations in poly(ethylene terephthalate) produced by crystallization and drawing as determined by neutron scattering. *Polymer* 27 (1986), 1391–1395.

172 Glatter, O. and Kratky, O. *Small Angle X-ray Scattering*. Academic Press, London, UK, 1982.

173 Gregory, P. and Huglin, M. B. Viscosity of aqueous and alkaline solutions of poly(ethylene oxide). *Makromol. Chem.* 187 (1986), 1745–1755.

174 Greschner, G. S. Mathmatical properties of polydisperse systems, 1 General relations. *Makromol. Chem.* 168 (1973), 273–288.

175 Grimme, S. Semiempirical GGA-type density functional constructed with a long-range dispersion correction. *J. Comput. Chem.* 27 (2006), 1787–1799.

176 Grimme, S., Antony, J., Ehrlich, S., and Krieg, H. A consistent and accurate *ab initio* parameterization of density functional dispersion correction (DFT-D) for the 94 elements H-Pu. *J. Chem. Phys.* 132 (2010), 154104.

177 Grimme, S., Hansen, A., Brandenburg, J. G., and Bannwarth, C. Dispersion-corrected mean-field electronic structure methods. *Chem. Rev.* 116 (2016), 5105–5154.

178 Guerin, W., Diallo, A. K., Kirilov, E., Helou, M., Slawinski, M., Brusson, J.-M., Carpentier, J.-F., and Guillaume, S. M. Enantiopure isotactic PCHC synthesized by ring-opening polymerization of cyclohexene carbonate. *Macromolecules* 47 (2014), 4230–4235.

179 Guinier, A. La diffraction des rayons X aux très petits angles: application à l'étude de phénomènes ultramicroscopiques. *Ann. Phys.* 11 (1939), 161–237.

180 Gutowsky, H. S., Belford, G. G., and McMahon, P. E. NMR studies of conformational equilibria in substituted ethanes. *J. Chem. Phys.* 36 (1962), 3353–3368.

181 Guzman, J., Riande, E., Welsh, W. J., and Mark, J. E. Configurational characteristics of the polysulfides, 3. Dipole moments of poly(trimethylene sulfide) and comparisons between some polysulfides and the corresponding polyoxides. *Makromol. Chem.* 183 (1982), 2573–2581.

182 Hagarman, A., Measey, T. J., Mathieu, D., Schwalbe, H., and Schweitzer-Stenner, R. Intrinsic propensities of amino acid residue in GxG peptides inferred from amide I' band profiles and NMR scalar coupling constants. *J. Am. Chem. Soc.* 132 (2010), 540–551.

183 Hall, I. H. and Pass, M. G. Chain conformation of poly(tetramethylene terephthalate) and its change with strain. *Polymer* 17 (1976), 807–816.

184 Hancock, R. D. and Martell, A. E. Ligand design for selective complexation of metal ions in aqueous solution. *Chem. Rev.* 89 (1989), 1875–1914.

185 Hashimoto, K., Hamano, T., and Okada, M. Degradation of several polyamides in soils. *J. Appl. Polym. Sci.* 54 (1994), 1579–1583.

186 Hashimoto, K., Sudo, M., Ohta, K., Sugimura, T., Yamada, H., and Aoki, T. Biodegradation of nylon4 and its blend with nylon 6. *J. Appl. Polym. Sci.* 86 (2002), 2307–2311.

187 Hashimoto, K., Sudo, M., Sugimura, T., and Inagaki, Y. Synthesis of novel block copolymers containing polyamide4 segments and control of their biodegradability. *J. Appl. Polym. Sci.* 92 (2004), 3492–3498.

188 Hatada, K. and Kitayama, T. *NMR Spectroscopy of Polymers*. Springer-Verlag, Berlin, Germany, 2004.

189 Heatley, F., Luo, Y., Ding, J., Mobbs, R. H., and Booth, C. A carbon-13 nuclear magnetic resonance study of the triad sequence structure of block and statistical copolymers of ethylene oxide and propylene oxide. *Macromolecules* 21 (1988), 2713–2721.

190 Hehre, W. J., Ditchfield, R., and Pople, J. A. Self-consistent molecular-orbital methods. XII. Further extensions of Gaussian-type basis sets for use in molecular orbital studies of organic molecules. *J. Chem. Phys.* 56 (1972), 2257–2261.

191 Helfer, C. A. and Mattice, W. L. The rotational isomeric state model. In *Physical Properties of Polymers Handbook*, J. E. Mark, Ed. Springer, New York, USA, 2007, ch. 3.

192 Helgaker, T., Jaszuński, M., and Pecul, M. The quantum-chemical calculation of NMR indirect spin-spin coupling constants. *Chem. Phys. Lett.* 53 (2008), 249–268.

193 Helgaker, T., Jaszuński, M., and Ruud, K. Ab initio methods for the calculation of NMR shielding and indirect spin-spin coupling constants. *Chem. Rev.* 99 (1999), 293–352.

194 Helgaker, T., Klopper, W., Koch, H., and Noga, J. Basis-set convergence of correlated calculations on water. *J. Chem. Phys.* 106 (1997), 9639–9646.

195 Helgaker, T., Watson, M., and Handy, N. C. Analytical calculation of nuclear magnetic resonance indirect spin-spin coupling constants at the generalized gradient approximation and hybrid levels of density-functional theory. *J. Chem. Phys.* 113 (2000), 9402–9409.

196 Higgins, J. S. and Benoit, H. C. *Polymers and Neutron Scattering*. Clarendon, Oxford, UK, 1994.

197 Hirano, T., Khanh, P. H., Tsuji, K., Sato, A., Tsuruta, T., Abe, A., Shimozawa, T., Kotera, A., Yamaguchi, N., and Kitahara, S. Dipole moments of poly(oxypropylene), poly[(*S*)-oxy(1-isopropylethylene)] and their monomeric model compounds. *Polym. J. (Tokyo, Jpn.)* 11 (1979), 905–907.

198 Hirano, T., Khanh, P. H., and Tsuruta, T. Deuterated epoxides and their polymers. III. NMR study of poly(*trans*-propylene oxide-1-d) in various solvents; a preliminary report. *Makromol. Chem.* 153 (1972), 331–336.

199 Hirosye, T., Einaga, Y., and Fujita, H. Excluded-volume effects in dilute polymer solutions. VIII. Poly(D,L-β-methyl β-propiolactone) in several solvents and reanalysis of data on poly(D-β-hydroxybutyrate). *Polym. J. (Tokyo, Jpn.)* 11 (1979), 819–826.

200 Hisano, T., Kasuya, K., Tezuka, Y., Ishii, N., Kobayashi, T., Shiraki, M., Oroudjev, E., Hansma, H., Iwata, T., Doi, Y., Saito, T., and Miki, K. The crystal structure of polyhydroxybutyrate depolymerase from *Penicillium funiculosum* provides insights into the recognition and degradation of biopolyesters. *J. Mol. Biol.* 356 (2006), 993–1004.

201 Hohenberg, P. and Kohn, W. Inhomogeneous electron gas. *Phys. Rev.* 136 (1964), B864–B871.

202 Holliday, L. and White, J. W. The stiffness of polymers in relation to their structure. *Pure Appl. Chem.* 26 (1971), 545–582.

203 Hori, Y., Pei, N., Kumagai, R., and Sasanuma, Y. Poly(*N*-protected ethylene imine-*alt*-ethylene sulfide) block to functionalize polymeric materials. *Polym. Chem.* 2 (2011), 2183–2185.

204 Huang, Y., Xu, Z., Huang, Y., Ma, D., Yang, J., and Mays, J. W. Characterization of poly(ϵ-caprolactone) via size exclusion chromatography with online right-angle laser-light scattering and viscometric detectors. *Int. J. Polym. Anal. Charact.* 8 (2003), 383–394.

205 Huggins, M. L. Solutions of long chain compounds. *J. Chem. Phys.* 9 (1941), 440–440.

206 Huggins, M. L. Some properties of solutions of long-chain compounds. *J. Phys. Chem.* 46 (1942), 151–158.

207 Huggins, M. L. The viscosity of dilute solutions of long-chain molecules. IV. Dependence on concentration. *J. Am. Chem. Soc.* 64 (1942), 2716–2718.

208 Hwang, M. J., Stockfisch, T. P., and Hagler, A. T. Derivation of class II force fields. 2. Derivation and characterization of a class II force field, CFF93, for the alkyl functional group and alkane molecules. *J. Am. Chem. Soc.* 116 (1994), 2515–2525.

209 Ichikawa, Y., Kondo, H., Igarashi, Y., Noguchi, K., Okuyama, K., and Washiyama, J. Crystal structures of α and β forms of poly(tetramethylene succinate). *Polymer* 41 (2000), 4719–4727.

210 Ichikawa, Y., Suzuki, J., Washiyama, J., Moteki, Y., Noguchi, K., and Okuyama, K. Strain-induced crystal modification in poly(tetramethylene succinate). *Polymer* 35 (1994), 3338–3339.

211 Ichikawa, Y., Washiyama, J., Moteki, Y., Noguchi, K., and Okuyama, K. Crystal transition mechanisms in poly(tetramethylene succinate). *Polym. J. (Tokyo, Jpn.)* 27 (1995), 1230–1238.

212 Iguchi, M. Growth of needle-like crystals of polyoxymethylene during polymerization. *Br. Polym. J.* 5 (1973), 195–198.

213 Iguchi, M. Memoir: a polymer whisker - the needle-shaped single crystal of polyoxymethylene. *Polymer* 168 (2019), 255–268.

214 Iguchi, M., Suehiro, T., Watanabe, Y., Nishi, Y., and Uryu, M. Composite materials reinforced with polyoxymethylene whiskers. *J. Mater. Sci.* 17 (1982), 1632–1638.

215 Imai, M., Kaji, K., Kanaya, T., and Sakai, Y. Ordering process in the induction period of crystallization of poly(ethylene terephthalate). *Phys. Rev. B.* 52 (1995), 12696–12704.

216 Inomata, K. and Abe, A. Conformation of 1,2-dimethoxyethane in the gas phase: a rotational isomeric state simulation of NMR vicinal coupling constants. *J. Phys. Chem.* 96 (1992), 7934–7937.

217 Inoue, S., Koinuma, H., and Tsuruta, T. Copolymerization of carbon dioxide and epoxide. *J. Polym. Sci., Part B* 7 (1969), 287–292.

218 Inoue, S., Koinuma, H., and Tsuruta, T. Copolymerization of carbon dioxide and epoxide. *Makromol. Chem.* 130 (1969), 210–220.

219 Inoue, S., Koinuma, H., Yokoo, Y., and Tsuruta, T. Stereochemistry of copolymerization of carbon dioxide with epoxycyclohexane. *Makromol. Chem.* 143 (1971), 97–104.

220 Ishida, R. Elucidation of structure-property relationships of poly(propylene oxide) and poly(propylene sulfide) crystals, March 2021. Graduation work, Department of Applied Chemistry and Biotechnology, Chiba University.

221 Jaffe, R. L., Smith, G. D., and Yoon, D. Y. Conformation of 1,2-dimethoxyethane from ab initio electronic structure calculations. *J. Phys. Chem.* 97 (1993), 12745–12751.

222 Jambe, B., Jonas, A., and Devaux, J. Crystalline structure of poly(methyl-*n*-propylsilane). *J. Polym. Sci., Part B: Polym. Phys.* 35 (1997), 1533–1543.

223 Jeffrey, G. A. *An Introduction to Hydrogen Bonding*. Oxford University Press, New York, USA, 1997.

224 Jones, A. A., Stockmayer, W. H., and Molinari, R. J. Low-frequency dielectric response of poly(ϵ-caprolactone) in dilute solution. *J. Polym. Sci. Polym. Symp.* 54 (1976), 227–235.

225 Joziasse, C. A. P., Veenstra, H., Grijpma, D. W., and Pennings, A. J. On the chain stiffness of poly(lactide)s. *Macromol. Chem. Phys.* 197 (1996), 2219–2229.

226 Juaristi, E. *Introduction to Stereochemistry and Conformational Analysis*. Wiley, New York, USA, 1991.

227 Juaristi, E., and Cuevas, G. *The Anomeric Effect*. CRC Press, Boca Raton, FL, USA, 1995.

228 Kaidu, R. Elucidation of structure-property relationships of poly(methylene sulfide) and poly(methylene selenide) crystals, March 2021. Graduation work, Department of Applied Chemistry and Biotechnology, Chiba University.

229 Kamiya, N., Inoue, Y., Yamamoto, Y., Chûjô, R., and Doi, Y. Conformational analysis of poly(3-hydroxybutyrate-*co*-3-hydroxyvalerate) in solution by proton NMR spectroscopy. *Macromolecules* 23 (1990), 1313–1317.

230 Kang, S., Zhang, G., Aou, K., Hsu, S. L., Stidham, H. D., and Yang, X. An analysis of poly(lactic acid) with varying regio regularity. *J. Chem. Phys.* 118 (2003), 3430–3436.

231 KariKari, E. K., Greso, A. J., Farmer, B. L., Miller, R. D., and Rabolt, J. F. Studies of the conformation and packing of polysilanes. *Macromolecules* 26 (1993), 3937–3945.

232 Karplus, M. Contact electron-spin coupling of nuclear magnetic moments. *J. Chem. Phys.* 30 (1959), 11–15.

233 Karplus, M. Vicinal proton coupling in nuclear magnetic resonance. *J. Am. Chem. Soc.* 85 (1963), 2870–2871.

234 Kasperczyk, J. E. Microstructure analysis of poly(lactic acid) obtained by lithium *tert*-butoxide as initiator. *Macromolecules* 28 (1995), 3937–3939.

235 Kato, H., Sasanuma, Y., Kaito, A., Tanigaki, N., Tanabe, Y., and Kinugasa, S. Unperturbed chain dimensions of poly(di-*n*-hexylsilane), poly(methyl-*n*-propylsilane), and poly(di-*n*-butylsilane). *Macromolecules* 34 (2001), 262–268.

236 Kawaguchi, Y. and Doi, Y. Structure of native poly(3-hydroxybutyrate) granules characterized by X-ray diffraction. *FEMS Microbiol. Lett.* 70 (1990), 151–156.

237 Kawaguchi, S., Imai, G., Suzuki, J., Miyahara, A., Kitano, T., and Ito, K. Aqueous solution properties of oligo- and poly(ethylene oxide) by static light scattering and intrinsic viscosity. *Polymer* 38 (1997), 2885–2891.

238 Kawai, A., Hamamoto, N., and Sasanuma, Y. Conformational characteristics and conformation-dependent properties of poly(ϵ-caprolactone). *Phys. Chem. Chem. Phys.* 44 (2022), 11382–11394.

239 Kawai, T., Uno, K., Oya, M., Komoto, T., Tsuchida, E., Nakajima, T., Fukada, E., Yamazaki, N., and Yamada, N., Eds. *Design of Polymers (Koubunshi no Bunshi Sekkei)*. Baifukan, Tokyo, Japan, 1972 (written in Japanese).

240 Kawasaki, N., Nakayama, A., Yamano, N., Takeda, S., Kawata, Y., Yamamoto, N., and ichi Aiba, S. Synthesis, thermal and mechanical properties and biodegradation of branched polyamide 4. *Polymer* 46 (2005), 9987–9993.

241 Keeler, J. *Understanding NMR Spectroscopy*, 2nd ed. Wiley, Chichester, West Sussex, UK, 2010.

242 Keith, T. A. and Bader, R. F. W. Calculation of magnetic response properties using atoms in molecules. *Chem. Phys. Lett.* 194 (1992), 1–8.

243 Keith, T. A. and Bader, R. F. W. Calculation of magnetic response properties using a continuous set of gauge transformations. *Chem. Phys. Lett.* 210 (1993), 223–231.

244 Kelly, K. M., Patterson, G. D., and Tonelli, A. E. Kerr effect studies of the poly(oxyethylenes). *Macromolecules* 10 (1977), 859–862.

245 Kittel, C. *Introduction to Solid State Physics*, 8th ed. Wiley, Hoboken, NJ, USA, 2005.

246 Klamt, A., Moya, C., and Palomar, J. A comprehensive comparison of the IEFPCM and SS(V)PE continuum solvation methods with the COSMO approach. *J. Chem. Theory Comput.* 11 (2015), 4220–4225.

247 Klamt, A. and Schüürmann, G. COSMO: a new approach to dielectric screening in solvents with explicit expressions for the screening energy and its gradient. *J. Chem. Soc., Perkin Trans. 2* (1993), 799–805.

248 Knecht, M. R. and Elias, H.-G. Zur flexibilität aliphatischer Polyester. *Makromol. Chem.* 157 (1972), 1–12.

249 Kobayashi, M., Adachi, T., Matsumoto, Y., Morishita, H., Takahashi, T., Ute, K., and Hatada, K. Polarized Raman and infrared studies of single crystals of orthorhombic modification of polyoxymethylene and its linear oligomer: crystal structures and vibrational assignments. *J. Raman Spectrosc.* 24 (1993), 533–538.

250 Kobayashi, S. and Kadokawa, J. Ring-opening polymerization of 1-(2,4,6-tri-*tert*-butylphenyl)-phosphirane: direct synthesis of a polyphosphine derivative. *Macromol. Rapid Commun.* 15 (1994), 567–571.

251 Kobayashi, M., Morishita, H., Shimomura, M., and Iguchi, M. Vibrational spectroscopic study on the solid-state phase transition of poly(oxymethylene) single crystals from the orthorhombic to the trigonal phase. *Macromolecules* 20 (1987), 2453–2456.

252 Kobayashi, S., Suzuki, M., and Saegusa, T. Cationic ring-opening polymerization of 2-phenyl-1,2-oxaphospholane (deoxophostone). *Polym. Bull.* 4 (1981), 315–321.

253 Koch, W. and Holthausen, M. C. *A Chemist's Guide to Density-Functional Theory*. Wiley-VCH, Weinheim, Germany, 2000.

254 Kohn, W. and Sham, L. J. Self-consistent equations including exchange and correlation effects. *Phys. Rev.* 140 (1965), A1133–A1138.

255 Koleske, J. V. and Lundberg, R. D. Lactone polymers. II. Hydrodynamic properties and unperturbed dimensions of poly-ϵ-caprolactone. *J. Polym. Sci., Part A-2* 7 (1969), 897–907.

256 Krigbaum, W. R. Relationships between $[\eta]$ or $\langle r^2\rangle^{3/2}$ and the second virial coefficient. *J. Polym. Sci.* 18 (1955), 315–320.

257 Krigbaum, W. R. Estimating the unperturbed dimensions of polymer molecules. *J. Polym. Sci.* 28 (1958), 213–221.

258 Krishnamoorti, R., Graessley, W. W., Zirkel, A., Richter, D., Hadjichristidis, N., Fetters, L. J., and Lohse, D. J. Melt-state polymer chain dimensions as a function of temperature. *J. Polym. Sci., Part B: Polym. Phys.* 40 (2002), 1768–1776.

259 Kubo, K., Ando, I., Shiibashi, T., Yamanobe, T., and Komoto, T. Conformations and ^{13}C NMR chemical shifts of some polyamides in the solid state as studied by high-resolution ^{13}C NMR spectroscopy. *J. Polym. Sci., Part B: Polym. Phys. Ed.* 29 (1991), 57–66.

260 Kugler, J., Fischer, E. W., Peuscher, M., and Eisenbach, C. D. Small angle neutron scattering studies of poly(ethylene oxide) in the melt. *Makromol. Chem.* 184 (1983), 2325–2334.

261 Kugler, J., Gilmer, J. W., Wiswe, D., Zachmann, H. G., Hahn, K., and Fischer, E. W. Study of transesterification in poly(ethylene terephthalate) by small-angle neutron scattering. *Macromolecules* 20 (1987), 1116–1119.

262 Kurata, M. and Stockmayer, W. H. Intrinsic viscosities and unperturbed dimensions of long chain molecules. *Fortschr. Hochpolym. Forsch.* 3 (1963), 196–312.

263 Kurata, M. and Tsunashima, Y. Viscosity-molecular weight relationships and unperturbed dimensions of linear chain molecules. In *Polymer Handbook*, J. Brandrup, E. H. Immergut, and E. A. Grulke, Eds. Wiley-Interscience, New York, USA, 1999, p. VII/48.

264 Kurata, M., Utiyama, H., and Kamada, K. Unperturbed dimensions of poly(tetrahydrofuran). *Makromol. Chem.* 88 (1965), 281–293.

265 Kurata, M. and Yamakawa, H. Theory of dilute polymer solution. II. Osmotic pressure and frictional properties. *J. Chem. Phys.* 29 (1958), 311–325.

266 Kurita, T., Fukuda, Y., Takahashi, M., and Sasanuma, Y. Crystalline moduli of polymers, evaluated from density functional theory calculations under periodic boundary conditions. *ACS Omega* 3 (2018), 4824–4835.

267 Kutzelnigg, W. Theory of magnetic susceptibilities and NMR chemical shifts in terms of localized quantities. *Isr. J. Chem.* 19 (1980), 193–200.

268 Kuzmany, H., Rabolt, J. F., Farmer, B. L., and Miller, R. D. Studies of chain conformational kinetics in poly(di-*n*-alkylsilanes) by spectroscopic methods 2. Conformation and packing of poly(di-*n*-hexylsilane). *J. Chem. Phys.* 85 (1986), 7413–7422.

269 Lar, J. Notes. Poly(methylene sulfide). *J. Org. Chem.* 26 (1961), 971–972.

270 Law, R. V. and Sasanuma, Y. Nature of the non-bonded (C–H)···O interaction of ethers $CH_3O–(CH_2)_n–OCH_3$ (n = 4–8). *J. Chem. Soc., Faraday Trans.* 92 (1996), 4885–4888.

271 Law, R. V. and Sasanuma, Y. Conformational characteristics of poly(tetramethylene oxide). *Macromolecules* 31 (1998), 2335–2342.

272 Lax, M., Barrett, A. J., and Domb, C. Polymer chain statistics and universality I. *J. Phys. A: Math. Gen.* 11 (1978), 361–374.

273 Leach, A. R. *Molecular Modelling*, 2nd ed. Pearson Education, Harlow, UK, 2001.

274 Leblanc, C. and Brisse, F. Structural study in the oligomethylene dithiobenzoate series, *n*DBS, where *n*, the number of methylene groups, is odd. *Can. J. Chem.* 70 (1992), 900–909.

275 Lee, E. C., Kim, D., Jurečka, P., Tarakeshwar, P., Hobza, P., and Kim, K. S. Understanding of assembly phenomena by aromatic-aromatic interactions: benzene dimer and the substituted systems. *J. Phys. Chem. A* 111 (2007), 3446–3457.

276 Lee, C., Yang, W., and Parr, R. G. Development of the Colle-Salvetti correlation-energy formula into a functional of the electron density. *Phys. Rev. B* 37 (1988), 785–789.

277 Lemoigne, M. Études sur láutolyse microbienne acidification par formation dácide *β*-oxybutyrique. *Ann. Inst. Pasteur* 39 (1925), 144–156.

278 Lemoigne, M. Produit de deshydratation et de polymerisation de lácide *β*-oxybutyrique. *Bull. Soc. Chim. Biol.* 8 (1926), 770–782.

279 Li, P., Ge, B., Zhang, S., Chen, S., Zhang, Q., and Zhao, Y. CO_2 capture by polyethylenimine-modified fibrous adsorbent. *Langmuir* 24 (2008), 6567–6574.

280 Li, J., Uzawa, J., and Doi, Y. Conformational behavior of methyl (3*R*)-3-[(3'*R*)-3'-hydroxybutanoyl]oxybutanoate in solutions: Effect of intramolecular hydrogen bond. *Bull. Chem. Soc. Jpn.* 70 (1997), 1887–1893.

281 Li, J., Uzawa, J., and Doi, Y. Conformational analysis of oligomers of (*R*)-3-hydroxybutanoic acid in solutions by ^{1}H NMR spectroscopy. *Bull. Chem. Soc. Jpn.* 71 (1998), 1683–1689.

282 Lieser, G., Fischer, E. W., and Ibel, K. Conformation of polyethylene molecules in the melt as revealed by small-angle neutron scattering. *J. Polym. Sci., Part C: Polym. Lett.* 13 (1975), 39–43.

283 Liu, Z., Gosser, Y., Baker, P. J., Ravee, Y., Lu, Z., Alemu, G., Li, H., Butterfoss, G. L., Kong, X.-P., Gross, R., and Montclare, J. K. Structural and functional studies of *Aspergillusoryzae* cutinase: enhanced thermostability and hydrolytic activity of synthetic ester and polyester degradation. *J. Am. Chem. Soc.* 131 (2009), 15711–15716.

284 Liu, Y., Ren, W.-M., Liu, J., and Lu, X.-B. Asymmetric copolymerization of CO_2 with meso-epoxides mediated by dinuclear cobalt(III) complexes: unprecedented enantioselectivity and activity. *Angew. Chem., Int. Ed.* 52 (2013), 11594–11598.

285 Liu, Y., Ren, W.-M., Zhang, W.-P., Zhao, R.-R., and Lu, X.-B. Crystalline CO_2-based polycarbonates prepared from *racemic* catalyst through intramolecularly interlocked assembly. *Nature Communications* 30 (2012), 487–492.

286 Longo, J. M., Sanford, M. J., and Coates, G. W. Ring-opening copolymerization of epoxides and cyclic anhydrides with discrete metal complexes: structure-property relationships. *Chem. Rev.* 116 (2016), 15167–15197.

287 Lovinger, A. J., Davis, D. D., Schilling, F. C., Padden, Jr., F. J., Bovey, F. A., and Zeigler, J. M. Solid-state structure and phase transitions of poly(dimethylsilylene). *Macromolecules* 24 (1991), 132–139.

288 Lovinger, A. J., Schilling, F. C., Bovey, F. A., and Zeigler, J. M. Characterization of poly(di-*n*-hexylsilane) in the solid state. I. X-ray and electron diffraction studies. *Macromolecules* 19 (1986), 2657–2660.

289 Lu, X.-B., Ren, W.-M., and Wu, G.-P. CO_2 copolymers from epoxides: catalyst activity, product selectivity, and stereochemistry control. *Acc. Chem. Res.* 45 (2012), 1721–1735.

290 Lüftl, S. and Visakh, P. M. Polyoxymethylene: state of art, new challenges and opportunities. In *Polyoxymethylene Handbook: Structure, Properties, Applications, and Their Nanocomposites*, S. Lüftl, P. M. Visakh, and S. Chandran, Eds. Scrivener Publishing, Salem, MA, USA, 2014, ch. 1.

291 Luinstra, G. A. and Borchardt, E. Material properties of poly(propylene carbonate). *Adv. Polym. Sci.* 245 (2012), 29–48.

292 Madkour, T. M. Polycarbonate. In *Polymer Data Handbook*, J. E. Mark, Ed. Oxford University Press, New York, USA, 1999, pp. 363–367.

293 Mandelkern, L. *Crystallization of Polymers Volume 1. Equilibrium Concepts*, 2nd ed. Cambridge University Press, Cambridge, UK, 2002.

294 Mandelkern, L. and Alamo, R. G. Polyethylene, linear high-density. In *Polymer Data Handbook*, J. E. Mark, Ed. Oxford University Press, New York, 1999, pp. 493–507.

295 Mark, J. E. Thermoelastic properties of rubberlike networks and their thermodynamic and molecular interpretation. *Rubber Chem. Technol.* 46 (1973), 593–618.

296 Mark, J. E. Thermoelastic results on rubberlike networks and their bearing on the foundations of elastic theory. *J. Polym. Sci., Macromol. Rev.* 11 (1976), 135–159.

297 Mark, J. E. Configurational partition functions and entropies for polyoxide chains having the repeat unit [$(CH_2)_y$–O–]. *J. Chem. Phys.* 67 (1977), 3300–3302.

298 Mark, J. E. Rubber elasticity. *J. Chem. Educ.* 58 (1981), 898–903.

299 Mark, J. E. and Chiu, D. S. Dipole moments of poly(trimethylene oxide) chains. *J. Chem. Phys.* 66 (1977), 1901–1904.

300 Mark, J. E. and Flory, P. J. The configuration of the polyoxyethylene chain. *J. Am. Chem. Soc.* 87 (1965), 1415–1423.

301 Mark, J. E. and Flory, P. J. Dipole moments of chain molecules. I. Oligomers and polymers of oxyethylene. *J. Am. Chem. Soc.* 88 (1966), 3702–3707.

302 Mark, J., Ngai, K., Graessley, W., Mandelkern, L., Samulski, E., Koenig, J., and Wignall, G. *Physical Properties of Polymers*, 3rd ed. Cambridge University Press, New York, 2004.

303 Masamoto, J. Poly(ethylene sulfide). In *Polymer Data Handbook*, J. E. Mark, Ed. Oxford University Press, New York, USA, 1999, pp. 553–557.

304 Maschio, L., Kirtman, B., Oriando, R., and Rèrat, M. Ab initio analytical infrared intensities for periodic systems through a coupled perturbed Hartree-Fock/Kohn-Sham method. *J. Chem. Phys.* 137 (2012), 204113.

305 Mattice, W. L. Conformational properties of a molecule consisting of three branches emanating from a common atom, evaluated using rotational isomeric state theory. *Macromolecules* 9 (1976), 48–52.

306 Mattice, W. L., Helfer, C. A., and Sokolov, A. P. Persistence length and finite chain length effect on characteristic ratios. *Macromolecules* 37 (2004), 4711–4717.

307 Mattice, W. L. and Suter, U. W. *Conformational Theory of Large Molecules: The Rotational Isomeric State Model in Macromolecular Systems.* Wiley-Interscience, New York, USA, 1994.

308 McNaught, A. D. and Wilkinson, A., Eds. *IUPAC. Compendium of Chemical Terminology*, 2nd ed. (the "Gold Book"). Blackwell Scientific Publications, Oxford, UK, 1997.

309 McQuarrie, D. A. *Statistical Mechanics.* University Science Books, Sausalito, CA, USA, 2000.

310 Mendicuti, F., Patel, B., Viswanadhan, V. N., and Mattice, W. L. Identification of conformations conducive to intramolecular excimer formation in polyesters with different numbers of methylene units between aromatic rings. *Polymer* 29 (1988), 1669–1674.

311 Mendicuti, F., Rodrigo, M. M., Tarazona, M. P., and Saiz, E. Dipole moments and kerr constants of dibenzoates of some diols with different numbers of methylene units as model compounds for polyesters. *Macromolecules* 23 (1990), 1139–1144.

312 Mennucci, B., Cammi, R., and Tomasi, J. Excited states and solvatochromic shifts within a nonequilibrium solvation approach: a new formulation of the integral equation formalism method at the self-consistent field, configuration interaction, and multiconfiguration self-consistent field level. *J. Chem. Phys.* 109 (1998), 2798–2807.

313 Mennucci, B., Cancès, E., and Tomasi, J. Evaluation of solvent effects in isotropic and anisotropic dielectrics and in ionic solutions with a unified integral equation method: theoretical bases, computational implementation, and numerical applications. *J. Phys. Chem. B* 101 (1997), 10506–10517.

314 Merrick, J. P., Moran, D., and Radom, L. An evaluation of harmonic vibrational frequency scale factors. *J. Phys. Chem. A* 111 (2007), 11683–11700.

315 Meyerhoff, V. G., and Shimotsuma, S. Das lösungsverhalten von polyäthylenterephthalat. *Makromol. Chem.* 135 (1970), 195–217.

316 Michl, J. and West, R. Conformations of linear chains. Systematics and suggestions for nomenclature. *Acc. Chem. Res.* 33 (2000), 821–823.

317 Middleton, J. C. and Tipton, A. J. Synthetic biodegradable polymers as orthopedic devices. *Biomaterials* 21 (2000), 2335–2346.

318 Miertuš, S., Scrocco, E., and J. Tomasi. Electrostatic interaction of a solute with a continuum. A direct utilizaion of ab initio molecular potentials for the prevision of solvent effects. *Chem. Phys.* 55 (1981), 117–129.

319 Miller, W. G., Brant, D. A., and Flory, P. J. Random coil configurations of polypeptide copolymers. *J. Mol. Biol.* 23 (1967), 67–80.

320 Miller, R. D., Hofer, D., Rabolt, J., and Fickes, G. N. Anomalously long-wavelength electronic transition in conformationally locked organosilane high polymers. *J. Am. Chem. Soc.* 107 (1985), 2172–2174.

321 Miller, R. D. and Michl, J. Polysilane high polymers. *Chem. Rev.* 89 (1989), 1359–1410.

322 Miyaki, Y., Einaga, Y., Hirosye, T., and Fujita, H. Solution properties of poly(D-β-hydroxybutyrate). 2. Light scattering and viscosity in trifluoroethanol and behavior of highly expanded polymer coils. *Macromolecules* 10 (1977), 1356–1364.

323 Miyasaka, T., Kinai, Y., and Imamura, Y. Conformational analysis of poly(oxymethylene). *Makromol. Chem.* 182 (1981), 3533–3543.

324 Møller, C. and Plesset, M. S. Note on an approximation treatment for many-electron systems. *Phys. Rev.* 46 (1934), 618–622.

325 Momma, K. and Izumi, F. VESTA 3 for three-dimensional visualization of crystal, volumetric and morphology data. *J. Appl. Crystallogr.* 44 (2011), 1272–1276.

326 Moreland, J. L., Gramada, A., Buzko, O. V., Zhang, Q., and Bourne, P. E. The molecular biology toolkit (MBT): a modular platform for developing molecular visualization applications. *BMC Bioinformatics* 6 (2005), 21.

327 Morgado, C. A., Jurečka, P., Svozil, D., Hobza, P., and Šponer, J. Reference MP2/CBS and CCSD(T) quantum-chemical calculations on stacked adenine dimers. Comparison with DFT-D, MP2.5, SCS(MI)-MP2, M06-2X, CBS(SCS-D) and force field descriptions. *Phys. Chem. Chem. Phys.* 12 (2010), 3522–3534.

328 Mortillaro, L., Credali, L., Russo, M., and Checchi, C. D. Polyselenomethylene. *J. Polym. Sci., Polym. Lett. Ed.* 3 (1965), 581–585.

329 Nagai, K. Even moments of the end-to-end distance of polymeric chains. *J. Chem. Phys.* 48 (1968), 5646–5655.

330 Nagai, K. and Ishikawa, T. Excluded-volume effect on dipole moments of polymer chains. *Polym. J. (Tokyo, Jpn.)* 2 (1971), 416–421.

331 Nagasawa, M., Ishii, T., Abe, D., and Sasanuma, Y. Structure–property relationships of aromatic polyamides and polythioamides: comparative consideration with those of analogous polyesters, polythioesters and polydithioesters. *RSC Advances* 5 (2015), 96611–96622.

332 Nagasawa, M., Sasanuma, Y., and Masu, H. N,N'-(Ethane-1,3-diyl) dibenzothioamide. *Acta Crystallogr., Sect. E* 70 (2014), o586.

333 Nagasawa, M., Sasanuma, Y., and Masu, H. N,N'-(Propane-1,3-diyl) dibenzothioamide. *Acta Crystallogr., Sect. E* 70 (2014), o639.

334 Nakafuku, C., and Yoshimura, H. Melting parameters of poly(glycolic acid). *Polymer* 45 (2004), 3583–3585.

335 Nakamae, K., Kameyama, M., Yoshikawa, M., and Matsumoto, T. Elastic modulus of crystalline regions of poly(butylene terephthalate) in the direction parallel to the chain axis. *J. Polym. Sci., Polym. Phys. Ed.* 20 (1982), 319–326.

336 Nash, L. K. *Elements of Statistical Thermodynamics*, 2nd ed. Dover Publications, Mineola, NY, USA, 2006.

337 Nash, D. W. and Pepper, D. C. Solution properties of poly(propylene sulphide). *Polymer* 16 (1975), 105–109.

338 Nelder, J. A. and Mead, R. A simplex method for function minimization. *Comput. J.* 7 (1965), 308–313.

339 Neumann, F., Teramae, H., Downing, J. W., and Michl, J. Gauche, ortho, and anti conformations of saturated A_4X_{10} chains: When will all six conformers exist? *J. Am. Chem. Soc.* 120 (1998), 573–582.

340 Nishino, T., Okamoto, T., and Sakurai, H. Cryogenic mechanical behavior of poly(trimethylene terephthalate). *Macromolecules* 44 (2011), 2106–2111.

341 Nozaki, K., Nakano, K., and Hiyama, T. Optically active polycarbonates: asymmetric alternating copolymerization of cyclohexene oxide and carbon dioxide. *J. Am. Chem. Soc.* 121 (1999), 11008–11009.

342 Nye, J. F. *Physical Properties of Crystals: Their Representation by Tensors and Matrices*. Oxford University Press, New York, USA, 1985.

343 Ochterski, J. W. *Vibrational analysis in Gaussian. White papers and technical notes*. Gaussian, Inc., 1999. https://gaussian.com/vib/.

344 Ochterski, J. W. *Thermochemistry in Gaussian. White papers and technical notes*. Gaussian, Inc., 2000. https://gaussian.com/thermo/.

345 Oguni, N., Maeda, S., and Tani, H. Structure analysis of poly(propylene-β-d oxide) by proton nuclear magnetic resonance spectroscopy. *Macromolecules* 6 (1973), 459–465.

346 Onsager, L. Electric moments of molecules in liquids. *J. Am. Chem. Soc.* 58 (1936), 1486–1493.

347 Otsuka, S., Kuwajima, I., Hosoya, J., Xu, Y., and Yamazaki, M. PoLyInfo: Polymer database for polymeric materials design. In *2011 International Conference on Emerging Intelligent Data and Web Technologies* (2011), National Institute for Materials Science Japan, pp. 22–29.

348 Ottosson, C.-H. and Michl, J. Conformers of n-Si_6Me_{14}: ab initio, molecular mechanics, and additive increment methods. *J. Phys. Chem. A* 104 (2000), 3367–3380.

349 Ovitt, T. M. and Coates, G. W. Stereoselective ring-opening polymerization of *rac*-lactide with a single-site, racemic aluminum alkoxide catalyst: synthesis of stereoblock poly(lactic acid). *J. Polym. Sci., Part A: Polym. Chem.* 38 (2000), 4686–4692.

350 Ovitt, T. M. and Coates, G. W. Stereochemistry of lactide polymerization with chiral catalysts: new opportunities for stereocontrol using polymer exchange mechanisms. *J. Am. Chem. Soc.* 124 (2002), 1316–1326.

351 Pérez, S. and Brisse, F. Conformations studies on oligomethylene glycol derivatives and related compounds. I. The cystal and molecular structure of ethylene glycol dibenzoate, $C_{16}H_{14}O_4$. *Acta Crystallogr., Sect. B* 32 (1976), 470–474.

352 Pérez, S. and Brisse, F. Trimethylene glycol dibenzoate. *Acta Crystallogr., Sect. B* 33 (1977), 3259–3262.

353 Palmer, A. and Brisse, F. La structure du N,N'-ethylene-dibenzamide. *Acta Crystallogr.* B36 (1980), 1447–1452.

354 Papageorgiou, G. Z. and Bikiaris, D. N. Crystallization and melting behavior of three biodegradable poly(alkylene succinates). A comparative study. *Polymer* 46 (2005), 12081–12092.

355 Parchaňský, V., Kapitán, J., Kaminský, J., Šebestík, J., and Bouř, P. Ramachandran plot for alanine dipeptide as determined from Raman optical activity. *J. Phys. Chem. Lett.* 4 (2013), 2763–2768.

356 Parr, R. G. and Yang, W. *Density-Functional Theory of Atoms and Molecules.* Oxford University Press, New York, USA, 1989.

357 Pauling, L. *The Nature of the Chemical Bond*, 3rd ed. Cornell University, Ythaca, NY, USA, 1960.

358 Pearson, R. G. Hard and soft acids and bases. *J. Am. Chem. Soc.* 85 (1963), 3533–3539.

359 Pell, A. J. and Keeler, J. Two-dimensional J-spectra with absorption-mode lineshapes. *J. Magn. Reson.* 189 (2007), 293–299.

360 Peng, C., Ayala, P. Y., Schlegel, H. B., and Frisch, M. J. Using redundant internal coordinates to optimize equilibrium geometries and transition states. *J. Comput. Chem.* 17 (1996), 49–56.

361 Peng, C. and Schlegel, H. B. Combining synchronous transit and quasi-Newton methods to find transition states. *Isr. J. Chem.* 33 (1993), 449–454.

362 Perdew, J. P., Chevary, J. A., Vosko, S. H., Jackson, K. A., Pederson, M. R., Singh, D. J., and Fiolhais, C. Atons, molecules, solids, and surfaces: applications of the generalized gradient approximation for exchange and correlation. *Phys. Rev. B* 46 (1992), 6671–6687.

363 Perevyazko, I., Gubarev, A. S., Tauhardt, L., Dobrodumov, A., Pavlov, G. M., and Schubert, U. S. Linear poly(ethylene imine)s: true molar masses, solution properties and conformation. *Polym. Chem.* 8 (2017), 7169–7179.

364 Perger, W., Criswell, J., Civalleri, B., and Dovesi, R. Ab-initio calculation of elastic constants of crystalline systems with the CRYSTAL code. *Comput. Phys. Commun.* 180 (2009), 1753–1759.

365 Pihlaja, K. and Kleinpeter, E. *Carbon-13 NMR Chemical Shifts in Structural and Stereochemical Analysis.* VCH Publishers, New York, USA, 1994.

366 Pisani, C., Dovesi, R., Roetti, C., Causà, M., Orlando, R., Casassa, S., and Saunders, V. R. CRYSTAL and EMBED, two computational tools for the ab initio study of electronic properties of crystals. *Int. J. Quantum Chem.* 77 (2000), 1032–1048.

367 Pitoňák, M., Neogrády, P., Černý, J., Grimme, S., and Hobza, P. Scaled MP3 non-covalent interaction energies agree closely with accurate CCSD(T) benchmark data. *ChemPhysChem* 10 (2009), 282–289.

368 Pople, J. A. and Beveridge, D. L. *Approximate Molecular Orbital Theory.* MaGraw-Hill, New York, USA, 1970.

369 Pople, J. A., Seeger, R., and Krishnan, R. Variational configuration interaction methods and comparison with perturbation theory. *Int. J. Quantum Chem.* 12 (1977), 149–163.

370 Poulin-Dandurand, S., Pérez, S., Revol, J.-F., and Brisse, F. The crystal structures of poly(trimethylene terephthalate) by X-ray and electron diffraction. *Polymer* 20 (1979), 419–426.

371 Pranamuda, H. and Tokiwa, Y. Degradation of poly(L-lactide) by strains belonging to genus *Amycolatopsis. Biotechnol. Lett.* 21 (1999), 901–905.

372 Pranamuda, H., Tsuchii, A., and Tokiwa, Y. Poly(L-lactide)-degrading enzyme produced by *Amycolatopsis* sp. *Macromol. Biosci.* 1 (2001), 25–29.

373 Provencher, S. W. A constrained regularization method for inverting data represented by linear algebraic or integral equations. *Comput. Phys. Commun.* 27 (1982), 213–227.

374 Provencher, S. W. CONTIN: A general purpose constrained regularization program for inverting noisy linear algebraic and integral equations. *Comput. Phys. Commun.* 27 (1982), 229–242.

375 Quarti, C., Milani, A., Civalleri, B., Orlando, R., and Castiglioni, C. Ab initio calculation of the crystalline structure and ir spectrum of polymers: nylon 6 polymorphs. *J. Phys. Chem. B* 116 (2012), 8299–8311.

376 Rabolt, J. F., Hofer, D., Miller, R. D., and Fickes, G. N. Studies of chain conformational kinetics in poly(di-*n*-alkylsilanes) by spectroscopic methods. 1. Poly(di-*n*-hexylsilane), poly(di-*n*-heptylsilane), and poly(di-*n*-octylsilane). *Macromolecules* 19 (1986), 611–616.

377 Radano, C. P., Baker, G. L., and Milton, R. Smith, I. Stereoselective polymerization of a racemic monomer with a racemic catalyst: direct preparation of the polylactic acid stereocomplex from racemic lactide. *J. Am Chem. Soc.* 122 (2000), 1552–1553.

378 Rahalkar, R. R., Mark, J. E., Boileau, S., Hemery, P., and Riande, E. Thermoelastic properties of rubberlike networks and their thermodynamic and molecular interpretation. *J. Polym. Sci., Polym. Phys. Ed.* 17 (1979), 1623–1625.

379 Ramsey, N. F. Electron coupled interactions between nuclear spins in molecules. *Phys. Rev.* 91 (1953), 303–307.

380 Rappe, A. K. and Goddard, W. A. Charge equilibration for molecular dynamics simulations. *J. Phys. Chem.* 95 (1991), 3358–3363.

381 Rassolov, V. A., Ratner, M. A., Pople, J. A., Redfern, P. C., and Curtiss, L. A. 6-31g* basis set for third-raw atoms. *J. Comput. Chem.* 22 (2001), 976–984.

382 Reed, A. E., Curtiss, L. A., and Weinhold, F. Intermolecular interactions from a natural bond orbital, donor-acceptor viewpoint. *Chem. Rev.* 88 (1988), 899–926.

383 Rehahn, M., Mattice, W. L., and Suter, U. W. *Rotational Isomeric State Models in Macromolecular Systems (Advances in Polymer Science)*, vols. 131/132. Springer-Verlag, Berlin, 1997.

384 Riande, E. Dipole moments of poly(oxytetramethylene) and poly(oxydecamethylene) chains. *Makromol. Chem.* 178 (1977), 2001–2010.

385 Riande, E., Boileau, S., Hemery, P., and Mark, J. E. An experimental study of the dipole moments of isotactic and atactic poly(propylene sulfide). *Macromolecules* 12 (1979), 702–704.

386 Riande, E., Boileau, S., Hemery, P., and Mark, J. E. Dipole moments of poly(propylene sulfide) chains. *J. Chem. Phys.* 71 (1979), 4206–4208.

387 Riande, E. and Guzmán, J. Statistical properties of alternating copolymers. 1. Dipole moments of poly(thiodiethylene glycol) chains. *Macromolecules* 12 (1979), 952–956.

388 Riande, E. and Saiz, E. *Dipole Moments and Birefringence of Polymers*. Prentice Hall, Englewood Cliffs, NJ, USA, 1992.

389 Roberts, J. D. *ABCs of FT-NMR*. University Science Books, Sausalito, CA, USA, 2000.

390 Roe, R.-J. *Methods of X-ray and Neutron Scattering in Polymer Science*. Oxford University Press, New York, USA, 2000.

391 Rossi, C. and Magnasco, V. Poly(ethylene oxide) in solution. *J. Polym. Sci.* 58 (1962), 977–989.

392 Saegusa, T., Ikeda, H., and Fujii, H. Isomerization polymerization of 2-oxazoline. I. Preparation of unsubstituted 2-oxazoline polymer. *Polym. J. (Tokyo, Jpn.)* 3 (1972), 35–39.

393 Saito, Y., Izuka, D., Kaneko, N., Togashi, D., Narumi, A., and Kawaguchi, S. Molecular characterization of poly(L-lactic acid) isolated chain. *Kobunshi Ronbunshu* 69 (2012), 416–423.

394 Sakakihara, H., Takahashi, Y., Tadokoro, H., Sigwalt, P., and Spassky, N. Structural studies of the optically active and racemic poly(propylene sulfides). *Macromolecules* 2 (1969), 515–520.

395 Salmeia, K. A., Vagin, S., Anderson, C. E., and Rieger, B. Poly(propylene carbonate): insight into the microstructure and enantioselective ring-opening mechanism. *Macromolecules* 45 (2012), 8604–8613.

396 Sánchez, A., Bello, A., Marco, C., and Fatou, J. G. Solution properties and chain dimensions of poly(trimethylene sulfide). *Makromol. Chem.* 189 (1988), 399–408.

397 Sasanuma, Y. Solvent effect on the conformation of 1,2-dimethoxypropane. *J. Phys. Chem.* 98 (1994), 13486–13488.

398 Sasanuma, Y. Conformational analysis of poly(propylene oxide) and its model compound 1,2-dimethoxypropane. *Macromolecules* 28 (1995), 8629–8638.

399 Sasanuma, Y. Intramolecular interactions of polyethers and polysulfides, investigated by NMR, ab initio molecular orbital calculations, and rotational isomeric state scheme: an advanced analysis of NMR data. In *Annual Reports* *on NMR Spectroscopy*, G. A. Webb, Ed., vol. 49. Academic Press (Elsevier Science), Oxford, UK, 2003, ch. 5, pp. 213–280.

400 Sasanuma, Y. Conformational characteristics, configurational properties, and thermodynamic characteristics of poly(ethylene terephthalate) and poly(ethylene-2,6-naphthalate). *Macromolecules* 42 (2009), 2854–2862.

401 Sasanuma, Y. Structure-property relationships of polymers, unraveled by molecular orbital, RIS, and periodic density functional theory calculations. In *Modern Applications of Flory's "Statistical Mechanics of Chain Molecules"*, A. E. Tonelli and G. Patterson, Eds. American Chemical Society / Oxford University Press, Oxford, UK, 2021, Chapter 10: 161–208.

402 Sasanuma, Y., Asai, S., and Kumagai, R. Conformational characteristics and configurational properties of poly(ethylene oxide-*alt*-ethylene sulfide). *Macromolecules* 40 (2007), 3488–3497.

403 Sasanuma, Y., Hattori, S., Imazu, S., Ikeda, S., Kaizuka, T., Iijima, T., Sawanobori, M., Azam, M. A., Law, R. V., and Steinke, J. H. G. Conformational analysis of poly(ethylene imine) and its model compounds: rotational and inversional isomerizations and intramolecular and intermolecular hydrogen bonds. *Macromolecules* 37 (2004), 9169–9183.

404 Sasanuma, Y., Hayashi, Y., Matoba, H., Touma, I., Ohta, H., Sawanobori, M., and Kaito, A. Conformational analysis of poly(propylene sulfide). *Macromolecules* 35 (2002), 8216–8226.

405 Sasanuma, Y., Iwata, T., Kato, Y., Kato, H., Yarita, T., Kinugasa, S., and Law, R. V. Carbon-13 NMR chemical shifts of dimeric model compounds of poly(propylene oxide): a proof of existence of the (C–H)···O attraction. *J. Phys. Chem.* 105 (2001), 3277–3283.

406 Sasanuma, Y., Kato, H., and Kaito, A. Conformational analysis of poly(di-*n*-butylsilane), poly(di-*n*-hexylsilane), and poly(methyl-*n*-propylsilane) by a rotational isomeric state scheme with molecular dynamics simulations. *J. Phys. Chem. B* 107 (2003), 11852–11860.

407 Sasanuma, Y. and Katsumata, S. Elucidation of conformational characteristics and configurational properties of poly((*R*)-3-hydroxybutyrate) by *ab initio* statistical mechanics. *Polym. J. (Tokyo, Jpn.)* 45 (2013), 727–737.

408 Sasanuma, Y. and Kumagai, R. Conformational characteristics and configurational properties of poly(ethylene imine-*alt*-ethylene sulfide) and the role of the secondary amine group as a junction of attractive interactions. *Macromolecules* 40 (2007), 7393–7399.

409 Sasanuma, Y., Kumagai, R., and Nakata, K. Prediction of structures, properties, and functions of alternating copolymers of ethylene imine and ethylene oxide as an example of molecular design for polymers. *Macromolecules* 39 (2006), 6752–6764.

410 Sasanuma, Y., Nonaka, Y., and Yamaguchi, Y. Conformational characteristics and configurational properties of poly(ethylene succinate) and poly(butylene succinate) and structure-property-function relationships of representative biodegradable polyesters. *Polymer* 56 (2015), 327–339.

411 Sasanuma, Y., Ogawa, Y., and Matsumoto, M. Predictive elucidation of conformational characteristics and configurational properties of poly(1-methylphosphirane) and poly(1-phenylphosphirane) as a molecular design. *Phys. Chem. Chem. Phys.* 12 (2010), 14619–14628.

412 Sasanuma, Y., Ohta, H., Touma, I., Matoba, H., Hayashi, Y., and Kaito, A. Conformational characteristics of poly(ethylene sulfide) and poly(ethylene oxide): solvent dependence of attractive and repulsive gauche effects. *Macromolecules* 35 (2002), 3748–3761.

413 Sasanuma, Y. and Sugita, K. The attractive gauche effect of ethylene oxides. *Polym. J. (Tokyo, Jpn.)* 38 (2006), 983–988.

414 Sasanuma, Y. and Suzuki, N. Influence of weak attractive interactions on structures and properties of poly(trimethylene terephthalate). *Macromolecules* 42 (2009), 7203–7212.

415 Sasanuma, Y. and Takahashi, Y. Structure-property relationships of poly(ethylene carbonate) and poly(propylene oxide). *ACS Omega* 2 (2017), 4808–4819.

416 Sasanuma, Y. and Tanaka, S. Molecular design of aromatic polythionoesters. *ACS Omega* 5 (2020), 3016–3029.

417 Sasanuma, Y., Teramae, F., Yamashita, H., Hamano, I., and Hattori, S. Conformational analysis of poly(trimethylene imine) and poly(*N*-methyltrimethylene imine) by the rotational isomeric state scheme with up to fourth-order intramolecular interactions. *Macromolecules* 38 (2005), 3519–3532.

418 Sasanuma, Y. and Touge, D. Configurational statistics of poly(L-lactide) and poly(DL-lactide) chains. *Polymer* 55 (2014), 1901–1911.

419 Sasanuma, Y., Wagai, Y., Suzuki, N., and Abe, D. Conformational characteristics and configurational properties of poly(butylene terephthalate) and structure-property relationships of aromatic polyesters. *Polymer* 54 (2013), 3904–3913.

420 Sasanuma, Y. and Watanabe, A. Conformational characteristics of poly(trimethylene sulfide) and structure-property relationships of representative polysulfides and polyethers. *Macromolecules* 39 (2006), 1646–1656.

421 Sasanuma, Y., Watanabe, A., and Tamura, K. Structure-property relationships of polyselenoethers $[\text{-}(CH_2)_y Se\text{-}]_x$ (y =1, 2, and 3) and related polyethers and polysulfides. *J. Phys. Chem. B* 112 (2008), 9613–9624.

422 Sasanuma, Y., Yamamoto, H., and Choi, S. Structure-property relationships of poly(glycolic acid) and poly(2-hydroxybutyrate). *Macromolecules* 52 (2019), 3730–3746.

423 Sawanobori, M., Sasanuma, Y., and Kaito, A. Conformational analysis of poly(methylene sulfide) and its oligomeric model compounds: anomeric effect and electron flexibility of polythioacetal. *Macromolecules* 34 (2001), 8321–8329.

424 Schelten, J., Ballard, D., Wignall, G., Longman, G., and Schmatz, W. Small-angle neutron scattering studies of molten and crystalline polyethylene. *Polymer* 17 (1976), 751–757.

425 Schilling, F. C., Bovey, F. A., Lovinger, A. J., and Zeigler, J. M. Characterization of poly(di-*n*-hexylsilane) in the solid state. II. Carbon-13 and silicon-29 magic-angle spinning NMR studies. *Macromolecules* 19 (1986), 2660–2663.

426 Schilling, F. C., Lovinger, A. J., Zeigler, J. M., Davis, D. D., and Bovey, F. A. Solid-state structures and thermochromism of poly(di-*n*-butylsilylene) and poly(di-*n*-pentylsilylene). *Macromolecules* 22 (1989), 3055–3063.

427 Schilling, F. C., Lovinger, A. J., Zeigler, J. M., Davis, D. D., and Bovey, F. A. Solid-state structures and thermochromism of poly(di-*n*-butylsilylene) and poly(di-*n*-pentylsilylene). *Macromolecules* 22 (1989), 3055–3063.

428 Schilling, F. C. and Tonelli, A. E. Carbon-13 NMR determination of poly(propylene oxide) microstructure. *Macromolecules* 19 (1986), 1337–1343.

429 Schmitz, K. S. *An Introduction to Dynamic Light Scattering by Macromolecules.* Academic Press, London, UK, 1990.

430 Schweitzer-Stenner, R. Distribution of conformations sampled by the central amino acid residue in tripeptides inferred from amide I band profiles and NMR scalar coupling constants. *J. Phys. Chem. B* 113 (2009), 2922–2932.

431 Sepulchre, M., Spassky, N., Ooteghem, D. V., and Goethals, E. J. 300 MHz ^{1}H-NMR study of poly(propylene sulfide). *J. Polym. Sci., Polym. Chem. Ed.* 12 (1974), 1683–1693.

432 Shao, H., Reddi, Y., and Cramer, C. J. Modeling the mechanism of CO_2/cyclohexene oxide copolymerization catalyzed by chiral zinc β-diiminates: factors affecting reactivity and isotacticity. *ACS Catal.* 10 (2020), 8870–8879.

433 Shi, Z., Olson, C. A., Rose, G. D., Baldwin, R. L., and Kallenbach, N. R. Polyproline II structure in a sequence of seven alanine residues. *Proc. Natl. Acad. Sci. U.S.A.* 99 (2002), 9190–9195.

434 Shimomura, M., Iguchi, M., and Kobayashi, M. Vibrational spectroscopic study on trigonal polyoxymethylene and polyoxymethylene-d_2 crystals. *Polymer* 29 (1988), 351–357.

435 Silla, E., naki Tu nón, I., and Pascual-Ahuir, J. L. GEPOL: An improved description of molecular surfaces II. Computing the molecular area and volume. *J. Comput. Chem.* 12 (1991), 1077–1088.

436 Slater, J. C. A simplification of the Hartree-Fock method. *Phys. Rev.* 81 (1951), 385–390.

437 Smith, G. D., Bedrov, D., and Borodin, O. Conformations and chain dimensions of poly(ethylene oxide) in aqueous solution: a molecular dynamics simulation study. *J. Am. Chem. Soc.* 122 (2000), 9548–9549.

438 Smith, G. D., Jaffe, R. L., and Yoon, D. Y. Conformational characteristics of dimethoxymethane based upon ab initio electronic structure calculations. *J. Phys. Chem.* 36 (1994), 9072–9077.

439 Smith, G. D., Jaffe, R. L., and Yoon, D. Y. Conformational characteristics of poly(oxymethylene) based upon ab initio electronic structure calculations. *J. Phys. Chem.* 36 (1994), 9078–9082.

440 Smith, G. D., Yoon, D. Y., Jaffe, R. L., Colby, R. H., Krishnamoorti, R., and Fetters, L. J. Conformations and structures of poly(oxyethylene) melts from molecular dynamics simulations and small-angle neutron scattering experiments. *Macromolecules* 29 (1996), 3462–3469.

441 Spassky, N., Wisniewski, M., Pluta, C., and Borgne, A. L. Highly stereoelective polymerization of *rac*-(D,L)-lactide with a chiral Schiff's base/aluminium alkoxide initiator). *Macromol. Chem. Phys.* 197 (1996), 2627–2637.

442 Spek, A. L. Structure validation in chemical crystallography. *Acta Crystallogr.* D65 (2009), 148–155.
443 Staudinger, H. Über Polymerisation. *Ber. Dtsch. Chem. Ges. B* 53 (1920), 1073–1085.
444 Stephens, P. J., Devlin, F. J., Chabalowski, C. F., and Frisch, M. J. Ab initio calculation of vibrational absorption and circular dichroism spectra using density functional force fields. *J. Phys. Chem.* 98 (1994), 11623–11627.
445 Stockmayer, W. H. Chain dimensions near the Flory temperature. *J. Polym. Sci. (Hoboken, NJ, U. S. A.)* 15 (1955), 595–598.
446 Stockmayer, W. H. Problems of the statistical thermodynamics of dilute polymer solutions. *Makromol. Chem.* 35 (1960), 54–74.
447 Stockmayer, W. H. and Fixman, M. On the estimation of unperturbed dimensions from intrinsic viscositiesxcin. *J. Polym. Sci., Part C: Polym. Symp.* 1 (1963), 137–141.
448 Sulley, G. S., Gregory, G. L., Chen, T. T. D., Carrodeguas, L. P., Trott, G., Santmarti, A., Lee, K.-Y., Terrill, N. J., and Williams, C. K. Switchable catalysis improves the properties of CO_2-derived polymers: poly(cyclohexene carbonate-*b*-ϵ-decalactone-*b*-cyclohexene carbonate) adhesives, elastomers, and toughened plastics. *J. Am. Chem. Soc.* 142 (2020), 4367–4378.
449 Sun, H. Ab initio calculations and force field development for computer simulation of polysilanes. *Macromolecules* 28 (1995), 701–712.
450 Sure, R., Brandenburg, J. G., and Grimme, S. Small atomic orbital basis set first-principles quantum chemical methods for large molecular and periodic systems: a critical analysis of error sources. *ChemistryOpen* 5 (2016), 94–109.
451 Surhone, L. M., Timpledon, M. T., and Marseken, S. F., Eds. *Nylon. List of synthetic polymers, polyamide, thermoplastic, peptide bond, condensation polymer, fiber, nylon 6-6, nylon 6, ballistic nylon, ripstop nylon, nylon-eating bacteria* (Beau Bassin, Mauritius, 2010), Betascript Publishing: VDM Publishing House.
452 Suter, U. W. and Flory, P. J. Conformational energy and configurational statistics of polypropylene. *Macromolecules* 8 (1975), 765–776.
453 Sychrovský, V., Gräfenstein, J., and Cremer, D. Nuclear magnetic resonance spin-spin coupling constants from coupled perturbed density functional theory. *J. Chem. Phys.* 113 (2000), 3530–3547.
454 Tachibana, K., Hashimoto, K., Yoshikawa, M., and Okawa, H. Isolation and characterization of microorganisms degrading nylon 4 in the composted soil. *Polym. Degrad. Stab.* 95 (2010), 912–917.
455 Tadokoro, H. *Structure of Crystalline Polymers*. Wiley-Interscience, New York, USA, 1979.
456 Tadokoro, H., Chatani, Y., Yoshihara, T., Tahara, S., and Murahashi, S. Structural studies on polyethers, $[\text{-}(CH_2)_m\text{-}O\text{-}]_n$. II. Molecular structure of polyethylene oxide. *Makromol. Chem.* 73 (1964), 109–127.
457 Tadokoro, H., Takahashi, Y., Chatani, Y., and Kakida, H. Structural studies of polyethers, $[-(CH_2)_m-O-]_n$, V. Polyoxacyclobutane. *Makromol. Chem.* 109 (1967), 96–111.
458 Tadokoro, H., Yasumoto, T., Murahashi, S., and Nitta, I. Molecular configuration of polyoxymethylene. *J. Polym. Sci.* 44 (1960), 266–269.

459 Taha, H. A., Castillo, N., Sears, D. N., Wasylishen, R. E., Lowary, T. L., and Roy, P.-N. Conformational analysis of arabinofuranosides: prediction of $^{3}J_{\mathrm{H,H}}$ using MD simulations with DFT-derived spin-spin coupling profiles. *J. Chem. Theory Comput.* 5 (2010), 212–222.

460 Taherimehr, M. and Pescarmona, P. P. Green polycarbonates prepared by the copolymerization of CO_2 with epoxides. *J. Appl. Polym. Sci.* 131 (2014), 41141.

461 Takahashi, Y. Neutron structure analysis of polyethylene-d_4. *Macromolecules* 31 (1988), 3868–3871.

462 Takahashi, Y. and Mark, J. E. Random-coil dimensions of poly(trimethylene oxide), an unusually compact chain molecule. *J. Am. Chem. Soc.* 98 (1976), 3756–3760.

463 Takahashi, Y., Sumita, I., and Tadokoro, H. Structural studies of polyethers. IX. Planar zigzag modification of poly(ethylene oxide). *J. Polym. Sci., Polym. Phys. Ed.* 11 (1973), 2113–2122.

464 Takahashi, Y. and Tadokoro, H. Structural studies of polyethers, $(\text{-}(CH_2)_m\text{-O-})_n$. X. Crystal structure of poly(ethylene oxide). *Macromolecules* 6 (1973), 672–675.

465 Takahashi, Y. and Tadokoro, H. Electron density calculation for polyoxymethylene in cylindrical coordinates by a new fourier method. *J. Polym. Sci.: Polym. Phys. Ed.* 16 (1978), 1219–1225.

466 Takahashi, Y. and Tadokoro, H. Least-squares refinement of molecular structure of polyoxymethylene. *J. Polym. Sci.: Polym. Phys. Ed.* 17 (1979), 123–130.

467 Takahashi, Y., Tadokoro, H., and Chatani, Y. Structure of polyethylene sulfide. *J. Macromol. Sci., Part B: Phys.* 2 (1968), 361–367.

468 Tanaka, S., Masu, H., and Sasanuma, Y. Crystal structures of 2-(benzenecarbothioyloxy)ethyl benzenecarbothioate and 2-(benzenecarbothioyloxy)ethyl benzoate. *Acta Crystallogr., Sect. E* 73 (2017), 1430–1433.

469 Tasaki, K. and Abe, A. NMR studies and conformational energy calculations of 1,2-dimethoxyethane and poly(oxyethylene). *Polym. J. (Tokyo, Jpn.)* 17 (1985), 641–655.

470 Tashiro, K., Hanesaka, M., Ohhara, T., Ozeki, T., Kitano, T., Nishu, T., Kurihara, K., Tamada, T., Kuroki, R., Fujiwara, S., Tanaka, I., and Niimura, N. Structural refinement and extraction of hydrogen atomic positions in polyoxymethylene crystal based on the first successful measurements of 2-dimensional high-energy synchrotron X-ray diffraction and wide-angle neutron diffraction patterns of hydrogenated and deuterated species. *Polym. J. (Tokyo, Jpn.)* 39 (2007), 1253–1273.

471 Taylor, H. M. and Karlin, S. *An Introduction to Stochastic Modeling*, 3rd ed. Academic Press, San Diego, CA, USA, 1998.

472 Teraoka, I. *Polymer Solutions: An Introduction to Physical Properties*. Wiley, New York, USA, 2002.

473 Thakur, K. A. M., Kean, R. T., Hall, E. S., Kolstad, J. J., and Munson, E. J. Stereochemical aspects of lactide stereo-copolymerization investigated by ^{1}H NMR: a case of changing stereospecificity. *Macromolecules* 31 (1998), 1487–1494.

474 Thakur, M., Majid, I., Hussain, S., and Nanda, V. Poly(ϵ-caprolactone): a potential polymer for biodegradable food packaging applications. *Packag. Technol. Sci.* 34 (2021), 449–461.

475 Thatcher, G. R. J. *The Anomeric Effect and Associated Stereoelectronic Effects*, vol. 539 of *ACS Symposium Series*. American Chemical Society, Washington, DC, USA, 1993.

476 Thomas, L. H. The calculation of atomic fields. *Math. Proc. Camb. Philos. Soc.* 23 (1927), 542–548.

477 Thomas, L. L., Christakis, T. J., and Jorgensen, W. L. Conformation of alkanes in the gas phase and pure liquids. *J. Phys. Chem. B* 110 (2006), 21198–21204.

478 Thorat, S. D., Phillips, P. J., Semenov, V., and Gakh, A. Physical properties of aliphatic polycarbonates made from CO_2 and epoxides. *J. Appl. Polym. Sci.* 89 (2003), 1163–1176.

479 Tokiwa, Y. and Calabia, B. P. Degradation of microbial polyesters. *Biotechnol. Lett.* 26 (2004), 1181–1189.

480 Tokiwa, Y., Calabia, B. P., Ugwu, C. U., and Aiba, S. Biodegradability of plastics. *Int. J. Mol. Sci.* 10 (2009), 3722–3742.

481 Tokiwa, Y. and Jarerat, A. Microbial degradation of aliphatic polyesters. *Macromol. Symp.* 201 (2003), 283–289.

482 Tomalia, D. A. and Ojha, N. D. Homopolymers of *N*-(2-hydroxyethyl)aziridines and *N*-(2-thioethyl)aziridines, August 18, 1973. U.S. Patent 3,752,854, The Dow Chemical Corporation, Midland, Mic.

483 Tomasi, J., Mennucci, B., and Cammi, R. Quantum mechanical continuum solvation models. *Chem. Rev.* 105 (2005), 2999–3093.

484 Tonelli, A. E. Calculation of the intramolecular contribution to the entropy of fusion in crystalline polymers. *J. Chem. Phys.* 52 (1970), 4749–4751.

485 Tonelli, A. E. *NMR Spectroscopy and Polymer Microstructure: The Conformational Connection*. VCH Publishers, New York, USA, 1989.

486 Tonelli, A. E. and Flory, P. J. The configuration statistics of random poly(lactic acid) chains. I. Experimental results. *Macromolecules* 2 (1969), 225–227.

487 Tonelli, A. and Shen, J. *Conformations. Connecting the Chemical Structures and Material Behaviors of Polymers*. CRC Press, Boca Raton, FL, USA, 2020.

488 Towler, M. D., Zupan, A., and Causà, M. Density functional theory in periodic systems using local Gaussian basis sets. *Comput. Phys. Commun.* 98 (1996), 181–205.

489 Truhlar, D. G. Basis-set extrapolation. *Chem. Phys. Lett.* 294 (1998), 45–48.

490 Tsuji, H. and Hayakawa, T. Hetero-stereocomplex formation between substituted poly(lactic acid)s with linear branched side chains, poly(L-2-hydroxybutanoic acid) and poly(D-2-hydroxy-3-methylbutanoic acid). *Polymer* 55 (2014), 721–726.

491 Tsuji, H., Hosokawa, M., and Sakamoto, Y. Ternary stereocomplex crystallization of poly(L-2-hydroxybutanoic acid), poly(D-2-hydroxybutanoic acid), and poly(D-lactic acid) from the melt. *Polymer* 54 (2013), 2190–2198.

492 Tsuji, H. and Ikeda, Y. Crystallization from the melt of poly(lactide)s with different optical purities and their blends. *Macromol. Chem. Phys.* 197 (1996), 3483–3499.

493 Tsuji, H. and Shimizu, S. Stereocomplex crystallization and homo-crystallization of enantiomeric poly(2-hydroxybutyrate)s: effects of molecular weight and crystallization conditions. *Polymer* 53 (2012), 5385–5392.

494 Tsuji, H. and Sobue, T. Cocrystallization of monomer units in lactic acid-based biodegradable copolymers, poly(L-lactic acid-*co*-L-2-hydroxybutanoic acid)s. *Polymer* 72 (2015), 202–211.

495 Tsuji, H., Takai, H., and Saha, S. K. Isothermal and non-isothermal crystallization behavior of poly(L-lactic acid): effects of stereocomplex as nucleating agent. *Polymer* 47 (2006), 3826–3837.

496 Tsuzuki, S., Honda, K., Uchimaru, T., Mikami, M., and Tanabe, K. Origin of attraction and directionality of the π/π interaction: model chemistry calculations of benzene dimer interaction. *J. Am. Chem. Soc.* 124 (2002), 104–112.

497 Tsuzuki, S., Uchimaru, T., and Tanabe, K. Conformational analysis of *n*-alkanes using density functional theory. Comparison with ab initio calculations. *Chem. Phys. Lett.* 246 (1995), 9–12.

498 Tsuzuki, S., Uchimaru, T., Tanabe, K., and Hirano, T. Conformational analysis of 1,2-dimethoxyethane by ab initio molecular orbital and molecular mechanics calculations: stabilization of the TGG' rotamer by the 1,5 CH_3/O nonbonding attractive interaction. *J. Phys. Chem.* 97 (1993), 1346–1350.

499 Tuzar, Z., Vošický, V., Bohdanecký, M., and Jalovecký, J. Unperturbed dimensions of poly(ethylene terephthalate). *Makromol. Chem.* 180 (1979), 1399–1402.

500 Tvaroška, I., Hricovíni, M., and Petráková, E. An attempt to derive a new Karplus-type equation of vicinal proton-carbon coupling constants for C-O-C-H segments of bonded atoms. *Carbohydr. Res.* 189 (1989), 359–362.

501 Twisleton, J. F., White, J. W., and Reynolds, P. A. Dynamical studies of fully oriented deuteropolyethylene by inelastic neutron scattering. *Polymer* 23 (1982), 578–588.

502 Uchida, T., Kurita, Y., and Kubo, M. The dipole moments and the structure of polyoxymethylene dimethyl ethers. *J. Polym. Sci.* 19 (1956), 365–372.

503 Uchida, T. and Tadokoro, H. Structural studies of polyethers. IV. Structure analysis of the polyoxymethylene molecule by three-dimensional fourier syntheses. *J. Polym. Sci. Part A-2* 5 (1967), 63–81.

504 Udipi, K. and Gillham, J. K. Poly(ethylene carbonate) and poly(propylene carbonate): transitions and thermomechanical spectra. *J. Appl. Polym. Sci.* 18 (1974), 1575–1580.

505 Ueda, A. S., Chatani, Y., and Tadokoro, H. Structural studies of polyesters. IV. Molecular and crystal structures of poly(ethylene succinate) and poly(ethylene oxalate). *Polym. J. (Tokyo, Jpn.)* 2 (1971), 387–397.

506 Ueno, Y., Bahry, M., and Okawara, M. Tetrathioquinodimethane chemistry. A new approach to the preparation of charge-transfer complex by lithium iodide reduction. *Tetrahedron Lett.* 18 (1977), 4607–4610.

507 van de Ven, F. J. M. *Multidimensional NMR in Liquids: Basic Principles and Experimental Methods*. VCH, New York, USA, 1995.

508 van Duijneveldt, F. B., van Duijneveldt-van de Rijdt, J. G. C. M., and van Lenthe, J. H. State of the art in counterpoise theory. *Chem. Rev.* 94 (1994), 1873–1885.

509 Varma-Nair, M., Cheng, J., Jin, Y., and Wunderlich, B. Thermal properties of polysilylenes. *Macromolecules* 24 (1991), 5442–5450.

510 Vasanthakumari, R., and Pennings, A. Crystallization kinetics of poly(L-lactic acid). *Polymer* 24 (1983), 175–178.

511 Vennemann, N., Lechner, M. D., and Oberthür, R. C. Thermodynamics and conformation of polyoxyethylene in aqueous solution under high pressure: 1. Small-angle neutron scattering and densitometric measurements at room temperature. *Polymer* 28 (1987), 1738–1748.

512 Veszprémi, T. and Fehér, M. *Quantum Chemistry: Fundamentals to Applications.* Kluwer Academic/Plenum Publishing, New York, USA, 1999.

513 Vos, L. D., de Voorde, B. V., Daele, L. V., Dubruel, P., and Vlierberghe, S. V. Poly(alkylene terephthalate)s: from current developments in synthetic strategies towards applications. *Eur. Polym. J.* 161 (2021), 110840.

514 Vosko, S. H., Wilk, L., and Nusair, M. Accurate spin-dependent electron liquid correlation energies for local spin density calculations: a critical analysis. *Can. J. Phys.* 58 (1980), 1200–1211.

515 Wallach, M. L. Viscosity-molecular weight relation and unperturbed dimensions of polyethylene terephthalate. *Makromol. Chem.* 103 (1967), 19–26.

516 Ward, I. M. and Sweeney, J. *Mechanical Properties of Solid Polymers.* Wiley, Chichester, West Sussex, UK, 2013.

517 Waser, P., Rueping, M., Seebach, D., Duchardt, E., and Schwalbe, H. On the solution structure of PHB: preparation and NMR analysis of isotopically labeled oligo[(*R*)-3-hydroxybutanoic acids] (OHBs). *Helv. Chim. Acta* 84 (2001), 1821–1845.

518 Watts, J. D., Gauss, J., and Bartlett, R. J. Coupled-cluster methods with noniterative triple excitations for restricted open-shell Hartree-Fock and other general single determinant reference functions. Energies and analytical gradients. *J. Chem. Phys.* 98 (1993), 8718–8733.

519 Welsh, W. J., DeBolt, L., and Mark, J. E. Conformational energies and unperturbed chain dimensions of polysilane and poly(dimethylsilylene). *Macromolecules* 19 (1986), 2978–2983.

520 Whinfield, J. R. and Dickson, J. T. Improvements relating to the manufacture of highly polymeric substances, 1941. British Patent 578,079.

521 Wignall, G. D. Small angle neutron and X-ray scattering. In *Physical Properties of Polymers Handbook*, J. E. Mark, Ed., 2nd ed. Springer, New York, USA, 2007, Chapter 23: 407–420.

522 Wignall, G. D., Hendricks, R. W., Koehler, W. C., Lin, J. S., Wai, M. P., Thomas, E. L., and Stein, R. S. Measurements of single chain form factors by small-angle neutron scattering from polystyrene blends containing high concentrations of labelled molecules. *Polymer* 22 (1981), 886–889.

523 Williams, C. E., Nierlich, M., Cotton, J. P., Jannink, G., Boué, F., Daoud, M., Farnoux, B., Picot, C., DeGennes, P. G., Rinaudo, M., Moan, M., and Wolff, C. Polyelectrolyte solutions: intrachain and interchain correlations observed by SANS. *J. Polym. Sci., Polym. Lett. Ed.* 17 (1979), 379–384.

524 Wilson, P. J. Density functional theory and its application to nuclear magnetic resonance shielding constants. In *Annual Reports on NMR Spectroscopy*, G. A. Webb, Ed., vol. 49. Academic Press / Elsevier, Oxford, UK, 2003.
525 Wolinski, K., Hinton, J. F., and Pulay, P. Efficient implementation of the gauge-independent atomic orbital method for NMR chemical shift calculations. *J. Am. Chem. Soc.* 112 (1990), 8251–8260.
526 Wong, M. W., Frisch, M. J., and Wiberg, K. B. Solvent effects. 1. The mediation of electrostatic effects by solvents. *J. Am. Chem. Soc.* 113 (1991), 4776–4782.
527 Woodruff, M. A. and Hutmacher, D. W. The return of a forgotten polymer – polycaprolactone in the 21th century. *Prog. Polym. Sci.* 35 (2010), 1217–1256.
528 Wu, G.-P., Jiang, S.-D., Lu, X.-B., Ren, W.-M., and Yan, S.-K. Stereoregular poly(cyclohexene carbonate)s: unique crystallization behavior. *Chin. J. Polym. Sci.* 30 (2012), 487–492.
529 Wu, X., Jin, J., Zhang, L., and Xu, J. Configuration-dependent properties of the poly(dimethylsilmethylene) chain in the third-order interaction approximation. *J. Polym. Sci., Part B: Phys. Ed.* 31 (1993), 455–459.
530 Wu, G.-P., Ren, W.-M., Luo, Y., Li, B., Zhang, W.-Z., and Lu, X.-B. Enhanced asymmetric induction for the copolymerization of CO_2 and cyclohexene oxide with unsymmetric enantiopure salenCo(III) complexes: synthesis of crystalline CO_2-based polycarbonate. *J. Am. Chem. Soc.* 134 (2012), 5682–5688.
531 Wu, G., Tashiro, K., Kobayashi, M., Komatsu, T., and Nakagawa, K. A study on mechanical deformation of highly oriented poly(oxymethylene) by vibrational spectroscopy and X-ray diffraction: stress and temperature dependence of Young's modulus. *Macromolecules* 22 (1989), 758–765.
532 Xu, J., Song, X., Zhou, Z., and Yan, D. Third-order interaction approximation for linear polymer chains. *J. Polym. Sci., Part B: Phys. Ed.* 29 (1991), 877–882.
533 Yamakawa, H. *Modern Theory of Polymer Solutions*. Harper & Row, New York, USA, 1971.
534 Yamane, K., Sato, H., Ichikawa, Y., Sunagawa, K., and Shigaki, Y. Development of an industrial production technology for high-molecular-weight polyglycolic acid. *Polym. J. (Tokyo, Jpn.)* 46 (2014), 769–775.
535 Yamano, N., Kawasaki, N., Takeda, S., and Nakayama, A. Production of 2-pyrrolidone from biobased glutamate by using *Escherichiacoli*. *J. Polym. Environ.* 21 (2013), 528–533.
536 Yamano, N., Nakayama, A., Kawasaki, N., Yamamoto, N., and Aiba, S. Mechanism and characterization of polyamide 4 degradation by *Pseudomonas* sp. *J. Polym. Environ.* 16 (2008), 141–146.
537 Yang, X., Kang, S., Hsu, S. L., Stidham, H. D., Smith, P. B., and Leugers, A. A spectroscopic analysis of chain flexibility of poly(lactic acid). *Macromolecules* 34 (2001), 5037–5041.
538 Yarita, T., Nomura, A., Abe, K., and Takeshita, Y. Supercritical fluid chromatographic determination of tocopherols on an ODS-silica gel column. *J. Chromatogr. A* 679 (1994), 329–334.
539 Yokouchi, M., Chatani, Y., Tadokoro, H., Teranishi, K., and Tani, H. Structural studies of polyesters: 5. Molecular and crystal structures of optically active and racemic poly(β-hydroxybutyrate). *Polymer* 14 (1973), 267–272.

540 Yokouchi, M., Sakakibara, Y., Chatani, Y., Tadokoro, H., Tanaka, T., and Yoda, K. Structures of two crystalline forms of poly(butylene terephthalate) and reversible transition between them by mechanical deformation. *Macromolecules* 9 (1976), 266–273.

541 Yoshida, Y. Elucidation of structure-property relationships of poly(ethylene oxide) and poly(ethylene sulfide) crystals, March 2021. Graduation work, Department of Applied Chemistry and Biotechnology, Chiba University.

542 Yoshida, N., Aoki, D., and Sasanuma, Y. Configurational statistics of poly(cyclohexene carbonate). *Macromolecules* 53 (2020), 9362–9374.

543 Yoshida, N., Aoki, D., and Sasanuma, Y. Correction to configurational statistics of poly(cyclohexene carbonate). *Macromolecules* 53 (2020), 10299–10299.

544 Yoshida, H., Kaneko, I., Matsuura, H., Ogawa, Y., and Tasumi, M. Importance of an intramolecular 1,5-CH...O interaction and intermolecular interactions as factors determining conformational equilibria in 1,2-dimethoxyethane: a matrix-isolation infrared spectroscopic study. *Chem. Phys. Lett.* 196 (1992), 601–606.

545 Yuan, Q. W. Poly(ethylene oxide). In *Polymer Data Handbook*, J. E. Mark, Ed. Oxford University Press, New York, USA, 1999, p. 542.

546 Zacharopoulos, N. and Economou, I. G. Morphology and organization of poly(propylene imine) dendrimers in the melt from molecular dynamics simulation. *Macromolecules* 35 (2002), 1814–1821.

547 Zhang, J. Study of poly(trimethylene terephthalate) as an engineering thermoplastics material. *J. Appl. Polym. Sci.* 91 (2004), 1657–1666.

548 Zhang, L., Wang, S., Li, J., Liu, X., Chen, P., Zhao, T., and Zhang, L. A nitrogen-containing all-solid-state hyperbranched polymer electrolyte for superior performance lithium batteries. *J. Mater. Chem. A* 7 (2019), 6801–6808.

549 Zimm, B. H. Application of the methods of molecular distribution to solutions of large molecules. *J. Chem. Phys.* 14 (1946), 164–179.

550 Zimm, B. H. Apparatus and methods for measurement and interpretation of the angular variation of light scattering; preliminary results on polystyrene solutions. *J. Chem. Phys.* 16 (1948), 1099–1116.

551 Zimm, B. H. The scattering of light and the radial distribution function of high polymer solutions. *J. Chem. Phys.* 16 (1948), 1093–1099.

552 Zimm, B. H. Excluded volume in polymer chains. *J. Chem. Phys.* 21 (1953), 1716–1723.

Index

Conformational Analysis of Polymers: Methods and Techniques for Structure-Property Relationships and Molecular Design, First Edition. Yuji Sasanuma.

c

d

e

t

Printed and bound by CPI Group (UK) Ltd, Croydon, CR0 4YY
17/09/2026
14964321-0005